Basic Concepts of Linear Algebra

Basic Concepts of Linear Algebra

S. Isaak
and
M. N. Manougian
University of South Florida

W · W · Norton & Company · Inc · New York

W. W. Norton & Company, Inc. 500 Fifth Avenue,
New York, N. Y. 10110
W. W. Norton & Company Ltd, 37 Great Russell Street,
London WC1B 3NU

Library of Congress Cataloging in Publication Data
Isaak, S
Basic concepts of linear algebra.
Includes index.
1. Algebras, Linear. I. Manougian, M. N., joint author. II. Title.
QA184.I8 1976 512'.5 75-31789
ISBN 0-393-09199-6

Printed in the United States of America
7 8 9

To Edahn
Eilam
and
Michael

Contents

Preface

The purpose of this text is to introduce students of mathematics, the sciences, and engineering to the basic concepts of linear algebra. It is our intent to set forth lucidly some of the main concepts, language, and techniques of linear algebra for students at the freshman and sophomore levels who have completed a course in college algebra and trigonometry. The text is suitable for a one-semester or one-quarter course.

Mathematical concepts are introduced with simple examples. Following a precise mathematical formulation, many worked-out examples are given to illustrate the concepts discussed. Explanations of the methods of linear algebra are supplemented with relevant theorems. Proofs of the simpler theorems are detailed so that the student can follow the arguments. The more difficult theorems are clearly stated and their proofs omitted. References to other texts which supply these omitted proofs have been supplied in the margin for the energetic and well-prepared student.

Each page is divided into two parts, with the marginal portion used for illustrations, notes, comments, and remarks to aid the student.

For better understanding, historical developments of some of the basic concepts are presented. Furthermore, the text contains numerous historical comments and biographical sketches of some of the famous mathematicians who contributed to the development of linear algebra. The purpose here is to familiarize the student with the human aspect, life, and character of those who created and developed linear algebra in particular and mathematics in general.

We have covered the material in Chapters 1 through 7 of this text in a one-quarter course meeting four hours per week. This is equivalent to a one-semester course meeting three hours per week. In a well-prepared class portions of Chapter 8 can be included also. Chapter 0 is a review chapter for the student who lacks the required algebraic background. Students who have had a course in college algebra may skip Chapter 0 and start with Chapter 1.

Chapter 0 through Section 1 of Chapter 8 require no calculus. For the benefit of the student with a course in elementary calculus, some sections contain problems that relate the concepts of linear algebra to those of calculus. These problems are clearly marked and may be omitted with no loss of continuity. The last two sections of Chapter 8 illustrate the use of linear algebra in the applications of differential equations.

Each section contains many worked-out examples and numerous

problems ranging from the elementary to the more challenging. The problems are designed to enhance the student's understanding of the concepts introduced in each of the sections. Furthermore, each chapter concludes with a chapter review. These serve as a self-test for the student, emphasizing the main concepts introduced in the chapter.

Answers to all computational problems are included at the end of the book. Hints for solving some of the more challenging problems are given there, as well as in the marginal column.

We are grateful to the reviewers of the manuscript: Joseph L. Ercolano of Baruch College, Robert B. Herrera of Los Angeles City College, Bobby H. Jones of Pasadena City College, Cleon R. Yohe of Washington University, and Leonard Gillman of the University of Texas, Austin. Their comments and suggestions were most helpful. Also, we wish to thank the many students at the University of South Florida who read and commented on the original manuscript, and we are grateful to Janelle Fortson and Terry Luth for their excellent typing. Finally, we wish to thank Christopher P. Lang and the staff of W. W. Norton & Company for their assistance and encouragement.

S. Isaak
M. N. Manougian

Tampa, Florida
May 1975

Basic Concepts of Linear Algebra

Preliminaries

0

0-1 Sets

Objectives

1 Introduce the concept of a set and discuss universal sets, subsets, and equality of sets.
2 Introduce the following operations involving sets: intersection, union, and complement.

The concept of a **set** is very natural to our thought processes. The idea of a set of objects has come naturally to the attention of philosophers and mathematicians throughout history. The basic difficulty in the concept of a set that interested philosophers was the notion of an **infinite set.** Zeno (fifth century B.C.) and Aristotle (fifth century B.C.) considered the problem, and philosophers throughout the Middle Ages tackled the question of whether there can be an infinite collection of objects. Galileo (1564–1642) rejected the idea of an infinite set, and Gauss (1777–1855) protested against its use in mathematics. Bernhard Bolzano (1781–1848) defended the existence of infinite sets. His work, however, was philosophical rather than mathematical.

The men responsible for the beginnings of our theory of sets are the nineteenth-century mathematicians Georg Cantor and George Boole. The former is considered the creator of set theory.

A set is a **well-defined** collection of distinct objects. The objects are called the **elements** of the set.

We shall not concern ourselves with the definition of the word "set." Rather, we shall accept the concept of a set as an undefined term. A set is well-defined. This is the basic property that describes a set. It means that if A is a set and a is an object then it must be unambiguously clear whether or not a belongs to A. Thus "the set of all good teachers at your college" is ill-defined. The statement does not tell us what a good teacher is. On the other hand, "all mathematics instructors," is a well-defined collection and therefore is a set.

The student can produce readily many examples of sets. For example:

1 The set of colors of the American flag
2 The set of students at your college
3 The set of all women on the 1972 Miami Dolphins football team
4 The set of all former presidents of the United States
5 The set of integers divisible by 2
6 The set of all island states of the United States in 1973

From these examples we see that a set may have one element (as in 6) no element (as in 3), more than one but a finite number of elements (as in 1, 2, 4), or an infinite number of elements (as in 5). The set described in 5 is called an **infinite** set, where the other sets are called **finite.** The set that has no element (as in 3) is called the **null** (or **empty**) set and is denoted by $\varnothing$.

In general, capital letters are used to denote sets and lowercase letters are used to denote the elements of a set. We use the notation

$a \in A$

which means that the object designated by a is an element of the set whose name is A. If the object b is not an element of A we write

$b \notin A$

There are two general methods for describing sets. A set may be defined by listing all of its elements. For example,

$S = \{$Jane, Sam, Jill$\}$

This reads "S is the set whose elements are the people Jane, Sam, and Jill." Note the use of the commas to separate the elements which are enclosed with braces $\{\ \}$.

If A is the set of all integers strictly between 1 and 100 it will be cumbersome to list all the elements of A. In this case we describe in words a property possessed by each element of the set. Thus

$A = \{x : x$ is an integer strictly between 1 and $100\}$

An alternate notation is

$A = \{2,3,4, \ldots ,99\}$

The three dots indicate that the rest of the elements are included and each is calculated by using the rule indicated by the first three elements. Care should be exercised in using this notation. The first three elements should make it absolutely and unambiguously clear what the fourth, fifth, etc., elements are.

Here x denotes an element; the notation : reads "such that," and is followed by a description of the property possessed by the elements of A.

Suppose we wish to consider the various species of animals in the city zoo. Each species is a set of animals. In our discussion there is an overall set to which the elements of all other sets must belong. Such a set is called a *universal set.*

Definition 0-1 A **universal set** is a set to which the elements of all other sets in a given discussion must belong. The universal set is denoted by $\mathbf{U}$.

A universal set for our discussion above could be the set of all animals in the zoo. Or we could pick as our universal set the set of all animals on this earth. Thus it is possible to have more than one universal set for a particular discussion. It is important to specify what universal set we do select.

Suppose $\mathbf{U} = \{x : x$ is an animal in the city zoo$\}$. We find that the set $F = \{y : y$ is a four-legged animal in the city zoo$\}$ has the property that for every $y \in F$, $y \in \mathbf{U}$. This leads to the following definition.

Definition 0-2 If every element of set A is an element of set B, then A is said to be a **subset** of B. In symbols, we write

$A \subset B$

(Read "A is a subset of B.") If A is not a subset of B we write $A \not\subset B$.

Example 1 Let

$U = \{1,2,3,4,5,6,7,8,9\}$

a Find the set A whose elements are multiples of 3.

b Find all the subsets of A.

Solution **a** For this discussion the set A is given by

$A = \{3,6,9\}$

Note that 12 is a multiple of 3 but $12 \notin A$ because $12 \notin U$.

b The subsets of $\{3,6,9\}$ are as follows:

$\{3\}$, $\{6\}$, $\{9\}$, $\{3,6\}$, $\{3,9\}$, $\{6,9\}$, $\varnothing$, and $\{3,6,9\}$

Some authors write $B \subset A$ to mean "B is a proper subset of A" and $B \subseteq A$ to mean "B is a subset of A."

It is by convention that the null set is regarded as a subset of every set. The set $\{3,6,9\}$ is called an **improper subset** of A. The rest of the subsets are called **proper subsets.**

Definition 0-3 The sets A and B are said to be **equal** if and only if $A \subset B$ and $B \subset A$. In symbols we write $A = B$.

If sets A and B are not equal we write $A \neq B$.

Example 2 $\{2,4,6\} = \{4,2,6\}$

Example 3 $\{2,4,6\} = \{1 + 1, 2 + 2, 6\}$

Example 4 $\{2,4,6\} \neq \{2,4\}$

Consider the two sets

Note: $A \subset B$; however, $B \not\subset A$.

$A = \{\text{red, blue}\}$
$B = \{\text{white, red, blue, green}\}$

Obviously the two sets are not equal. The sets A and B have elements in common; that is, the colors red and blue are elements of A and of B. The set consisting of elements that belong to both A and B is called the *intersection* of A and B.

Definition 0-4 The **intersection** of two sets A and B is the set of all elements that belong to both A and B. In symbols we write $A \cap B$.

Example 5 Let $A = \{\text{red, blue, purple}\}$, $B = \{\text{white, red, blue, green}\}$. Find $A \cap B$.

Solution $A \cap B = \{\text{red, blue}\}$.

Example 6 Let A be the set of all even natural numbers and let B be the set of all odd natural numbers. Find $A \cap B$.

Solution $A \cap B = \varnothing$. This is the case because there is no element in A that is also an element of B. Hence, the intersection of these two sets is the null set.

Whenever the intersection of two sets A and B is the null set, then A and B are said to be **disjoint.**

Definition 0-5 The **union** of two sets A and B is the set consisting of elements that belong to either A, or B, or both. In symbols we write $A \cup B$.

Example 7 Let $A = \{x : x \text{ is a girl with blonde hair}\}$ and let $B = \{y : y \text{ is a girl with blue eyes}\}$. Find $A \cup B$.

Solution $A \cup B = \{z : z \text{ is a girl having either blonde hair, or blue eyes, or both}\}$.

Example 8 Let $A = \{1,2,3,4,5\}$ and $B = \{2,4,6\}$.

a Find $A \cup B$
b Find $A \cap B$

Solution
a $A \cup B = \{1,2,3,4,5,6\}$
b $A \cap B = \{2,4\}$

A third basic operation involving sets is given in the following:

Definition 0-6 Let A be a set and $\mathbf{U}$ a universal set such that $A \subset \mathbf{U}$. Then the **complement** of A, denoted by A^c, is the set of all elements in $\mathbf{U}$ that are not in A.

Example 9 Let $\mathbf{U} = \{a,b,c,d,e,f\}$, $A = \{a,b,c\}$, and $B = \{a,c,e\}$. Find A^c, B^c, $(A \cup B)^c$, $(A \cap B)^c$, $A^c \cup B^c$, and $A^c \cap B^c$.

Solution
a $A^c = \{d,e,f\}$
b $B^c = \{b,d,f\}$
c $A \cup B = \{a,b,c,e\}$ and so $(A \cup B)^c = \{d,f\}$
d $A \cap B = \{a,c\}$ and so $(A \cap B)^c = \{b,d,e,f\}$
e $A^c \cup B^c = \{b,d,e,f\}$
f $A^c \cap B^c = \{d,f\}$

Biographical Sketch

GEORG CANTOR (1845–1898) *was born in St. Petersburg, Russia, to a well-to-do Danish merchant. The family, which though of Jewish descent had converted to Christianity, moved to Frankfurt in 1856. Cantor's talents in mathematics were recognized at an early age. He received his university education at Zurich and at the University of Berlin, and among his instructors were the famous mathematicians Weierstrass and Kronecker. He received his Ph.D. in 1867. His dissertation, involving a problem stated by Gauss, was a fine piece of work, but where there was talent no one saw genius. The same could be said for all the papers he wrote up to the age of 29. At that age Cantor published his first revolutionary paper on the theory of infinite sets. Instead of being given an influential position he was continually bypassed by less eminent mathematicians at the University of Berlin; a position there was his cherished ambition. He spent his entire professional career at the University of Hall, a third-rate institution where he was appointed an (unpaid) lecturer in 1868; he lived on whatever he could get from his students. In 1875 he was appointed full professor.*

Cantor blamed Kronecker for his failure to obtain a position at the University of Berlin. Kronecker attacked Cantor's work, and the viciousness of these attacks might have contributed to Cantor's first breakdown in 1884. Cantor eventually died in a mental institution.

Cantor's theory of infinite sets—one of the most original contributions to mathematics in the last 2,500 years—deserved better than the ridicule it received. After his death it was universally agreed that Cantor had come up with something fundamentally new in mathematics.

Exercises

1 Which of the following collections are sets?

a The tall students at your college
b The students enrolled in a mathematics course at your college
c The cold countries of Europe
d The known planets of our solar system
e The number zero
f Persons under the age of 35 who have been presidents of the United States

2 List the elements in each of the following sets:

$A = \{x : x$ is an even number less than $10\}$
$B = \{y : y$ is a letter of the word "follow"$\}$
$C = \{p : p$ is a planet of our solar system$\}$

$D = \{n : n$ is a natural number less than $13\}$
$E = \{x : x$ is a factor of $50\}$
$F = \{x : x$ is a positive prime number less than $14\}$

In problems 3–20, let $\mathbf{U} = \{1,2,3,4,5,6,7,8,9\}$, $A = \{1,3,5,7,9\}$, $B = \{1,2,3,4\}$, and $C = \{2,4,6,8\}$. List the elements of each of the following sets.

3 A^c **4** B^c **5** C^c
6 $A \cup B$ **7** $B \cup C$ **8** $A \cup C$
9 $A \cap C$ **10** $B \cap C$ **11** $A^c \cap C^c$
12 $A^c \cap C$ **13** $A^c \cap B^c$ **14** $(A \cap B)^c$
15 $(A \cup B) \cap C$ **16** $(A \cap B) \cap C$ **17** $(A^c \cap B)^c$
18 $\varnothing^c$ **19** $(A^c)^c$ **20** $(\varnothing^c)^c$

21 List the non-empty subsets of the set $\{i,o,u\}$.

22 Are the following sets equal?
$A = \{x : x$ is a letter of the word "tease"$\}$
$B = \{y : y$ is a letter of the word "state"$\}$
$C = \{z : z$ is a letter of the word "seat"$\}$

23 Find all possible solutions of x and y such that
a $\{3x, y\} = \{6,9\}$ **b** $\{3x\} = \{0\}$
c $\{4x,3\} = \{10,3\}$ **d** $\{x, y^2\} = \{9,3\}$
e $\{x,x^2\} = \{9,3\}$ **f** $\{x,x^2\} = \{4,-3\}$

24 Given that
$A \cup B = \{1,2,3,4\}$ $A \cup C = \{1,2,3,4,5,7\}$
$A \cap B = \{2,4\}$ $A \cap C = \varnothing$ $B \cap C = \{1,3\}$

find A, B, and C.

25 Give conditions under which each of the following statements is true.
a $A \cup B = B$ **b** $A \cup B = \varnothing$ **c** $A \cap \mathbf{U} = \mathbf{U}$
d $B^c \cap \mathbf{U} = \mathbf{U}$ **e** $A \cap B = B$ **f** $A \cap B = A \cup B$

26 In a survey of 100 freshman students it was found that the number of students taking the various subjects were as follows:

45—mathematics 18—mathematics and philosophy
62—English 9—English and philosophy
25—philosophy 23—mathematics and English
6—all three subjects

a How many students had mathematics as the only subject?
b How many students had mathematics and English but not philosophy?

c How many students had mathematics and philosophy but not English?

d How many students were studying none of these subjects?

0-2 Functions

Objectives

1 Introduce the concept of a function.

2 Discuss the domain of a function and operations on functions.

The concept of a **function** or a relation between two sets of objects had its beginnings in the seventeenth century. Galileo used the notion of a function in his works. By the eighteenth century the concept of a function had been developed and many of the elementary functions had been recognized. Research on problems concerning the motion of objects along curved paths led to the function concept. Among the outstanding mathematicians of the eighteenth century who contributed to the development of this concept are Leibnitz, L'Hôpital, Huygens, Euler, and members of the Bernoulli family. The word "function" was introduced into the mathematical language by Leibnitz.

Biographical Sketch

GALILEO GALILEI (1564–1642) *was born in Pisa to a cloth merchant. He entered the University of Pisa to study medicine but at the age of seventeen switched to mathematics. He was refused a position at the University of Bologna because he was not "sufficiently distinguished." Galileo became a professor at Pisa and later at the University of Padua. In 1610 he was appointed Chief Mathematician in Florence at the court of the Grand Duke Cosimo II de Medici who set him up comfortably.*

Galileo was an active mathematician, physicist, and astronomer, and a talented writer. He discovered the satellites of Jupiter, independently invented the microscope, and designed the first pendulum clock. He also designed and sold a popular compass.

Galileo's teaching of the Copernican system was condemned by the Roman Inquisition. In 1633 he was forbidden to publish and was put under house arrest. However, he continued to work and his book Dialogues Concerning Two New Sciences *was smuggled to Holland, where it was published in 1638.*

We shall illustrate the concept of a function with some examples. A precise definition of functions will follow.

Example 1 Let A be the set of persons $p_1, p_2, \ldots, p_n$ and let B be the set of all birthdays. Now consider the relation in which each person's birthday is associated with that person. The important feature here is that each person has only one birthday. We illustrate this in Figure 0-1. Here we say that f is a "rule" that assigns to each element of the set A one and only one element of the set B. Thus, f is said to be a **function.** The set A is called the **domain** of the function f. For each p_i in A, the corresponding unique element in B is called the **image** of p_i; and the set of all images of the elements of A is called the **range** of the function f.

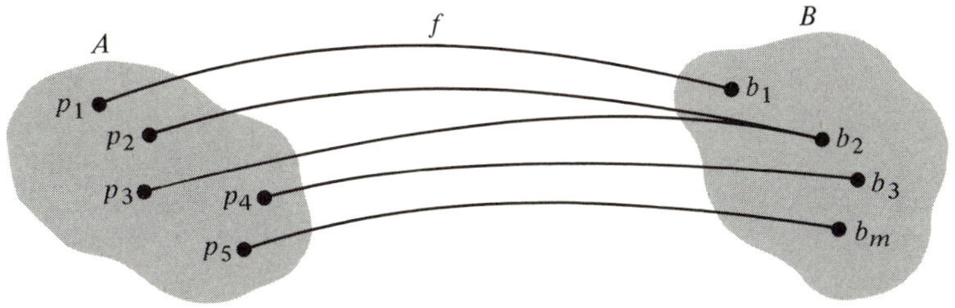

Figure 0-1

Observe that associated with each $p_i \in A$ is a unique element in B. However, a given element in B may be associated with more than one element in A.

Example 2 Consider the two commonly used temperature scales, the centigrade scale and the Fahrenheit scale. Let the centigrade scale be represented by the set $C = \{x : x \geq -273.15\}$ and let the Fahrenheit scale be represented by the set $F = \{y : y \geq -459.67\}$. We find that 0°C (0 degrees centigrade) is associated with 32°F (32 degrees Fahrenheit) and 100°C is associated with 212°F. In general, the association between the elements of the sets C and F is given by the "rule"

$$y = \tfrac{9}{5}x + 32 \qquad [1]$$

Thus for each $x \in C$ there exists a unique $y \in F$. Here we have a function f defined by equation 1. The set C is the domain of f and the set F is the range of f.

The following notation, which is due to Euler, is commonly used. For each x in the domain of f the image in F is designated by $f(x)$, read "f of x." Thus equation 1 may be written as

$$f(x) = \tfrac{9}{5}x + 32 \qquad [2]$$

Here $f(x)$ plays the same role as y. Thus we write

$$\begin{aligned} f(0) &= 32 \\ f(100) &= 212 \\ f(-5) &= 23 \\ f(-40) &= -40 \end{aligned}$$

and so on.

In the examples above we see that we are dealing with

a two sets, A and B, and

b A "rule" or formula that assigns to each element of A a single element in B.

We now give a precise definition of a function.

Definition 0-7 Let A and B denote two nonempty sets. Then f is a **function** from A to B, written $f: A \longrightarrow B$, provided that f is a rule that assigns to each element of A a single element of B.

The set A is called the domain of f and the set of all images of the elements of A, a subset of B, is called the range of f.

The fact that f assigns $x \in A$ to $y \in B$ is denoted by $f(x) = y$.

When the sets A and B are real numbers, the function $f: A \longrightarrow B$ is called a **real-valued function** (or simply a real function).

For a real function defined by a formula, if the domain is not specified we shall assume that the domain consists of all real num-

Biographical Sketch **LEONHARD EULER (1707–1783)** *was born near Basel, Switzerland. He received his early mathematical education from his father, Paul Euler, a Calvinist pastor who had studied mathematics with Jacob Bernoulli. Leonhard Euler entered the University of Basel to study theology. There he studied mathematics under Johannes Bernoulli. When Euler was 17 he received his master's degree and at 19 he presented his first research paper to the French Academy of Sciences. Euler succeeded Daniel Bernoulli to the Chair of Mathematics at the Academy of Sciences in St. Petersburg (now Leningrad), Russia. Later, at the invitation of the King of Prussia, he became Director of the Department of Mathematics at the Academy of Berlin. In 1735 he had lost an eye following an illness. Beginning 17 years before his death, Euler gradually became blind, but he continued working even after he had lost his sight entirely.*

Euler, one of the master analysts of the eighteenth century, was one of the most productive mathematicians of all time. He wrote on nearly every branch of mathematics then known. Euler's major fields were calculus, differential equations, differential geometry, theory of numbers, series, and the calculus of variations. Over 70 large volumes of Euler's works have appeared so far. He wrote more than 700 papers on various branches of mathematics. Besides his work in mathematics, Euler did important research in astronomy, hydrodynamics, and optics. Euler was alert to the end. On his last day he was calculating the orbit of the planet Uranus. He suffered a stroke that evening.

bers for which the formula is meaningful. We illustrate this in the following example.

Example 3 Give the domain of each of the functions defined by the following formulas.

Note: If f is a function and $f(x_1) = y_1$ and $f(x_1) = y_2$ then $y_1 = y_2$

Notation: In general, functions are denoted by letters such as f, g, h, F, H, etc.

a $f(x) = x^2 + 1$

b $g(t) = \dfrac{1}{t(t-2)}$

c $h(u) = \sqrt{u+1}$

d $f(s) = \dfrac{1}{\sqrt{s^2 - s - 2}}$

Solution **a** Since for every real number x, $x^2 + 1$ is a real number,

$\mathbf{R}$ is the set of real numbers.

domain $f = \{x : x \in \mathbf{R}\}$

Note that in this case the range of f is $\{y : y \geq 1\}$.

b For the equation in example b to be meaningful we must have $t(t-2) \neq 0$. Therefore

domain $g = \{t : t \in \mathbf{R}$ and $t \neq 0$ and $t \neq 2\}$

c Since $\sqrt{u+1}$ is a real number if and only if $u + 1 \geq 0$, $u \geq -1$ and

domain $h = \{u : u \geq -1\}$

Note: The solution of the inequality $s^2 - s - 2 > 0$ is $\{s : s < -1\} \cup \{s : s > 2\}$

d Here $f(s)$ is a real number for values of s for which $s^2 - s - 2 > 0$. Therefore

domain $f = \{s : s < -1\} \cup \{s : s > 2\}$

Functions may be combined with the operations of addition, subtraction, multiplication, and division. If f and g are functions whose domains are A and B, respectively, then the sum $f + g$, the difference $f - g$, the product fg, and quotient f/g are defined by

$(f + g)(x) = f(x) + g(x)$ for each $x \in A \cap B$ **[3]**

$(f - g)(x) = f(x) - g(x)$ for each $x \in A \cap B$ **[4]**

$(fg)(x) = f(x)g(x)$ for each $x \in A \cap B$ **[5]**

$(f/g)(x) = f(x)/g(x)$ for each $x \in A \cap B$ and $g(x) \neq 0$ **[6]**

Example 4 Let the functions f and g be defined by

$f(x) = x^2 + 4 \qquad g(x) = \sqrt{x-2}$

Describe the functions $f + g$, $f - g$, fg, and f/g.

Solution Here the domain of f is the set of all real numbers. The domain of g is $\{x : x \geq 2\}$. Therefore, we have

$$\begin{aligned}(f + g)(x) &= f(x) + g(x) \\ &= x^2 + 4 + \sqrt{x - 2}\end{aligned}$$

and

$$\text{domain } (f + g) = \{x : x \geq 2\}$$

Secondly,

$$\begin{aligned}(f - g)(x) &= f(x) - g(x) \\ &= x^2 + 4 - \sqrt{x - 2}\end{aligned}$$

and

$$\text{domain } (f - g) = \{x : x \geq 2\}$$

Next we have

$$\begin{aligned}(fg)(x) &= f(x)g(x) \\ &= (x^2 + 4)(\sqrt{x - 2})\end{aligned}$$

and

$$\text{domain } (fg) = \{x : x \geq 2\}$$

Finally,

$$\begin{aligned}(f/g)(x) &= \frac{f(x)}{g(x)} \\ &= \frac{x^2 + 4}{\sqrt{x - 2}}\end{aligned}$$

and

Notice the strict inequality in the domain of f/g. Why?

$$\text{domain } (f/g) = \{x : x > 2\}$$

Another useful operation involving two functions f and g is their **composition.**

Definition 0-8 Let f and g be two functions. The composite function of f and g, denoted by $f \circ g$, is defined by

$$(f \circ g)(x) = f(g(x)) \qquad \textbf{[7]}$$

The domain of $f \circ g$ is $\{x : x \in \text{domain } g \text{ and } g(x) \in \text{domain } f\}$. See the diagram in Figure 0-2.

Example 5 Let the functions f and g be defined by

$$f(x) = x^2 + 4, \qquad g(x) = \sqrt{x - 2}$$

Find $f \circ g$ and $g \circ f$.

Solution From equation 7 we have

$$\begin{aligned}(f \circ g)(x) &= f(g(x)) \\ &= f(\sqrt{x-2}) \\ &= (\sqrt{x-2})^2 + 4 \\ &= x + 2\end{aligned}$$

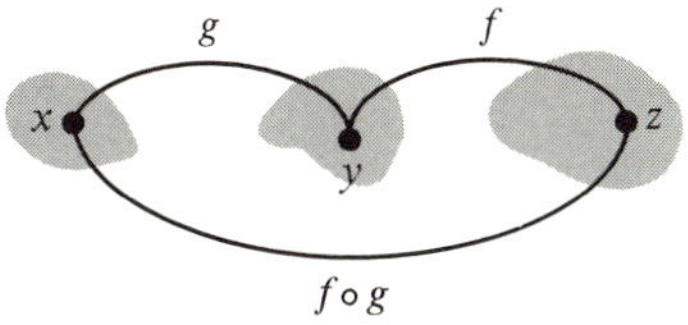

Figure 0-2

Now,

$$\text{domain } f = \{x : x \in \mathbf{R}\}$$
$$\text{domain } g = \{x : x \geq 2\}$$

Therefore,

$$\begin{aligned}\text{domain } f \circ g &= \{x : x \in \text{domain } g \text{ and } g(x) \in \text{domain } f\} \\ &= \{x : x \geq 2 \text{ and } \sqrt{x-2} \text{ is real}\} \\ &= \{x : x \geq 2\}\end{aligned}$$

For $g \circ f$ we have

$$\begin{aligned}(g \circ f)(x) &= g(f(x)) \\ &= g(x^2 + 4) \\ &= \sqrt{(x^2+4) - 2} \\ &= \sqrt{x^2 + 2}\end{aligned}$$

and

Note that in this example $f \circ g \neq g \circ f$.

$$\begin{aligned}\text{domain } g \circ f &= \{x : x \in \text{domain } f \text{ and } f(x) \in \text{domain } g\} \\ &= \{x : x \in \mathbf{R} \text{ and } x^2 + 4 \geq 2\} \\ &= \{x : x \in \mathbf{R}\}\end{aligned}$$

Exercises In problems 1–15, determine the domain of the functions defined by the given equations.

1 $f(x) = 3x + 5$

2 $g(x) = \dfrac{1}{x-1}$

3 $h(t) = t^2 + 1$

4 $g(u) = \dfrac{1}{u^2+1}$

5 $f(t) = \dfrac{3}{t(t+1)}$

6 $h(r) = \dfrac{1}{r^2-1}$

7 $f(x) = \sqrt{x+5}$

8 $f(x) = \sqrt[3]{x+5}$

9 $g(x) = \sqrt{9 - x^2}$

10 $h(u) = \sqrt{u(u - 2)}$

11 $h(t) = \dfrac{4t^2 - 1}{3t + 1}$

12 $g(s) = \sqrt{s^2 - 3s - 4}$

13 $f(x) = \sqrt{\dfrac{1}{2 - x}}$

14 $g(x) = \dfrac{\sqrt{x + 2}}{\sqrt{4 - x^2}}$

15 $h(x) = \dfrac{1}{\sqrt{x^2 - x + 2}}$

16 If the function f is defined by $f(x) = x^2 - 3$, find

a $f(1)$ **b** $f(0)$ **c** $f(-2)$
d $f(\frac{1}{2})$ **e** $f(r + 1)$ **f** $f(a + h) - f(a)$

17 If the function g is defined by $g(x) = \sqrt{2x + 1}$, find

a $g(0)$ **b** $g(3)$ **c** $g(\frac{1}{2})$ **d** $g(3x + 1)$

18 If the function g is defined by $g(x) = 2x^2 - 5$, find the number x in the domain of g such that

a $g(x) = 0$ **b** $g(x) = 3$ **c** $g(x) = 5$

In problems 19–24 the functions f and g are defined. In each problem, define the following functions and determine the domain of the resulting function.

a $f + g$ **b** $f - g$ **c** fg
d f/g **e** $f \circ g$ **f** $g \circ f$

19 $f(x) = x - 4,\quad g(x) = x^2 + 1$

20 $f(x) = 2x + 1,\quad g(x) = x^2$

21 $f(x) = x^2 + 4,\quad g(x) = \sqrt{x}$

22 $f(x) = \dfrac{x - 1}{x + 1},\quad g(x) = \dfrac{2}{x}$

23. $f(x) = \dfrac{2}{x},\quad g(x) = \sqrt{x - 3}$

24 $f(x) = \sqrt{x^2 - 4},\quad g(x) = \sqrt{x - 2}$

A function f is said to be an **even** function if for every x and $-x$ in the domain of f, $f(x) = f(-x)$; and f is said to be an **odd** function if for every x and $-x$ in the domain of f, $f(-x) = -f(x)$.

In problems 25–30, determine whether f is even or odd.

Note:

a If $f(x) = x^2$, then
$f(-x) = (-x)^2 = x^2 = f(x)$
Hence,
$f(x) = x^2$ is an even function.

b If $f(x) = x$, then
$f(-x) = -x = -f(x)$.
Hence,
$f(x) = x$ is an odd function.

25 $f(x) = x^2 - 3x^2 + 1$

26 $f(x) = 2x^3 - 5x$

27 $f(t) = t^4 - 1$

28 $f(s) = 2s^5 + 3$

29 $f(x) = |x|$

30 $f(x) = \dfrac{x + 1}{x - 1}$

31 If f and g are odd functions, show that

a $f + g$ and $f - g$ are both odd functions

b fg and f/g are both even functions

Biographical Sketch

GOTTFRIED WILHELM LEIBNITZ (1646–1716) *was born in Leipzig to a scholarly family. He was only 6 when his father, a professor of philosophy, died. At the age of 15 he entered the University of Leipzig as a law student. His first two years were devoted to philosophy. His introduction to Cardan, Campanelli, Kepler, Galileo, and Descartes led him to be interested in the new mathematical sciences. In his third year, he studied all that the University of Jena could offer in mathematics; his last two years were devoted to law. In 1666 he wrote "On the Art of Combination," and that same year he presented his doctorate thesis in law at the University of Altdorf. In 1676 he accepted the position of Director of the Library and legal advisor to the Duke of Brunswick at Hanover. There he spent the remaining 40 years of his life.*

Leibnitz is one of the greatest thinkers of modern times. He was the builder of a remarkable system of modern philosophy. As a mathematician he was responsible for the development of the differential and integral calculus and symbolic logic. He invented determinants and designed a calculating machine. As a jurist he simplified and codified law and was a pioneer in international law. As a theologian he tried to reconcile the Protestant and Catholic churches. He was also a diplomat, historian, geologist, economist, and linguist. His last years were saddened by ill health and controversy over the legitimacy of his discovery of the calculus. Newton had come to a similar discovery by different means a few years before Leibnitz (1650), though it was published after Leibnitz's. If chronological priority belonged to Newton, nevertheless Leibnitz's discovery was perfectly independent, and his method and notations have been universally recognized in preference to Newton's. Leibnitz died alone and forgotten, no one attending his funeral except his faithful servant. In scientific circles his death was not known until a year later.

Review of Chapter 0

1 A *well-defined* collection of distinct objects is called a

________________.

2 The *objects* in the collection are called ________________.

3 A set that has no elements is called the ________________ set.

4 Give an example of an infinite set:

5 Give an example of a finite set:

6 Define the term universal set:

7 Define the term subset:

8 If $A = \{2,4,6,8\}$ and $B = \{2,4\}$, then B________A.

9 List the proper subsets of B above:

10 Define the intersection of two sets:

11 Define the union of two sets.

12 Define the complement of a set.

13 Define the term function:

14 What is the domain of a function?

15 If $f(x) = x^2 - 6x + 5$, then $f(0) =$ ________________, $f(1) =$ ________________, $f(2) =$ ________________. For what values of x is $f(x) = 0$?

16 **a** If $f(x) = \sqrt{1 - x}$, then domain $f =$ ________________.

b If $f(x) = \dfrac{x + 1}{x\sqrt{x + 2}}$, then domain $f =$ ________________.

17 Define the composition of two functions:

18 Define an even function:

19 Define an odd function:

Matrix Algebra

1

1-1 Addition and Scalar Multiplication of Matrices

Objectives

1. Introduce the concept of a matrix.
2. Discuss addition of matrices and scalar multiplication of a matrix.

In this chapter the reader will be introduced to the concept of a **matrix** (plural: matrices). This mathematical object was invented by Cayley in the nineteenth century. The word matrix was first used in 1858 by James J. Sylvester (1814–1897). The concept of a matrix does not constitute a great innovation in mathematics; it is an innovation in language. Nevertheless, it is a highly useful tool in mathematics. Today the concept of a matrix has become a tool in all branches of mathematics, the sciences, and engineering. In particular, matrices are essential in the study of linear algebra. In this section we introduce matrices and define two operations involving them.

Let us consider the following examples.

Example 1 Sam goes to the grocery store with the following list of items to buy:

4 candles
1 box of napkins
2 loaves of wheat germ bread
1 12-lb. turkey
3 bottles of rosé wine

At the store Sam finds that the prices are as follows:

candles:	2 for \$0.25
box of napkins:	\$0.36
loaf of bread	\$0.42
turkey, per lb.:	\$0.55
wine, per bottle	\$2.85

How much is Sam's bill excluding sales taxes?

Solution To find the total cost of the items bought, Sam multiplies the number of units bought times the cost per unit and adds the results, obtaining an answer of \$16.85.

$$\begin{aligned} 0.25 \times 2 &= 0.50 \\ 0.36 \times 1 &= 0.36 \\ 0.42 \times 2 &= 0.84 \\ 0.55 \times 12 &= 6.60 \\ 2.85 \times 3 &= \underline{8.55} \\ & \ 16.85 \end{aligned}$$

Here we have multiplication between two lists of numbers. Addition and multiplication of two or more such lists of numbers occur frequently in very many applications of mathematics today. This has led to the idea of "matrix arithmetic."

Example 2 Ted and Alice are members of a linear algebra class. In the midterm examination their scores were 84 and 96, while on the final examination they scored 96 and 84, respectively. Since the final examination carried more weight than the midterm examination, Ted ended up with an average of 92 while Alice finished with an average of 88.

Find the weights that were used by the instructor to arrive at the last two averages.

Solution If we denote by x the weight belonging to the midterm examination and by y the weight associated with the final examination, we have

For Ted $84x + 96y = 92$
For Alice $96x + 84y = 88$

In this system it is clear that the coefficients of x and y and the numbers 92 and 88 on the right-hand side of the equations play a significant role in determining the solution (x, y). Clearly the *position* of the numbers 84, 96 and 92, 88 relative to x and y is important. With this in mind we ask: Is it possible to separate the known quantities from the unknown? If so, could we then perform certain operations to determine the unknown? The answer is in the affirmative. One way of doing this is to write the system in the form

The reader could compare this equation with the single algebraic equation

$ax = b$

$$\begin{bmatrix} 84 & 96 \\ 96 & 84 \end{bmatrix} \begin{bmatrix} x \\ y \end{bmatrix} = \begin{bmatrix} 92 \\ 88 \end{bmatrix}$$

Each array of numbers in brackets is called a matrix.

We call the arrangement of numbers $\begin{bmatrix} 84 & 96 \\ 96 & 84 \end{bmatrix}$ the **matrix of coefficients** of the system of equations. The so-called solution vector $\begin{bmatrix} x \\ y \end{bmatrix}$ may be obtained by proper manipulations with the matrix $\begin{bmatrix} 84 & 96 \\ 96 & 84 \end{bmatrix}$ and the ordered pair $\begin{bmatrix} 92 \\ 88 \end{bmatrix}$. It can easily be checked that $x = \frac{1}{3}$ and $y = \frac{2}{3}$ is a solution of the system above. The study of matrices in this chapter will enable us to obtain this solution and solve such systems (and even more complicated ones) in an efficient way.

Example 3 Consider the two lines given by the equations

$5x - 3y = 3$ and $x + 2y = 11$

To find the points of intersection we have to solve the system

$$\begin{aligned} 5x - 3y &= 3 \\ x + 2y &= 11 \end{aligned}$$

This system may be written, using matrices, as follows:

Note carefully that in order to write the system in matrix form, we have displaced only the unknowns x and y, leaving all other numbers in their original relative positions.

$$\begin{bmatrix} 5 & -3 \\ 1 & 2 \end{bmatrix} \begin{bmatrix} x \\ y \end{bmatrix} = \begin{bmatrix} 3 \\ 11 \end{bmatrix}$$

Once again the solution given by $x = 3$, $y = 4$ can be found by simple matrix manipulations which will be discussed later in this chapter.

Many problems in the sciences lead to systems of equations containing more unknowns than the examples above. The general situation can be described by m equations and n unknowns.

m and n are used to denote positive integers.

$$\begin{aligned} a_{11}x_1 + a_{12}x_2 + \cdots + a_{1n}x_n &= b_1 \\ a_{21}x_1 + a_{22}x_2 + \cdots + a_{2n}x_n &= b_2 \\ \vdots \qquad\qquad\qquad\qquad & \quad \vdots \\ a_{m1}x_1 + a_{m2}x_2 + \cdots + a_{mn}x_n &= b_m \end{aligned}$$

Such a system may be written as

$$\begin{bmatrix} a_{11} & a_{12} & \cdots & a_{1n} \\ a_{21} & a_{22} & \cdots & a_{2n} \\ \vdots & & & \vdots \\ a_{m1} & a_{m2} & \cdots & a_{mn} \end{bmatrix} \begin{bmatrix} x_1 \\ x_2 \\ \vdots \\ x_n \end{bmatrix} = \begin{bmatrix} b_1 \\ b_2 \\ \vdots \\ b_m \end{bmatrix}$$

Now we define the concept of a matrix.

Definition 1-1 A **matrix** is a rectangular array of numbers; the generalized form is shown below:

$$A = \begin{bmatrix} a_{11} & a_{12} & \cdots & a_{1n} \\ a_{21} & a_{22} & \cdots & a_{2n} \\ \vdots & \vdots & & \vdots \\ a_{m1} & a_{m2} & \cdots & a_{mn} \end{bmatrix}$$

We say that the matrix A is $m \times n$ if it has m rows and n columns; the matrix is said to have **order** $m \times n$ (read "m by n").

For example, the matrix

Here the first row consists of the numbers 2, 1, and 3. The order in which these numbers appear is important.

$$A = \begin{bmatrix} 2 & 1 & 3 \\ 1 & 5 & 4 \end{bmatrix}$$

Thus, the (i,j) entry is the number in the ith row from the top and in the jth column from the left.

Note: For a 3×2 matrix B we have

$$B = (b_{ij}) = \begin{bmatrix} b_{11} & b_{12} \\ b_{21} & b_{22} \\ b_{31} & b_{32} \end{bmatrix}$$

is 2×3 because it has two rows and three columns. The number that appears in the ith row and jth column will be designated by a_{ij} and will be referred to as the (i,j) **entry** of the matrix A. Unless stated otherwise, we shall assume that all the entries of any matrix are real numbers.

We shall often use the notation $A = (a_{ij})$ for the matrix A, indicating that a_{ij} is the (i,j) entry in the matrix A.

Whenever new mathematical objects are introduced, it must be made clear what is meant by the statement that object A is *equal* to object B. The following definition will make this statement clear when the objects are matrices.

Definition 1-2 Let $A = (a_{ij})$ be an $m \times n$ matrix and $B = (b_{ij})$ a $p \times q$ matrix. We say that A is **equal** to B, and write $A = B$, if and only if $m = p$, $n = q$, and $a_{ij} = b_{ij}$ for $i = 1, 2, \ldots, m$ and $j = 1, 2, \ldots, n$.

Equality of matrices.

It follows from this definition that if $A = B$ then A and B have the same number of rows and the same number of columns, and their corresponding (i,j) entries are equal for all possible i and j. If A is not equal to B we write $A \neq B$.

Example 4 Let

$$A = \begin{bmatrix} 1 & 2 & 4 & 0 \\ 3 & 1 & 0 & 1 \end{bmatrix} \quad \text{and} \quad B = \begin{bmatrix} 1 & 2 & 4 \\ 3 & 1 & 0 \end{bmatrix}$$

Is $A = B$?

Solution Since A is a 2×4 matrix and B is a 2×3 matrix, A and B have different numbers of columns. It follows from definition 1-2 that $A \neq B$.

Example 5 Let

$$C = \begin{bmatrix} 0 & 0 & 1 \\ 0 & 0 & 1 \end{bmatrix} \quad \text{and} \quad D = \begin{bmatrix} 0 & 1 & 0 \\ 0 & 0 & 1 \end{bmatrix}$$

Is $C = D$?

Solution Both matrices are 2×3 and contain entries which are either 1 or 0. A simple check reveals that $c_{13} = 1$ while the corresponding entry $d_{13} = 0$ so that $c_{13} \neq d_{13}$. Further checking is unnecessary, and we conclude that $C \neq D$.

We now introduce two basic operations involving matrices, the *sum* and the *scalar multiplication* of matrices.

Definition 1-3 Let A and B be $m \times n$ matrices. The **sum** of A and B, written $A + B$, is the $m \times n$ matrix C for which $c_{ij} = a_{ij} + b_{ij}$ where $i = 1, 2, \ldots, m$ and $j = 1, 2, \ldots, n$.

We write $A + B = C$.

Notice that the sum of A and B is defined only when the two matrices have the same number of rows and the same number of columns.

Example 6 Let

$$A = \begin{bmatrix} 0 & 2 & 4 \\ 1 & 3 & 5 \end{bmatrix} \quad \text{and} \quad B = \begin{bmatrix} \sqrt{2} & 0 & \sqrt{3} \\ 1 & 5 & -1 \end{bmatrix}$$

Find $A + B$.

Solution We have

$$A + B = \begin{bmatrix} 0+\sqrt{2} & 2+0 & 4+\sqrt{3} \\ 1+1 & 3+5 & 5+(-1) \end{bmatrix} = \begin{bmatrix} \sqrt{2} & 2 & 4+\sqrt{3} \\ 2 & 8 & 4 \end{bmatrix}$$

Definition 1-4 Let A be an $m \times n$ matrix and α be a real number. The **scalar** multiple of A by α, denoted by αA, is the matrix B for which $b_{ij} = \alpha a_{ij}$ where $i = 1, 2, \ldots, m$ and $j = 1, 2, \ldots, n$.

Scalar multiplication.

The real number α in αA is called a scalar.

In other words, to obtain the matrix $B = \alpha A$, one multiplies each entry of the matrix A by the same number α.

Example 7 Let

$$A = \begin{bmatrix} 1 & -2 \\ 3 & 5 \end{bmatrix} \quad \text{and} \quad \alpha = \tfrac{1}{2}$$

Find αA.

Solution We have

$$\alpha A = \begin{bmatrix} (\frac{1}{2})(1) & (\frac{1}{2})(-2) \\ (\frac{1}{2})(3) & (\frac{1}{2})(5) \end{bmatrix} = \begin{bmatrix} \frac{1}{2} & -1 \\ \frac{3}{2} & \frac{5}{2} \end{bmatrix}$$

The set of all $m \times n$ matrices under the operations of sum and scalar multiplication satisfies algebraic properties which will remind us of the system of real numbers. The proofs of these properties depend on properties of the real numbers. Here we shall assume that the reader is familiar with the basic algebraic properties of the real number system.

Theorem 1-1 Let A, B, and C be $m \times n$ matrices. Then

Matrix addition is **commutative** and **associative.**

a $A + B = B + A$

b $(A + B) + C = A + (B + C)$

Proof **a** Here we have to show that the (i,j) entries of $A + B$ and $B + A$ are the same. Now, if $A = (a_{ij})$ and $B = (b_{ij})$, then the (i,j) entries of $A + B$ and $B + A$ are $a_{ij} + b_{ij}$ and $b_{ij} + a_{ij}$, respectively. Since addition of real numbers is commutative, we have

$$a_{ij} + b_{ij} = b_{ij} + a_{ij}$$

for $i = 1, 2, \ldots, m$ and $j = 1, 2, \ldots, n$. Therefore, by Definition 1-2 we have

$$A + B = B + A$$

b The proof of this part is based on the associativity of the real numbers with respect to addition, and the details are left for the reader as an exercise.

A matrix having only zero entries is called a **zero** matrix.

Let us denote by 0 a matrix such that all its entries are zero. If A is an $m \times n$ matrix then $A + 0$ will be meaningful only if 0 is also an $m \times n$ matrix. With this understanding we can state the following:

Theorem 1-2 Let 0 be the $m \times n$ zero matrix. Then

$$A + 0 = A$$

for every $m \times n$ matrix A.

Proof The (i,j) entry of $A + 0$ is $a_{ij} + 0$ while the (i,j) entry of A is a_{ij}. Since

$$a_{ij} + 0 = a_{ij}$$

the proof is complete.

The following question may be raised: Is there any matrix other than 0 with the property mentioned in Theorem 1-2? The answer to this question is provided by the next theorem.

Theorem 1-3 Let B be an $m \times n$ matrix such that $A + B = A$ for some $m \times n$ matrix A. Then $B = 0$.

For fixed m and n, 0 is the unique matrix satisfying the matrix equation $A + X = A$ for any $m \times n$ matrix A.

Proof The (i,j) entry of $A + B$ is $a_{ij} + b_{ij}$ and the corresponding entry of A is a_{ij}. Since $A + B = A$, we have

$$a_{ij} + b_{ij} = a_{ij}$$

for $i = 1, 2, \ldots, m$ and $j = 1, 2, \ldots, n$. It follows that $b_{ij} = 0$ and hence $B = 0$.

From the real number system we know that if a is any real number then there is a unique real number b such that

$$a + b = 0$$

Do matrices share a similar property? The answer is provided by the following theorem.

Theorem 1-4 If A is any given $m \times n$ matrix, then there exists a unique $m \times n$ matrix B such that

$$A + B = 0$$

Proof There are two parts to this theorem. First, we need to show that there exists a matrix B such that $A + B = 0$. This is a simple matter because for $B = (b_{ij})$ we can define $b_{ij} = -a_{ij}$. This choice implies that the (i,j) entry of $A + B$ is $a_{ij} + (-a_{ij}) = 0$, which proves the first part. The second part of the theorem deals with the uniqueness of B. We have to show that if B and C are $m \times n$ matrices such that $A + B = 0$ and $A + C = 0$, then $B = C$.

We consider the matrices $(C + A) + B$ and $C + (A + B)$. Theorem 1-1 implies that

$$(C + A) + B = C + (A + B)$$

From the left-hand side of this equation we have

The reader is urged to justify each of the steps taken.

$$\begin{aligned}(C + A) + B &= (A + C) + B \\ &= 0 + B \\ &= B + 0 \\ &= B\end{aligned}$$

And from the right-hand side of the equation we have

$$\begin{aligned}C + (A + B) &= C + 0 \\ &= C\end{aligned}$$

It follows that $B = C$ and uniqueness is proved.

We denote by $-A$ the unique matrix with the property $A + (-A) = 0$. Using the definition of scalar multiplication (see Definition 1-4), we see at once that $-A = (-1)A$.

$-A$ is called the **negative** of the matrix A.

Further algebraic properties of scalar multiplication of matrices are summarized in the following theorem.

Theorem 1-5 Let A and B be $m \times n$ matrices and let α and β be real numbers. Then

a $\alpha(A + B) = \alpha A + \alpha B$

b $(\alpha + \beta)A = \alpha A + \beta A$

c $\alpha(\beta A) = (\alpha\beta)A$

Proof **a** The (i,j) entry of $\alpha(A + B)$ is $\alpha(a_{ij} + b_{ij})$, while the (i,j) entry of $\alpha A + \alpha B$ is $\alpha a_{ij} + \alpha b_{ij}$. By the distributive law for real numbers we have

$$\alpha(a_{ij} + b_{ij}) = \alpha a_{ij} + \alpha b_{ij}$$

See problem 15.

which proves part a. The proofs of parts b and c are left for the reader as exercises.

Example 8 Let

$$\alpha = 5,\ \beta = 2,\ A = \begin{bmatrix} 2 & 1 & 0 \\ 0 & 3 & 5 \end{bmatrix}, \quad \text{and} \quad B = \begin{bmatrix} 0 & 1 & 3 \\ 2 & 0 & 1 \end{bmatrix}$$

Find $\alpha(A + B)$, $\alpha A + \alpha B$, $(\alpha + \beta)A$, $\alpha A + \beta A$, $\alpha(\beta A)$, and $(\alpha\beta)A$.

Solution

$$\alpha(A + B) = 5\left\{\begin{bmatrix} 2 & 1 & 0 \\ 0 & 3 & 5 \end{bmatrix} + \begin{bmatrix} 0 & 1 & 3 \\ 2 & 0 & 1 \end{bmatrix}\right\} = 5\begin{bmatrix} 2 & 2 & 3 \\ 2 & 3 & 6 \end{bmatrix}$$

$$= \begin{bmatrix} 10 & 10 & 15 \\ 10 & 15 & 30 \end{bmatrix}$$

$$\alpha A + \alpha B = 5\begin{bmatrix} 2 & 1 & 0 \\ 0 & 3 & 5 \end{bmatrix} + 5\begin{bmatrix} 0 & 1 & 3 \\ 2 & 0 & 1 \end{bmatrix} = \begin{bmatrix} 10 & 5 & 0 \\ 0 & 15 & 25 \end{bmatrix}$$

$$+ \begin{bmatrix} 0 & 5 & 15 \\ 10 & 0 & 5 \end{bmatrix} = \begin{bmatrix} 10 & 10 & 15 \\ 10 & 15 & 30 \end{bmatrix}$$

$$(\alpha + \beta)A = (5 + 2)\begin{bmatrix} 2 & 1 & 0 \\ 0 & 3 & 5 \end{bmatrix} = \begin{bmatrix} 7\cdot 2 & 7\cdot 1 & 7\cdot 0 \\ 7\cdot 0 & 7\cdot 3 & 7\cdot 5 \end{bmatrix}$$

$$= \begin{bmatrix} 14 & 7 & 0 \\ 0 & 21 & 35 \end{bmatrix}$$

$$\alpha A + \beta A = 5\begin{bmatrix} 2 & 1 & 0 \\ 0 & 3 & 5 \end{bmatrix} + 2\begin{bmatrix} 2 & 1 & 0 \\ 0 & 3 & 5 \end{bmatrix} = \begin{bmatrix} 10 & 5 & 0 \\ 0 & 15 & 25 \end{bmatrix}$$

$$+ \begin{bmatrix} 4 & 2 & 0 \\ 0 & 6 & 10 \end{bmatrix} = \begin{bmatrix} 14 & 7 & 0 \\ 0 & 21 & 35 \end{bmatrix}$$

$$\alpha(\beta A) = 5\left\{2\begin{bmatrix} 2 & 1 & 0 \\ 0 & 3 & 5 \end{bmatrix}\right\} = 5\begin{bmatrix} 4 & 2 & 0 \\ 0 & 6 & 10 \end{bmatrix} = \begin{bmatrix} 20 & 10 & 0 \\ 0 & 30 & 50 \end{bmatrix}$$

$$(\alpha\beta)A = (5\cdot 2)\begin{bmatrix} 2 & 1 & 0 \\ 0 & 3 & 5 \end{bmatrix} = 10\begin{bmatrix} 2 & 1 & 0 \\ 0 & 3 & 5 \end{bmatrix} = \begin{bmatrix} 20 & 10 & 0 \\ 0 & 30 & 50 \end{bmatrix}$$

Biographical Sketch

ARTHUR CAYLEY (1821–1895). *Arthur Cayley's father was an English merchant involved in trade with Russia, and Arthur was born during one of his parents' frequent visits to England. When Arthur was 8 years old his father retired permanently in England and the boy was sent to private schools. His mathematical talent was immediately spotted and his teachers all agreed he should make mathematics his career.*

Cayley entered Trinity College at the age of 17. By the time he was 20 he was far ahead of his fellow students in mathematics. He was elected Fellow of Trinity and assistant tutor. During the first 3 years his duties were light, allowing him to spend much time on the mathematical researches he had begun as an undergraduate. The 25 papers he published in this 3-year period map out much of the work that was to occupy him for the rest of his life. Included were his study of geometry of n *dimensions (which he originated) his development of the theory of algebraic invariants, his invention of the theory of matrices, and his outstanding contributions to the theory of elliptic functions.*

Cayley's interests were not solely confined to mathematics, and sandwiched between his hard work were vacations on the Continent, where mountaineering and watercolor sketching became a passion. He had always loved good literature and read classical Greek, French, German, and Italian works in the original.

Cayley left Cambridge in 1846 at the age of 25, unable to find a position as a mathematician because he could not in good conscience take the holy orders which would guarantee him such a job. He became attracted to law and began to prepare himself for a legal career. After 3 years of studies he was admitted to the bar in 1849. For 14 dreary years of law work he managed, by taking enough work, to make a decent living, spending the rest of his time writing mathematical papers, publishing between 200 and 300, many of which are now classic. It was during Cayley's legal career that he became acquainted with James J. Sylvester.

In 1863 Cayley accepted the newly established professorship of mathematics at Cambridge University and was married that same year. He now devoted his life to mathematical research and university administration where his business training, even temper, and legal experience were invaluable. He was instrumental in having women finally admitted as students at Cambridge.

Cayley's creative activity did not diminish up to the day he died, after a long and painful illness.

Exercises

In problems 1–4, find the indicated sums.

1 $\begin{bmatrix} 1 & 3 \\ 5 & 0 \end{bmatrix} + \begin{bmatrix} -2 & 0 \\ 1 & 1 \end{bmatrix}$

2 $\begin{bmatrix} 3 & 1 & -3 \\ 0 & 7 & 2 \end{bmatrix} + \begin{bmatrix} -2 & 4 & 4 \\ 1 & 1 & 3 \end{bmatrix}$

3 $\begin{bmatrix} 1 & 2 \\ 3 & 4 \\ 5 & 6 \end{bmatrix} + \begin{bmatrix} 6 & 5 \\ 4 & 3 \\ 2 & 1 \end{bmatrix}$

4 $\begin{bmatrix} 2 & 0 & 2 \\ 1 & 3 & 1 \\ 1 & -1 & 4 \end{bmatrix} + \begin{bmatrix} 10 & 3 & 4 \\ 1 & 0 & 1 \\ 3 & 0 & 3 \end{bmatrix}$

In each of problems 5–7, find the matrix A that satisfies the given equation.

5 $A + \begin{bmatrix} 1 & -3 \\ 2 & 1 \end{bmatrix} = \begin{bmatrix} 5 & 4 \\ -2 & 1 \end{bmatrix}$

6 $A - \begin{bmatrix} 0 & -1 \\ 0 & 0 \end{bmatrix} = \begin{bmatrix} 3 & -1 \\ 2 & 1 \end{bmatrix}$

7 $A + \begin{bmatrix} -1 & 3 & 2 \\ 1 & 0 & 0 \end{bmatrix} = \begin{bmatrix} 4 & -1 & 1 \\ 2 & 1 & -4 \end{bmatrix}$

8 Let A be a 3×2 matrix. Let the (i,j) entry of A be defined by $a_{ij} = i + j$. Find the actual numerical values of all the entries of A.

9 Let B be a 3×3 matrix. Let the (i,j) entry of B be defined by $b_{ij} = i - j$. Find all the entries of B.

10 Let C be a 2×2 matrix. Let the (i,j) entry of C be defined by $c_{ij} = 2i + 3j$. Find all the entries of C.

11 Let D be a 4×4 matrix. Let the (i,j) entry of D be defined by $d_{ij} = \delta_{ij}$ where δ_{ij} is the Kronecker delta. Find all the entries of D.

The Kronecker delta, denoted δ_{ij}, is defined by

$$\delta_{ij} = \begin{cases} 1 & \text{if } i = j \\ 0 & \text{if } i \neq j \end{cases}$$

12 Let A and B be two 2×2 matrices. Let $a_{ij} = 2 \cdot i \cdot j$ and $b_{ij} = i^2 + j^2$. Find A, B, $A + B$, and $A - B$. Note: $A - B = A + (-B)$.

13 Let $\alpha = -2$, $A = \begin{bmatrix} 3 & -2 \\ 1 & 5 \end{bmatrix}$, and $B = \begin{bmatrix} 0 & 1 \\ -3 & 4 \end{bmatrix}$. Find $\alpha(A + B)$ and $\alpha A + \alpha B$.

14 Let $\alpha = 3$, $\beta = -2$, and $A = \begin{bmatrix} 7 & -3 \\ 8 & 1 \end{bmatrix}$. Find $(\alpha + \beta)A$ and $\alpha A + \beta A$. Also find $\alpha(\beta A)$ and $(\alpha\beta)A$.

15 Prove parts b and c of Theorem 1-5.

16 Let A be a 3×3 matrix. Given that $a_{12} = -4$, $a_{13} = -7$, and $a_{ij} = \alpha i + \beta j$, find all the remaining entries of A. (Hint: first find α and β.)

LEOPOLD KRONECKER (1823-1891) *was born in Liegnitz, Prussia, to a prosperous family. His performance at school was brilliant and he showed talent in mathematics at an early age. Kronecker studied at the University of Berlin and was greatly influenced by the outstanding mathematician Dirichlet. Kronecker's Ph.D. dissertation, "On Complex Units," is in the field of algebraic numbers. Kronecker was a staunch algebraist and had a prolonged mathematical war against the father of modern analysis, Weierstrass. He declared, "God made the integers; all the rest is the work of man."*

1-2 Product of Matrices

Objectives

1 Discuss the product of two matrices.
2 Establish associative and distributive laws for multiplication of matrices.

In the previous section we discussed some basic properties associated with matrix addition and scalar multiplication. We now introduce another important operation involving matrices—the *product.*

Definition 1-5 Let $A = (a_{ij})$ be an $m \times n$ matrix and $B = (b_{ij})$ be an $n \times k$ matrix.

The **product** of A and B, denoted by AB, is the $m \times k$ matrix C such that

$$c_{ij} = a_{i1}b_{1j} + a_{i2}b_{2j} + \cdots + a_{in}b_{nj}$$

and we write $AB = C$.

Observation 1 To obtain the (i,j) entry of AB one has to multiply the entries of the ith row of A by the corresponding entries of the jth column of B and sum up these products.

Observation 2 It should be clear from the definition that the product AB is defined only if the number of columns in A is equal to the number of rows in B. In short, the "length" of every row in A must equal the "length" of every column in B.

Example 1 Let

$$A = \begin{bmatrix} a_{11} & a_{12} \\ a_{21} & a_{22} \end{bmatrix} \quad \text{and} \quad B = \begin{bmatrix} b_{11} & b_{12} & b_{13} \\ b_{21} & b_{22} & b_{23} \end{bmatrix}$$

Then

$$AB = \begin{bmatrix} a_{11}b_{11} + a_{12}b_{21} & a_{11}b_{12} + a_{12}b_{22} & a_{11}b_{13} + a_{12}b_{23} \\ a_{21}b_{11} + a_{22}b_{21} & a_{21}b_{12} + a_{22}b_{22} & a_{21}b_{13} + a_{22}b_{23} \end{bmatrix}$$

Notice that the number of rows in AB is equal to the number of rows in A and that the number of columns in AB is equal to the number of columns in B. Observe also that BA is not defined because B is a 2×3 matrix and A is a 2×2 matrix.

In general, if A is an $m \times n$ matrix and B is a $p \times q$ matrix, then in order for both AB and BA to be defined we must have $n = p$ and $q = m$.

Thus, if A is a 2×3 matrix and B is a 3×2 matrix, both AB and BA will be defined. The matrix AB will be of order 2×2 and the matrix BA will be of order 3×3. In this case it is obvious that $AB \neq BA$. There still remains the possibility that AB and BA are both $m \times m$ matrices. This will be the case if and only if A and B are both $m \times m$ matrices. A matrix having the same number of rows as columns is called a **square matrix.**

We ask the following question: If A and B are square matrices of order m, that is, $m \times m$, is it true that $AB = BA$?

The answer to this question is negative. Matrix multiplication is not commutative, as the following example shows.

Consider the effect of two successive transformations

$$x_1 = a_{11}x + a_{12}y$$
$$y_1 = a_{21}x + a_{22}y$$

and

$$x_2 = b_{11}x_1 + b_{12}y_1$$
$$y_2 = b_{21}x_1 + b_{22}y_1$$

Then the relation between x_2, y_2 and x, y is given by

$$x_2 = (b_{11}a_{11} + b_{12}a_{12})x + (b_{11}a_{12} + b_{12}a_{22})y$$
$$y_2 = (b_{21}a_{11} + b_{22}a_{21})x + (b_{21}a_{12} + b_{22}a_{22})y$$

Cayley used this result and defined the product of the matrices to be

$$\begin{bmatrix} b_{11} & b_{12} \\ b_{21} & b_{22} \end{bmatrix}\begin{bmatrix} a_{11} & a_{12} \\ a_{21} & a_{22} \end{bmatrix} = \begin{bmatrix} b_{11}a_{11} + b_{12}a_{21} & b_{11}a_{12} + b_{12}a_{22} \\ b_{21}a_{11} + b_{22}a_{21} & b_{21}a_{12} + b_{22}a_{22} \end{bmatrix}$$

For brevity, a matrix of order $m \times m$ is called a **square matrix of order** m.

Is matrix multiplication commutative?

Example 2 Let

$$A = \begin{bmatrix} 1 & 1 \\ 0 & 1 \end{bmatrix} \quad \text{and} \quad B = \begin{bmatrix} 0 & 0 \\ 1 & 1 \end{bmatrix}$$

Show that $AB \neq BA$.

Solution We find that

$$AB = \begin{bmatrix} 1 & 1 \\ 1 & 1 \end{bmatrix} \quad \text{and} \quad BA = \begin{bmatrix} 0 & 0 \\ 1 & 2 \end{bmatrix}$$

so that $AB \neq BA$.

It is important to note that there exist square matrices C and D such that $CD = DC$. The following example exhibits such a possibility.

Example 3 Let

$$C = \begin{bmatrix} \alpha & 0 \\ 0 & \alpha \end{bmatrix} \quad \text{and} \quad D = \begin{bmatrix} d_{11} & d_{12} \\ d_{21} & d_{22} \end{bmatrix}$$

where α and $d_{ij}(i = 1, 2; j = 1, 2)$ are any real numbers. Show that $CD = DC$.

Solution

$$CD = \begin{bmatrix} \alpha d_{11} & \alpha d_{12} \\ \alpha d_{21} & \alpha d_{22} \end{bmatrix} \quad \text{and} \quad DC = \begin{bmatrix} d_{11}\alpha & d_{12}\alpha \\ d_{21}\alpha & d_{22}\alpha \end{bmatrix}$$

Now for real numbers $\alpha d_{ij} = d_{ij}\alpha$, so that $CD = DC$.

Is matrix multiplication associative?

The next problem we wish to examine is whether matrix multiplication is associative or not. We first consider a particular example.

Example 4 Let

$$A = \begin{bmatrix} 2 & 1 & -1 \\ 0 & -1 & 2 \end{bmatrix}, \quad B = \begin{bmatrix} 1 \\ 3 \\ 2 \end{bmatrix}, \quad \text{and} \quad C = [-2 \quad 1]$$

Find $(AB)C$ and $A(BC)$. Is $(AB)C = A(BC)$?

Solution

$$(AB)C = \left\{ \begin{bmatrix} 2 & 1 & -1 \\ 0 & -1 & 2 \end{bmatrix} \begin{bmatrix} 1 \\ 3 \\ 2 \end{bmatrix} \right\} [-2 \quad 1]$$

$$= \begin{bmatrix} (2)(1) + (1)(3) + (-1)(2) \\ (0)(1) + (-1)(3) + (2)(2) \end{bmatrix} [-2 \quad 1]$$

$$= \begin{bmatrix} 3 \\ 1 \end{bmatrix} [-2 \quad 1] = \begin{bmatrix} (3)(-2) & (3)(1) \\ (1)(-2) & (1)(1) \end{bmatrix} = \begin{bmatrix} -6 & 3 \\ -2 & 1 \end{bmatrix}$$

$$A(BC) = \begin{bmatrix} 2 & 1 & -1 \\ 0 & -1 & 2 \end{bmatrix} \left\{ \begin{bmatrix} 1 \\ 3 \\ 2 \end{bmatrix} [-2 \quad 1] \right\}$$

$$= \begin{bmatrix} 2 & 1 & -1 \\ 0 & -1 & 2 \end{bmatrix} \begin{bmatrix} (1)(-2) & (1)(1) \\ (3)(-2) & (3)(1) \\ (2)(-2) & (2)(1) \end{bmatrix}$$

$$= \begin{bmatrix} 2 & 1 & -1 \\ 0 & -1 & 2 \end{bmatrix} \begin{bmatrix} -2 & 1 \\ -6 & 3 \\ -4 & 2 \end{bmatrix}$$

$$= \begin{bmatrix} (2)(-2) + (1)(-6) + (-1)(-4) & (2)(1) + (1)(3) + (-1)(2) \\ (0)(-2) + (-1)(-6) + (2)(-4) & (0)(1) + (-1)(3) + (2)(2) \end{bmatrix}$$

$$= \begin{bmatrix} -6 & 3 \\ -2 & 1 \end{bmatrix}$$

Thus, we have $(AB)C = A(BC) = \begin{bmatrix} -6 & 3 \\ -2 & 1 \end{bmatrix}$

In fact this last result holds for all matrices A, B, and C, provided that both $(AB)C$ and $A(BC)$ are defined. We shall prove this in the next theorem. However, for the general case, the expressions that we get for the (i,j) entry of $(AB)C$ and $A(BC)$ are fairly complicated. It is therefore advantageous to introduce at this point the **Σ notation** and some of its properties. For instance,

$$a_1b_1 + a_2b_2 + \cdots + a_nb_n = \sum_{i=1}^{n} a_ib_i$$

The letter i, called the **index,** is a "dummy" and can be replaced by another letter. Thus,

$$\sum_{k=1}^{n} a_kb_k = \sum_{i=1}^{n} a_ib_i$$

Using the Σ notation, we can write

$$a_{i1}b_{1j} + a_{i2}b_{2j} + \cdots + a_{in}b_{nj} = \sum_{k=1}^{n} a_{ik}b_{kj}$$

Some simple properties of the Σ notation are:

1 $\sum_{i=1}^{n} (a_i + b_i)c_i = \sum_{i=1}^{n} a_ic_i + \sum_{i=1}^{n} b_ic_i$

2 $\sum_{i=1}^{n} \alpha a_ib_i = \alpha\left(\sum_{i=1}^{n} a_ib_i\right)$

3 $\sum_{j=1}^{n}\sum_{i=1}^{m} d_{ij} = \sum_{i=1}^{m}\sum_{j=1}^{n} d_{ij}$

Notice that the summation runs over the index k only. We now prove the following theorem.

Theorem 1-6 Associative law for multiplication.

Let A, B, and C be $m \times n$, $n \times k$, and $k \times p$ matrices, respectively. Then

$$(AB)C = A(BC)$$

Proof Let $AB = D$, $(AB)C = E$, $BC = F$, and $A(BC) = G$. Then

$$e_{ij} = \sum_{r=1}^{k} d_{ir}c_{rj} \qquad \text{and} \qquad g_{ij} = \sum_{s=1}^{n} a_{is}f_{sj}$$

where

$$d_{ir} = \sum_{s=1}^{n} a_{is}b_{sr} \qquad \text{and} \qquad f_{sj} = \sum_{r=1}^{k} b_{sr}c_{rj}$$

Thus we have

$$e_{ij} = \sum_{r=1}^{k} \left(\sum_{s=1}^{n} a_{is}b_{sr} \right) c_{rj}$$

and

$$g_{ij} = \sum_{s=1}^{n} a_{is} \left(\sum_{r=1}^{k} b_{sr}c_{rj} \right)$$

Now we must show that $e_{ij} = g_{ij}$ for $i = 1, 2, \ldots, m$ and $j = 1, 2, \ldots, p$. Using properties of the Σ notation, we have

$$\begin{aligned}
e_{ij} &= \sum_{r=1}^{k} (a_{i1}b_{1r} + \cdots + a_{in}b_{nr})c_{rj} \\
&= \sum_{r=1}^{k} (a_{i1}b_{1r}c_{rj} + \cdots + a_{in}b_{nr}c_{rj}) \\
&= \sum_{r=1}^{k} a_{i1}b_{1r}c_{rj} + \cdots + \sum_{r=1}^{k} a_{in}b_{nr}c_{rj} \\
&= a_{i1} \sum_{r=1}^{k} b_{1r}c_{rj} + \cdots + a_{in} \sum_{r=1}^{k} b_{nr}c_{rj} \\
&= \sum_{s=1}^{n} a_{is} \sum_{r=1}^{k} b_{sr}c_{rj} \\
&= g_{ij}
\end{aligned}$$

Since e_{ij} is the (i,j) entry of $(AB)C$ and g_{ij} is the (i,j) entry of $A(BC)$, it follows that $(AB)C = A(BC)$.

Another useful result is given in the following theorem.

Theorem 1-7 Let A be an $m \times n$ matrix, B an $n \times k$ matrix, and α any real number. Then

$$(\alpha A)B = A(\alpha B) = \alpha(AB)$$

See problem 6.

The proof is left for the reader as an exercise.

Example 5 Let

$$\alpha = 2,\ A = \begin{bmatrix} 1 & 0 & 3 \\ 0 & 4 & 1 \end{bmatrix}, \quad \text{and} \quad B = \begin{bmatrix} -1 & 2 \\ 0 & 1 \\ 1 & -1 \end{bmatrix}$$

Find $(\alpha A)B$, $A(\alpha B)$, and $\alpha(AB)$.

Solution

$$(\alpha A)B = \left\{2\begin{bmatrix} 1 & 0 & 3 \\ 0 & 4 & 1 \end{bmatrix}\right\}\begin{bmatrix} -1 & 2 \\ 0 & 1 \\ 1 & -1 \end{bmatrix}$$

$$= \begin{bmatrix} 2 & 0 & 6 \\ 0 & 8 & 2 \end{bmatrix}\begin{bmatrix} -1 & 2 \\ 0 & 1 \\ 1 & -1 \end{bmatrix} = \begin{bmatrix} 4 & -2 \\ 2 & 6 \end{bmatrix}$$

$$A(\alpha B) = \begin{bmatrix} 1 & 0 & 3 \\ 0 & 4 & 1 \end{bmatrix}\left\{2\begin{bmatrix} -1 & 2 \\ 0 & 1 \\ 1 & -1 \end{bmatrix}\right\}$$

$$= \begin{bmatrix} 1 & 0 & 3 \\ 0 & 4 & 1 \end{bmatrix}\begin{bmatrix} -2 & 4 \\ 0 & 2 \\ 2 & -2 \end{bmatrix} = \begin{bmatrix} 4 & -2 \\ 2 & 6 \end{bmatrix}$$

$$\alpha(AB) = 2\left\{\begin{bmatrix} 1 & 0 & 3 \\ 0 & 4 & 1 \end{bmatrix}\begin{bmatrix} -1 & 2 \\ 0 & 1 \\ 1 & -1 \end{bmatrix}\right\} = 2\begin{bmatrix} 2 & -1 \\ 1 & 3 \end{bmatrix} = \begin{bmatrix} 4 & -2 \\ 2 & 6 \end{bmatrix}$$

Matrices also satisfy distributive laws, as indicated by the next two theorems.

Theorem 1-8 Distributive laws for multiplication.

Let A and B be $m \times n$ matrices and C an $n \times k$ matrix. Then $(A + B)C = AC + BC$.

Proof Let $(A + B)C = D$ and $AC + BC = E$. We must show that $d_{ij} = e_{ij}$ for $i = 1, 2, \ldots, m$ and $j = 1, 2, \ldots, k$. We have

$$d_{ij} = \sum_{p=1}^{n} (a_{ip} + b_{ip})c_{pj}$$

and

$$e_{ij} = \sum_{p=1}^{n} a_{ip}c_{pj} + \sum_{p=1}^{n} b_{ip}c_{pj}$$

But

$$\sum_{p=1}^{n} (a_{ip} + b_{ip})c_{pj} = \sum_{p=1}^{n} (a_{ip}c_{pj} + b_{ip}c_{pj})$$

$$= \sum_{p=1}^{n} a_{ip}c_{pj} + \sum_{p=1}^{n} b_{ip}c_{pj}$$

and the proof is complete.

Theorem 1-9 Let C be an $m \times n$ matrix and let A and B be $n \times k$ matrices. Then $C(A + B) = CA + CB$.

The proof follows the pattern of Theorem 1-8 and is left to the student as an exercise.

The necessity for proving two separate distributive laws is dictated by the fact that in general $(A + B)C \neq C(A + B)$: matrix multiplication is not commutative.

Example 6 Let

See problem 8.

$$A = \begin{bmatrix} 1 & 0 \\ -1 & 1 \end{bmatrix}, \quad B = \begin{bmatrix} 0 & 2 \\ 1 & -1 \end{bmatrix}, \quad \text{and} \quad C = \begin{bmatrix} 1 & 1 \\ 1 & 0 \end{bmatrix}$$

Find $(A + B)C$, $AC + BC$, $C(A + B)$, and $CA + CB$.

Solution

$$(A + B)C = \left\{\begin{bmatrix} 1 & 0 \\ -1 & 1 \end{bmatrix} + \begin{bmatrix} 0 & 2 \\ 1 & -1 \end{bmatrix}\right\}\begin{bmatrix} 1 & 1 \\ 1 & 0 \end{bmatrix}$$

$$= \begin{bmatrix} 1 & 2 \\ 0 & 0 \end{bmatrix}\begin{bmatrix} 1 & 1 \\ 1 & 0 \end{bmatrix} = \begin{bmatrix} 3 & 1 \\ 0 & 0 \end{bmatrix}$$

$$AC + BC = \begin{bmatrix} 1 & 0 \\ -1 & 1 \end{bmatrix}\begin{bmatrix} 1 & 1 \\ 1 & 0 \end{bmatrix} + \begin{bmatrix} 0 & 2 \\ 1 & -1 \end{bmatrix}\begin{bmatrix} 1 & 1 \\ 1 & 0 \end{bmatrix}$$

$$= \begin{bmatrix} 1 & 1 \\ 0 & -1 \end{bmatrix} + \begin{bmatrix} 2 & 0 \\ 0 & 1 \end{bmatrix} = \begin{bmatrix} 3 & 1 \\ 0 & 0 \end{bmatrix}$$

Thus, $(A + B)C = AC + BC$.

$$C(A+B) = \begin{bmatrix} 1 & 1 \\ 1 & 0 \end{bmatrix} \left\{ \begin{bmatrix} 1 & 0 \\ -1 & 1 \end{bmatrix} + \begin{bmatrix} 0 & 2 \\ 1 & -1 \end{bmatrix} \right\}$$

$$= \begin{bmatrix} 1 & 1 \\ 1 & 0 \end{bmatrix} \begin{bmatrix} 1 & 2 \\ 0 & 0 \end{bmatrix} = \begin{bmatrix} 1 & 2 \\ 1 & 2 \end{bmatrix}$$

$$CA + CB = \begin{bmatrix} 1 & 1 \\ 1 & 0 \end{bmatrix} \begin{bmatrix} 1 & 0 \\ -1 & 1 \end{bmatrix} + \begin{bmatrix} 1 & 1 \\ 1 & 0 \end{bmatrix} \begin{bmatrix} 0 & 2 \\ 1 & -1 \end{bmatrix}$$

$$= \begin{bmatrix} 0 & 1 \\ 1 & 0 \end{bmatrix} + \begin{bmatrix} 1 & 1 \\ 0 & 2 \end{bmatrix} = \begin{bmatrix} 1 & 2 \\ 1 & 2 \end{bmatrix}$$

Thus, $C(A + B) = CA + CB$.

Exercises

1 Write each product as a single matrix.

a $\begin{bmatrix} 1 & 2 & -2 \end{bmatrix} \begin{bmatrix} 1 \\ 2 \\ -1 \end{bmatrix}$ **b** $\begin{bmatrix} 2 & 1 \end{bmatrix} \begin{bmatrix} -2 \\ 3 \end{bmatrix}$

c $\begin{bmatrix} 1 & -3 \\ 1 & 2 \end{bmatrix} \begin{bmatrix} -1 & 4 \\ 2 & 1 \end{bmatrix}$ **d** $\begin{bmatrix} 1 & 0 \\ -2 & 3 \end{bmatrix} \begin{bmatrix} 5 & 0 \\ -1 & -2 \end{bmatrix}$

e $\begin{bmatrix} 3 & 2 & 1 \\ 0 & 1 & -1 \end{bmatrix} \begin{bmatrix} -1 & 0 \\ 1 & -3 \\ 2 & 4 \end{bmatrix}$ **f** $\begin{bmatrix} 3 & 0 & -2 \\ 1 & 4 & 5 \end{bmatrix} \begin{bmatrix} -1 & 2 \\ 1 & 0 \\ 2 & 1 \end{bmatrix}$

g $\begin{bmatrix} 1 & -1 & 2 \\ 0 & 2 & -1 \\ 1 & 0 & 3 \end{bmatrix} \begin{bmatrix} -2 & 0 & -1 \\ 0 & 1 & 1 \\ -1 & 2 & 1 \end{bmatrix}$ **h** $\begin{bmatrix} 1 & 4 & 7 \\ 2 & 5 & 8 \\ 3 & 6 & 9 \end{bmatrix} \begin{bmatrix} 1 & 0 & 0 \\ 0 & 1 & 0 \\ 0 & 0 & 1 \end{bmatrix}$

2 Let $A = \begin{bmatrix} 1 & 0 & 2 \\ 3 & 1 & 0 \end{bmatrix}$ and $B = \begin{bmatrix} -1 & -3 \\ 0 & -1 \\ -2 & 0 \end{bmatrix}$. Find AB and BA.

3 Let $A = \begin{bmatrix} -1 & 2 & 0 \\ 1 & 3 & 2 \end{bmatrix}$, $B = \begin{bmatrix} 1 & 0 & -1 & 0 \\ 0 & 2 & -3 & 1 \end{bmatrix}$,

$C = \begin{bmatrix} 5 & -1 \\ 1 & 0 \\ 2 & -3 \end{bmatrix}$, and $D = \begin{bmatrix} 1 & 0 & 3 & 0 & 5 \\ 0 & 2 & 0 & 4 & 0 \\ -1 & 2 & 0 & 4 & 0 \\ 0 & -2 & 0 & -4 & 0 \end{bmatrix}$.

Determine which of the following products are meaningful and compute those that are: AB, BA, AC, CA, AD, DA, BC, CB, CD, and DC.

4 Let $A = \begin{bmatrix} 1 & 0 & -1 \\ 2 & 0 & -2 \end{bmatrix}$, $B = \begin{bmatrix} 1 & 0 & -2 & 0 \\ 0 & 2 & 0 & -3 \\ -1 & 0 & 3 & 0 \end{bmatrix}$, and

$C = \begin{bmatrix} 4 & 0 & -1 & 0 \\ -1 & 1 & 2 & 3 \\ 2 & 0 & -3 & 0 \\ -3 & 3 & 4 & 4 \end{bmatrix}$. Compute $(AB)C$ and $A(BC)$.

5 Let $A = \begin{bmatrix} 1 & 2 & 3 \\ 1 & 0 & -1 \\ 0 & 4 & 0 \end{bmatrix}$ and $B = \begin{bmatrix} -3 & 0 \\ -2 & 4 \\ -1 & 0 \end{bmatrix}$. Let $\alpha = 5$. Find $(\alpha A)B$, $A(\alpha B)$, and $\alpha(AB)$.

6 Prove Theorem 1-7.

7 Let $A = \begin{bmatrix} -1 & 2 & -3 \\ -4 & 5 & -6 \end{bmatrix}$, $B = \begin{bmatrix} 3 & -2 & 1 \\ 6 & -5 & 4 \end{bmatrix}$, and $C = \begin{bmatrix} 2 & 0 \\ -3 & 1 \\ 0 & 1 \end{bmatrix}$.

Find $(A + B)C$ and $AC + BC$. Find $C(A + B)$ and $CA + CB$.

8 Prove Theorem 1-9.

Note: Do not use Example 2 in this section.

9 Find two 2×2 matrices A and B such that $AB \neq BA$.

Note: $A^2 = AA$, $B^2 = BB$, $A^3 = AAA$, and so on.

10 Let A and B be square matrices. Under what condition do the following equations hold?

a $(A + B)(A + B) = A^2 + 2AB + B^2$

b $(A + B)(A - B) = A^2 - B^2$

11 Find matrices A and B satisfying the following simultaneous conditions: A and B are 2×2 matrices, $A \neq B$, $B \neq \begin{bmatrix} 0 & 0 \\ 0 & 0 \end{bmatrix}$, $A \neq \begin{bmatrix} 0 & 0 \\ 0 & 0 \end{bmatrix}$, and $AB = \begin{bmatrix} 0 & 0 \\ 0 & 0 \end{bmatrix}$.

12 Find a matrix $A \neq 0$ such that $A^2 = 0$.

13 Find a matrix $B \neq 0$ such that $B^2 \neq 0$ but $B^3 = 0$.

14 Find a matrix $C \neq 0$ such that $C^2 \neq 0$, $C^3 \neq 0$, but $C^4 = 0$.

Remember:

$$A - B = A + (-B) = A + (-1)B$$

15 Let A and B be $m \times n$ matrices and C an $n \times k$ matrix. Show that $(A - B)C = AC - BC$.

16 Find matrices A, B, and C satisfying the following simultaneous conditions: A, B, and C are 2×2 matrices; A, B, and C are *different* from $\begin{bmatrix} 0 & 0 \\ 0 & 0 \end{bmatrix}$; $A \neq B$; and $AC = BC$. (This shows that, in general, one may *not* cancel the matrix C in an equation $AC = BC$.)

1-3 Elementary Matrices

Objective
Introduce the concept of an elementary matrix of first, second, and third kinds.

This section will be devoted to the introduction of the *identity matrix* and the *elementary matrices* associated with it. These matrices play a significant role in applications discussed later in this chapter.

Let A be an $n \times n$ matrix. We say that A is a **square matrix** of **order** n. We sometimes denote such a matrix by A_n to emphasize the dimension. The entries a_{ii} $(i = 1, 2, \ldots, n)$ are called the elements of the **main diagonal** of the matrix A_n.

Definition 1-6
The identity matrix of order n.

The matrix I_n defined by

$$I_n = (\delta_{ij})$$

Remember:

$$\delta_{ij} = \begin{cases} 1 & \text{if } i = j \\ 0 & \text{if } i \neq j \end{cases}$$

where δ_{ij} is the Kronecker delta, is called the **identity matrix** of order n.

We may describe I_n by saying that all the entries of its main diagonal are 1 while all other entries of I_n are 0. Thus,

$$I_1 = [1],\ I_2 = \begin{bmatrix} 1 & 0 \\ 0 & 1 \end{bmatrix},\ I_3 = \begin{bmatrix} 1 & 0 & 0 \\ 0 & 1 & 0 \\ 0 & 0 & 1 \end{bmatrix},\ I_4 = \begin{bmatrix} 1 & 0 & 0 & 0 \\ 0 & 1 & 0 & 0 \\ 0 & 0 & 1 & 0 \\ 0 & 0 & 0 & 1 \end{bmatrix}$$

and so on.

Let A be a 2×3 matrix. Then we have

$$AI_3 = \begin{bmatrix} a_{11} & a_{12} & a_{13} \\ a_{21} & a_{22} & a_{23} \end{bmatrix} \begin{bmatrix} 1 & 0 & 0 \\ 0 & 1 & 0 \\ 0 & 0 & 1 \end{bmatrix} = \begin{bmatrix} a_{11} & a_{12} & a_{13} \\ a_{21} & a_{22} & a_{23} \end{bmatrix}$$

and

$$I_2A = \begin{bmatrix} 1 & 0 \\ 0 & 1 \end{bmatrix} \begin{bmatrix} a_{11} & a_{12} & a_{13} \\ a_{21} & a_{22} & a_{23} \end{bmatrix} = \begin{bmatrix} a_{11} & a_{12} & a_{13} \\ a_{21} & a_{22} & a_{23} \end{bmatrix}$$

Thus, the effect of multiplying the matrix A by an identity matrix I (of proper order) from the left or from the right is to leave A unchanged.

It follows from the definition of matrix multiplication and the definition of the identity matrix I that we have

$$BI_n = B$$

and

$$I_m B = B$$

for every $m \times n$ matrix B.

In the following examples we shall examine the effect of multiplying a given matrix A (from the left) by special matrices derived from the identity matrix I.

Example 1 Let

$$A = \begin{bmatrix} 1 & 4 \\ 2 & 5 \\ 3 & 6 \end{bmatrix} \quad \text{and} \quad E = \begin{bmatrix} 0 & 0 & 1 \\ 0 & 1 & 0 \\ 1 & 0 & 0 \end{bmatrix}$$

Find EA.

Solution

$$EA = \begin{bmatrix} 0 & 0 & 1 \\ 0 & 1 & 0 \\ 1 & 0 & 0 \end{bmatrix} \begin{bmatrix} 1 & 4 \\ 2 & 5 \\ 3 & 6 \end{bmatrix} = \begin{bmatrix} 3 & 6 \\ 2 & 5 \\ 1 & 4 \end{bmatrix}.$$

A close examination of the resulting matrix shows that one gets the matrix $\begin{bmatrix} 3 & 6 \\ 2 & 5 \\ 1 & 4 \end{bmatrix}$ from the matrix $\begin{bmatrix} 1 & 4 \\ 2 & 5 \\ 3 & 6 \end{bmatrix}$ by interchanging the first and the third rows. Also, one can easily see that interchanging the first and the third rows of I_3 produces the matrix E. Thus, we may think of E as a "row changer" when it multiplies a matrix A from the left. This leads to the following definition.

Definition 1-7 We call E an **elementary matrix of the first kind** if E is obtained from the identity matrix I_n by interchanging any two of its rows.

Elementary matrix of the first kind.

An important property of an elementary matrix E of the first kind is illustrated in the next example.

Example 2 Let

$$A = \begin{bmatrix} 1 & 4 \\ 2 & 5 \\ 3 & 6 \end{bmatrix} \quad \text{and} \quad E = \begin{bmatrix} 1 & 0 & 0 \\ 0 & 0 & 1 \\ 0 & 1 & 0 \end{bmatrix}$$

Find $E(EA)$.

Solution

$$E(EA) = \begin{bmatrix} 1 & 0 & 0 \\ 0 & 0 & 1 \\ 0 & 1 & 0 \end{bmatrix} \left\{ \begin{bmatrix} 1 & 0 & 0 \\ 0 & 0 & 1 \\ 0 & 1 & 0 \end{bmatrix} \begin{bmatrix} 1 & 4 \\ 2 & 5 \\ 3 & 6 \end{bmatrix} \right\}$$

$$= \begin{bmatrix} 1 & 0 & 0 \\ 0 & 0 & 1 \\ 0 & 1 & 0 \end{bmatrix} \begin{bmatrix} 1 & 4 \\ 3 & 6 \\ 2 & 5 \end{bmatrix} = \begin{bmatrix} 1 & 4 \\ 2 & 5 \\ 3 & 6 \end{bmatrix} = A$$

Actually, there was no need for any numerical computation because we could argue this way: Since E is obtained from I_3 by interchanging the second and third rows of I_3, the matrix EA is obtained from A by interchanging the second and the third rows in A. Multiplying EA by E to get $E(EA)$ means that we interchange the second and the third rows in EA, thus restoring the rows of A to their original position. In short, if we multiply a matrix A twice from the left by a "row changer," E, the net effect is to leave the matrix A unchanged. Notice in particular that

$$EE = E(EI_3) = I_3$$

Before formally introducing another type of elementary matrix, let us consider the following example.

Example 3 Let

$$A = \begin{bmatrix} 1 & 4 \\ 2 & 5 \\ 3 & 6 \end{bmatrix} \quad \text{and} \quad G = \begin{bmatrix} 1 & 0 & 0 \\ 0 & 1 & 0 \\ 0 & 0 & \alpha \end{bmatrix} \quad \text{where} \quad \alpha \neq 0$$

Find GA.

Solution

$$GA = \begin{bmatrix} 1 & 0 & 0 \\ 0 & 1 & 0 \\ 0 & 0 & \alpha \end{bmatrix} \begin{bmatrix} 1 & 4 \\ 2 & 5 \\ 3 & 6 \end{bmatrix} = \begin{bmatrix} 1 & 4 \\ 2 & 5 \\ 3\alpha & 6\alpha \end{bmatrix}$$

We describe G by saying that it is obtained from I_3 by multiplying the third row of I_3 by $\alpha \neq 0$. A simple inspection reveals that GA is obtained from A by multiplying the third row of A by α. Hence, we may think of G as a "row multiplier."

Definition 1-8
Elementary matrix of the second kind.

We call G an **elementary matrix of the second kind** if G is obtained from the identity matrix I_n by multiplying one of its rows by a real number $\alpha \neq 0$.

The reason for the requirement $\alpha \neq 0$ will become apparent in the following example.

Example 4 Let

$$A = \begin{bmatrix} 1 & 4 \\ 2 & 5 \\ 3 & 6 \end{bmatrix} \quad \text{and} \quad G = \begin{bmatrix} 1 & 0 & 0 \\ 0 & 1 & 0 \\ 0 & 0 & \alpha \end{bmatrix} \quad \text{where} \quad \alpha \neq 0$$

Find a matrix $\bar{G}$ such that $\bar{G}(GA) = A$.

Solution We may describe the problem here by saying that we seek a matrix $\bar{G}$ which will cancel the effect that the "row multiplier" G had on the matrix A. Since the effect of G on A is to multiply its third row by $\alpha \neq 0$, it follows that $\bar{G}$ will cancel this effect if it multiplies the same row by $\frac{1}{\alpha}$. This is possible only if $\alpha \neq 0$. If we define $\bar{G}$ by the equation

$$\bar{G} = \begin{bmatrix} 1 & 0 & 0 \\ 0 & 1 & 0 \\ 0 & 0 & \frac{1}{\alpha} \end{bmatrix}$$

then we have

$$\bar{G}(GA) = \begin{bmatrix} 1 & 0 & 0 \\ 0 & 1 & 0 \\ 0 & 0 & \frac{1}{\alpha} \end{bmatrix} \left\{ \begin{bmatrix} 1 & 0 & 0 \\ 0 & 1 & 0 \\ 0 & 0 & \alpha \end{bmatrix} \begin{bmatrix} 1 & 4 \\ 2 & 5 \\ 3 & 6 \end{bmatrix} \right\} = \begin{bmatrix} 1 & 0 & 0 \\ 0 & 1 & 0 \\ 0 & 0 & \frac{1}{\alpha} \end{bmatrix} \begin{bmatrix} 1 & 4 \\ 2 & 5 \\ 3\alpha & 6\alpha \end{bmatrix}$$

$$= \begin{bmatrix} 1 & 4 \\ 2 & 5 \\ 3 & 6 \end{bmatrix} = A$$

In particular, we have $\bar{G}G = I_3$.

The third type of elementary matrix is of a somewhat more involved nature, as the following example will show.

Example 5 Let

$$A = \begin{bmatrix} 1 & 4 \\ 2 & 5 \\ 3 & 6 \end{bmatrix} \quad \text{and} \quad H = \begin{bmatrix} 1 & 0 & 0 \\ 0 & 1 & 0 \\ \beta & 0 & 1 \end{bmatrix}$$

where β is a real number. Find HA.

Solution

$$HA = \begin{bmatrix} 1 & 0 & 0 \\ 0 & 1 & 0 \\ \beta & 0 & 1 \end{bmatrix} \begin{bmatrix} 1 & 4 \\ 2 & 5 \\ 3 & 6 \end{bmatrix} = \begin{bmatrix} 1 & 4 \\ 2 & 5 \\ 3 + (1)(\beta) & 6 + (4)(\beta) \end{bmatrix}$$

We may describe H by saying that it is obtained from I_3 by adding β times the first row of I_3 to the third row. The effect of H on A upon multiplication of A from the left is of a similar nature. We describe the matrix HA by saying that it is obtained from the matrix A by adding β times the first row of A to the third row. Note that we did *not* stipulate the condition $\beta \neq 0$ as in Examples 3 and 4.

We are now ready to introduce the third type of elementary matrix.

Definition 1-9 We call H an **elementary matrix of the third kind** if H is obtained from the identity matrix I_n by adding β times one row to *another* row of I_n.

Elementary matrix of the third kind.

We have already seen in Examples 2 and 4 that the effect of multiplication of a matrix by elementary matrices of the first and second kind can be "undone." The same is true also for elementary matrices of the third kind, as the following example shows.

Example 6 Let

$$A = \begin{bmatrix} 1 & 4 \\ 2 & 5 \\ 3 & 6 \end{bmatrix} \quad \text{and} \quad H = \begin{bmatrix} 1 & 0 & 0 \\ 0 & 1 & 0 \\ \beta & 0 & 1 \end{bmatrix}$$

where β is a real number. Find a matrix $\bar{H}$ such that $\bar{H}(HA) = A$.

Solution We have seen that the effect of H on A in obtaining the product HA is to add β times the first row to the third row of A. Thus, in order to cancel this effect we must find a matrix $\bar{H}$ such that performing the product $\bar{H}(HA)$ will result in subtracting β times the first row of HA (which is also the first row of A) from the third row of HA. Such a matrix $\bar{H}$ is given by

$$\bar{H} = \begin{bmatrix} 1 & 0 & 0 \\ 0 & 1 & 0 \\ -\beta & 0 & 1 \end{bmatrix}$$

We find

$$\bar{H}(HA) = \begin{bmatrix} 1 & 0 & 0 \\ 0 & 1 & 0 \\ -\beta & 0 & 1 \end{bmatrix} \begin{bmatrix} 1 & 0 & 0 \\ 0 & 1 & 0 \\ \beta & 0 & 1 \end{bmatrix} \begin{bmatrix} 1 & 4 \\ 2 & 5 \\ 3 & 6 \end{bmatrix}$$

$$= \begin{bmatrix} 1 & 0 & 0 \\ 0 & 1 & 0 \\ -\beta & 0 & 1 \end{bmatrix} \begin{bmatrix} 1 & 4 \\ 2 & 5 \\ 3+\beta & 6+4\beta \end{bmatrix} = \begin{bmatrix} 1 & 4 \\ 2 & 5 \\ 3 & 6 \end{bmatrix} = A$$

Note in particular that $\bar{H}H = I_3$.

Exercises

1 Let A be a 3×2 matrix.

a Find an elementary matrix E such that EA is the matrix obtained from A by interchanging the first and third row of A.

b Find an elementary matrix G such that GA is the matrix obtained from A by multiplying the third row of A by -3.

c Find an elementary matrix H such that HA is the matrix obtained from A by adding 4 times the first row of A to the second row.

2 Determine which of the following are elementary matrices.

a $\begin{bmatrix} 0 & 1 \\ 1 & 0 \end{bmatrix}$ **b** $\begin{bmatrix} 1 & 2 \\ 2 & 1 \end{bmatrix}$ **c** $\begin{bmatrix} 1 & 0 \\ 0 & 1 \end{bmatrix}$

d $\begin{bmatrix} 1 & 0 \\ 8 & 1 \end{bmatrix}$ **e** $\begin{bmatrix} 0 & 0 \\ 0 & 1 \end{bmatrix}$

f $\begin{bmatrix} 0 & 0 & 1 \\ 1 & 0 & 0 \\ 0 & 1 & 0 \end{bmatrix}$ **g** $\begin{bmatrix} 0 & 0 & 1 \\ 0 & 1 & 0 \\ 1 & 0 & 0 \end{bmatrix}$

h $\begin{bmatrix} 1 & 0 & 0 \\ 0 & -5 & 0 \\ 0 & 0 & 1 \end{bmatrix}$ **i** $\begin{bmatrix} 3 & 0 & 0 \\ 0 & 1 & 0 \\ 0 & 0 & -1 \end{bmatrix}$

j $\begin{bmatrix} 1 & 0 & 0 \\ -1 & 1 & 0 \\ 0 & 0 & 1 \end{bmatrix}$ **k** $\begin{bmatrix} 1 & 0 & 0 \\ 3 & 1 & 0 \\ 1 & 0 & 1 \end{bmatrix}$

l $\begin{bmatrix} 1 & 0 & 0 & 0 \\ 0 & 1 & 0 & 0 \\ 0 & 4 & 1 & 0 \\ 0 & 0 & 0 & 1 \end{bmatrix}$ **m** $\begin{bmatrix} 3 & 0 & 0 & 0 \\ 0 & 1 & 0 & 0 \\ 0 & -2 & 1 & 0 \\ 0 & 0 & 0 & 1 \end{bmatrix}$

3 Let $A = \begin{bmatrix} 0 & 0 & 1 & 0 \\ 0 & 1 & 0 & 0 \\ 1 & 0 & 0 & 0 \\ 0 & 0 & 0 & 1 \end{bmatrix}$, $B = \begin{bmatrix} 1 & 0 & 0 & 0 \\ 0 & 1 & 0 & 0 \\ 0 & 0 & 5 & 0 \\ 0 & 0 & 0 & 1 \end{bmatrix}$, and $C = \begin{bmatrix} 1 & 0 & 0 & 0 \\ 0 & 1 & 0 & 0 \\ 0 & 3 & 1 & 0 \\ 0 & 0 & 0 & 1 \end{bmatrix}$. Find matrices D, E, and F such that $AD = I_4$, $BE = I_4$, and $CF = I_4$.

4 Let $A = \begin{bmatrix} 1 & 6 & 7 \\ 2 & 5 & 8 \\ 3 & 4 & 9 \end{bmatrix}$, $B = \begin{bmatrix} 1 & 6 & 7 \\ 2 & 5 & 8 \\ -3 & -4 & -9 \end{bmatrix}$, $C = \begin{bmatrix} 1 & 6 & 7 \\ 3 & 4 & 9 \\ 2 & 5 & 8 \end{bmatrix}$,

and $D = \begin{bmatrix} 1 & 6 & 7 \\ 3 & 11 & 15 \\ 3 & 4 & 9 \end{bmatrix}$.

a Find the elementary matrix E_1 such that $E_1A = B$.
b Find the elementary matrix E_2 such that $E_2B = A$.
c Find the elementary matrix E_3 such that $E_3A = C$.
d Find the elementary matrix E_4 such that $E_4C = A$.
e Find the elementary matrix E_5 such that $E_5A = D$.
f Find the elementary matrix E_6 such that $E_6D = A$.
g Find elementary matrices E_7 and E_8 such that $E_7(E_8B) = C$.
h Find elementary matrices E_9 and E_{10} such that $E_9(E_{10}B) = D$.

5 a Let A be a 3×4 matrix. Find AI_4 and I_3A.
b Let B be an $m \times n$ matrix. Find BI_n and I_mB.

6 Let A be an $n \times n$ matrix satisfying the matrix equation $5A + A^2 = I_n$. Find a matrix B (expressed in terms of A) such that $AB = I_n$.

7 Let $E = \begin{bmatrix} 0 & 0 & 1 \\ 0 & 1 & 0 \\ 1 & 0 & 0 \end{bmatrix}$, $G = \begin{bmatrix} 1 & 0 & 0 \\ 0 & 2 & 0 \\ 0 & 0 & 1 \end{bmatrix}$, and $H = \begin{bmatrix} 1 & 0 & 0 \\ 5 & 1 & 0 \\ 0 & 0 & 1 \end{bmatrix}$. Find E^7, G^7, and H^7.

8 Let E be an elementary $m \times m$ matrix. Show that E^n is also an elementary matrix for every positive integer n.

1-4 Nonsingular Matrices and Linear Systems

Objectives

1 Introduce the concept of a nonsingular matrix.
2 Discuss the solution of the system $A\mathbf{x} = \mathbf{b}$ where A is a nonsingular matrix,

$$\mathbf{x} = \begin{bmatrix} x_1 \\ x_2 \\ \vdots \\ x_n \end{bmatrix}, \quad \text{and} \quad \mathbf{b} = \begin{bmatrix} b_1 \\ b_2 \\ \vdots \\ b_n \end{bmatrix}.$$

3 Discuss the fact that every nonsingular $n \times n$ matrix is a product of elementary matrices.

Matrix methods provide a simple and systematic procedure for analyzing mathematical problems involving systems of equations. Systems of equations arise frequently in many branches of mathematics, the physical and biological sciences, engineering, economics, and other disciplines.

Let us consider an electric network. A schematic representation of the network is given in Figure 1-1. The network involves two shunt circuits and a series circuit (see Figures 1-2 and 1-3). Here, i_1 and

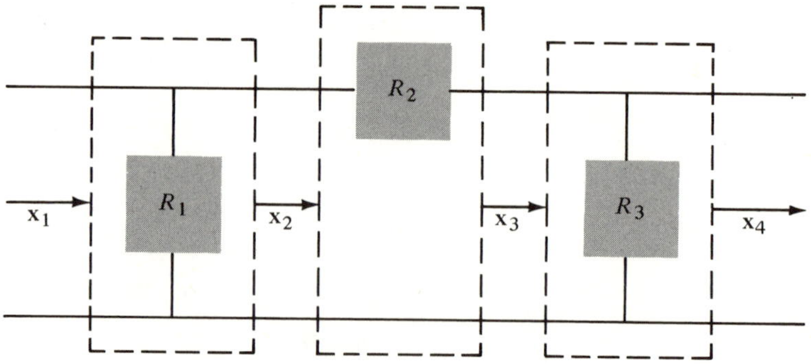

Figure 1-1

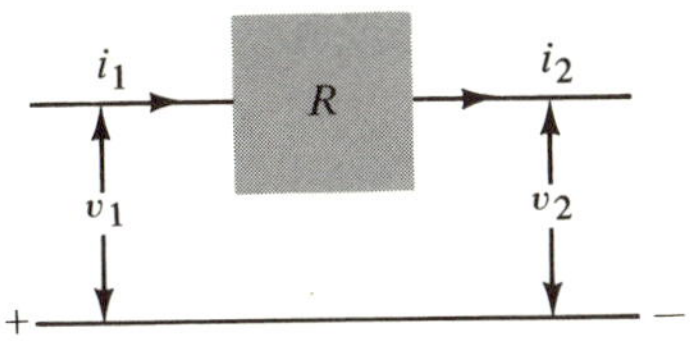

Figure 1-2 Series circuit

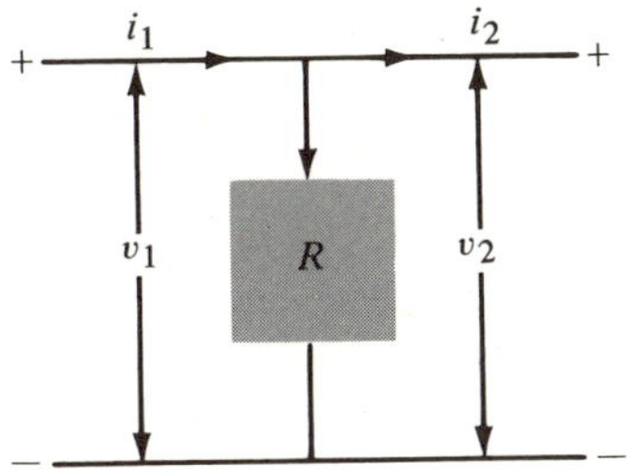

Figure 1-3 Shunt

i_2 represent input and output currents, respectively; v_1 and v_2 represent input and output voltages, respectively. The basic principles of electricity that govern such networks are

a Ohm's Law: The potential drop across a resistor is equal to the current times the resistance; $v = iR$.

b Kirchhoff's Law: The algebraic sum of the currents flowing into a junction point is zero.

In the case of the series circuit (see Figure 1-2), application of the laws above yields the equations

$$i_2 = i_1$$

and

$$v_2 = v_1 - i_1 R$$

In matrix notation, we have

$$\begin{bmatrix} i_2 \\ v_2 \end{bmatrix} = \begin{bmatrix} 1 & 0 \\ -R & 1 \end{bmatrix} \begin{bmatrix} i_1 \\ v_1 \end{bmatrix} \tag{1}$$

In the case of the shunt we find

$$\frac{v_2}{R} = \frac{v_1}{R}$$

or

$$v_2 = v_1$$

and

$$i_2 = \frac{-v_1}{R} + i_1$$

In matrix notation,

$$\begin{bmatrix} i_2 \\ v_2 \end{bmatrix} = \begin{bmatrix} 1 & -1/R \\ 0 & 1 \end{bmatrix} \begin{bmatrix} i_1 \\ v_1 \end{bmatrix} \tag{2}$$

In order to simplify, we define the following matrices,

$$\mathbf{x}_j = \begin{bmatrix} i_j \\ v_j \end{bmatrix}, \quad A = \begin{bmatrix} 1 & 0 \\ -R & 1 \end{bmatrix}, \quad \text{and} \quad B = \begin{bmatrix} 1 & -1/R \\ 0 & 1 \end{bmatrix}$$

where $j = 1, 2$; then equations 1 and 2 can be written in the more concise form

The student who is not familiar with simple electrical circuits should accept the validity of these equations and not worry about the derivations.

$$\mathbf{x}_2 = A\mathbf{x}_1 \tag{1'}$$
$$\mathbf{x}_2 = B\mathbf{x}_1 \tag{2'}$$

We now return to the network illustrated in Figure 1-1. We find

$$\mathbf{x}_2 = M_1\mathbf{x}_1$$
$$\mathbf{x}_3 = M_2\mathbf{x}_2 \qquad \textbf{[3]}$$
$$\mathbf{x}_4 = M_3\mathbf{x}_3$$

where M_i $(i = 1, 2, 3)$ is either matrix A or matrix B with the appropriate value for the resistance involved. Thus for given R_1, R_2, and R_3 and known input $\mathbf{x}_1$, the value of $\mathbf{x}_4$ can easily be determined from the equations in 3:

$$\mathbf{x}_4 = M_3M_2M_1\mathbf{x}_1 \qquad \textbf{[4]}$$

With matrix multiplication as defined in Section 2-2, equation 4 can be solved readily. This example illustrates how matrix multiplication provides a simple and systematic approach to solving problems in connection with electrical networks.

Unfortunately, many systems of equations are not as simple to solve as those indicated above. However, the matrix approach is simpler and more systematic than other methods.

Let us consider the general form of a linear system of m equations in n unknowns:

$$\begin{aligned} a_{11}x_1 + a_{12}x_2 + \cdots + a_{1n}x_n &= b_1 \\ a_{21}x_1 + a_{22}x_2 + \cdots + a_{2n}x_n &= b_2 \\ \vdots \qquad\qquad\qquad\qquad & \quad \vdots \\ a_{m1}x_1 + a_{m2}x_2 + \cdots + a_{mn}x_n &= b_m \end{aligned} \qquad \textbf{[5]}$$

Systems of equations like system 5 arise frequently in many disciplines. It is therefore vital to know how to solve such a system—assuming a solution does exist.

A **solution** of system 5 is an n-tuple of real numbers $x_1, x_2, \ldots, x_n$ satisfying each of the equations in system 5. It is possible that system 5 may *not* have a solution. In this case the system is called **inconsistent.** If $b_1 = b_2 = \cdots = b_m = 0$, then the system is called a **homogeneous system of linear equations.** If at least one of the b's is different from zero, say $b_i \neq 0$, then the system is called **nonhomogeneous.**

Using matrix multiplication, we may write system 5 in the following form:

$$\begin{bmatrix} a_{11} & a_{12} & \cdots & a_{1n} \\ a_{21} & a_{22} & \cdots & a_{2n} \\ \vdots & & & \vdots \\ a_{m1} & a_{m2} & \cdots & a_{mn} \end{bmatrix} \begin{bmatrix} x_1 \\ x_2 \\ \vdots \\ x_n \end{bmatrix} = \begin{bmatrix} b_1 \\ b_2 \\ \vdots \\ b_m \end{bmatrix} \qquad \textbf{[6]}$$

The matrix A given by

$$A = \begin{bmatrix} a_{11} & a_{12} & \cdots & a_{1n} \\ a_{21} & a_{22} & \cdots & a_{2n} \\ \vdots & & & \vdots \\ a_{m1} & a_{m2} & \cdots & a_{mn} \end{bmatrix}$$

is called the **matrix of coefficients.** If we let

$$\mathbf{x} = \begin{bmatrix} x_1 \\ x_2 \\ \vdots \\ x_n \end{bmatrix} \quad \text{and} \quad \mathbf{b} = \begin{bmatrix} b_1 \\ b_2 \\ \vdots \\ b_m \end{bmatrix}$$

then equation 6 can be expressed by

$$A\mathbf{x} = \mathbf{b}$$

One must remember that A is an $m \times n$ matrix, $\mathbf{x}$ is an $n \times 1$ matrix, and $\mathbf{b}$ is an $m \times 1$ matrix. In order to obtain a solution $\mathbf{x}$ of the system $A\mathbf{x} = \mathbf{b}$, we must somehow eliminate the coefficient matrix A. One is tempted to try to divide by A. Unfortunately such an operation has not been defined for matrices. Nevertheless, when the matrix A possesses a special property, called *nonsingularity,* it is possible to use matrix multiplication to eliminate A from the left side of the system $A\mathbf{x} = \mathbf{b}$ and thus obtain a solution $\mathbf{x}$. In this section we confine our discussion to linear systems $A\mathbf{x} = \mathbf{b}$ where

Solutions of the system $A\mathbf{x} = \mathbf{b}$ may exist even though A does not possess the property of nonsingularity. This type of linear system will be discussed later in this chapter.

$$\mathbf{x} = \begin{bmatrix} x_1 \\ x_2 \\ \vdots \\ x_n \end{bmatrix}, \mathbf{b} = \begin{bmatrix} b_1 \\ b_2 \\ \vdots \\ b_n \end{bmatrix}$$

and A is an $n \times n$ matrix having the property of nonsingularity, which is defined as follows.

Definition 1-10 The square matrix A of order n is called **nonsingular** if there exists a square matrix B of order n such that $AB = BA = I_n$. We say that B is an **inverse** of A. If A does not have an inverse, we say that A is a **singular** matrix.

Example 1 Let

$$A = \begin{bmatrix} 1 & -1 \\ 1 & 1 \end{bmatrix} \quad \text{and} \quad B = \begin{bmatrix} \frac{1}{2} & \frac{1}{2} \\ -\frac{1}{2} & \frac{1}{2} \end{bmatrix}$$

Then $AB = BA = I_2$. Thus B is an inverse of A.

The student is encouraged to show this.

Example 2 Let

$$A = \begin{bmatrix} 1 & 1 \\ 1 & 1 \end{bmatrix}$$

Show that A does not have an inverse; i.e., that A is a singular matrix.

Solution Assume B is an inverse of A. Then $AB = I_2$. If

$$B = \begin{bmatrix} b_{11} & b_{12} \\ b_{21} & b_{22} \end{bmatrix}$$

then $AB = I_2$ may be expressed as follows:

$$\begin{bmatrix} 1 & 1 \\ 1 & 1 \end{bmatrix}\begin{bmatrix} b_{11} & b_{12} \\ b_{21} & b_{22} \end{bmatrix} = \begin{bmatrix} 1 & 0 \\ 0 & 1 \end{bmatrix}$$

Performing the product on the left of this matrix equation yields

$$\begin{bmatrix} b_{11} + b_{21} & b_{12} + b_{22} \\ b_{11} + b_{21} & b_{12} + b_{22} \end{bmatrix} = \begin{bmatrix} 1 & 0 \\ 0 & 1 \end{bmatrix}$$

Thus we have a linear system of four equations in four unknowns

Note that the first two equations cannot both be true. Similarly, the last two equations cannot both be true.

$$\begin{aligned} b_{11} + b_{21} &= 1 \\ b_{11} + b_{21} &= 0 \\ b_{12} + b_{22} &= 0 \\ b_{12} + b_{22} &= 1 \end{aligned}$$

This system is obviously inconsistent; hence, the matrix A does not possess an inverse.

We emphasize that the concept of an inverse matrix is defined for *square matrices only.* We may still have two matrices A and B of orders $m \times n$ and $n \times m$, respectively, with $m \neq n$, satisfying the equation $AB = I_m$. It would be *wrong* to conclude from such an equation that A is a nonsingular matrix and that B is its inverse. The condition $m \neq n$ means that neither matrix qualifies for either of these titles. The following example illustrates such a situation.

See Definition 1-10.

Example 3 Let

$$A = \begin{bmatrix} 1 & 0 & 0 \\ 0 & 1 & 0 \end{bmatrix} \quad \text{and} \quad B = \begin{bmatrix} 1 & 0 \\ 0 & 1 \\ 0 & 0 \end{bmatrix}$$

Then

$$AB = \begin{bmatrix} 1 & 0 \\ 0 & 1 \end{bmatrix} = I_2$$

Obviously A is not a nonsingular matrix nor is B its inverse, because neither matrix is a square matrix.

Now, suppose A is an $n \times n$ nonsingular matrix. How many different inverses does A possess? In other words, how many different solutions does the matrix equation $AX = I_n$ have where X denotes an unknown $n \times n$ matrix? The answer to this question is provided by the following theorem.

Theorem 1-10 Let A be a nonsingular $n \times n$ matrix. If B and C are inverses of A, then $B = C$.

The inverse of a nonsingular matrix A is unique.

Proof Assuming that B and C are inverses of A implies that

(See Definition 1-10.)

$$AB = BA = I_n \qquad \text{and} \qquad AC = CA = I_n$$

The key to the proof is to examine the products

$$(BA)C \qquad \text{and} \qquad B(AC)$$

By associativity (see Theorem 1-6),

$$(BA)C = B(AC)$$

But

$$(BA)C = I_nC = C$$

and

$$B(AC) = BI_n = B$$

We conclude that $B = C$.

Since a nonsingular matrix A has a *unique* inverse, we call it *the* inverse of A. We denote it by A^{-1}. The matrix A^{-1} satisfies the equations

$$AA^{-1} = A^{-1}A = I_n$$

As an immediate application of the uniqueness of an inverse matrix, we prove the following fundamental theorem.

Theorem 1-11 Let A be a nonsingular matrix of order n. Then the linear system $A\mathbf{x} = \mathbf{b}$ of n equations and n unknowns possesses a unique solution given by $\mathbf{x} = A^{-1}\mathbf{b}$.

Proof Since A is nonsingular it has a unique inverse A^{-1}. Multiplying $A\mathbf{x} = \mathbf{b}$ by A^{-1} from the left yields

$$A^{-1}(A\mathbf{x}) = A^{-1}\mathbf{b}$$

But

$$A^{-1}(A\mathbf{x}) = (A^{-1}A)\mathbf{x} = I_n\mathbf{x} = \mathbf{x}$$

so that

$$\mathbf{x} = A^{-1}\mathbf{b}$$

is a solution of $A\mathbf{x} = \mathbf{b}$.

We now must show the uniqueness of the solution just obtained. Let us assume that $\mathbf{y}$ and $\mathbf{z}$ are arbitrary solutions of $A\mathbf{x} = \mathbf{b}$. Then we have

$$A\mathbf{y} = \mathbf{b}$$

and

$$A\mathbf{z} = \mathbf{b}$$

We conclude that $A\mathbf{y} = A\mathbf{z}$. Upon multiplication by A^{-1} from the left we get

$$A^{-1}(A\mathbf{y}) = A^{-1}(A\mathbf{z})$$

But

$$A^{-1}(A\mathbf{y}) = (A^{-1}A)\mathbf{y} = I_n\mathbf{y} = \mathbf{y}$$

and

$$A^{-1}(A\mathbf{z}) = (A^{-1}A)\mathbf{z} = I_n\mathbf{z} = \mathbf{z}$$

so that $\mathbf{y} = \mathbf{z}$. Since $\mathbf{x} = A^{-1}\mathbf{b}$ is a solution of $A\mathbf{z} = \mathbf{b}$, it must be the only one.

Example 4 Solve the system

$$\begin{aligned} x_1 - x_2 &= 3 \\ x_1 + x_2 &= 5 \end{aligned}$$

Solution The system may be written as

$$\begin{bmatrix} 1 & -1 \\ 1 & 1 \end{bmatrix}\begin{bmatrix} x_1 \\ x_2 \end{bmatrix} = \begin{bmatrix} 3 \\ 5 \end{bmatrix}$$

In Example 1 we found that the matrix $\begin{bmatrix} 1 & -1 \\ 1 & 1 \end{bmatrix}$ has an inverse given by

$$\begin{bmatrix} 1 & -1 \\ 1 & 1 \end{bmatrix}^{-1} = \begin{bmatrix} \frac{1}{2} & \frac{1}{2} \\ -\frac{1}{2} & \frac{1}{2} \end{bmatrix}$$

Hence we may write

$$\begin{bmatrix} \frac{1}{2} & \frac{1}{2} \\ -\frac{1}{2} & \frac{1}{2} \end{bmatrix}\left\{\begin{bmatrix} 1 & -1 \\ 1 & 1 \end{bmatrix}\begin{bmatrix} x_1 \\ x_2 \end{bmatrix}\right\} = \begin{bmatrix} \frac{1}{2} & \frac{1}{2} \\ -\frac{1}{2} & \frac{1}{2} \end{bmatrix}\begin{bmatrix} 3 \\ 5 \end{bmatrix}$$

But

$$\begin{bmatrix} \frac{1}{2} & \frac{1}{2} \\ -\frac{1}{2} & \frac{1}{2} \end{bmatrix}\left\{\begin{bmatrix} 1 & -1 \\ 1 & 1 \end{bmatrix}\begin{bmatrix} x_1 \\ x_2 \end{bmatrix}\right\} = \left\{\begin{bmatrix} \frac{1}{2} & \frac{1}{2} \\ -\frac{1}{2} & \frac{1}{2} \end{bmatrix}\begin{bmatrix} 1 & -1 \\ 1 & 1 \end{bmatrix}\right\}\begin{bmatrix} x_1 \\ x_2 \end{bmatrix}$$

$$= \begin{bmatrix} 1 & 0 \\ 0 & 1 \end{bmatrix}\begin{bmatrix} x_1 \\ x_2 \end{bmatrix} = \begin{bmatrix} x_1 \\ x_2 \end{bmatrix}$$

and

$$\begin{bmatrix} \frac{1}{2} & \frac{1}{2} \\ -\frac{1}{2} & \frac{1}{2} \end{bmatrix}\begin{bmatrix} 3 \\ 5 \end{bmatrix} = \begin{bmatrix} \frac{3}{2} + \frac{5}{2} \\ -\frac{3}{2} + \frac{5}{2} \end{bmatrix} = \begin{bmatrix} 4 \\ 1 \end{bmatrix}$$

so that

$$\begin{bmatrix} x_1 \\ x_2 \end{bmatrix} = \begin{bmatrix} 4 \\ 1 \end{bmatrix}$$

This means that $x_1 = 4$ and $x_2 = 1$.

In order to be able to make effective use of Theorem 1-11, we must have a method of computing the inverse of the matrix A in the given system $A\mathbf{x} = \mathbf{b}$.

It is a simple problem if the matrix of coefficients A happens to be one of the three types of elementary matrices which we encountered in the previous section. The following examples will show that elementary matrices are invertible matrices; that is, they possess inverse matrices in accordance with Definition 1-10.

Example 5 Solve the system $A\mathbf{x} = \mathbf{b}$ if

$$A = \begin{bmatrix} 1 & 0 & 0 \\ 0 & 0 & 1 \\ 0 & 1 & 0 \end{bmatrix} \quad \text{and} \quad \mathbf{b} = \begin{bmatrix} 1 \\ 2 \\ 3 \end{bmatrix}$$

Solution Since A is an elementary matrix obtained from the identity matrix I_3 by interchanging the second and third rows, it is obvious that A is its own inverse; that is, $A^{-1} = A$. Computation of $A^{-1}A$ yields

$$A^{-1}A = AA = \begin{bmatrix} 1 & 0 & 0 \\ 0 & 0 & 1 \\ 0 & 1 & 0 \end{bmatrix}\begin{bmatrix} 1 & 0 & 0 \\ 0 & 0 & 1 \\ 0 & 1 & 0 \end{bmatrix} = \begin{bmatrix} 1 & 0 & 0 \\ 0 & 1 & 0 \\ 0 & 0 & 1 \end{bmatrix} = I_3$$

Therefore, the solution of the system $A\mathbf{x} = \mathbf{b}$ is given by:

$$\mathbf{x} = A^{-1}\mathbf{b} = \begin{bmatrix} 1 & 0 & 0 \\ 0 & 0 & 1 \\ 0 & 1 & 0 \end{bmatrix}\begin{bmatrix} 1 \\ 2 \\ 3 \end{bmatrix} = \begin{bmatrix} 1 \\ 3 \\ 2 \end{bmatrix}$$

Thus $x_1 = 1$, $x_2 = 3$, and $x_3 = 2$.

Example 6 Solve the system $A\mathbf{x} = \mathbf{b}$ if

$$A = \begin{bmatrix} 1 & 0 & 0 \\ 0 & 2 & 0 \\ 0 & 0 & 1 \end{bmatrix} \quad \text{and} \quad \mathbf{b} = \begin{bmatrix} 1 \\ 2 \\ 3 \end{bmatrix}$$

Solution By inspection we find that $A^{-1} = \begin{bmatrix} 1 & 0 & 0 \\ 0 & \frac{1}{2} & 0 \\ 0 & 0 & 1 \end{bmatrix}$. Hence, we have

See Example 4 in Section 1-3.

$$\mathbf{x} = A^{-1}\mathbf{b} = \begin{bmatrix} 1 & 0 & 0 \\ 0 & \frac{1}{2} & 0 \\ 0 & 0 & 1 \end{bmatrix}\begin{bmatrix} 1 \\ 2 \\ 3 \end{bmatrix} = \begin{bmatrix} 1 \\ 1 \\ 3 \end{bmatrix}$$

Therefore, $x_1 = 1$, $x_2 = 1$, and $x_3 = 3$.

Example 7 Solve the system $A\mathbf{x} = \mathbf{b}$ if

$$A = \begin{bmatrix} 1 & 0 & 0 \\ 0 & 1 & 0 \\ 5 & 0 & 1 \end{bmatrix} \quad \text{and} \quad \mathbf{b} = \begin{bmatrix} 1 \\ 2 \\ 3 \end{bmatrix}$$

Solution Since A is obtained from I_3 by adding 5 times the first row to the third row, we know that A^{-1} will have the form

See Example 6, Section 1-3. Observe that:

$$A^{-1}A = \begin{bmatrix} 1 & 0 & 0 \\ 0 & 1 & 0 \\ -5 & 0 & 1 \end{bmatrix}\begin{bmatrix} 1 & 0 & 0 \\ 0 & 1 & 0 \\ 5 & 0 & 1 \end{bmatrix} = \begin{bmatrix} 1 & 0 & 0 \\ 0 & 1 & 0 \\ 0 & 0 & 1 \end{bmatrix} = I_3$$

$$A^{-1} = \begin{bmatrix} 1 & 0 & 0 \\ 0 & 1 & 0 \\ -5 & 0 & 1 \end{bmatrix}$$

Thus

$$\mathbf{x} = A^{-1}\mathbf{b} = \begin{bmatrix} 1 & 0 & 0 \\ 0 & 1 & 0 \\ -5 & 0 & 1 \end{bmatrix}\begin{bmatrix} 1 \\ 2 \\ 3 \end{bmatrix} = \begin{bmatrix} 1 \\ 2 \\ -2 \end{bmatrix}$$

Therefore, $x_1 = 1$, $x_2 = 2$, and $x_3 = -2$.

See Section 1-3.

It is important to emphasize that each one of the three different types of elementary matrices is an invertible matrix. In fact, each of the inverses is of the same type exhibited in Examples 5, 6, and 7. Thus, the inverses of the matrices

$$\begin{bmatrix} 1 & 0 & 0 \\ 0 & 0 & 1 \\ 0 & 1 & 0 \end{bmatrix}, \quad \begin{bmatrix} 1 & 0 & 0 \\ 0 & \alpha & 0 \\ 0 & 0 & 1 \end{bmatrix}, \quad \text{and} \quad \begin{bmatrix} 1 & 0 & 0 \\ 0 & 1 & 0 \\ \beta & 0 & 1 \end{bmatrix}$$

are given by

$$\begin{bmatrix} 1 & 0 & 0 \\ 0 & 0 & 1 \\ 0 & 1 & 0 \end{bmatrix}^{-1} = \begin{bmatrix} 1 & 0 & 0 \\ 0 & 0 & 1 \\ 0 & 1 & 0 \end{bmatrix}$$

$$\begin{bmatrix} 1 & 0 & 0 \\ 0 & \alpha & 0 \\ 0 & 0 & 1 \end{bmatrix}^{-1} = \begin{bmatrix} 1 & 0 & 0 \\ 0 & \frac{1}{\alpha} & 0 \\ 0 & 0 & 1 \end{bmatrix}, \quad \text{where} \quad \alpha \neq 0$$

and

$$\begin{bmatrix} 1 & 0 & 0 \\ 0 & 1 & 0 \\ \beta & 0 & 1 \end{bmatrix}^{-1} = \begin{bmatrix} 1 & 0 & 0 \\ 0 & 1 & 0 \\ -\beta & 0 & 1 \end{bmatrix}$$

We now pose a more advanced problem. Suppose $A\mathbf{x} = \mathbf{b}$ is a given linear system and suppose that $A = E_1E_2$ where E_1 and E_2 are known elementary matrices. We wish to find A^{-1} so that we may solve the system $A\mathbf{x} = \mathbf{b}$. The following example illustrates the procedure involved.

Example 8 Solve the system $A\mathbf{x} = \mathbf{b}$ if $A = E_1E_2$ where

$$E_1 = \begin{bmatrix} 1 & 0 & 0 \\ 0 & 3 & 0 \\ 0 & 0 & 1 \end{bmatrix}, \quad E_2 = \begin{bmatrix} 1 & 0 & 0 \\ 0 & 1 & 0 \\ 4 & 0 & 1 \end{bmatrix}, \quad \text{and} \quad \mathbf{b} = \begin{bmatrix} 1 \\ 0 \\ -1 \end{bmatrix}$$

Solution The system $A\mathbf{x} = \mathbf{b}$ may be written as

$$(E_1E_2)\mathbf{x} = \mathbf{b}$$

or

Here we use the associativity property of matrix multiplication.

$$E_1(E_2\mathbf{x}) = \mathbf{b}$$

Since E_1^{-1} can be easily determined, we have

$$(E_1^{-1}E_1)(E_2\mathbf{x}) = E_1^{-1}\mathbf{b}$$

or

$$I_3(E_2\mathbf{x}) = E_1^{-1}\mathbf{b}$$

or

$$E_2\mathbf{x} = E_1^{-1}\mathbf{b}$$

Multiplying by E_2^{-1} from the left, we obtain

$$(E_2^{-1}E_2)\mathbf{x} = E_2^{-1}E_1^{-1}\mathbf{b}$$

or

$$\mathbf{x} = E_2^{-1}E_1^{-1}\mathbf{b}$$

Since

$$E_1^{-1} = \begin{bmatrix} 1 & 0 & 0 \\ 0 & \frac{1}{3} & 0 \\ 0 & 0 & 1 \end{bmatrix}$$

and

$$E_2^{-1} = \begin{bmatrix} 1 & 0 & 0 \\ 0 & 1 & 0 \\ -4 & 0 & 1 \end{bmatrix}$$

we get

$$\mathbf{x} = \begin{bmatrix} 1 & 0 & 0 \\ 0 & 1 & 0 \\ -4 & 0 & 1 \end{bmatrix}\begin{bmatrix} 1 & 0 & 0 \\ 0 & \frac{1}{3} & 0 \\ 0 & 0 & -1 \end{bmatrix} = \begin{bmatrix} 1 \\ 0 \\ -5 \end{bmatrix}$$

We note that $(E_1E_2)^{-1} = E_2^{-1}E_1^{-1}$. This is true for any two invertible matrices, as the following theorem indicates.

Theorem 1-12 Let A and B be nonsingular matrices of order n. Then AB is also a nonsingular matrix of order n and its inverse is given by $(AB)^{-1} = B^{-1}A^{-1}$.

The inverse of the product is the product of the inverses in reverse order.

Proof In order to prove this theorem we have to show that

$$(B^{-1}A^{-1})(AB) = I_n \qquad \text{and} \qquad (AB)(B^{-1}A^{-1}) = I_n$$

We have

$$(B^{-1}A^{-1})(AB) = B^{-1}(A^{-1}A)B = B^{-1}(I_nB) = B^{-1}B = I_n$$

and

$$(AB)(B^{-1}A^{-1}) = A(BB^{-1})A^{-1} = A(I_nA^{-1}) = AA^{-1} = I_n$$

Judging from the structure of $(AB)^{-1} = B^{-1}A^{-1}$ one can easily guess the appropriate formulas for inverses of matrices expressed as products like ABC, $ABCD$, etc., provided all the factor matrices are invertible. The general result is expressed in the following.

Theorem 1-13 Let $A_1, A_2, \ldots, A_k$ be nonsingular matrices of order n. Then the product $A_1A_2 \ldots A_k$ is also a nonsingular matrix of order n, and its inverse is given by $(A_1A_2 \ldots A_k)^{-1} = A_k^{-1}A_{(k-1)}^{-1} \ldots A_2^{-1}A_1^{-1}$.

See problem 15.

The proof follows the pattern of the previous theorem, and is left for the student as an exercise.

The use of Theorem 1-13 will enable us to solve the system $A\mathbf{x} = \mathbf{b}$ provided A is given as a product of k elementary matrices E_1, $E_2, \ldots, E_k$.

We outline the procedure for the case where $k = 4$. In this case the system $A\mathbf{x} = \mathbf{b}$ may be written as $(E_1E_2E_3E_4)\mathbf{x} = \mathbf{b}$ where each E_i $(i = 1, 2, 3, 4)$ is a known elementary matrix of order n.

We find E_i^{-1} for $i = 1, 2, 3, 4$ and compute the product $E_4^{-1}E_3^{-1}E_2^{-1}E_1^{-1}$. Theorem 1-13 guarantees that

$$(E_1E_2E_3E_4)^{-1} = E_4^{-1}E_3^{-1}E_2^{-1}E_1^{-1}$$

and therefore the solution $\mathbf{x}$ will be given by

$$\mathbf{x} = (E_4^{-1}E_3^{-1}E_2^{-1}E_1^{-1})\mathbf{b}$$

In practice, when we seek solutions of the system $A\mathbf{x} = \mathbf{b}$ of n equations and n unknowns, the matrix A is given in terms of n^2 entries rather than as a product of elementary matrices. If we can express the matrix A as a product of the k elementary matrices E_1, $E_2, \ldots, E_k$ then the equation $A = E_1E_2 \ldots E_k$ will imply that $A^{-1} = E_k^{-1} \ldots E_2^{-1}E_1^{-1}$ and the unique solution of $A\mathbf{x} = \mathbf{b}$ will be given by

See Theorem 1-13.

$$\mathbf{x} = E_k^{-1} \ldots E_2^{-1}E_1^{-1}\mathbf{b}$$

It is fortunate that every nonsingular matrix can be expressed as a product of elementary matrices, as indicated by the following theorem.

Theorem 1-14 Let A be an $n \times n$ matrix. Then A is invertible (nonsingular) if and only if there exist elementary matrices $E_1, E_2, \ldots, E_k$ such that $A = E_1E_2 \ldots E_k$.

There are two statements in the last theorem which should be clearly understood. The first statement says that if A is a product of k elementary matrices $E_1, E_2, \ldots, E_k$, then A is invertible. This in essence is a corollary of Theorem 1-13. The second statement says that if A is nonsingular, then there exist k elementary matrices $E_1, E_2, \ldots, E_k$ such that

$$A = E_1E_2 \ldots E_k$$

For proof of Theorem 1-14 see *Linear Algebra* by K. Hoffman and R. Kunze (2nd ed., Prentice-Hall, 1971), p. 23.

The proof of Theorem 1-14 is beyond the scope of this book, but we will make use of the theorem in the next section where we develop a method of finding the inverse (if it exists) for a given matrix A of order n.

Exercises

1 For each of the given linear systems, find the matrix of coefficients A and express the system in the form $A\mathbf{x} = \mathbf{b}$.

a $$\begin{aligned} 2x - 4y - z &= 1 \\ x + 5y &= 6 \\ 2y + 3z &= 7 \end{aligned}$$

b $$\begin{aligned} x - y &= 2 \\ 2x + 3y &= 7 \\ x + 2y &= 4 \end{aligned}$$

c $$\begin{aligned} 5x_1 - 2x_2 - 4x_3 - x_4 &= 0 \\ -x_1 + x_2 - 2x_3 &= 0 \\ 2x_1 - 7x_2 + x_3 + 6x_4 &= 0 \\ 5x_2 - 2x_3 - x_4 &= 0 \end{aligned}$$

d $$\begin{aligned} 2x_1 + 3x_2 - 4x_3 - 5x_4 + 7x_5 &= 12 \\ -x_1 - 5x_2 + 6x_3 - x_4 - x_5 &= 7 \end{aligned}$$

2 Let $A = \begin{bmatrix} 0 & 0 \\ 0 & 0 \end{bmatrix}$. Does A^{-1} exist? Prove your answer. (Hint: Assume $A^{-1} = \begin{bmatrix} a & b \\ c & d \end{bmatrix}$ and check the equation $\begin{bmatrix} a & b \\ c & d \end{bmatrix}\begin{bmatrix} 0 & 0 \\ 0 & 0 \end{bmatrix} = \begin{bmatrix} 1 & 0 \\ 0 & 1 \end{bmatrix}$ for possible solutions for a, b, c, and d.)

3 Let $B = \begin{bmatrix} 0 & 0 \\ 1 & 2 \end{bmatrix}$. Does B^{-1} exist? Prove your answer.

4 Let $C = \begin{bmatrix} 0 & 2 \\ 0 & 3 \end{bmatrix}$. Does C^{-1} exist? Prove your answer.

5 Let $D = \begin{bmatrix} 2 & 1 & 0 & -1 \\ 3 & 2 & 0 & 1 \\ 1 & 0 & 0 & 1 \\ -2 & 1 & 0 & 3 \end{bmatrix}$. Does D^{-1} exist? Prove your answer.

6 Let $E = \begin{bmatrix} 1 & 2 & 3 & 4 \\ 0 & 0 & 0 & 0 \\ 3 & 2 & 1 & 4 \\ 1 & 0 & -1 & 0 \end{bmatrix}$. Does E^{-1} exist? Prove your answer.

7 Let A be an $n \times n$ matrix such that its entire ith row (or jth column) consists of zeros. Does A^{-1} exist? Prove your answer.

8 Let A and B be $n \times n$ matrices such that $A = EB$ where E is an elementary matrix. Show that A is invertible if and only if B is invertible.

9 Let A and B be $n \times n$ matrices such that $A = E_1E_2B$ where E_1 and E_2 are some given elementary matrices. Show that A is invertible if and only if B is invertible.

10 Generalize the results obtained in problems 8 and 9.

11 Let $A\mathbf{x} = \mathbf{0}$ where A is nonsingular. Show that the so-called trivial solution $\mathbf{x} = \mathbf{0}$ is a unique solution.

12 Let A be an $n \times n$ matrix and B an $n \times k$ matrix. Assume that $B \neq 0$ and $AB = 0$. Show that A is a singular matrix.

13 Let A be an $n \times n$ matrix and let B and C be $n \times k$ matrices. Assume that $AB - AC = 0$. Show that if A is nonsingular then $B = C$.

14 Let A be an invertible matrix. Show that A^{-1} is also an invertible matrix and that $(A^{-1})^{-1} = A$.

15 Prove Theorem 1-13.

1-5 Inversion of Matrices

Objectives

1 Establish a method of checking the invertibility of a matrix A and computing A^{-1} whenever it exists by use of elementary row operations.
2 Solve linear systems of equations with an invertible matrix of coefficients.

In this section we establish a method which will enable us to determine whether or not a given matrix A of order $n \times n$ is invertible and to compute A^{-1} whenever it exists. In describing this method we speak about elementary row operations in accordance with the following definition.

Definition 1-11 Let E be an elementary $n \times n$ matrix and let A be an $n \times m$ matrix.

a If E is obtained from I_n by interchanging any two rows of I_n, then we say that EA is obtained from A by performing an **elementary row operation of type I** on A.

b If E is obtained from I_n by multiplying one row of I_n by a number α, $\alpha \neq 0$, then we say that EA is obtain from A by performing an **elementary row operation of type II** on A.

Elementary row operations.

c If E is obtained from I_n by multiplying one row of I_n by β and then adding it to *another* row of I_n, then we say that EA is obtained from A by performing an **elementary row operation of type III** on A.

Thus, every elementary row operation on A is equivalent to multiplying A from the left by a suitable elementary matrix E.

It follows from Theorem 1-14 that if a matrix A of order $n \times n$ is nonsingular, then it may be expressed as a product of k elementary matrices. By Theorem 1-13, A^{-1} may be expressed in a similar fashion. Thus, if A is nonsingular then A^{-1} is nonsingular too and there exist k elementary matrices $E_1, E_2, \ldots, E_k$ such that

$$A^{-1} = E_k E_{(k-1)} \cdots E_2 E_1$$

Since we have $A^{-1}A = I_n$, we may write

$$(E_k E_{(k-1)} \cdots E_2 E_1)A = I_n \qquad \textbf{[1]}$$

Equation 1 may be interpreted also as follows: If we can find a sequence of k elementary row operations (represented by k elementary matrices $E_1, E_2, \ldots, E_k$) such that the matrix A will be transformed into the identity matrix I_n as a result of performing this sequence on A, then

$$A^{-1} = E_k E_{(k-1)} \cdots E_2 E_1$$

This suggests the following procedure for finding A^{-1} (if it exists): Form a partitioned matrix of order $n \times 2n$ by putting the matrices A and I_n next to each other as follows:

$$[A \vdots I_n] = \begin{bmatrix} a_{11} & a_{12} & \cdots & a_{1n} & \vdots & 1 & 0 & \cdots & 0 \\ a_{21} & a_{22} & \cdots & a_{2n} & \vdots & 0 & 1 & \cdots & 0 \\ \vdots & \vdots & & \vdots & \vdots & \vdots & \vdots & & \vdots \\ a_{n1} & a_{n2} & \cdots & a_{nn} & \vdots & 0 & 0 & \cdots & 1 \end{bmatrix}$$

Now, apply row operations to $[A \vdots I_n]$ in order to transform it. If the sequence $E_1, E_2, \ldots, E_k$ transforms A into I_n, then the same sequence transforms I_n (from the partitioned matrix $[A \vdots I_n]$) into $E_k E_{k-1} \cdots E_2 E_1$. It follows from equation 1 that $E_k E_{(k-1)} \cdots E_2 E_1 = A^{-1}$. Thus, if we transform the partitioned matrix $[A \vdots I_n]$ into the matrix $[I_n \vdots B]$ by means of elementary row operations, then $B = A^{-1}$.

We still need to show how one chooses the elementary row operations so that when they are applied to A, I_n is produced.

We illustrate this method by applying it to a matrix of order 3×3.

Example 1 Let

$$A = \begin{bmatrix} 2 & 4 & 6 \\ 3 & 1 & 2 \\ 0 & 1 & -1 \end{bmatrix}$$

Find A^{-1}.

Solution We put I_3 next to A and perform row operations on the partitioned matrix

$$[A \vdots I_3] = \left[\begin{array}{ccc:ccc} 2 & 4 & 6 & 1 & 0 & 0 \\ 3 & 1 & 2 & 0 & 1 & 0 \\ 0 & 1 & -1 & 0 & 0 & 1 \end{array}\right]$$

This gives us the (1,1) entry of I_3.

We multiply the first row by $\frac{1}{2}$ and obtain

$$\left[\begin{array}{ccc:ccc} 1 & 2 & 3 & \frac{1}{2} & 0 & 0 \\ 3 & 1 & 2 & 0 & 1 & 0 \\ 0 & 1 & -1 & 0 & 0 & 1 \end{array}\right]$$

Elementary row operation of type III.

Multiplying the first row by -3 and adding it to the second row, we have

$$\left[\begin{array}{ccc:ccc} 1 & 2 & 3 & \frac{1}{2} & 0 & 0 \\ 0 & -5 & -7 & -\frac{3}{2} & 1 & 0 \\ 0 & 1 & -1 & 0 & 0 & 1 \end{array}\right]$$

Elementary row operation of type I.

Observe that we have already succeeded in producing, on the left side, the first column of I_3. Now we turn our attention to producing the second column of I_3. We interchange the second and third rows and get

$$\left[\begin{array}{ccc:ccc} 1 & 2 & 3 & \frac{1}{2} & 0 & 0 \\ 0 & 1 & -1 & 0 & 0 & 1 \\ 0 & -5 & -7 & -\frac{3}{2} & 1 & 0 \end{array}\right]$$

Elementary row operation of type III.

Multiplying the second row by 5 and adding it to the third row we obtain

$$\left[\begin{array}{ccc:ccc} 1 & 2 & 3 & \frac{1}{2} & 0 & 0 \\ 0 & 1 & -1 & 0 & 0 & 1 \\ 0 & 0 & -12 & -\frac{3}{2} & 1 & 5 \end{array}\right]$$

Elementary row operation of type III.

Now multiply the second row by -2 and add it to the first row to get

$$\left[\begin{array}{ccc:ccc} 1 & 0 & 5 & \frac{1}{2} & 0 & -2 \\ 0 & 1 & -1 & 0 & 0 & 1 \\ 0 & 0 & -12 & -\frac{3}{2} & 1 & 5 \end{array}\right]$$

Elementary row operation of type II.

So far we have produced the first two columns of I_3. Multiply the third row by $-\frac{1}{12}$ to get

$$\begin{bmatrix} 1 & 0 & 5 & \vdots & \frac{1}{2} & 0 & -2 \\ 0 & 1 & -1 & \vdots & 0 & 0 & 1 \\ 0 & 0 & 1 & \vdots & \frac{3}{24} & -\frac{1}{12} & -\frac{5}{12} \end{bmatrix}$$

Elementary row operation of type III.

Now add to the second row 1 times the third row to get

$$\begin{bmatrix} 1 & 0 & 5 & \vdots & \frac{1}{2} & 0 & -2 \\ 0 & 1 & 0 & \vdots & \frac{3}{24} & -\frac{1}{12} & \frac{7}{12} \\ 0 & 0 & 1 & \vdots & \frac{3}{24} & -\frac{1}{12} & -\frac{5}{12} \end{bmatrix}$$

Elementary row operation of type III.

Finally, add -5 times the third row to the first row to obtain

$$[I_n \vdots A^{-1}] = \begin{bmatrix} 1 & 0 & 0 & \vdots & -\frac{3}{24} & \frac{5}{12} & \frac{1}{12} \\ 0 & 1 & 0 & \vdots & \frac{3}{24} & -\frac{1}{12} & \frac{7}{12} \\ 0 & 0 & 1 & \vdots & \frac{3}{24} & -\frac{1}{12} & -\frac{5}{12} \end{bmatrix}$$

Hence

$$A^{-1} = \begin{bmatrix} -\frac{3}{24} & \frac{5}{12} & \frac{1}{12} \\ \frac{3}{24} & -\frac{1}{12} & \frac{7}{12} \\ \frac{3}{24} & -\frac{1}{12} & -\frac{5}{12} \end{bmatrix}$$

Students are advised to check their results by computing the product $A^{-1}A$ (or AA^{-1}), and to show that in fact $A^{-1}A = I_n$.

It is important to notice the methodical way in which we proceeded to transform $[A \vdots I_n]$ into $[I_n \vdots A^{-1}]$. By means of row operations, one generates one column after the other of the matrix I_n. If at any stage of this operation we find it impossible to generate any further columns of I_n, we conclude that an inverse of the matrix A does not exist. This is illustrated in the next example.

Example 2 Let

$$A = \begin{bmatrix} 0 & 0 & 2 \\ 1 & 2 & 6 \\ 3 & 6 & 9 \end{bmatrix}$$

Find A^{-1} if it exists.

Solution We start again with the partitioned matrix

$$[A \vdots I_n] = \begin{bmatrix} 0 & 0 & 2 & \vdots & 1 & 0 & 0 \\ 1 & 2 & 6 & \vdots & 0 & 1 & 0 \\ 3 & 6 & 9 & \vdots & 0 & 0 & 1 \end{bmatrix}$$

Classify the type of operations performed in solving this problem.

In order to generate the left column of I_n we interchange the first and second rows of $[A \vdots I_n]$, obtaining

$$\begin{bmatrix} 1 & 2 & 6 & \vdots & 0 & 1 & 0 \\ 0 & 0 & 2 & \vdots & 1 & 0 & 0 \\ 3 & 6 & 9 & \vdots & 0 & 0 & 1 \end{bmatrix}$$

We add -3 times the first row to the third row and get

$$\begin{bmatrix} 1 & 2 & 6 & \vdots & 0 & 1 & 0 \\ 0 & 0 & 2 & \vdots & 1 & 0 & 0 \\ 0 & 0 & -9 & \vdots & 0 & -3 & 0 \end{bmatrix}$$

Now multiply the second row by $\frac{1}{2}$, obtaining

$$\begin{bmatrix} 1 & 2 & 6 & \vdots & 0 & 1 & 0 \\ 0 & 0 & 1 & \vdots & \frac{1}{2} & 0 & 0 \\ 0 & 0 & -9 & \vdots & 0 & -3 & 1 \end{bmatrix}$$

We now add 9 times the second row to the third row and get

$$\begin{bmatrix} 1 & 2 & 6 & \vdots & 0 & 1 & 0 \\ 0 & 0 & 1 & \vdots & \frac{1}{2} & 0 & 0 \\ 0 & 0 & 0 & \vdots & \frac{9}{2} & -3 & 0 \end{bmatrix}$$

At this stage we stop and conclude that A^{-1} does not exist, because the left half of the partitioned matrix, namely

$$\begin{bmatrix} 1 & 2 & 6 \\ 0 & 0 & 1 \\ 0 & 0 & 0 \end{bmatrix}$$

contains a row of zeros. Such a matrix is not invertible (see Problem 7 of Section 1-4), and therefore we cannot transform A into I_3 by any elementary row operations.

The method of finding the inverse of a given matrix A can be used also to factor A in terms of elementary matrices. Let us use Example 1 in performing such a factorization. We examine closely the row operations performed on the matrix A given by

$$A = \begin{bmatrix} 2 & 4 & 6 \\ 3 & 1 & 2 \\ 0 & 1 & -1 \end{bmatrix}$$

If we express each row operation as an elementary matrix, we get the following equation:

The coefficient of A is A^{-1}.

$$\left\{\begin{bmatrix}1 & 0 & -5\\0 & 1 & 0\\0 & 0 & 1\end{bmatrix}\begin{bmatrix}1 & 0 & 0\\0 & 1 & 1\\0 & 0 & 1\end{bmatrix}\begin{bmatrix}1 & 0 & 0\\0 & 1 & 0\\0 & 0 & -\frac{1}{12}\end{bmatrix}\begin{bmatrix}1 & -2 & 0\\0 & 1 & 0\\0 & 0 & 1\end{bmatrix}\begin{bmatrix}1 & 0 & 0\\0 & 1 & 0\\0 & 5 & 1\end{bmatrix}\begin{bmatrix}1 & 0 & 0\\0 & 0 & 1\\0 & 1 & 0\end{bmatrix}\begin{bmatrix}1 & 0 & 0\\-3 & 1 & 0\\0 & 0 & 1\end{bmatrix}\begin{bmatrix}\frac{1}{2} & 0 & 0\\0 & 1 & 0\\0 & 0 & 1\end{bmatrix}\right\}A = I_3$$

We now find A by taking $(A^{-1})^{-1}$ (see problem 14, Section 1-4) and using Theorem 1-13:

Remember the inversion of elementary matrices.

$$A = (A^{-1})^{-1} = \begin{bmatrix}\frac{1}{2} & 0 & 0\\0 & 1 & 0\\0 & 0 & 1\end{bmatrix}^{-1}\begin{bmatrix}1 & 0 & 0\\-3 & 1 & 0\\0 & 0 & 1\end{bmatrix}^{-1}\begin{bmatrix}1 & 0 & 0\\0 & 0 & 1\\0 & 1 & 0\end{bmatrix}^{-1}\begin{bmatrix}1 & 0 & 0\\0 & 1 & 0\\0 & 5 & 1\end{bmatrix}^{-1}$$

$$\begin{bmatrix}1 & -2 & 0\\0 & 1 & 0\\0 & 0 & 1\end{bmatrix}^{-1}\begin{bmatrix}1 & 0 & 0\\0 & 1 & 0\\0 & 0 & -\frac{1}{12}\end{bmatrix}^{-1}\begin{bmatrix}1 & 0 & 0\\0 & 1 & 1\\0 & 0 & 1\end{bmatrix}^{-1}\begin{bmatrix}1 & 0 & -5\\0 & 1 & 0\\0 & 0 & 1\end{bmatrix}^{-1}$$

Thus

$$A = \begin{bmatrix}2 & 0 & 0\\0 & 1 & 0\\0 & 0 & 1\end{bmatrix}\begin{bmatrix}1 & 0 & 0\\3 & 1 & 0\\0 & 0 & 1\end{bmatrix}\begin{bmatrix}1 & 0 & 0\\0 & 0 & 1\\0 & 1 & 0\end{bmatrix}\begin{bmatrix}1 & 0 & 0\\0 & 1 & 0\\0 & -5 & 1\end{bmatrix}\begin{bmatrix}1 & 2 & 0\\0 & 1 & 0\\0 & 0 & 1\end{bmatrix}$$

$$\begin{bmatrix}1 & 0 & 0\\0 & 1 & 0\\0 & 0 & -12\end{bmatrix}\begin{bmatrix}1 & 0 & 0\\0 & 1 & -1\\0 & 0 & 1\end{bmatrix}\begin{bmatrix}1 & 0 & 5\\0 & 1 & 0\\0 & 0 & 1\end{bmatrix}$$

We now return to the problem of solving the system of n equations and n unknowns:

$$A\mathbf{x} = \mathbf{b}$$

Based on our work above, we can find a solution $\mathbf{x}$ provided A^{-1} exists. The solution is then

$$\mathbf{x} = A^{-1}\mathbf{b}$$

In practice, when we are confronted with the system $A\mathbf{x} = \mathbf{b}$ we actually do not need to find an explicit expression for A^{-1}. The reason for this stems from the fact that A^{-1} (if it exists) can be expressed as a product of k elementary matrices. Let

$$A^{-1} = E_1E_2 \dots E_k$$

where $E_i (i = 1, 2, \dots, k)$ is an elementary matrix. Multiplying $A\mathbf{x} = \mathbf{b}$ from the left by $E_1E_2 \dots E_k$ we obtain

$$(E_1E_2 \dots E_k)A\mathbf{x} = (E_1E_2 \dots E_k)\mathbf{b}$$

Let us interpret this equation in the light of

$$(E_1E_2 \dots E_k)A\mathbf{x} = A^{-1}A\mathbf{x} = I_n\mathbf{x} = \mathbf{x}$$

The elementary operations represented by $E_1, E_2, \dots, E_k$ and performed on $A\mathbf{x}$ are used also to transform $\mathbf{b}$ into $E_1E_2 \dots E_k\mathbf{b}$. This suggests the following procedure for solving the system $A\mathbf{x} = \mathbf{b}$: We form the so-called **augmented** matrix by adding to the matrix of coefficients A an additional column $\mathbf{b}$. We denote such an augmented matrix by $[A \vdots \mathbf{b}]$. Then we perform elementary row operations on the augmented matrix until we obtain $[I_n \vdots \mathbf{c}]$. Since $\mathbf{c} = A^{-1}\mathbf{b}$, it follows that $\mathbf{x} = \mathbf{c}$ is the solution. Obviously, there is no need here to find A^{-1} explicitly. We illustrate this method in the following example.

This is possible provided A^{-1} exists!

Example 3 Let a system of linear equations be given by

$$2x_1 - 3x_2 = 7$$
$$3x_1 + 4x_2 = 2$$

Find the solution vector $\begin{bmatrix} x_1 \\ x_2 \end{bmatrix}$.

Solution We form the augmented matrix

In matrix form, the system is

$$\begin{bmatrix} 2 & -3 \\ 3 & 4 \end{bmatrix}\begin{bmatrix} x_1 \\ x_2 \end{bmatrix} = \begin{bmatrix} 7 \\ 2 \end{bmatrix}$$

$$\left[\begin{array}{cc|c} 2 & -3 & 7 \\ 3 & 4 & 2 \end{array}\right]$$

Multiplying the first row by $\frac{1}{2}$, we get

$$\left[\begin{array}{cc|c} 1 & -\frac{3}{2} & \frac{7}{2} \\ 3 & 4 & 2 \end{array}\right]$$

We add -3 times the first row to the second row and obtain

$$\left[\begin{array}{cc|c} 1 & -\frac{3}{2} & \frac{7}{2} \\ 0 & \frac{17}{2} & -\frac{17}{2} \end{array}\right]$$

Multiplying the second row by $\frac{2}{17}$, we get

$$\left[\begin{array}{cc|c} 1 & -\frac{3}{2} & \frac{7}{2} \\ 0 & 1 & -1 \end{array}\right]$$

Finally, we multiply the second row by $\frac{3}{2}$ and add it to the first row, obtaining

$$\left[\begin{array}{cc|c} 1 & 0 & 2 \\ 0 & 1 & -1 \end{array}\right]$$

In this example

$A = \begin{bmatrix} 2 & -3 \\ 3 & 4 \end{bmatrix}$

Since the last matrix is of the form $[I_2 \vdots \mathbf{c}]$ we have solved the original system with the solution given by $\begin{bmatrix} x_1 \\ x_2 \end{bmatrix} = \begin{bmatrix} 2 \\ -1 \end{bmatrix}$. We emphasize again that in using this procedure no explicit form of A^{-1} was needed to solve for $\mathbf{x}$.

Exercises

1 Determine which of the following matrices are singular. Find the inverse for each nonsingular matrix.

a $\begin{bmatrix} 1 & 2 \\ 2 & 1 \end{bmatrix}$ **b** $\begin{bmatrix} 2 & 6 \\ 3 & 9 \end{bmatrix}$ **c** $\begin{bmatrix} 1 & 2 & 7 \\ -1 & 3 & 3 \\ 0 & 1 & 2 \end{bmatrix}$ **d** $\begin{bmatrix} 1 & -3 & 4 \\ 2 & -5 & 8 \\ 1 & 1 & 5 \end{bmatrix}$

2 Show that $A = \begin{bmatrix} 2 & -1 \\ 1 & 3 \end{bmatrix}$ is nonsingular by finding A^{-1}. Express A^{-1} and A as products of elementary matrices.

3 Show that $B = \begin{bmatrix} 1 & 2 & -6 \\ 2 & 5 & -12 \\ 1 & 4 & -3 \end{bmatrix}$ is nonsingular by finding B^{-1}. Express B^{-1} and B as products of elementary matrices.

4 Let $A = \begin{bmatrix} 1 & 2 & p \\ 3 & -1 & 1 \\ 5 & 3 & -5 \end{bmatrix}$. Determine p such that the matrix A is singular. How many solutions does this problem have? Explain.

5 Solve the following systems of linear equations:

a $\begin{aligned} 2x_1 - 3x_2 &= 2 \\ 4x_1 + x_2 &= 5 \end{aligned}$ **b** $\begin{aligned} 2x_1 - 3x_2 &= 0 \\ 4x_1 + x_2 &= 3 \end{aligned}$ **c** $\begin{aligned} 2x_1 - 3x_2 &= 0 \\ 4x_1 + x_2 &= 0 \end{aligned}$

d $\begin{aligned} 3x_1 + 4x_2 &= -2 \\ x_1 + 3x_2 &= 1 \end{aligned}$ **e** $\begin{aligned} 5x_1 + 2x_2 &= 0 \\ 15x_1 - 4x_2 &= 5 \end{aligned}$ **f** $\begin{aligned} 7x_1 + 5x_2 &= 0 \\ 5x_1 + 7x_2 &= 0 \end{aligned}$

6 Solve the systems of linear equations given by

a $\begin{aligned} -3x_1 - x_2 - 3x_3 &= 1 \\ 6x_1 + x_2 + 3x_3 &= 0 \\ 2x_1 \qquad + x_3 &= 1 \end{aligned}$ **b** $\begin{aligned} 2x_1 + 3x_2 + 4x_3 &= 11 \\ 3x_1 + 4x_2 - x_3 &= 2 \\ 4x_1 - x_2 + x_3 &= 1 \end{aligned}$

c $\begin{aligned} 2x_1 + 3x_2 + 4x_3 &= 0 \\ 3x_1 + 4x_2 - x_3 &= 0 \\ 4x_1 - x_2 + x_3 &= 0 \end{aligned}$

7 Solve the systems of linear equations given by

a $\begin{aligned} 2x_1 - 2x_2 + 6x_3 - 4x_4 &= 1 \\ x_2 + 3x_3 + 2x_4 &= 0 \\ 2x_1 - 2x_2 + 9x_3 \qquad &= 1 \\ 2x_1 - 2x_2 + 6x_3 \qquad &= 0 \end{aligned}$

b
$$\begin{aligned} x_1 - x_2 + 2x_3 + 3x_4 &= 8 \\ 2x_1 + 2x_2 \qquad + 2x_4 &= 8 \\ 4x_1 + x_2 - x_3 - x_4 &= 6 \\ x_1 + 2x_2 + 3x_3 \qquad &= -6 \end{aligned}$$

Hint: Obtain two equations in the two unknowns a and b.

8 The equation $ax_1 + bx_2 = 1$ represents a line L in the x_1x_2-plane. Find a and b given that the line L passes through the points (2,3) and (1,4).

9 The equation $ax_1 + bx_2 + cx_3 = 1$ represents a plane P in $x_1x_2x_3$-space. Find a, b, and c given that the plane P contains the points (2,1,0), (1,2,0), and (0,1,1).

10 Let $ad - bc \neq 0$. Show that $\begin{bmatrix} a & b \\ c & d \end{bmatrix}^{-1} = \dfrac{1}{ad-bc}\begin{bmatrix} d & -b \\ -c & a \end{bmatrix}$.

11 Let A be a 4×4 matrix defined by

$$a_{ij} = \begin{cases} 1 & \text{if } i \leq j \\ 0 & \text{if } i > j \end{cases}$$

a Show that A is nonsingular by finding A^{-1}.
b Find $(A^2)^{-1}$ in terms of A^{-1}.

1-6 Row Reduction Technique for Solving Linear Systems

Objectives
1. Establish the Gauss-Jordan elimination method for solving linear systems of m equations in n unknowns.
2. Show that a homogeneous system with more unknowns than equations has infinitely many solutions.

In the previous section we discussed only the problem of finding a solution to the linear system $A\mathbf{x} = \mathbf{b}$ when A was an invertible $n \times n$ matrix. In this section we consider the more general problem of finding solutions to the system $A\mathbf{x} = \mathbf{b}$ where A is an $m \times n$ matrix (and thus not necessarily invertible). We shall find out that there exist nonhomogeneous systems that have no solution. On the other hand, there are systems (homogeneous as well as nonhomogeneous) that have more than one solution, in fact, infinitely many solutions. The basic idea in searching for solutions of a linear system is the possibility of transforming the system into another one from which all solutions may be obtained (if they exist) by simple inspection.

The concept of equivalence of linear systems plays an important role in the process of transforming one linear system into another. The following definition clarifies this concept.

Definition 1-12

Equivalency of linear systems.

Let $A\mathbf{x} = \mathbf{b}$ be a system of m linear equations in n unknowns. Let $B\mathbf{x} = \mathbf{c}$ be a system of r linear equations in n unknowns. We say that the two systems are **equivalent** if they both have exactly the same solutions.

Example 1 The system of equations

$$3x_1 - x_2 = 1$$
$$2x_1 + 3x_2 = 8$$

has the unique solution $x_1 = 1$ and $x_2 = 2$.

The system of equations

$$x_1 + x_2 = 3$$
$$5x_1 - x_2 = 3$$
$$9x_1 - 4x_2 = 1$$

also has the unique solution $x_1 = 1$ and $x_2 = 2$. Thus, the two systems are equivalent.

The system

$$3x_1 - x_2 = 1$$
$$2x_1 + 3x_2 = 8$$

may be written as

$$\begin{bmatrix} 3 & -1 \\ 2 & 3 \end{bmatrix}\begin{bmatrix} x_1 \\ x_2 \end{bmatrix} = \begin{bmatrix} 1 \\ 8 \end{bmatrix}$$

Since $\begin{bmatrix} 3 & -1 \\ 2 & 3 \end{bmatrix}$ is nonsingular, the solution must be unique.

The system

$$x_1 + x_2 = 3$$
$$5x_1 - x_2 = 3$$

has the unique solution $x_1 = 1$, $x_2 = 2$. Since the equation $9x_1 - 4x_2 = 1$ is also satisfied by $x_1 = 1$, $x_2 = 2$ it follows that the system

$$x_1 + x_2 = 3$$
$$5x_1 - x_2 = 3$$
$$9x_1 - 4x_2 = 1$$

has a unique solution.

We have seen in the previous section that by performing elementary row operations on the augmented matrix $[A \vdots \mathbf{b}]$ of the linear system $A\mathbf{x} = \mathbf{b}$ we obtained the augmented matrix $[I_n \vdots \mathbf{c}]$ of the system $I_n\mathbf{x} = \mathbf{c}$ with $\mathbf{c} = A^{-1}\mathbf{b}$ (provided A^{-1} exists). In this case, the elementary row operations produced two equivalent linear systems of equations, $A\mathbf{x} = \mathbf{b}$ and $I_n\mathbf{x} = \mathbf{c}$, and the solution was readily available from the latter. It can be proved that elementary row operations applied to the augmented matrix of one linear system of equations produce an augmented matrix of an equivalent linear system of equations. This fact will help us in establishing a method for solving a system $A\mathbf{x} = \mathbf{b}$ of m equations with n unknowns provided a solution exists.

The reader may recall from the previous section that when we searched for solutions of the linear system $A\mathbf{x} = \mathbf{b}$, we attempted to transform the augmented matrix $[A \vdots \mathbf{b}]$ into $[I_n \vdots \mathbf{c}]$ where possible. Whenever we succeeded in such an attempt, we actually transformed the system $A\mathbf{x} = \mathbf{b}$ into the simple system $I_n\mathbf{x} = \mathbf{c}$ where the solution $\mathbf{x} = \mathbf{c}$ was readily available.

We need to establish a similar process of simplification for the linear system $A\mathbf{x} = \mathbf{b}$ of m equations and n unknowns. We must therefore transform the system $A\mathbf{x} = \mathbf{b}$ into an equivalent system $B\mathbf{x} = \mathbf{c}$ from which solutions are easy to obtain. The remainder of this section is devoted to this task.

Definition 1-13 An $m \times n$ matrix is said to be in **row-reduced echelon** form if it satisfies the following properties:

Row-reduced echelon matrix.

1 If a row does not consist entirely of zeros, then the first nonzero entry (from the left) in the row is 1. This entry is called the **leading entry** of the row.

2 If there are rows that consist entirely of zeros, then they are below all rows which do not consist entirely of zeros.

3 If row i and row $i+1$ are two successive rows that do not consist entirely of zeros, then the leading entry 1 in row i occurs to the left of the leading entry 1 in row $i+1$.

4 If a column contains a leading entry of some row, then the remaining entries in that column must be zeros.

Example 2 The following matrices satisfy properties 1–4 of Definition 1-13 and thus are in row-reduced echelon form:

$$\begin{bmatrix} 0 & 0 & 0 \\ 0 & 0 & 0 \end{bmatrix}, \begin{bmatrix} 1 & 0 \\ 0 & 1 \end{bmatrix}, \begin{bmatrix} 1 & 0 & 0 & 4 \\ 0 & 1 & 0 & 3 \\ 0 & 0 & 1 & 2 \end{bmatrix}, \begin{bmatrix} 0 & 1 & 3 & 2 & 0 \\ 0 & 0 & 0 & 0 & 1 \\ 0 & 0 & 0 & 0 & 0 \\ 0 & 0 & 0 & 0 & 0 \end{bmatrix}$$

Example 3 The following matrices fail to satisfy some of the properties 1–4 of Definition 1-13 and therefore are not in row-reduced echelon form:

These matrices fail to satisfy properties 2, 3, 4 and 1 respectively.

$$\begin{bmatrix} 0 & 0 & 0 \\ 1 & 0 & 0 \end{bmatrix}, \begin{bmatrix} 0 & 1 \\ 1 & 0 \end{bmatrix}, \begin{bmatrix} 1 & 2 & 0 & 4 \\ 0 & 1 & 0 & 3 \\ 1 & 0 & 1 & 2 \end{bmatrix}, \begin{bmatrix} 0 & 4 & 3 & 2 & 0 \\ 0 & 0 & 0 & 0 & 3 \\ 0 & 0 & 0 & 0 & 0 \\ 0 & 0 & 0 & 0 & 0 \end{bmatrix}$$

Example 4 Each of the following row-reduced echelon matrices is an augmented matrix of a corresponding linear system of equations. Solve each of the systems.

a $\left[\begin{array}{cc:c} 1 & 0 & 2 \\ 0 & 1 & 3 \end{array}\right]$ **b** $\left[\begin{array}{ccc:c} 1 & 0 & 2 & 4 \\ 0 & 1 & 3 & 5 \end{array}\right]$

c $\left[\begin{array}{ccccc:c} 1 & 2 & 0 & 3 & 4 & -1 \\ 0 & 0 & 1 & 0 & 2 & -2 \\ 0 & 0 & 0 & 0 & 0 & 0 \end{array}\right]$ **d** $\left[\begin{array}{ccccc:c} 1 & 2 & 0 & 3 & 4 & 0 \\ 0 & 0 & 1 & 0 & 2 & 0 \\ 0 & 0 & 0 & 0 & 0 & 1 \end{array}\right]$

Solution **a** The corresponding system of equations is given by

Note: Here the system is
$1 \cdot x_1 + 0 \cdot x_2 = 2$
$0 \cdot x_1 + 1 \cdot x_2 = 3$

$$\begin{aligned} x_1 \quad &= 2 \\ x_2 &= 3 \end{aligned}$$

Hence the solution is $x_1 = 2$ and $x_2 = 3$.

b The corresponding system of equations is given by

$$\begin{aligned} x_1 \quad + 2x_3 &= 4 \\ x_2 + 3x_3 &= 5 \end{aligned}$$

Since the coefficients of x_1 and x_2 are the leading entries of the rows in the matrix in b, we refer to x_1 and x_2 as the **leading unknowns.**

We now solve for the leading unknowns in terms of the unknown x_3 in the system of equations above, obtaining

$$\begin{aligned} x_1 &= 4 - 2x_3 \\ x_2 &= 5 - 3x_3 \end{aligned}$$

By assigning completely arbitrary values to x_3, we determine the values of x_1 and x_2. Thus the system above possesses infinitely many solutions.

c The corresponding system of equations is given by

$$\begin{aligned} x_1 + 2x_2 \quad\; + 3x_4 + 4x_5 &= -1 \\ x_3 \qquad + 2x_5 &= -2 \end{aligned}$$

The last row of the matrix corresponds to the equation

$$0 \cdot x_1 + 0 \cdot x_2 + 0 \cdot x_3 + 0 \cdot x_4 + 0 \cdot x_5 = 0$$

and is therefore omitted.

Solving for the leading unknowns x_1 and x_3 above, we get

$$\begin{aligned} x_1 &= -1 - 2x_2 - 3x_4 - 4x_5 \\ x_3 &= -2 - 2x_5 \end{aligned}$$

We are now free to choose arbitrary values for x_2, x_4, and x_5. Every such choice will in turn determine x_1 and x_3. Thus we have infinitely many solutions to the system of equations.

d The change we introduced in the matrix in c to produce the matrix in d resulted in an inconsistent system of equations. To see this, we need only to write down the equation corresponding to the third row of matrix d. We have

$$0 \cdot x_1 + 0 \cdot x_2 + 0 \cdot x_3 + 0 \cdot x_4 + 0 \cdot x_5 = 1,$$

which obviously cannot be satisfied regardless of the choice of numerical values for x_1, x_2, x_3, x_4, and x_5. Thus system d does not have a solution.

The examples above made it clear that if we seek solutions of a linear system of equations $A\mathbf{x} = \mathbf{b}$, our first step is to transform the augmented matrix $[A \vdots \mathbf{b}]$ into a row-reduced echelon matrix. Once this has been achieved, the resulting equivalent system of equations is simple to solve, as we have seen in the examples above.

The procedure of transforming the system of equations $A\mathbf{x} = \mathbf{b}$ into the system $B\mathbf{x} = \mathbf{c}$ where $[B \vdots \mathbf{c}]$ is in row-reduced echelon form is called **Gauss-Jordan elimination.**

To help the reader attain proficiency in solving linear systems of equations we illustrate the procedure of row-reduction of matrices by solving the following system.

Example 5 Solve the system of equations

$$\begin{aligned} 2x_1 - x_2 + 4x_3 + x_4 - x_5 &= 4 \\ 3x_1 + 2x_2 - 2x_3 - 5x_4 + 2x_5 &= 9 \\ x_1 + x_2 + 6x_3 + 2x_4 + x_5 &= 7 \end{aligned}$$

Solution The augmented matrix of the system is given by

$$\left[\begin{array}{rrrrr:r} 2 & -1 & 4 & 1 & -1 & 4 \\ 3 & 2 & -2 & -5 & 2 & 9 \\ 1 & 1 & 6 & 2 & 1 & 7 \end{array}\right]$$

Motivated by Definition 1-13, we find a leading entry in the third row. Therefore we interchange row 1 and row 3 and obtain

$$\left[\begin{array}{rrrrr:r} 1 & 1 & 6 & 2 & 1 & 7 \\ 3 & 2 & -2 & -5 & 2 & 9 \\ 2 & -1 & 4 & 1 & -1 & 4 \end{array}\right]$$

Having a leading entry located in the first column, we need to produce zeros elsewhere in that column. We do it by multiplying the first row by -3 and adding it to the second row, and then multiplying the first row by -2 and adding it to the third row. We obtain

$$\left[\begin{array}{rrrrr:r} 1 & 1 & 6 & 2 & 1 & 7 \\ 0 & -1 & -20 & -11 & -1 & -12 \\ 0 & -3 & -8 & -3 & -3 & -10 \end{array}\right]$$

We need to produce a leading entry in any one of the remaining rows. This leading entry should be located in a column as close as possible to the first one (which is already in its final form). If we multiply the second row by -1 we produce a leading entry there that is also located in the second column. Thus we have

$$\left[\begin{array}{rrrrr:r} 1 & 1 & 6 & 2 & 1 & 7 \\ 0 & 1 & 20 & 11 & 1 & 12 \\ 0 & -3 & -8 & -3 & -3 & -10 \end{array}\right]$$

We now produce two zeros in the second column by multiplying the second row by -1 and adding it to the first row, and then multiplying the second row by 3 and adding it to the third row. We obtain

$$\left[\begin{array}{rrrrr:r} 1 & 0 & -14 & -9 & 0 & -5 \\ 0 & 1 & 20 & 11 & 1 & 12 \\ 0 & 0 & 52 & 30 & 0 & 26 \end{array}\right]$$

Having produced leading entries in rows 1 and 2, we turn our attention to row 3. The number 52 is the first nonzero entry from the left. Thus, multiplying the third row by $\frac{1}{52}$ will produce a leading entry in that row:

$$\left[\begin{array}{rrrrr:r} 1 & 0 & -14 & -9 & 0 & -5 \\ 0 & 1 & 20 & 11 & 1 & 12 \\ 0 & 0 & 1 & \frac{30}{52} & 0 & \frac{26}{52} \end{array}\right]$$

CARL F. GAUSS (1777-1855) *was born in Brunswick, Germany, to an extremely poor family. Gauss's talent and genius were manifested at an early age. His remarkable achievements in school earned him the attention of the Duke of Brunswick, who undertook his education and put him through the University of Göttingen. Gauss was a prolific writer, and his early fame was made in the theory of numbers. He worked in many areas of mathematics: He developed the theory of surfaces and the normal laws of distribution in the theory of probability, and invented the method of least squares. Gauss is one of the greatest of all mathematicians. Among his famous students are Riemann and Dedekind. Gauss considered teaching a waste of time but when he taught he did so superbly.*

In order to produce the desired two zeros in the third column, we multiply the third row by 14 and add it to the first row, and then multiply the third row by -20 and add it to the second row. Simplifying fractions, we obtain

$$\begin{bmatrix} 1 & 0 & 0 & -\frac{12}{13} & 0 & \vdots & 2 \\ 0 & 1 & 0 & -\frac{7}{13} & 1 & \vdots & 2 \\ 0 & 0 & 1 & \frac{15}{26} & 0 & \vdots & \frac{1}{2} \end{bmatrix}$$

The last matrix is in row-reduced echelon form. The corresponding system of equations is given by

$$\begin{aligned} x_1 \qquad + (-\tfrac{12}{13})x_4 \qquad &= 2 \\ x_2 \quad + (-\tfrac{7}{13})x_4 + x_5 &= 2 \\ x_3 + \quad \tfrac{15}{26}\, x_4 \qquad &= \tfrac{1}{2} \end{aligned}$$

Solving for the leading unknowns x_1, x_2, and x_3, we obtain

$$\begin{aligned} x_1 &= 2 + \tfrac{12}{13}x_4 \\ x_2 &= 2 + \tfrac{7}{13}x_4 - x_5 \\ x_3 &= \tfrac{1}{2} - \tfrac{15}{26}x_4 \end{aligned}$$

Thus, we see now that the original system of equations has infinitely many solutions depending on our arbitrary choice of values for x_4 and x_5.

We like to emphasize again that there are linear nonhomogeneous systems of equations which possess no solutions at all. We recognize such a system also by transforming its augmented matrix into row-reduced echelon form. The following example illustrates such a system.

Example 6 Consider the system of linear equations

$$\begin{aligned} x_1 + x_2 + 2x_3 + 3x_4 &= 1 \\ 2x_1 + x_2 + 3x_3 + 4x_4 &= 1 \\ 3x_1 + x_2 + 4x_3 + 5x_5 &= 2 \end{aligned}$$

Show that the system has no solution.

Solution First, we transform the augmented matrix of the system into row-reduced echelon form. We leave it to the reader to show that the augmented matrix of the system,

$$\begin{bmatrix} 1 & 1 & 2 & 3 & \vdots & 1 \\ 2 & 1 & 3 & 4 & \vdots & 1 \\ 3 & 1 & 4 & 5 & \vdots & 2 \end{bmatrix}$$

will be transformed into the matrix

$$\begin{bmatrix} 1 & 0 & 1 & 1 & \vdots & 0 \\ 0 & 1 & 1 & 2 & \vdots & 0 \\ 0 & 0 & 0 & 0 & \vdots & 1 \end{bmatrix}$$

which is in the row-reduced echelon form. The corresponding system of equations is given by

$$\begin{aligned} x_1 \qquad\qquad + \quad x_3 + \quad x_4 &= 0 \\ x_2 + \quad x_3 + \quad 2x_4 &= 0 \\ 0 \cdot x_1 + 0 \cdot x_2 + 0 \cdot x_3 + 0 \cdot x_4 &= 1 \end{aligned}$$

We wrote the third equation in that form to make it clear that there exist no x_1, x_2, x_3, and x_4 which satisfy the equation. Thus, the system cannot have a solution.

The discussion above indicates that for a linear nonhomogeneous system of equations one and only one of the following must hold:

1 The system has a unique solution.

2 The system has infinitely many solutions.

3 The system has no solution.

We wish to point out that the homogeneous system $A\mathbf{x} = \mathbf{0}$ will never possess property 3 above. The reason, of course, is that the system $A\mathbf{x} = \mathbf{0}$ always has at least one solution, the so-called **trivial** solution, where all the unknown x_i's are zero. We have seen that if A^{-1} exists, then the system $A\mathbf{x} = \mathbf{0}$ has the unique solution $\mathbf{x} = \mathbf{0}$. A natural question is therefore the following: Which linear homogeneous systems $A\mathbf{x} = \mathbf{0}$ will also have nontrivial solutions, that is, solutions which are different from the solution $\mathbf{x} = \mathbf{0}$? The answer is given in the following simple theorem.

The trivial solution is $\mathbf{x} = \mathbf{0}$

Theorem 1-15 If the homogeneous system $A\mathbf{x} = \mathbf{0}$ has more unknowns than equations, then the system has infinitely many solutions.

Note: All the solutions except one are nontrivial solutions.

Proof Let the system $A\mathbf{x} = \mathbf{0}$ consist of m equations and n unknowns with the condition $m < n$. We now transform the augmented matrix $[A \vdots \mathbf{0}]$ into the augmented matrix $[B \vdots \mathbf{0}]$ where $[B \vdots \mathbf{0}]$ is in row-reduced echelon form. If the number of leading entries in the matrix $[B \vdots \mathbf{0}]$ is r, then obviously $r \leq m$. Now, since we assumed that the number of unknowns n is greater than m, we have $r < n$. This implies that in the system $B\mathbf{x} = \mathbf{0}$ (which is equivalent to $A\mathbf{x} = \mathbf{0}$) we can solve for r leading unknowns in terms of the remaining $n - r$ unknowns. Since we can choose those $n - r$ unknowns quite arbitrarily it fol-

Note: $r \leq m$ because one cannot have more leading entries than the number of rows in the matrix.

lows that the system $B\mathbf{x} = \mathbf{0}$ (and therefore also $A\mathbf{x} = \mathbf{0}$) has infinitely many solutions. This completes the proof.

The following example will illustrate the argument used in the proof of Theorem 1-15.

Example 7 Let a linear homogeneous system be given by

$$\begin{aligned} x_1 + 2x_2 + 3x_3 \phantom{{}+x_4} + 4x_5 &= 0 \\ 2x_1 + 4x_2 + 6x_3 + x_4 + 9x_5 &= 0 \\ x_1 + 2x_2 + 3x_3 + x_4 + 5x_5 &= 0 \end{aligned}$$

The augmented matrix of the system is given by

$$\left[\begin{array}{ccccc:c} 1 & 2 & 3 & 0 & 4 & 0 \\ 2 & 4 & 6 & 1 & 9 & 0 \\ 1 & 2 & 3 & 1 & 5 & 0 \end{array}\right]$$

Transforming this matrix into row-reduced echelon form, we get the matrix

The student is encouraged to do this!

$$\left[\begin{array}{ccccc:c} 1 & 2 & 3 & 0 & 4 & 0 \\ 0 & 0 & 0 & 1 & 1 & 0 \\ 0 & 0 & 0 & 0 & 0 & 0 \end{array}\right]$$

The corresponding system of equations is given by

We omitted the last equation because it imposes no condition on the unknowns.

$$\begin{aligned} x_1 + 2x_2 + 3x_3 \quad + 4x_5 &= 0 \\ x_4 + x_5 &= 0 \end{aligned}$$

Solving for the leading unknowns x_1 and x_4 in terms of the remaining unknowns, we get

$$\begin{aligned} x_1 &= -2x_2 - 3x_3 - 4x_5 \\ x_4 &= -x_5 \end{aligned}$$

Thus, besides the choice $x_2 = x_3 = x_5 = 0$ which leads to $x_1 = x_4 = 0$ and results in the trivial solution $\mathbf{x} = \mathbf{0}$, we can choose for x_2, x_3, and x_5 arbitrary values (not all zero) and obtain infinitely many nontrivial solutions.

An important observation which should be brought to the attention of the reader is that each attempt we made in this section to transform a matrix into row-reduced echelon form always ended with success. The reason for this fact is given in the following theorem.

Theorem 1-16 Every matrix A can be transformed by means of elementary row operations into a matrix B which is in row-reduced echelon form.

We omit the tedious proof of the theorem.

Exercises

1 Determine which of the following matrices are in row-reduced echelon form.

a $\begin{bmatrix} 1 & 1 \\ 0 & 0 \end{bmatrix}$ **b** $\begin{bmatrix} 0 & 1 \\ 1 & 0 \end{bmatrix}$

c $\begin{bmatrix} 0 & 1 \\ 0 & 0 \end{bmatrix}$ **d** $\begin{bmatrix} 0 & 0 \\ 2 & 1 \end{bmatrix}$

e $\begin{bmatrix} 1 & 1 & 0 \\ 0 & 0 & 1 \\ 0 & 0 & 0 \end{bmatrix}$ **f** $\begin{bmatrix} 0 & 0 & 1 \\ 0 & 1 & 0 \\ 1 & 0 & 0 \end{bmatrix}$

g $\begin{bmatrix} 1 & 2 & 3 \\ 0 & 0 & 0 \\ 0 & 0 & 0 \end{bmatrix}$ **h** $\begin{bmatrix} 1 & 3 & 0 & 0 \\ 0 & 0 & 1 & 2 \\ 0 & 0 & 0 & 0 \end{bmatrix}$

i $\begin{bmatrix} 1 & 0 & 0 & 2 \\ 0 & 0 & 1 & 3 \\ 0 & 1 & 0 & 4 \end{bmatrix}$ **j** $\begin{bmatrix} 1 & 0 & 2 & 4 & 6 \\ 0 & 1 & 3 & 5 & 7 \end{bmatrix}$

k $\begin{bmatrix} 0 & 1 & -1 & 0 & 0 \\ 0 & 0 & 0 & 1 & 0 \\ 0 & 0 & 0 & 0 & 1 \\ 0 & 0 & 0 & 0 & 0 \end{bmatrix}$ **l** $\begin{bmatrix} 1 & 0 & 0 & 0 & 0 \\ 0 & 2 & 0 & 0 & 0 \\ 0 & 0 & 3 & 0 & 0 \\ 0 & 0 & 0 & 0 & 1 \end{bmatrix}$

2 Construct a 3×4 matrix A such that conditions 1 and 3 of Definition 1-13 are satisfied while conditions 2 and 4 are not satisfied.

3 Each of the following row-reduced echelon matrices is an augmented matrix of a corresponding linear system of equations. Solve each of the systems.

a $\left[\begin{array}{cccc:c} 1 & 0 & -1 & 2 & 3 \\ 0 & 1 & -2 & 1 & 4 \end{array}\right]$ **b** $\left[\begin{array}{ccccc:c} 1 & -1 & 0 & -2 & 0 & 1 \\ 0 & 0 & 1 & 3 & 0 & 1 \\ 0 & 0 & 0 & 0 & 1 & 1 \end{array}\right]$

c $\left[\begin{array}{ccccc:c} 1 & 2 & 3 & 0 & 4 & 0 \\ 0 & 0 & 0 & 1 & 5 & 0 \\ 0 & 0 & 0 & 0 & 0 & 1 \end{array}\right]$

4 Solve each of the following systems by using Gauss-Jordan elimination (see Example 5).

a
$$\begin{aligned} 3x_1 + 2x_2 + x_3 &= 10 \\ 2x_1 + x_2 - x_3 &= 1 \end{aligned}$$

b
$$\begin{aligned} -x_1 + 2x_2 - 3x_3 + 4x_4 &= 8 \\ 2x_1 - 4x_2 - x_3 + 2x_4 &= -3 \\ 5x_1 - 4x_2 - x_3 + 2x_4 &= -3 \end{aligned}$$

c
$$\begin{aligned} x_1 + x_2 + x_3 - x_4 &= 0 \\ 2x_1 - 4x_2 - 4x_3 + x_4 &= 0 \\ 4x_1 + 2x_2 + 2x_3 - 3x_4 &= 0 \\ 7x_1 - x_2 - x_3 - 3x_4 &= 0 \end{aligned}$$

Remember, a system of equations is called inconsistent if it has *no* solution.

5 Determine which of the following systems are inconsistent.

a
$$\begin{aligned} 2x_1 + 3x_2 &= 12 \\ 3x_1 - x_2 &= 7 \\ 4x_1 - 5x_2 &= 2 \end{aligned}$$

b
$$\begin{aligned} x_1 + 2x_2 &= 7 \\ -x_1 + 3x_2 &= 3 \\ 3x_1 - 4x_2 &= 0 \end{aligned}$$

c
$$\begin{aligned} -x_1 + 3x_2 &= 5 \\ 3x_1 - x_2 &= 1 \end{aligned}$$

d
$$\begin{aligned} 2x_1 - 3x_2 &= -1 \\ -2x_1 + 3x_2 &= 2 \end{aligned}$$

e
$$\begin{aligned} x_1 - 3x_2 + 4x_3 &= 2 \\ 2x_1 + 5x_2 - 3x_3 &= 4 \\ 4x_1 + x_2 + 5x_3 &= 9 \end{aligned}$$

6 Find all values of k for which the following systems have no solution.

a
$$\begin{aligned} x_1 + 2x_2 + 3x_3 &= 5 \\ 2x_1 - x_2 - 2x_3 &= 1 \\ 3x_1 - x_2 - x_3 &= k \end{aligned}$$

b
$$\begin{aligned} 2x_1 - 3x_2 - x_3 &= 2 \\ x_1 + 3x_2 - 2x_3 &= k \\ 7x_1 + 3x_2 - 8x_3 &= k^2 \end{aligned}$$

7 Let $A = \begin{bmatrix} a & b \\ c & d \end{bmatrix}$. Show that if $ad - bc \neq 0$ then A can be row-reduced to the matrix $I_2 = \begin{bmatrix} 1 & 0 \\ 0 & 1 \end{bmatrix}$.

8 Show that if $ad - bc \neq 0$ then the system

$$\begin{bmatrix} a & b \\ c & d \end{bmatrix} \begin{bmatrix} x_1 \\ x_2 \end{bmatrix} = \begin{bmatrix} 0 \\ 0 \end{bmatrix}$$

has the unique (trivial) solution $x_1 = 0$, $x_2 = 0$.

9 Let the equations $ax + by = 0$ and $cx + dy = 0$ represent the lines L_1 and L_2, respectively, in the xy-plane. Describe the relation between L_1 and L_2 if the system

$$\begin{aligned} ax + by &= 0 \\ cx + dy &= 0 \end{aligned}$$

has

a a unique solution (which must be the trivial one)

b more than one solution.

10 Let the equations $ax + by + cz = 0$, $dx + ey + fz = 0$, and $gx + hy + kz = 0$ represent the planes p_1, p_2, and p_3, respectively, in xyz-space. Describe the relation between p_1, p_2, and p_3 if the system

$$ax + by + cz = 0$$
$$dx + ey + fz = 0$$
$$gx + hy + kz = 0$$

has

a a unique solution

b more than one solution.

Review of Chapter 1

1 A matrix is a ______________ array of numbers.

2 An $m \times n$ matrix has ______________ rows and ______________ columns.

3 The (i,j) entry of the matrix A, denoted by a_{ij}, is the number that appears in the ______________ row and ______________ column of A.

4 Let A and B be matrices. What do we mean by the equation $A = B$?

5 Under what condition is the sum of the matrices A and B defined?

6 To obtain the scalar multiple of A by α, we multiply ______________ entry of the matrix A by the same number α.

7 Let A be a 3×2 matrix. What does 0 represent in the equation $A + 0 = A$?

8 We denote by $-A$ the unique matrix with the property that ______________.

9 If A and B are $m \times n$ matrices and α, β are real numbers, then

$\alpha(A + B) =$ ______________

$(\alpha + \beta)A =$ ______________

$\alpha(\beta A) =$ ______________

10 Under what condition is the product of two matrices A and B defined?

11 Let $A = \begin{bmatrix} a_{11} & a_{12} & a_{13} \\ a_{21} & a_{22} & a_{23} \end{bmatrix}$ and $B = \begin{bmatrix} b_{11} \\ b_{21} \\ b_{31} \end{bmatrix}$. Find AB. Is BA defined?

12 A matrix having the same number of rows as columns is called a ______________ matrix.

13 Give an example which shows that matrix multiplication is not commutative ($AB \neq BA$).

14 State the associative law for matrix multiplication.

15 Let A, B, and C be square matrices of order n. Under what condition do the following equations hold?

a $(A - B)^2 = A^2 - 2AB + B^2$

b $(A - B)C = CA - CB$

16 Let A and B be matrices such that $AB = 0$. Does it follow that $A = 0$ or $B = 0$?

17 Let A be a 3×4 matrix. Describe the effect of multiplying A by the elementary matrix $E = \begin{bmatrix} 0 & 0 & 1 \\ 0 & 1 & 0 \\ 1 & 0 & 0 \end{bmatrix}$ from the left.

18 Let A be a 3×4 matrix. Describe the effect of multiplying A by the elementary matrix $G = \begin{bmatrix} 1 & 0 & 0 \\ 0 & 1 & 0 \\ 0 & 0 & 3 \end{bmatrix}$ from the left.

19 Let A be a 3×4 matrix. Describe the effect of multiplying A by the elementary matrix $H = \begin{bmatrix} 1 & 0 & 0 \\ 4 & 1 & 0 \\ 0 & 0 & 1 \end{bmatrix}$ from the left.

20 Let A be an $n \times n$ matrix satisfying the matrix equation $3A + 2A^2 + A^3 = I_n$. Find the inverse A^{-1} (if it exists) and express it in terms of the matrix A.

21 Does the matrix $A = \begin{bmatrix} 1 & 0 & 0 \\ 0 & 1 & 0 \end{bmatrix}$ have an inverse? Why or why not?

22 Let A be a nonsingular matrix. How many different inverses does A have?

23 Let A be a square matrix of order n. Under what condition does the linear system $A\mathbf{x} = \mathbf{b}$ of n equations and n unknowns have a unique solution?

24 Let A, B, and C be nonsingular $n \times n$ matrices. Express $(ABC)^{-1}$ in terms of A^{-1}, B^{-1}, and C^{-1}.

25 Let E_1, E_2, and E_3 be elementary matrices. Is it possible for the product $E_1E_2E_3$ to be a singular matrix? Why?

26 Write down the matrix of coefficients for the linear system

$$\begin{aligned} 3x - 2y &= 1 \\ y + 4z &= 0 \\ 2x - \quad z &= 2 \end{aligned}$$

27 Express the matrix $A = \begin{bmatrix} 0 & 3 \\ 1 & -12 \end{bmatrix}$ as a product of elementary matrices.

28 Solve the linear system of equations

$$\begin{aligned} 8x + 3y + 3z &= 1 \\ 4x + \ \ y \qquad &= 2 \\ 2x + \ \ y + \ \ z &= 3 \end{aligned}$$

How many solutions does the system have?

29 Does there exist a linear homogeneous system $A\mathbf{x} = \mathbf{0}$ which has no solution?

30 Does there exist a linear nonhomogeneous system $A\mathbf{x} = \mathbf{b}$ which has no solution?

31 Under what condition does the homogeneous system $A\mathbf{x} = \mathbf{0}$ have infinitely many solutions?

32 Explain the role played by row-reduced echelon matrices in finding solutions for a linear system $A\mathbf{x} = \mathbf{b}$.

Vector Spaces

2

2-1 Definition of General Vector Spaces

Objectives

1. Introduce the concept of a real vector space.
2. Give a variety of examples of vector spaces.

We focus our attention again on the system consisting of the set M of all $m \times n$ matrices together with the operations of addition and scalar multiplication defined on it. In Chapter 1 we proved some basic properties of this system. Our aim here is to show that there exist other systems that possess the same basic properties although the nature of their elements is quite different. To illustrate, let us take the set P_1 of all polynomials of degree not exceeding 1. Thus

$$P_1 = \{ax + b : a \text{ and } b \text{ are any real numbers}\}$$

We now define addition and scalar multiplication on P_1 as follows: Let $\mathbf{p}_1 = a_1x + b_1$ and $\mathbf{p}_2 = a_2x + b_2$ be any two elements of P_1. We define addition by the equation

$$\mathbf{p}_1 + \mathbf{p}_2 = (a_1 + a_2)x + (b_1 + b_2)$$

and we define scalar multiplication by the equation

$$\alpha\mathbf{p}_1 = (\alpha a_1)x + (\alpha b_1)$$

where α is any real number. The system consisting of the set P_1 together with addition and scalar multiplication as defined has many properties which we found to be associated with matrices. A list of ten such properties for the two systems is given in the following table. The striking resemblance of the properties for the two systems should be noted by the reader.

P_1 is the set of all polynomials of degree ≤ 1	M is the set of all $m \times n$ matrices
1 If $\mathbf{p}_1 \in P_1$ and $\mathbf{p}_2 \in P_1$, then $\mathbf{p}_1 + \mathbf{p}_2 \in P_1$	**1** If $\mathbf{A} \in M$ and $\mathbf{B} \in M$, then $\mathbf{A} + \mathbf{B} \in M$
2 $\mathbf{p}_1 + \mathbf{p}_2 = \mathbf{p}_2 + \mathbf{p}_1$ for all $\mathbf{p}_1$, $\mathbf{p}_2 \in P_1$	**2** $\mathbf{A} + \mathbf{B} = \mathbf{B} + \mathbf{A}$ for all $\mathbf{A}$, $\mathbf{B} \in M$
3 $(\mathbf{p}_1 + \mathbf{p}_2) + \mathbf{p}_3 = \mathbf{p}_1 + (\mathbf{p}_2 + \mathbf{p}_3)$ for all $\mathbf{p}_1, \mathbf{p}_2, \mathbf{p}_3 \in P_1$	**3** $(\mathbf{A} + \mathbf{B}) + \mathbf{C} = \mathbf{A} + (\mathbf{B} + \mathbf{C})$ for all $\mathbf{A}, \mathbf{B}, \mathbf{C} \in M$
4 There exists a unique element in P_1, denoted by $\mathbf{0}$, such that $\mathbf{p} + \mathbf{0} = \mathbf{p}$ for all $\mathbf{p} \in P_1$ ($\mathbf{0}$ is called the zero polynomial)	**4** There exists a unique element in M, denoted by $\mathbf{0}$, such that $\mathbf{A} + \mathbf{0} = \mathbf{A}$ for all $\mathbf{A} \in M$ ($\mathbf{0}$ is called the zero matrix)
5 For each $\mathbf{p} \in P_1$ there exists a unique element $\mathbf{q} \in P_1$ such that $\mathbf{p} + \mathbf{q} = \mathbf{0}$	**5** For each $\mathbf{A} \in M$ there exists a unique element $\mathbf{B} \in M$ such that $\mathbf{A} + \mathbf{B} = \mathbf{0}$
6 If $\mathbf{p} \in P_1$ and α is any real number, then $\alpha\mathbf{p} \in P_1$	**6** If $\mathbf{A} \in M$ and α is any real number, then $\alpha\mathbf{A} \in M$
7 $\alpha(\mathbf{p}_1 + \mathbf{p}_2) = \alpha\mathbf{p}_1 + \alpha\mathbf{p}_2$ for	**7** $\alpha(\mathbf{A} + \mathbf{B}) = \alpha\mathbf{A} + \alpha\mathbf{B}$ for all

all $\mathbf{p}_1$, $\mathbf{p}_2 \in P_1$ and any real number α	$\mathbf{A}$, $\mathbf{B} \in M$ and any real number α
8 $(\alpha + \beta)\mathbf{p} = \alpha\mathbf{p} + \beta\mathbf{p}$ for all $\mathbf{p} \in P_1$ and any real numbers α and β	**8** $(\alpha + \beta)\mathbf{A} = \alpha\mathbf{A} + \beta\mathbf{A}$ for all $\mathbf{A} \in M$ and any real numbers α and β
9 $\alpha(\beta\mathbf{p}) = (\alpha\beta)\mathbf{p}$ for all $\mathbf{p} \in P_1$ and any real numbers α and β	**9** $\alpha(\beta\mathbf{A}) = (\alpha\beta)\mathbf{A}$ for all $\mathbf{A} \in M$ and any real numbers α and β
10 $1\mathbf{p} = \mathbf{p}$ for all $\mathbf{p} \in P_1$	**10** $1\mathbf{A} = \mathbf{A}$ for all $\mathbf{A} \in M$

Let us analyze some of the ten properties mentioned above. Property 1 simply states that adding two polynomials $\mathbf{p}_1$ and $\mathbf{p}_2$ of degree not exceeding 1 results in a polynomial which is also of degree not exceeding 1. We sometimes say that the set P_1 of all polynomials of degree not exceeding 1 is **closed under addition** of polynomials. Property 1 as related to $m \times n$ matrices states that addition of two matrices A and B which are of order $m \times n$ results in a matrix which is also of order $m \times n$. Thus, the set of all $m \times n$ matrices is **closed under addition** of matrices.

Property 6 deals with the operation of scalar multiplication. It states that if $\mathbf{p}$ is a polynomial of degree not exceeding 1 then so is $\alpha\mathbf{p}$ for each real number α. We say in this case that P_1 is **closed under scalar multiplication.** Property 6 as related to $m \times n$ matrices states that the set M is **closed under scalar multiplication.** The reader should recall that properties 2–5 and 7–9 were proved for $m \times n$ matrices in Section 1-1. Property 10 is a simple consequence of the definition of scalar multiplication. We wish to emphasize that in order to establish the validity of the properties listed in the table, one makes heavy use of similar properties of real numbers. A careful check of the ten properties reveals that the system consisting of the set $\mathbf{R}$ of real numbers together with the usual addition and multiplication defined on it satisfies them too.

The student is encouraged to establish the properties listed in the table for the set P_1.

This leads us to suspect that there might be other such systems which satisfy all 10 properties (and possibly more). Rather than investigate each of the systems separately, we shall consider them simultaneously. This can be done with the aid of the procedure of **abstraction.**

A mathematical system which possesses properties similar to the ones that we listed in the table will be called a real *vector space.* The exact definition is as follows:

Definition 2-1 A real vector space.

A **real vector space** is a mathematical system consisting of a set V together with two operations, addition and scalar multiplication, such that the following properties are satisfied.

1 If $\mathbf{v}_1 \in V$ and $\mathbf{v}_2 \in V$ then $\mathbf{v}_1 + \mathbf{v}_2 \in V$.

2 $\mathbf{v}_1 + \mathbf{v}_2 = \mathbf{v}_2 + \mathbf{v}_1$ for all $\mathbf{v}_1, \mathbf{v}_2 \in V$.

3 $(\mathbf{v}_1 + \mathbf{v}_2) + \mathbf{v}_3 = \mathbf{v}_1 + (\mathbf{v}_2 + \mathbf{v}_3)$ for all $\mathbf{v}_1, \mathbf{v}_2, \mathbf{v}_3 \in V$.

4 There exists a unique element $\mathbf{0}$ in V such that $\mathbf{v} + \mathbf{0} = \mathbf{v}$ for all $\mathbf{v} \in V$. The element $\mathbf{0}$ is called the **zero vector.**

5 For each $\mathbf{v} \in V$ there exists a unique element $\mathbf{u} \in V$ such that $\mathbf{v} + \mathbf{u} = \mathbf{0}$. We denote $\mathbf{u}$ by $-\mathbf{v}$.

6 If $\mathbf{v} \in V$ and α is any real number then $\alpha\mathbf{v} \in V$.

7 $\alpha(\mathbf{v}_1 + \mathbf{v}_2) = \alpha\mathbf{v}_1 + \alpha\mathbf{v}_2$ for all $\mathbf{v}_1, \mathbf{v}_2 \in V$ and any real number α.

8 $(\alpha + \beta)\mathbf{v} = \alpha\mathbf{v} + \beta\mathbf{v}$ for all $\mathbf{v} \in V$ and any real numbers α and β.

9 $\alpha(\beta\mathbf{v}) = (\alpha\beta)\mathbf{v}$ for all $\mathbf{v} \in V$ and any real numbers α and β.

10 $1\mathbf{v} = \mathbf{v}$ for all $\mathbf{v} \in V$.

Properties 1–10 are sometimes called the axioms of a vector space.

Every element of V is called a **vector,** while the real numbers used in the operation of scalar multiplication are called **scalars.** Given a set V together with addition and scalar multiplication defined on it, one must go through the tedious chore of checking whether all the axioms for a vector space are satisfied before declaring V to be a vector space. The table in this section brings out the fact that the set M of all $m \times n$ matrices, together with addition and scalar multiplication defined on it, is a vector space. The table also indicates that the set P_1 of all polynomials of degree ≤ 1 is a vector space under the operations of addition and scalar multiplication defined on it.

We caution the reader about the misconception that the term vector refers only to a directed line segment. Rather, the set of all directed line segments in the xy-plane, together with the operations of addition and scalar multiplication as defined in the next section, will be found to constitute just another example of a vector space with some geometrical flavor attached to it. Thus the word vector might designate a matrix, a polynomial, or any other mathematical object, depending on the nature of the vector space under discussion.

We emphasize here that some mathematical systems, consisting of a set V and two operations defined on it, fail to be vector spaces because one or more of the axioms in Definition 2-1 are not satisfied. The following two examples will exhibit such systems.

Example 1 Let M_2 be the set of all 2×2 matrices. We define addition on M_2 by the equation

This is the usual addition of matrices.

$$\begin{bmatrix} a_{11} & a_{12} \\ a_{21} & a_{22} \end{bmatrix} + \begin{bmatrix} b_{11} & b_{12} \\ b_{21} & b_{22} \end{bmatrix} = \begin{bmatrix} a_{11} + b_{11} & a_{12} + b_{12} \\ a_{21} + b_{21} & a_{22} + b_{22} \end{bmatrix}$$

We define scalar multiplication on M_2 by the equation

$$\alpha \begin{bmatrix} a_{11} & a_{12} \\ a_{21} & a_{22} \end{bmatrix} = \begin{bmatrix} 0 & 0 \\ 0 & 0 \end{bmatrix} \quad \text{for all real } \alpha$$

The last operation is certainly quite different from the usual scalar multiplication of matrices as we encountered it in Chapter 1. One can easily check that the first nine axioms for a vector space are satisfied for M_2 under the two operations defined on it. However, the tenth axiom is violated, because for

$$\alpha = 1 \quad \text{and} \quad \begin{bmatrix} a_{11} & a_{12} \\ a_{21} & a_{22} \end{bmatrix} = \begin{bmatrix} 5 & 3 \\ 1 & -1 \end{bmatrix}$$

we have by the definition of scalar multiplication above

$$1\begin{bmatrix} 5 & 3 \\ 1 & -1 \end{bmatrix} = \begin{bmatrix} 0 & 0 \\ 0 & 0 \end{bmatrix}$$

while axiom 10 requires that

$$1\begin{bmatrix} 5 & 3 \\ 1 & -1 \end{bmatrix} = \begin{bmatrix} 5 & 3 \\ 1 & -1 \end{bmatrix}$$

Thus, M_2 under the two operations defined on it here does not constitute a vector space.

Example 2 Let S_2 be the set of all ordered pairs of real numbers. Thus, $S_2 = \{(x_1,x_2)\colon x_1 \text{ and } x_2 \text{ are real}\}$.

We define addition on S_2 by the equation

Notice the term $a_2 - b_2$.

$$(a_1,a_2) + (b_1,b_2) = (a_1 + b_1, a_2 - b_2)$$

We define scalar multiplication by the equation

$$\alpha(a_1,a_2) = (\alpha a_1,\alpha a_2) \text{ for all real } \alpha$$

Let us check which of the axioms for a vector space are satisfied (and which are not) for S_2 together with the two operations defined above.

To describe the fact that S_2 satisfies axioms 1 and 6 of Definition 2-1 we sometimes say that S_2 has the closure property.

S_2 is obviously closed under addition and scalar multiplication, so that axioms 1 and 6 of Definition 2-1 are satisfied. But axioms 2 and 3 do not hold for our example. To see this, we find

$$(a_1,a_2) + (b_1,b_2) = (a_1 + b_1, a_2 - b_2)$$

and

$$(b_1,b_2) + (a_1,a_2) = (b_1 + a_1, b_2 - a_2)$$

Since $a_2 - b_2 \neq b_2 - a_2$ (unless $a_2 = b_2$) we have

$$(a_1,a_2) + (b_1,b_2) \neq (b_1,b_2) + (a_1,a_2)$$

Similarly,

$$\begin{aligned}[(a_1,a_2) + (b_1,b_2)] + (c_1,c_2) &= (a_1 + b_1, a_2 - b_2) + (c_1,c_2) \\ &= (a_1 + b_1 + c_1, a_2 - b_2 - c_2)\end{aligned}$$

while

$$\begin{aligned}(a_1,a_2) + [(b_1,b_2) + (c_1,c_2)] &= (a_1,a_2) + (b_1 + c_1, b_2 - c_2) \\ &= (a_1 + b_1 + c_1, a_2 - b_2 + c_2)\end{aligned}$$

which proves that

$$[(a_1,a_2) + (b_1,b_2)] + (c_1,c_2) \neq (a_1,a_2) + [(b_1,b_2) + (c_1,c_2)]$$

because $a_2 - b_2 - c_2 \neq a_2 - b_2 + c_2$ (unless $c_2 = 0$).

Axiom 4 is satisfied because $(0,0)$ satisfies the equation $(a_1,a_2) + (0,0) = (a_1 + 0, a_2 - 0) = (a_1,a_2)$ for all $(a_1,a_2) \in S_2$.

Axiom 5 is satisfied because for any $(a_1,a_2) \in S_2$ we have $(-a_1,a_2) \in S_2$ and

$$(a_1,a_2) + (-a_1,a_2) = (a_1 + (-a_1), a_2 - a_2) = (0,0)$$

Axiom 7 is satisfied, as one easily sees from the following equations:

$$\begin{aligned}\alpha[(a_1,a_2) + (b_1,b_2)] &= \alpha(a_1 + b_1, a_2 - b_2)\\ &= (\alpha(a_1 + b_1), \alpha(a_2 - b_2))\\ \alpha(a_1,a_2) + \alpha(b_1,b_2) &= (\alpha a_1,\alpha a_2) + (\alpha b_1,\alpha b_2)\\ &= (\alpha a_1 + \alpha b_1, \alpha a_2 - \alpha b_2)\end{aligned}$$

Since $\alpha(a_1 + b_1) = \alpha a_1 + \alpha b_1$ and $\alpha(a_2 - b_2) = \alpha a_2 - \alpha b_2$, it follows that

$$\alpha[(a_1,a_2) + (b_1,b_2)] = \alpha(a_1,a_2) + \alpha(b_1,b_2)$$

Axiom 8 does not hold. In order to see this, we write

$$\begin{aligned}(\alpha + \beta)(a_1,a_2) &= ((\alpha + \beta)a_1, (\alpha + \beta)a_2)\\ &= (\alpha a_1 + \beta a_1, \alpha a_2 + \beta a_2)\end{aligned}$$

and

$$\begin{aligned}\alpha(a_1,a_2) + \beta(a_1,a_2) &= (\alpha a_1,\alpha a_2) + (\beta a_1,\beta a_2)\\ &= (\alpha a_1 + \beta a_1, \alpha a_2 - \beta a_2)\end{aligned}$$

Since $\alpha a_2 + \beta a_2 \neq \alpha a_2 - \beta a_2$ (unless $\beta a_2 = -\beta a_2$) it follows that

$$(\alpha + \beta)(a_1,a_2) \neq (a_1,a_2) + (a_1,a_2)$$

Axiom 9 is satisfied because

$$\alpha[\beta(a_1,a_2)] = \alpha(\beta a_1,\beta a_2) = (\alpha(\beta a_1), \alpha(\beta a_2))$$

and

$$(\alpha\beta)(a_1,a_2) = ((\alpha\beta)a_1, (\alpha\beta)a_2)$$

so that

$$\alpha[\beta(a_1,a_2)] = (\alpha\beta)(a_1, a_2)$$

Axiom 10 is satisfied because $1(a_1,a_2) = (1 \cdot a_1, 1 \cdot a_2) = (a_1,a_2)$.

In conclusion we may say that S_2 together with addition and scalar multiplication as defined above is not a vector space because axioms 2, 3, and 8 fail to be satisfied.

See the table in this section.

At this point the reader has already seen examples of mathematical systems which are vector spaces as well as examples of systems which fail to satisfy one or more of the axioms. The point that we have tried to make by providing Examples 1 and 2 is that in determining whether a mathematical system is a vector space, the reader should carefully check the given system against each of the ten axioms in Definition 2-1.

The following theorem exhibits some properties which follow directly from the axioms in Definition 2-1 and therefore hold for every vector space.

Theorem 2-1 Let V be a vector space. Then

a $0\mathbf{v} = \mathbf{0}$ for every $\mathbf{v} \in V$.

b $(-1)\mathbf{v} = -\mathbf{v}$ for every $\mathbf{v} \in V$.

c $\alpha\mathbf{0} = \mathbf{0}$ for every real number α.

d The equation $\alpha\mathbf{v} = 0$ implies that either $\alpha = \mathbf{0}$ or $\mathbf{v} = \mathbf{0}$.

Proof In part **a** we state that the scalar product of the real number 0 and any vector $\mathbf{v} \in V$ is the zero vector $\mathbf{0}$ of V. The proof proceeds as follows. We have

Here we have used the property of the real number 0, namely

$0 = 0 + 0$

and then the distributive property for a vector space—axiom 8, in Definition 2-1.

$$0\mathbf{v} = (0 + 0)\mathbf{v} \qquad [1]$$

Thus we obtain

$$0\mathbf{v} = 0\mathbf{v} + 0\mathbf{v} \qquad [2]$$

By axiom 6, $0\mathbf{v}$ is an element of V. Therefore by axiom 5, V contains the unique element $-(0\mathbf{v})$ such that $0\mathbf{v} + (-0\mathbf{v}) = \mathbf{0}$. Thus by adding $-(0\mathbf{v})$ to both sides of equation 2 we get

$$0\mathbf{v} + (-0\mathbf{v}) = (0\mathbf{v} + 0\mathbf{v}) + (-0\mathbf{v})$$

or

Applying axioms 3–5.

$$\begin{aligned} \mathbf{0} &= 0\mathbf{v} + [0\mathbf{v} + (-0\mathbf{v})] \\ &= 0\mathbf{v} + \mathbf{0} \\ &= 0\mathbf{v} \end{aligned}$$

This completes the proof.

In order to prove part **b**, we must show that the element $(-1)\mathbf{v}$ has the same property as $-\mathbf{v}$, namely,

See axiom 5 of Definition 2-1.

$$(-1)\mathbf{v} + \mathbf{v} = -\mathbf{v} + \mathbf{v} = \mathbf{0}$$

Hence, if we prove that $(-1)\mathbf{v} + \mathbf{v} = \mathbf{0}$, by axiom 5 it follows that $(-1)\mathbf{v}$ and $-\mathbf{v}$ must be identical. We have

Note: by axiom 10 $1\mathbf{v} = \mathbf{v}$

$$(-1)\mathbf{v} + \mathbf{v} = (-1)\mathbf{v} + 1\mathbf{v}$$

But by axiom 8 and part **a**

$$(-1)\mathbf{v} + 1\mathbf{v} = ((-1) + 1)\mathbf{v} = 0\mathbf{v} = \mathbf{0}$$

Thus we have proved that $(-1)\mathbf{v} + \mathbf{v} = \mathbf{0}$ and hence $(-1)\mathbf{v} = -\mathbf{v}$.

See problem 2.

The proof of part **c** is similar to the proof of part **a** and is left for the reader as an exercise.

To prove part **d** we must show that if $\alpha \neq 0$ then $\mathbf{v}$ must be the $\mathbf{0}$ vector. Assume therefore that $\alpha \neq 0$. Then $\frac{1}{\alpha}$ is a real number which obviously satisfies the equation $\frac{1}{\alpha} \cdot \alpha = 1$. Multiplying both sides of the equation $\alpha\mathbf{v} = \mathbf{0}$ by $\frac{1}{\alpha}$ we get

$$\frac{1}{\alpha}(\alpha\mathbf{v}) = \frac{1}{\alpha}\mathbf{0}$$

Using axioms 9 and 10 gives us the following:

$$\frac{1}{\alpha}(\alpha\mathbf{v}) = \left(\frac{1}{\alpha} \cdot \alpha\right)\mathbf{v} = 1\mathbf{v} = \mathbf{v}$$

while $\frac{1}{\alpha}\mathbf{0} = \mathbf{0}$ by part **c**. Hence we get $\mathbf{v} = \mathbf{0}$.

The conclusions of Theorem 2-1 will be used quite often in the following chapters.

Biographical Sketch

JAMES JOSEPH SYLVESTER (1814–1897), *born in London, was the youngest child of a large Jewish family. His oldest brother emigrated to the United States and took the name of Sylvester—for reasons unknown. The whole family adopted the name, hence James Joseph Sylvester.*

Sylvester's genius for mathematics was apparent very early. After attending private schools he entered the Royal Institution at Liverpool. He entered St. John's College, Cambridge, in 1831 but had to interrupt his academic career almost immediately due to severe illness. Finally in 1837 he took the mathematics tripos and placed second. However, he was ineligible to compete for the Smith's prize because of his religion.

Sylvester's intellectual interests included both Greek and Latin classics, and many of his papers were peppered with quotations from these works. He was also well versed in the English, French, German, and Italian classics, which he read in the original. Sylvester was also very much interested in music.

In his intellectual pursuits Sylvester greatly resembled his friend Cayley. But there was a marked difference in their attitude towards other mathematicians' works. While Cayley could not read enough of all existing works, Sylvester was bored by others' works.

Sylvester's first job was that of Professor of Natural Philosophy at University College, London, at the age of 24. His next move was disastrous. He accepted a position as Professor of Mathematics at the University of Virginia. He lasted all of three months, resigning when the university refused to discipline a student who had insulted him. This experience killed his taste for teaching for the next decade. Instead he became an actuary for a life insurance company. In 1846 he began to prepare for a legal career, and became a lawyer in 1850. This brought him together with Cayley, and the encounter renewed Sylvester's fascination with mathematics. The two friends met often, discussing the theory of invariants, which both of them were creating.

Sylvester never married. He became a Professor of Mathematics at the Royal Military Academy, where he remained sixteen years until he was forcibly "retired" at the age of 56. His superiors tried to swindle him out of his pension, but he fought vigorously and retained it. After his retirement, Sylvester lived in London, doing next to no mathematics but plenty of versifying and chess playing. Then in 1876 he was offered a post at the newly established Johns Hopkins University. His years there (1876–1883) were his happiest and most productive. No longer forced to spend his energies fighting the world, he directed them full-strength towards his research and at the age of 63 began a second mathematical career. In 1878 he founded the American Journal of Mathematics.

In 1883 Sylvester was offered a chair at Oxford University. He arrived with all the zest and enthusiasm of his younger days to spring upon his students a brand new mathematical theory—differential invariants. In 1893 Sylvester's eyesight began to dim and he was forced little by little to relinquish all his duties. He died at the age of 83.

Exercises

1 Let $V = \{\mathbf{v}\}$ where $\mathbf{v}$ is a fixed object and define addition on V by $\mathbf{v} + \mathbf{v} = \mathbf{v}$ and scalar multiplication by $\alpha\mathbf{v} = \mathbf{v}$. Show that V together with the two operations defined on it is a vector space.

2 Prove part **c** of Theorem 2-1.

3 Let S_2 be the set of all ordered pairs of real numbers. Define addition on S_2 by: $(a_1,a_2) + (b_1,b_2) = (a_1 + b_1, a_2 + b_2)$. Define scalar multiplication on S_2 by: $\alpha(a_1,a_2) = (\alpha a_1,\alpha a_2)$. Determine whether S_2 is a vector space under these operations.

4 Let S_2 be the same set as in problem 3 above. Define addition on S_2 by: $(a_1,a_2) + (b_1,b_2) = (a_1 + 2b_1, a_2 + 2b_2)$. Define scalar multiplication on S_2 by: $\alpha(a_1,a_2) = (\alpha a_1,\alpha a_2)$ for all real α and all $(a_1,a_2) \in S_2$. Determine whether or not S_2 is a vector space under these operations.

5 Let P be the set of all polynomials of degree exactly 2. Thus, every element of P has the form $a_2x^2 + a_1x + a_0$ where $a_2 \neq 0$. Define addition on P by the equation

$$(a_2x^2 + a_1x + a_0) + (b_2x^2 + b_1x + b_0) = (a_2 + b_2)x^2 + (a_1 + b_1)x + (a_0 + b_0)$$

and scalar multiplication by

$$\alpha(a_2x^2 + a_1x + a_0) = (\alpha a_2)x^2 + (\alpha a_1)x + \alpha a_0$$

for all real α. Determine whether P is a vector space under these operations.

6 Let $P_2 = \{a_2x^2 + a_1x + a_0\colon a_0, a_1, \text{ and } a_2 \text{ are real numbers}\}$. (Notice that we no longer require $a_2 \neq 0$, and hence the set P_2 is different from the set P mentioned in problem 5 above.) Define addition and scalar multiplication on P_2 as in problem 5 above. Determine whether P_2 is a vector space under these operations.

2-2 Some Vector Spaces

Objectives

1 Introduce the vector spaces $\mathbf{R}^2$ and $\mathbf{R}^3$ and their geometric representation.
2 Discuss the vector space $\mathbf{R}^n$.
3 Study function spaces.

We devote this section to some special examples of vector spaces. The purpose of this is twofold. First, it will substantially increase the collection of vector spaces with which the reader is familiar so that further concepts of linear algebra may be tackled effectively. Second, some of the vector spaces can be illustrated geometrically, a fact which will certainly enhance the reader's understanding of the concepts involved and also suggest immediate applications.

Example 1 Let

$$\mathbf{R}^2 = \left\{\begin{bmatrix} x \\ y \end{bmatrix} : x \text{ and } y \text{ are real numbers}\right\}$$

Define addition on $\mathbf{R}^2$ by the equation

$$\begin{bmatrix} x_1 \\ y_1 \end{bmatrix} + \begin{bmatrix} x_2 \\ y_2 \end{bmatrix} = \begin{bmatrix} x_1 + x_2 \\ y_1 + y_2 \end{bmatrix}$$

Define scalar multiplication on $\mathbf{R}^2$ by the equation

$$\alpha \begin{bmatrix} x \\ y \end{bmatrix} = \begin{bmatrix} \alpha x \\ \alpha y \end{bmatrix}$$

We now state that $\mathbf{R}^2$ is a vector space under the operations of addition and scalar multiplication as defined above. Rather than prove separately that each of the 10 axioms of Definition 2-1 is satisfied, we take another approach here which the reader should study carefully.

We may look at $\mathbf{R}^2$ as being the set of all 2×1 matrices. It is obvious that the operations defined on $\mathbf{R}^2$ above are exactly the same as addition and scalar multiplication for matrices. Since we found that the set of all $m \times n$ matrices satisfies all 10 axioms of Definition 2-1 (and hence is a vector space), we may specialize to the case $m = 2$ and $n = 1$ to find that $\mathbf{R}^2$ with the operations defined on it above is indeed a vector space.

See the table in Section 2-1.

Despite the fact that the vector space $\mathbf{R}^2$ is purely algebraic in its appearance, we can give a geometrical representation of it at this point. To this end, let us take a rectangular coordinate system in the xy-plane. A vector $\begin{bmatrix} x \\ y \end{bmatrix}$ of $\mathbf{R}^2$ can be represented by a directed line segment having its **initial** point at the origin (0,0) and its **terminal** point at (x, y), or in a more general form, having its initial point at an arbitrary point (a,b) and its terminal point at $(a + x, b + y)$. The vector with initial point (x_1, y_1) and terminal point (x_2, y_2) and the vector with initial point (x_3, y_3) and terminal point (x_4, y_4) are considered **equal** if and only if $x_2 - x_1 = x_4 - x_3$ and $y_2 - y_1 = y_4 - y_3$. We may therefore speak of the arrow from (0,0) to (x, y), or from (a,b) to $(a + x,\ b + y)$, as representing the vector $\begin{bmatrix} x \\ y \end{bmatrix}$ of $\mathbf{R}^2$.

Let $\mathbf{a} = \begin{bmatrix} x_1 \\ y_1 \end{bmatrix}$ and $\mathbf{b} = \begin{bmatrix} x_2 \\ y_2 \end{bmatrix}$ be vectors represented by the arrows from the origin (0,0) to the points (x_1, y_1) and (x_2, y_2), respectively. We represent the sum of $\mathbf{a}$ and $\mathbf{b}$ by the arrow from (0,0) to the point $(x_1 + x_2, y_1 + y_2)$, as illustrated in Figure 2-1.

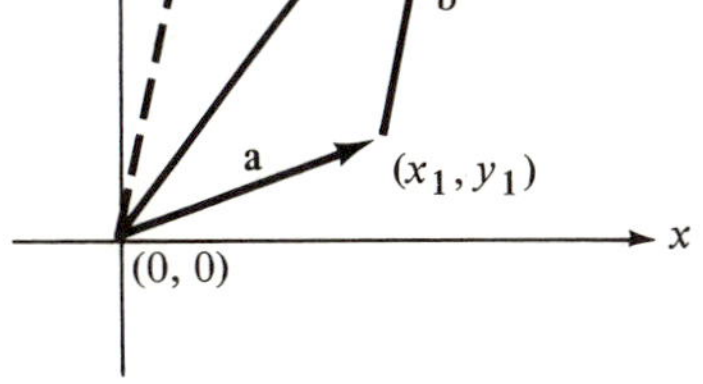

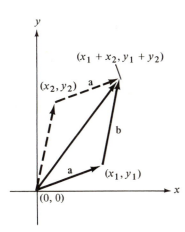

Figure 2-1

The geometrical construction of the sum $\mathbf{a} + \mathbf{b}$ may be described as follows: First, construct the arrow leading from (0,0) to (x_1, y_1), representing the vector $\mathbf{a} = \begin{bmatrix} x_1 \\ y_1 \end{bmatrix}$. Second, construct the

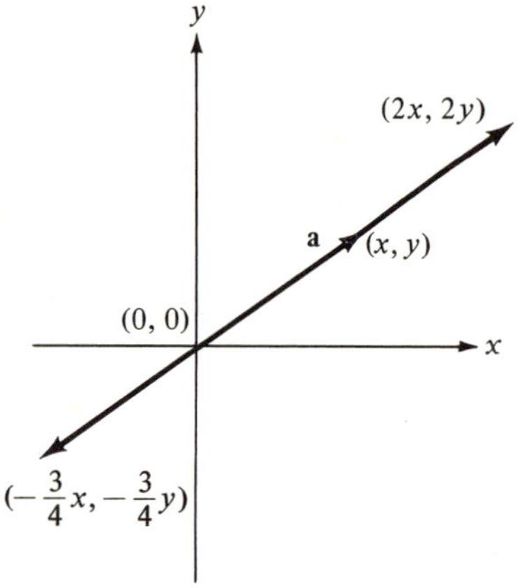

Figure 2-2

arrow leading from (x_1, y_1) to $(x_1 + x_2, y_1 + y_2)$, representing the vector $\mathbf{b} = \begin{bmatrix} x_2 \\ y_2 \end{bmatrix}$. Notice that we have chosen the terminal point of the vector $\mathbf{a}$ to be the initial point of the vector $\mathbf{b}$. This is in agreement with remarks made above.

We now construct the arrow leading from the origin (0,0) to the point $(x_1 + x_2, y_1 + y_2)$, representing the sum $\mathbf{a} + \mathbf{b}$.

Let α be any real number and let the vector $\mathbf{a} = \begin{bmatrix} x \\ y \end{bmatrix}$ be represented by the arrow from the origin (0,0) to the point (x, y). We represent the vector $\alpha\mathbf{a} = \begin{bmatrix} x \\ y \end{bmatrix}$ by the arrow from (0,0) to $(\alpha x, \alpha y)$. Figure 2-2 illustrates the construction of $\alpha\mathbf{a}$ for the cases $\alpha = 2$ and $\alpha = \frac{3}{4}$.

Example 2 Let

$$\mathbf{R}^3 = \left\{ \begin{bmatrix} x \\ y \\ z \end{bmatrix} : x, y, \text{ and } z \text{ are real numbers} \right\}$$

Define addition on $\mathbf{R}^3$ by the equation

$$\begin{bmatrix} x_1 \\ y_1 \\ z_1 \end{bmatrix} + \begin{bmatrix} x_2 \\ y_2 \\ z_2 \end{bmatrix} = \begin{bmatrix} x_1 + x_2 \\ y_1 + y_2 \\ z_1 + z_2 \end{bmatrix}$$

Define scalar multiplication on $\mathbf{R}^3$ by the equation

$$\alpha \begin{bmatrix} x \\ y \\ z \end{bmatrix} = \begin{bmatrix} \alpha x \\ \alpha y \\ \alpha z \end{bmatrix}$$

We may look upon $\mathbf{R}^3$ together with the operations defined above as being the same as the set of all 3×1 matrices under the usual addition and scalar multiplication defined on it. Thus, it follows that $\mathbf{R}^3$ is a vector space under the operations defined above.

As in Example 1, the vector space $\mathbf{R}^3$ has also a geometrical representation. We consider a rectangular coordinate system in xyz-space. See Figure 2-3. A vector $\begin{bmatrix} x \\ y \\ z \end{bmatrix}$ of $\mathbf{R}^3$ may be represented by a directed line segment having its initial point at the origin (0,0,0) and its terminal point at (x, y, z); or in a more general form, having an arbitrary initial point (a, b, c) and a terminal point $(a + x, b + y, c + z)$.

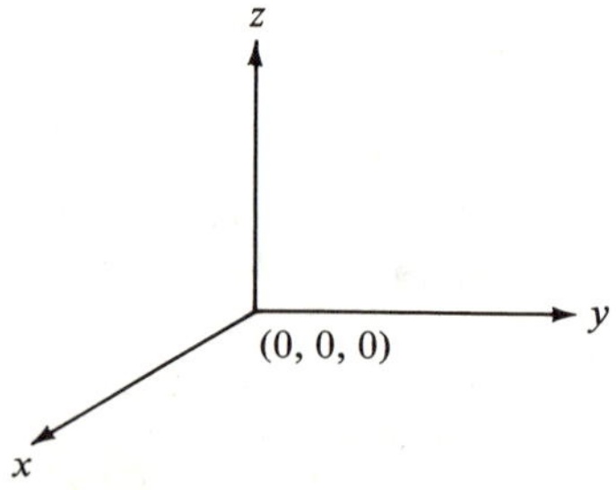

Figure 2-3

Thus, the vector with initial point (x_1, y_1, z_1) and terminal point

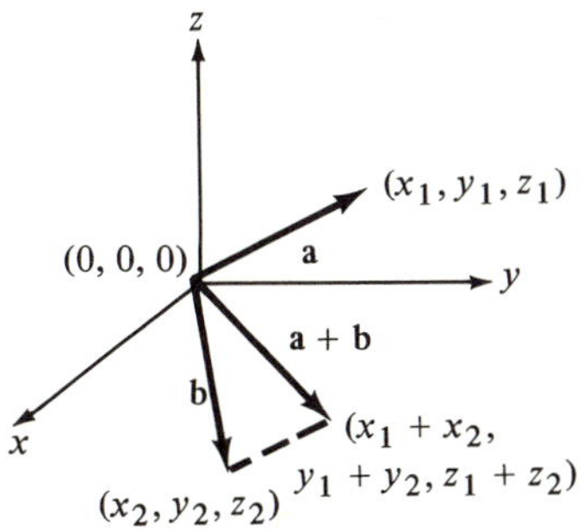

Figure 2-4

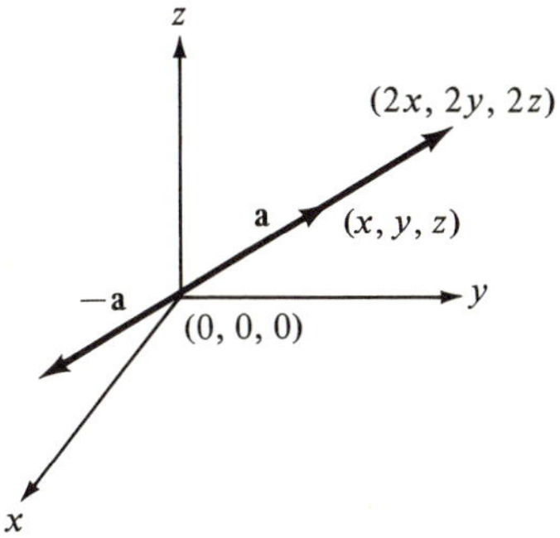

Figure 2-5

Note that the vectors $\begin{bmatrix}1\\0\\0\end{bmatrix}, \begin{bmatrix}0\\1\\0\end{bmatrix},$ and $\begin{bmatrix}0\\0\\1\end{bmatrix}$ lie on the x, y, and z axes, respectively.

(x_2,y_2,z_2) and the vector with initial point (x_3,y_3,z_3) and terminal point (x_4,y_4,z_4) are equal if and only if $x_2 - x_1 = x_4 - x_3$, $y_2 - y_1 = y_4 - y_3$, and $z_2 - z_1 = z_4 - z_3$. We speak of the arrow from $(0,0,0)$ to (x,y,z) or from (a,b,c) to $(a + x,\ b + y,\ c + z)$ as representing the vector $\begin{bmatrix}x\\y\\z\end{bmatrix}$ of $\mathbf{R}^3$.

Let the vectors

$$\mathbf{a} = \begin{bmatrix}x_1\\y_1\\z_1\end{bmatrix} \quad \text{and} \quad \mathbf{b} = \begin{bmatrix}x_2\\y_2\\z_2\end{bmatrix}$$

be represented by the arrows from $(0,0,0)$ to (x_1,y_1,z_1) and (x_2,y_2,z_2), respectively. We represent the sum of $\mathbf{a}$ and $\mathbf{b}$ by the arrow from $(0,0,0)$ to $(x_1 + x_2,\ y_1 + y_2,\ z_1 + z_2)$, as indicated in Figure 2-4.

Let α be any real number and let the vector $\mathbf{a} = \begin{bmatrix}x\\y\\z\end{bmatrix}$ be represented by the arrow from $(0,0,0)$ to (x,y,z). We represent the vector $\alpha\mathbf{a} = \alpha\begin{bmatrix}x\\y\\z\end{bmatrix}$ by the arrow from $(0,0,0)$ to $(\alpha x,\alpha y,\alpha z)$. Figure 2-5 illustrates the construction of $\alpha\mathbf{a}$ for the cases $\alpha = 2$ and $\alpha = -1$. The vector space $\mathbf{R}^3$ and its geometrical representation play a significant role in physical applications. The motion of a particle in space, its velocity, its acceleration, or the force acting on it may be viewed as vectors in $\mathbf{R}^3$, and one can manipulate with such vectors according to the rules which hold in $\mathbf{R}^3$.

In many physical and mathematical applications it is customary (and also useful) to express a vector $\begin{bmatrix}a\\b\\c\end{bmatrix}$ of $\mathbf{R}^3$ in the form

$$\begin{bmatrix}a\\b\\c\end{bmatrix} = a\begin{bmatrix}1\\0\\0\end{bmatrix} + b\begin{bmatrix}0\\1\\0\end{bmatrix} + c\begin{bmatrix}0\\0\\1\end{bmatrix}$$

If we take the origin $(0,0,0)$ as the initial point of the vectors $\begin{bmatrix}1\\0\\0\end{bmatrix}, \begin{bmatrix}0\\1\\0\end{bmatrix},$ and $\begin{bmatrix}0\\0\\1\end{bmatrix}$, then their terminal points are $(1,0,0)$, $(0,1,0)$, and $(0,0,1)$, respectively. We label the vectors thus:

$$\mathbf{i} = \begin{bmatrix} 1 \\ 0 \\ 0 \end{bmatrix}, \mathbf{j} = \begin{bmatrix} 0 \\ 1 \\ 0 \end{bmatrix}, \quad \text{and} \quad \mathbf{k} = \begin{bmatrix} 0 \\ 0 \\ 1 \end{bmatrix}$$

Thus we may write

$$\begin{bmatrix} a \\ b \\ c \end{bmatrix} = a\mathbf{i} + b\mathbf{j} + c\mathbf{k}$$

Hence, every vector $\begin{bmatrix} a \\ b \\ c \end{bmatrix} \in \mathbf{R}^3$ can be expressed as above in terms of the **basis** vectors $\mathbf{i}$, $\mathbf{j}$, and $\mathbf{k}$. We call the numbers a, b, c the x, y, and z **components** of the vector $\begin{bmatrix} a \\ b \\ c \end{bmatrix}$.

The mathematical meaning attached to the word basis will be defined rigorously in the next chapter.

Example 3 It is quite natural to generalize the previous two examples by considering the following space. Let n be a fixed positive integer. Let

$$\mathbf{R}^n = \left\{ \begin{bmatrix} x_1 \\ x_2 \\ \vdots \\ x_n \end{bmatrix} : x_i \text{ is a real number for } i = 1, 2, 3, \ldots, n \right\}$$

We refer to the operations defined in Example 3 as the **standard** operations on $\mathbf{R}^n$.

Define addition on $\mathbf{R}^n$ by the equation

$$\begin{bmatrix} x_1 \\ x_2 \\ \vdots \\ x_n \end{bmatrix} + \begin{bmatrix} y_1 \\ y_2 \\ \vdots \\ y_n \end{bmatrix} = \begin{bmatrix} x_1 + y_1 \\ x_2 + y_2 \\ \vdots \\ x_n + y_n \end{bmatrix}$$

Define scalar multiplication on $\mathbf{R}^n$ by the equation

$$\alpha \begin{bmatrix} x_1 \\ x_2 \\ \vdots \\ x_n \end{bmatrix} = \begin{bmatrix} \alpha x_1 \\ \alpha x_2 \\ \vdots \\ \alpha x_n \end{bmatrix}$$

Identifying $\mathbf{R}^n$ with the set of all $n \times 1$ matrices, it follows that we have constructed a vector space. The vector $\begin{bmatrix} 0 \\ 0 \\ \vdots \\ 0 \end{bmatrix}$ serves as the zero vector in $\mathbf{R}^n$. The vector

$$\begin{bmatrix} -x_1 \\ -x_1 \\ \vdots \\ -x_n \end{bmatrix} \text{ is the additive inverse of } \begin{bmatrix} x_1 \\ x_2 \\ \vdots \\ x_n \end{bmatrix}$$

because

$$\begin{bmatrix} x_1 \\ x_2 \\ \vdots \\ x_n \end{bmatrix} + \begin{bmatrix} -x_1 \\ -x_2 \\ \vdots \\ -x_n \end{bmatrix} = \begin{bmatrix} 0 \\ 0 \\ \vdots \\ 0 \end{bmatrix}$$

The space $\mathbf{R}^n$ for $n \geq 4$ has a similar structure to $\mathbf{R}^2$ and $\mathbf{R}^3$ and will be found to be very useful in later discussions. For the particular case $n = 1$, we have the vector space $\mathbf{R}^1$. In this space every vector is simply a real number. Hence, any real number in the vector space $\mathbf{R}^1$ plays a dual role, one as a vector and the other as a scalar.

Example 4 Let F be the set of all real-valued functions defined on some interval I. Let f and g be elements of F and let α be any real number. We define addition and scalar multiplication on F by the equations

$$(f + g)(x) = f(x) + g(x)$$

and

$$(\alpha f)(x) = \alpha f(x)$$

By a **function space** we mean a space whose elements are functions.

for all $x \in I$. Thus, the value of $f + g$ at x, denoted by $(f + g)(x)$, is obtained by adding the values of f and g at x. The value of αf at x is obtained by multiplying the value of f at x by α. It follows that the **function space** F is a vector space under the operations defined above. We emphasize that a vector in this space is nothing but a real-valued function defined on the interval I. The zero function $[f(x) \equiv 0$ for all $x \in I]$ serves as the zero vector in the space F. The additive inverse of f is $-f$, which satisfies the equation $(-f)(x) = -(f(x))$. The closure property of F (axioms 1 and 6 of Definition 2-1) with respect to the addition and the scalar multiplication defined above follows directly. All the remaining axioms are satisfied for F because their validity stems directly from similar properties of real numbers. Thus, $f + g = g + f$ because $(f + g)(x) = f(x) + g(x)$, $(g + f)(x) = g(x) + f(x)$, and $f(x) + g(x) = g(x) + f(x)$ for all $x \in I$.

The reader should check that the remaining axioms are also satisfied.

We conclude this section by an example that will bring out the fact that two vector spaces can have the same underlying set of elements and still be different. This will necessarily result from different operations defined on the underlying set.

Example 5 Let n be a fixed positive integer. Let

$$S^n = \left\{ \begin{bmatrix} x_1 \\ x_2 \\ \vdots \\ x_n \end{bmatrix} : x_i \text{ is a real number for } i = 1, 2, 3, \ldots, n \right\}$$

Define addition on S^n by the equation

$$\begin{bmatrix} x_1 \\ x_2 \\ \vdots \\ x_n \end{bmatrix} + \begin{bmatrix} y_1 \\ y_2 \\ \vdots \\ y_n \end{bmatrix} = \begin{bmatrix} (x_1^3 + y_1^3)^{\frac{1}{3}} \\ (x_2^3 + y_2^3)^{\frac{1}{3}} \\ \vdots \\ (x_n^3 + y_n^3)^{\frac{1}{3}} \end{bmatrix}$$

Define scalar multiplication on S^n by the equation

$$\alpha \begin{bmatrix} x_1 \\ x_2 \\ \vdots \\ x_n \end{bmatrix} = \begin{bmatrix} \alpha^{\frac{1}{3}} x_1 \\ \alpha^{\frac{1}{3}} x_2 \\ \vdots \\ \alpha^{\frac{1}{3}} x_n \end{bmatrix}$$

A simple check shows that S^n together with the operations defined above is a vector space. But a comparison with Example 3 shows that $S^n = \mathbf{R}^n$ while the resulting vector spaces in these two examples are different. For instance, the addition of the two vectors $\begin{bmatrix} 1 \\ 1 \\ \vdots \\ 1 \end{bmatrix}$ and $\begin{bmatrix} 1 \\ 1 \\ \vdots \\ 1 \end{bmatrix}$ in Example 3 will produce the vector $\begin{bmatrix} 2 \\ 2 \\ \vdots \\ 2 \end{bmatrix}$, while the addition of the same two vectors in Example 5 will produce the vector $\begin{bmatrix} 2^{\frac{1}{3}} \\ 2^{\frac{1}{3}} \\ \vdots \\ 2^{\frac{1}{3}} \end{bmatrix}$.

Exercises

In each of the exercises below a set of objects is given, together with operations of addition and scalar multiplication. Determine for each set whether or not it is a vector space under the operations defined on it. For each set which is not a vector space under the given operations, list all axioms that fail to be satisfied.

1 The set of all pairs of real numbers of the form $\begin{bmatrix} x \\ 0 \end{bmatrix}$ with the operations $\begin{bmatrix} x_1 \\ 0 \end{bmatrix} + \begin{bmatrix} x_2 \\ 0 \end{bmatrix} = \begin{bmatrix} x_1 + x_2 \\ 0 \end{bmatrix}$ and $\alpha \begin{bmatrix} x \\ 0 \end{bmatrix} = \begin{bmatrix} \alpha x \\ 0 \end{bmatrix}$.

2 The set of all pairs of real numbers $\begin{bmatrix} x \\ y \end{bmatrix}$ with the operations

$$\begin{bmatrix} x_1 \\ y_1 \end{bmatrix} + \begin{bmatrix} x_2 \\ y_2 \end{bmatrix} = \begin{bmatrix} x_1 + x_2 \\ y_1 + y_2 \end{bmatrix} \quad \text{and} \quad \alpha \begin{bmatrix} x \\ y \end{bmatrix} = \begin{bmatrix} \alpha x \\ y \end{bmatrix}.$$

See Example 3.

3 The set of all pairs of real numbers of the form $\begin{bmatrix} x \\ 2x \end{bmatrix}$ with the standard operations defined on $\mathbf{R}^2$.

4 The set of all pairs of real numbers $\begin{bmatrix} x \\ y \end{bmatrix}$ where $x < 0$ with the standard operations defined on $\mathbf{R}^2$.

5 The set of all triples of real numbers of the form $\begin{bmatrix} x \\ x \\ x \end{bmatrix}$ with the standard operations defined on $\mathbf{R}^3$.

6 The set of all triples of real numbers $\begin{bmatrix} x \\ y \\ z \end{bmatrix}$ with the operations

$$\begin{bmatrix} x_1 \\ y_1 \\ z_1 \end{bmatrix} + \begin{bmatrix} x_2 \\ y_2 \\ z_2 \end{bmatrix} = \begin{bmatrix} x_1 + x_2 \\ y_1 + y_2 \\ z_1 + z_2 \end{bmatrix} \text{ and } \alpha \begin{bmatrix} x \\ y \\ z \end{bmatrix} = \begin{bmatrix} x \\ \alpha y \\ z \end{bmatrix}.$$

7 The set of all triples of real numbers of the form $\begin{bmatrix} x \\ y \\ 0 \end{bmatrix}$ with the standard operations defined on $\mathbf{R}^3$.

8 The set of all real-valued functions defined on the real line whose graph contains the origin, with the operations defined in Example 4.

9 The set of all real-valued functions defined on the real line whose graph contains the point $(1,1)$ in the xy-plane, with the operations defined in Example 4.

10 The set of all pairs of real numbers $\begin{bmatrix} x \\ y \end{bmatrix}$ with the operations

$$\begin{bmatrix} x_1 \\ y_1 \end{bmatrix} + \begin{bmatrix} x_2 \\ y_2 \end{bmatrix} = \begin{bmatrix} (x_1{}^5 + x_2{}^5)^{\frac{1}{5}} \\ (y_1{}^5 + y_2{}^5)^{\frac{1}{5}} \end{bmatrix} \text{ and } \alpha \begin{bmatrix} x \\ y \end{bmatrix} = \begin{bmatrix} \alpha^{\frac{1}{5}} x \\ \alpha^{\frac{1}{5}} y \end{bmatrix}.$$

11 The set of all pairs of real numbers $\begin{bmatrix} x \\ y \end{bmatrix}$ with the operations

$$\begin{bmatrix} x_1 \\ y_1 \end{bmatrix} + \begin{bmatrix} x_2 \\ y_2 \end{bmatrix} = \begin{bmatrix} (x_1{}^2 + x_2{}^2)^{\frac{1}{2}} \\ (y_1{}^2 + y_2{}^2)^{\frac{1}{2}} \end{bmatrix} \text{ and } \alpha \begin{bmatrix} x \\ y \end{bmatrix} = \begin{bmatrix} |\alpha|^{\frac{1}{2}} x \\ |\alpha|^{\frac{1}{2}} y \end{bmatrix}.$$

12 The set of all matrices of the form $\begin{bmatrix} 0 & a & b \\ -a & 0 & c \\ -b & -c & 0 \end{bmatrix}$ with the usual addition and scalar multiplication defined on matrices.

2-3 Subspaces

Objectives
1 Introduce the concept of a subspace of a vector space.
2 Study a variety of examples of subspaces.

In this section we consider some special subsets of vector spaces. The subset of a vector space V consisting of the zero vector alone has the property that it is itself a vector space under the same operations defined on V. We can see this easily by checking that all ten axioms of Definition 2-1 are satisfied. Since V is also considered a subset of itself (although an improper subset) we can state that if V contains at least two different elements, then V has at least two subsets, the set V itself and the set consisting of the zero vector of V, which are vector spaces under the operations defined on V. It would be interesting to examine other subsets of V (provided V contains more elements than the zero vector) and find out whether or not they are vector spaces under the operations defined on V.

The following example will show that a vector space may have proper subsets which fail to be vector spaces.

Example 1 Let the vector space V be $\mathbf{R}^2$ together with the standard operations defined on it. Define two different subsets of $\mathbf{R}^2$ by

$$S = \left\{ \begin{bmatrix} 0 \\ y \end{bmatrix} : y \text{ is a real number} \right\}$$

and

$$T = \left\{ \begin{bmatrix} 1 \\ y \end{bmatrix} : y \text{ is a real number} \right\}$$

$S \cap T = \varnothing$

The sets S and T are proper subsets of $\mathbf{R}^2$ and they are disjoint. We assert that S is a vector space under the operations defined on $\mathbf{R}^2$ while T is not. One reason that T fails to be a vector space is that it does not contain a zero vector among its elements. Furthermore, the set T is not closed under addition and scalar multiplication. To see this, construct the sum of two elements of T, $\begin{bmatrix} 1 \\ y_1 \end{bmatrix}$

and $\begin{bmatrix} 1 \\ y_2 \end{bmatrix}$. We have

$$\begin{bmatrix} 1 \\ y_1 \end{bmatrix} + \begin{bmatrix} 1 \\ y_2 \end{bmatrix} = \begin{bmatrix} 2 \\ y_1 + y_2 \end{bmatrix}$$

But

$$\begin{bmatrix} 2 \\ y_1 + y_2 \end{bmatrix} \notin T$$

as we know from the definition of T.

Furthermore, for $\alpha = 3$ we have

$$\alpha \begin{bmatrix} 1 \\ y \end{bmatrix} = 3 \begin{bmatrix} 1 \\ y \end{bmatrix} = \begin{bmatrix} 3 \\ 3y \end{bmatrix}$$

This is because axioms 1, 4, and 6 of Definition 2-1 do not hold.

but $\begin{bmatrix} 3 \\ 3y \end{bmatrix} \notin T$. Thus T fails to be a vector space.

On the other hand, S satisfies the axioms that T has failed to satisfy. S contains the element $\begin{bmatrix} 0 \\ 0 \end{bmatrix}$ which serves as the zero vector. S is also closed under addition and scalar multiplication defined on $\mathbf{R}^2$. This follows from the definition of S and from the equations

$$\begin{bmatrix} 0 \\ y_1 \end{bmatrix} + \begin{bmatrix} 0 \\ y_2 \end{bmatrix} = \begin{bmatrix} 0 \\ y_1 + y_2 \end{bmatrix} \quad \text{and} \quad \alpha \begin{bmatrix} 0 \\ y \end{bmatrix} = \begin{bmatrix} 0 \\ \alpha y \end{bmatrix}$$

For every $\begin{bmatrix} 0 \\ y \end{bmatrix} \in S$ we have also $\begin{bmatrix} 0 \\ -y \end{bmatrix} \in S$; hence, S satisfies axiom 5 of Definition 2-1. All the remaining axioms are satisfied automatically for S because they hold for $\mathbf{R}^2$ (which includes S).

A set like S merits a special name to describe its properties in relation to $\mathbf{R}^2$. We call S a *subspace* of $\mathbf{R}^2$ and refer to the operations defined on S as being inherited from the "parent" vector space $\mathbf{R}^2$. The following is the general definition of a subspace.

Definition 2-2

Definition of subspace.

A subset S of a vector space V is called a **subspace** of V if S is itself a vector space under operations inherited from V.

As mentioned before, every vector space V has at least two subspaces: V itself and the subspace consisting of the zero vector of V. These are called the **trivial** subspaces of V.

Note: The two trivial subspaces might coincide.

At this point we wish to provide an answer to the following question: Given a nonempty subset S of a vector space V, how can one determine in the most efficient way whether or not S is a subspace of V? The answer is given in the following theorem.

Theorem 2-2 Let S be a nonempty subset of a vector space V. Let S be closed under addition and scalar multiplication inherited from V. Then S is a subspace of V.

Proof We have to show that S satisfies all ten axioms for a vector space. Axioms 1 and 6 are satisfied because we have assumed that S is closed under the operations inherited from V. Axioms 2 and 3 are satisfied because addition of vectors in V is commutative and associative and every element of S is also an element of V. Axioms 7, 8, 9, and 10 are satisfied for S because they hold for all elements of V, in particular for all elements of S. It remains to be shown that axioms 4 and 5 hold also.

First we show that S contains the zero vector $\mathbf{0}$ of V. Since S is not empty it contains at least one element, say $\mathbf{v} \in S$. From Theorem 2-1 it follows that $0\mathbf{v} = \mathbf{0}$. By assumption, S is closed under scalar multiplication, hence S contains the vector $\mathbf{0}$ from V. In order to prove uniqueness of $\mathbf{0}$ we assume that S contains an element $\widetilde{\mathbf{0}}$ with the property that $\mathbf{v} + \widetilde{\mathbf{0}} = \mathbf{v}$ for all $\mathbf{v} \in S$. We will show that $\mathbf{0} = \widetilde{\mathbf{0}}$. This follows immediately from the equations $\mathbf{0} + \widetilde{\mathbf{0}} = \widetilde{\mathbf{0}}$ and $\mathbf{0} + \widetilde{\mathbf{0}} = \mathbf{0}$. Thus $\mathbf{0}$ is the unique element of S which satisfies the equation $\mathbf{v} + \mathbf{0} = \mathbf{v}$ for every $\mathbf{v} \in S$. Finally, to show that axiom 5 holds for S, let us consider $(-1)\mathbf{v}$ where $\mathbf{v} \in S$. Since $(-1)\mathbf{v} = -\mathbf{v}$ it follows that the additive inverse of $\mathbf{v}$, $-\mathbf{v}$, belongs to S. The uniqueness of $-\mathbf{v}$ follows from the fact that the equation $\mathbf{v} + \mathbf{x} = \mathbf{0}$ had a unique solution in the vector space V, namely $-\mathbf{v}$, and S is a subset of V. This completes the proof.

The importance of this theorem follows from the simple and short criterion it provides for deciding whether or not a subset S of a vector space V is a subspace of V. We can also use this theorem in the following way. Suppose we are given a set S together with addition and scalar multiplication defined on it. Suppose we can find a vector space V that contains S as a subset and such that the operations on S can be looked upon as inherited from V. Then in order to show that S is a vector space all we have to show is that S is closed under addition and scalar multiplication defined on it. The following examples will clarify the use of Theorem 2-2.

Example 2 Let S be the set of all 2×2 matrices of the form $\begin{bmatrix} x & 0 \\ 0 & x \end{bmatrix}$ where x is an arbitrary real number. Show that S is a vector space under the standard operations defined on matrices.

Solution We already know that the set M_2 of all 2×2 matrices is a vector space under standard addition and scalar multiplication defined on

See Example 1 in Section 2-1.

matrices. Since M_2 obviously contains S, all we have to establish here is that S is closed under the operations defined on it. Let

$$\begin{bmatrix} x_1 & 0 \\ 0 & x_1 \end{bmatrix} \quad \text{and} \quad \begin{bmatrix} x_2 & 0 \\ 0 & x_2 \end{bmatrix}$$

be two elements of S. We have

$$\begin{bmatrix} x_1 & 0 \\ 0 & x_1 \end{bmatrix} + \begin{bmatrix} x_2 & 0 \\ 0 & x_2 \end{bmatrix} = \begin{bmatrix} x_1 + x_2 & 0 \\ 0 & x_1 + x_2 \end{bmatrix}$$

which again belongs to S. Also,

$$\alpha \begin{bmatrix} x & 0 \\ 0 & x \end{bmatrix} = \begin{bmatrix} \alpha x & 0 \\ 0 & \alpha x \end{bmatrix}$$

which is an element of S. Since S is closed under addition and scalar multiplication inherited from M_2, it follows that S is a subspace of M_2 and hence a vector space under the operations defined on it.

Example 3 Let

$\mathbf{R}$ is the set of all real numbers.

$$P_4 = \{a_4x^4 + a_3x^3 + a_2x^2 + a_1x + a_0,\ a_i \in \mathbf{R} \text{ for } i = 0, 1, 2, 3, 4\}$$

Define addition of two polynomials from P_4 by the equation

$$(a_4x^4 + a_3x^3 + a_2x^2 + a_1x + a_0) + (b_4x^4 + b_3x^3 + b_2x^2 + b_1x + b_0) = (a_4 + b_4)x^4 + (a_3 + b_3)x^3 + (a_2 + b_2)x^2 + (a_1 + b_1)x + (a_0 + b_0)$$

and scalar multiplication by

$$\alpha(a_4x^4 + a_3x^3 + a_2x^2 + a_1x + a_0) = (\alpha a_4)x^4 + (\alpha a_3)x^3 + (\alpha a_2)x^2 + (\alpha a_1)x + \alpha a_0$$

Show that P_4 is a vector space under the operations defined on it.

Solution

See Example 4 in the previous section.

We view polynomials in P_4 as being functions defined on the real line. We remind the reader that it has already been proved that the set F of all real-valued functions defined on the real line is a vector space under addition and scalar multiplication of functions. Since $P_4 \subset F$, we must ascertain only that P_4 is closed under the operations defined on it. But this follows directly from the definition of these operations by merely inspecting the right-hand side of each equation above. Hence P_4 is a subspace of F and therefore a vector space in its own right under the operations defined on it.

Example 4 Let

$$S = \left\{ \begin{bmatrix} x \\ y \\ z \end{bmatrix} : x + y = z, \text{ and } x, y, z \in \mathbf{R} \right\}$$

Show that S is a subspace of $\mathbf{R}^3$.

Solution We must show that S is closed under the standard operations defined on $\mathbf{R}^3$.

Let $\begin{bmatrix} x_1 \\ y_1 \\ z_1 \end{bmatrix}$ and $\begin{bmatrix} x_2 \\ y_2 \\ z_2 \end{bmatrix}$ be elements of S. Then we have

$$\begin{bmatrix} x_1 \\ y_1 \\ z_1 \end{bmatrix} + \begin{bmatrix} x_2 \\ y_2 \\ z_2 \end{bmatrix} = \begin{bmatrix} x_1 + x_2 \\ y_1 + y_2 \\ z_1 + z_2 \end{bmatrix}$$

In order for $\begin{bmatrix} x_1 + x_2 \\ y_1 + y_2 \\ z_1 + z_2 \end{bmatrix}$ to be an element of S, the equation

$$(x_1 + x_2) + (y_1 + y_2) = (z_1 + z_2)$$

Why? must be satisfied. Since $x_1 + y_1 = z_1$ and $x_2 + y_2 = z_2$, we obtain the following by adding the two equations:

$$(x_1 + y_1) + (x_2 + y_2) = z_1 + z_2$$

Because addition of real numbers is commutative and associative, we can rewrite the left-hand side of the equation as $(x_1 + x_2) + (y_1 + y_2)$, thus proving that S is closed under addition. To prove that S is closed under scalar multiplication, we examine

$$\alpha \begin{bmatrix} x \\ y \\ z \end{bmatrix} = \begin{bmatrix} \alpha x \\ \alpha y \\ \alpha z \end{bmatrix}$$

From $x + y = z$ we get

$$\alpha(x + y) = \alpha z$$

or

$$\alpha x + \alpha y = \alpha z$$

But the last equation means simply that $\begin{bmatrix} \alpha x \\ \alpha y \\ \alpha z \end{bmatrix}$ is an element of S

provided $\begin{bmatrix} x \\ y \\ z \end{bmatrix}$ is also an element of S. Thus S is closed under scalar multiplication. We conclude that S is a subspace of $\mathbf{R}^3$.

Exercises

1 Determine which of the following are subspaces of $\mathbf{R}^2$.

a all vectors $\begin{bmatrix} x \\ x \end{bmatrix}$ where $x \in \mathbf{R}$

b all vectors $\begin{bmatrix} x \\ y \end{bmatrix}$ where $x + y = 0$, $x, y \in \mathbf{R}$

c all vectors $\begin{bmatrix} x \\ y \end{bmatrix}$ where $x + y = 1$, $x, y \in \mathbf{R}$

d all vectors $\begin{bmatrix} 0 \\ y \end{bmatrix}$ where $y \in \mathbf{R}$

2 Determine which of the following are subspaces of $\mathbf{R}^3$.

a all vectors $\begin{bmatrix} x \\ 2x \\ 3x \end{bmatrix}$ where $x \in \mathbf{R}$

b all vectors $\begin{bmatrix} x \\ y \\ z \end{bmatrix}$ where $x - y = z$, x, y, and $z \in \mathbf{R}$

c all vectors $\begin{bmatrix} x \\ 1 \\ 0 \end{bmatrix}$ where $x \in \mathbf{R}$

d all vectors $\begin{bmatrix} x \\ y \\ z \end{bmatrix}$ where $x + y + 2 = z$, x, y, and $z \in \mathbf{R}$

See Example 3.

3 Determine which of the following are subspaces of P_4.

a all polynomials $a_4x^4 + a_3x^3 + a_2x^2 + a_1x + a_0$ for which $a_4 = a_3 = a_2 = a_1 = a_0$

b all polynomials $a_4x^4 + a_3x^3 + a_2x^2 + a_1x + a_0$ for which $a_1 + a_0 = 0$

c all polynomials $a_4x^4 + a_3x^3 + a_2x^2 + a_1x + a_0$ where $a_0 = 2$

d all polynomials $a_4x^4 + a_3x^3 + a_2x^2 + a_1x + a_0$ where $a_0 - a_1 = 1$

4 Determine which of the following are subspaces of the vector space M_2 of all 2×2 matrices.

a all matrices $\begin{bmatrix} a & b \\ c & d \end{bmatrix}$ where $a < b < c < d$

b all matrices $\begin{bmatrix} a & b \\ c & d \end{bmatrix}$ where $b = c$

c all matrices $\begin{bmatrix} a & b \\ c & d \end{bmatrix}$ where $a + b + c + d = 0$

d all matrices $\begin{bmatrix} a & b \\ c & d \end{bmatrix}$ where a, b, c, d are integers

5 Determine which of the following are subspaces of the function space F defined in Example 4 of the previous section, where the interval I is the whole real line.
a all f such that $f(1) = 0$
b all f such that $f(1) = 2$
c all f such that $f(-1) = f(1)$
d all f such that $f(1) = f(0) + f(2)$

6 Let V be a vector space and let S and T be subspaces of V.
a Show that $S \cap T \neq \varnothing$.
b Show that $S \cap T$ is a subspace of V.
c Construct an example to show that $S \cup T$ is not necessarily a subspace of V.

In problem 6a you need to show that $S \cap T$ contains at least one element.

Review of Chapter 2

In each of the problems 1 through 10 below a set of objects is given together with operations of addition and scalar multiplication. Determine for each set whether or not it is a vector space under the operations defined on it. For each set that is not a vector space under the given operations, list the axioms that fail to be satisfied.

1 The set of all pairs of real numbers $\begin{bmatrix} 0 \\ y \end{bmatrix}$ with the operations

$$\begin{bmatrix} 0 \\ y_1 \end{bmatrix} + \begin{bmatrix} 0 \\ y_2 \end{bmatrix} = \begin{bmatrix} 0 \\ y_1 + y_2 \end{bmatrix} \text{ and } \alpha \begin{bmatrix} 0 \\ y \end{bmatrix} = \begin{bmatrix} 0 \\ \alpha y \end{bmatrix}.$$

2 The set of all pairs of real numbers $\begin{bmatrix} x \\ y \end{bmatrix}$ with the operations

$$\begin{bmatrix} x_1 \\ y_2 \end{bmatrix} + \begin{bmatrix} x_2 \\ y_2 \end{bmatrix} = \begin{bmatrix} x_1 + x_2 \\ y_1 + y_2 \end{bmatrix} \text{ and } \alpha \begin{bmatrix} x \\ y \end{bmatrix} = \begin{bmatrix} x \\ \alpha y \end{bmatrix}.$$

3 The set of all pairs of real numbers $\begin{bmatrix} 3y \\ y \end{bmatrix}$ with the standard operations defined on $\mathbf{R}^2$.

4 The set of all pairs of real numbers $\begin{bmatrix} x \\ y \end{bmatrix}$ where $y > 0$ with the standard operations defined on $\mathbf{R}^2$.

5 The set of all pairs of real numbers $\begin{bmatrix} x \\ y \end{bmatrix}$ where $x \geq y$ with the standard operations defined on $\mathbf{R}^2$.

6 The set of all triples of real numbers $\begin{bmatrix} x \\ 0 \\ z \end{bmatrix}$ with the standard operations defined on $\mathbf{R}^3$.

7 The set of all triples of real numbers $\begin{bmatrix} x \\ y \\ z \end{bmatrix}$ with the operations

$$\begin{bmatrix} x_1 \\ y_1 \\ z_1 \end{bmatrix} + \begin{bmatrix} x_2 \\ y_2 \\ z_2 \end{bmatrix} = \begin{bmatrix} x_1 + 3x_2 \\ y_1 + 3y_2 \\ z_1 + 3z_2 \end{bmatrix} \text{ and } \alpha \begin{bmatrix} x \\ y \\ z \end{bmatrix} = \begin{bmatrix} \alpha x \\ \alpha y \\ \alpha z \end{bmatrix}.$$

8 The set of all triples of real numbers $\begin{bmatrix} x \\ y \\ z \end{bmatrix}$ with the operations

$$\begin{bmatrix} x_1 \\ y_1 \\ z_1 \end{bmatrix} + \begin{bmatrix} x_2 \\ y_2 \\ z_2 \end{bmatrix} = \begin{bmatrix} x_1 + x_2 \\ y_1 + y_2 \\ z_1 + z_2 \end{bmatrix} \text{ and } \alpha \begin{bmatrix} x \\ y \\ z \end{bmatrix} = \begin{bmatrix} x \\ y \\ \alpha z \end{bmatrix}.$$

9 The set of all triples of real numbers $\begin{bmatrix} x \\ 0 \\ -x \end{bmatrix}$ with the standard operations defined on $\mathbf{R}^3$.

10 The set of all matrices $\begin{bmatrix} a & b \\ c & d \end{bmatrix}$ with the operations $\begin{bmatrix} a_1 & b_1 \\ c_1 & d_1 \end{bmatrix} +$

$$\begin{bmatrix} a_2 & b_1 \\ c_2 & d_2 \end{bmatrix} = \begin{bmatrix} a_1 + a_2 & b_1 + b_2 \\ c_1 + c_2 & d_1 + d_2 \end{bmatrix} \text{ and } \alpha \begin{bmatrix} a & b \\ c & d \end{bmatrix} = \begin{bmatrix} \alpha a & b \\ c & \alpha d \end{bmatrix}.$$

11 Complete the following sentence:
A subset S of a vector space V is called a *subspace* of V if

____________________.

12 Determine which of the following are subspaces of $\mathbf{R}^2$.

a all vectors $\begin{bmatrix} x \\ y \end{bmatrix}$ where $x + 2y = 0,\ x, y \in \mathbf{R}$

b all vectors $\begin{bmatrix} x \\ y \end{bmatrix}$ where $y = x + 1,\ x, y \in \mathbf{R}$

c all vectors $\begin{bmatrix} x \\ y \end{bmatrix}$ where $x = 3y$, $x, y \in \mathbf{R}$

d all vectors $\begin{bmatrix} x \\ y \end{bmatrix}$ where $x \geq 0$, $y \leq 0$, $x, y \in \mathbf{R}$

13 Determine which of the following are subspaces of $\mathbf{R}^3$.

a all vectors $\begin{bmatrix} 2y \\ y \\ 3y \end{bmatrix}$ where $y \in \mathbf{R}$

b all vectors $\begin{bmatrix} 0 \\ y \\ 1 \end{bmatrix}$ where $y \in \mathbf{R}$

c all vectors $\begin{bmatrix} x \\ y \\ z \end{bmatrix}$ where $x + y + z = 0$, $x, y, z \in \mathbf{R}$

d all vectors $\begin{bmatrix} x \\ y \\ z \end{bmatrix}$ where $x \geq y \geq z$, $x, y, z \in \mathbf{R}$

14 Determine which of the following are subspaces of the vector space M_2 of all 2×2 matrices.

a all matrices $\begin{bmatrix} a & b \\ c & d \end{bmatrix}$ where $b = c = a + d$

b all matrices $\begin{bmatrix} a & b \\ c & d \end{bmatrix}$ where $a = d$ and $b = c$

c all matrices $\begin{bmatrix} a & b \\ c & d \end{bmatrix}$ where a, b, c, d are irrationals

d all matrices $\begin{bmatrix} a & b \\ c & d \end{bmatrix}$ where $a + b + d = c + 1$

15 Determine which of the following are subspaces of P_4.

a all polynomials $a_4x^4 + a_3x^3 + a_2x^2 + a_1x + a_0$ for which $a_4 \geq a_3 \geq a_2$

b all polynomials $a_4x^4 + a_3x^3 + a_2x^2 + a_1x + a_0$ for which $a_1 = a_0$

c all polynomials $a_4x^4 + a_3x^3 + a_2x^2 + a_1x + a_0$ for which $a_4 - a_3 + a_2 - a_1 + a_0 = 0$

d all polynomials $a_4x^4 + a_3x^3 + a_2x^2 + a_1x + a_0$ for which $a_4 = a_3 + 1$

Linear Independence and Finite-dimensional Vector Spaces

3

3-1 Linear Independence

Objectives

1. Introduce the concepts of linearly independent and linearly dependent sets of vectors.
2. Discuss examples of linearly independent sets and linearly dependent sets of vectors.

The concepts of **linear independence, spanning set,** and **basis** for a vector space play a major role in the investigation of the structure of any given vector space. These concepts will be introduced and discussed in this chapter.

First we shall discuss briefly the associativity of vector addition in a vector space. If $\mathbf{v}_1$, $\mathbf{v}_2$, and $\mathbf{v}_3$ are elements of a vector space V, then we know that

The common value is usually denoted by the parenthesis-free expression $\mathbf{v}_1 + \mathbf{v}_2 + \mathbf{v}_3$.

$$(\mathbf{v}_1 + \mathbf{v}_2) + \mathbf{v}_3 = \mathbf{v}_1 + (\mathbf{v}_2 + \mathbf{v}_3)$$

If $\mathbf{v}_1$, $\mathbf{v}_2$, $\mathbf{v}_3$, and $\mathbf{v}_4$ are elements of the vector space V, then we denote by $\mathbf{v}_1 + \mathbf{v}_2 + \mathbf{v}_3 + \mathbf{v}_4$ the common value of any of the equal expressions

$$\begin{aligned}(\mathbf{v}_1 + \mathbf{v}_2) + \mathbf{v}_3 + \mathbf{v}_4 &= (\mathbf{v}_1 + \mathbf{v}_2 + \mathbf{v}_3) + \mathbf{v}_4\\ &= \mathbf{v}_1 + (\mathbf{v}_2 + \mathbf{v}_3) + \mathbf{v}_4\\ &= \mathbf{v}_1 + (\mathbf{v}_2 + \mathbf{v}_3 + \mathbf{v}_4)\\ &= \mathbf{v}_1 + \mathbf{v}_2 + (\mathbf{v}_3 + \mathbf{v}_4)\end{aligned}$$

If $\mathbf{v}_1, \mathbf{v}_2, \ldots, \mathbf{v}_n$ are elements of V, then the expression $\mathbf{v}_1 + \mathbf{v}_2 + \cdots + \mathbf{v}_n$ is defined by induction. Thus, expressions like $\alpha_1\mathbf{v}_1 + \alpha_2\mathbf{v}_2 + \cdots + \alpha_n\mathbf{v}_n$ need no parentheses to determine their value, provided $\alpha_1, \alpha_2, \ldots, \alpha_n$ are scalars and $\mathbf{v}_1, \mathbf{v}_2, \ldots, \mathbf{v}_n$ are elements of a vector space V. We wish to emphasize also that since addition of vectors in a vector space is commutative, we may rearrange the sum $\alpha_1\mathbf{v}_1 + \alpha_2\mathbf{v}_2 + \cdots + \alpha_n\mathbf{v}_n$ in any order we desire without changing the value of the expression. Thus,

$$\begin{aligned}&\alpha_1\mathbf{v}_1 + \cdots + \alpha_i\mathbf{v}_i + \cdots + \alpha_j\mathbf{v}_j + \cdots + \alpha_n\mathbf{v}_n\\ &\qquad = \alpha_1\mathbf{v}_1 + \cdots + \alpha_j\mathbf{v}_j + \cdots + \alpha_i\mathbf{v}_i + \cdots + \alpha_n\mathbf{v}_n\end{aligned}$$

Definition 3-1 Let $S = \{\mathbf{v}_1, \mathbf{v}_2, \ldots, \mathbf{v}_n\}$ be a subset of a vector space V. Let $\alpha_1, \alpha_2, \ldots, \alpha_n$ be a system of scalars (real numbers). Then the vector $\alpha_1\mathbf{v}_1 + \alpha_2\mathbf{v}_2 + \cdots + \alpha_n\mathbf{v}_n$ is called a **linear combination** of the vectors $\mathbf{v}_1, \mathbf{v}_2, \ldots, \mathbf{v}_n$.

Linear combination.

Example 1 The vector $3\begin{bmatrix}1\\2\end{bmatrix}$ is a linear combination of the vector $\begin{bmatrix}1\\2\end{bmatrix}$ of $\mathbf{R}^2$ with $\alpha_1 = 3$. In this case there is only one vector involved, hence $n = 1$.

Example 2 The vector

$$2\begin{bmatrix}1\\3\end{bmatrix} + 3\begin{bmatrix}0\\1\end{bmatrix}$$

is a linear combination of the vectors $\begin{bmatrix}1\\3\end{bmatrix}$ and $\begin{bmatrix}0\\1\end{bmatrix}$ of $\mathbf{R}^2$ with $\alpha_1 = 2$, $\alpha_2 = 3$, $\mathbf{v}_1 = \begin{bmatrix}1\\3\end{bmatrix}$, and $\mathbf{v}_2 = \begin{bmatrix}0\\1\end{bmatrix}$. Note that by applying addition and scalar multiplication defined on $\mathbf{R}^2$ we get

$$2\begin{bmatrix}1\\3\end{bmatrix} + 3\begin{bmatrix}0\\1\end{bmatrix} = \begin{bmatrix}2\\9\end{bmatrix}$$

Thus, we may also say that the vector $\begin{bmatrix}2\\9\end{bmatrix}$ is a linear combination of $\begin{bmatrix}1\\3\end{bmatrix}$ and $\begin{bmatrix}0\\1\end{bmatrix}$ because there exist $\alpha_1 = 2$ and $\alpha_2 = 3$ such that

$$2\begin{bmatrix}1\\3\end{bmatrix} + 3\begin{bmatrix}0\\1\end{bmatrix} = \begin{bmatrix}2\\9\end{bmatrix}$$

We shall also express this relation by saying that the vector $\begin{bmatrix}2\\9\end{bmatrix}$ *is generated by* the vectors $\begin{bmatrix}1\\3\end{bmatrix}$ and $\begin{bmatrix}0\\1\end{bmatrix}$.

Example 3 Show that the vector $\begin{bmatrix}3\\7\end{bmatrix}$ is a linear combination of the vectors $\begin{bmatrix}1\\3\end{bmatrix}$ and $\begin{bmatrix}0\\1\end{bmatrix}$ of $\mathbf{R}^2$.

Solution We must show that there exist scalars α_1, α_2 such that

$$\alpha_1\begin{bmatrix}1\\3\end{bmatrix} + \alpha_2\begin{bmatrix}0\\1\end{bmatrix} = \begin{bmatrix}3\\7\end{bmatrix}$$

Since

$$\alpha_1\begin{bmatrix}1\\3\end{bmatrix} + \alpha_2\begin{bmatrix}0\\1\end{bmatrix} = \begin{bmatrix}\alpha_1 \cdot 1\\\alpha_1 \cdot 3\end{bmatrix} + \begin{bmatrix}\alpha_2 \cdot 0\\\alpha_2 \cdot 1\end{bmatrix} = \begin{bmatrix}\alpha_1\\3\alpha_1 + \alpha_2\end{bmatrix}$$

we get the vector equation

$$\begin{bmatrix}\alpha_1\\3\alpha_1 + \alpha_2\end{bmatrix} = \begin{bmatrix}3\\7\end{bmatrix}$$

The solution of this system is given by $\alpha_1 = 3$ and $\alpha_2 = -2$. Thus, we may write

$$\begin{bmatrix}3\\7\end{bmatrix} = 3\begin{bmatrix}1\\3\end{bmatrix} + (-2)\begin{bmatrix}0\\1\end{bmatrix}$$

Example 4 The matrix $\begin{bmatrix} a & b \\ c & d \end{bmatrix}$ may be expressed as a linear combination of the matrices $\begin{bmatrix} 1 & 0 \\ 0 & 0 \end{bmatrix}$, $\begin{bmatrix} 0 & 1 \\ 0 & 0 \end{bmatrix}$, $\begin{bmatrix} 0 & 0 \\ 1 & 0 \end{bmatrix}$, and $\begin{bmatrix} 0 & 0 \\ 0 & 1 \end{bmatrix}$.

Solution We have

$$\begin{bmatrix} a & b \\ c & d \end{bmatrix} = \begin{bmatrix} a & 0 \\ 0 & 0 \end{bmatrix} + \begin{bmatrix} 0 & b \\ 0 & 0 \end{bmatrix} + \begin{bmatrix} 0 & 0 \\ c & 0 \end{bmatrix} + \begin{bmatrix} 0 & 0 \\ 0 & d \end{bmatrix}$$

Since

$$\begin{bmatrix} a & 0 \\ 0 & 0 \end{bmatrix} = a\begin{bmatrix} 1 & 0 \\ 0 & 0 \end{bmatrix} \quad \text{and} \quad \begin{bmatrix} 0 & b \\ 0 & 0 \end{bmatrix} = b\begin{bmatrix} 0 & 1 \\ 0 & 0 \end{bmatrix}$$

$$\begin{bmatrix} 0 & 0 \\ c & 0 \end{bmatrix} = c\begin{bmatrix} 0 & 0 \\ 1 & 0 \end{bmatrix} \quad \text{and} \quad \begin{bmatrix} 0 & 0 \\ 0 & d \end{bmatrix} = d\begin{bmatrix} 0 & 0 \\ 0 & 1 \end{bmatrix}$$

we may write

$$\begin{bmatrix} a & b \\ c & d \end{bmatrix} = a\begin{bmatrix} 1 & 0 \\ 0 & 0 \end{bmatrix} + b\begin{bmatrix} 0 & 1 \\ 0 & 0 \end{bmatrix} + c\begin{bmatrix} 0 & 0 \\ 1 & 0 \end{bmatrix} + d\begin{bmatrix} 0 & 0 \\ 0 & 1 \end{bmatrix}$$

Thus, every element of the vector space M_2 of all 2×2 matrices can be generated by the four matrices

$$\begin{bmatrix} 1 & 0 \\ 0 & 0 \end{bmatrix}, \begin{bmatrix} 0 & 1 \\ 0 & 0 \end{bmatrix}, \begin{bmatrix} 0 & 0 \\ 1 & 0 \end{bmatrix}, \quad \text{and} \quad \begin{bmatrix} 0 & 0 \\ 0 & 1 \end{bmatrix}$$

Example 5 Let $P_2 = \{a_2x^2 + a_1x + a_0\colon a_2, a_1 \text{ and } a_0 \in \mathbf{R}\}$. Show that the element $3x^2 + 2x + 5$ of P_2 is a linear combination of $x^2 + 1$ and $x + 1$ (which also belong to P_2).

Solution Let

Note: In the previous chapter it has already been shown that P_2 is a vector space under the standard addition and scalar multiplication of polynomials.

$$\alpha_1 \cdot (x^2 + 1) + \alpha_2 \cdot (x + 1) = 3x^2 + 2x + 5$$

as polynomials in P_2.

After performing the operations on the left-hand side of the equation and rearranging the terms, we get

$$\alpha_1x^2 + \alpha_2x + (\alpha_1 + \alpha_2) = 3x^2 + 2x + 5$$

for all x.

Since two polynomials $a_2x^2 + a_1x + a_0$ and $b_2x^2 + b_1x + b_0$ are equal if and only if $a_2 = b_2$, $a_1 = b_1$, and $a_0 = b_0$, we get from the equation above the following conditions on α_1 and α_2:

Note that $\alpha_1 + \alpha_2 = 5$ follows from $\alpha_1 = 3$ and $\alpha_2 = 2$.

$$\alpha_1 = 3, \quad \alpha_2 = 2, \quad \text{and} \quad \alpha_1 + \alpha_2 = 5$$

Thus our solution may be written as

$$3x^2 + 2x + 5 = 3(x^2 + 1) + 2(x + 1)$$

We now proceed to introduce the concept of *linear independence* of vectors. We first illustrate the concept by some examples; a formal definition will follow.

Example 6 Consider the vectors $\begin{bmatrix}1\\2\end{bmatrix}$ and $\begin{bmatrix}2\\1\end{bmatrix}$ in the vector space $\mathbf{R}^2$. We wish to find an answer to the following question: How many different linear combinations of the vectors $\begin{bmatrix}1\\2\end{bmatrix}$ and $\begin{bmatrix}2\\1\end{bmatrix}$ will produce the zero vector $\begin{bmatrix}0\\0\end{bmatrix}$ of $\mathbf{R}^2$? In other words, we wish to find all different pairs of real numbers (α_1,α_2) such that

$$\alpha_1\begin{bmatrix}1\\2\end{bmatrix} + \alpha_2\begin{bmatrix}2\\1\end{bmatrix} = \begin{bmatrix}0\\0\end{bmatrix}$$

Solution This equation can be written as a system:

$$\begin{aligned} \alpha_1 + 2\alpha_2 &= 0 \\ 2\alpha_1 + \alpha_2 &= 0 \end{aligned}$$

Using matrix notation, we may write the system as

$$\begin{bmatrix}1 & 2\\2 & 1\end{bmatrix}\begin{bmatrix}\alpha_1\\\alpha_2\end{bmatrix} = \begin{bmatrix}0\\0\end{bmatrix}$$

The reader should have no difficulty in showing that $\begin{bmatrix}1 & 2\\2 & 1\end{bmatrix}^{-1}$ exists and is given by

$$\begin{bmatrix}1 & 2\\2 & 1\end{bmatrix}^{-1} = \begin{bmatrix}-\frac{1}{3} & \frac{2}{3}\\ \frac{2}{3} & -\frac{1}{3}\end{bmatrix}$$

If we multiply both sides of this equation from the left by $\begin{bmatrix}-\frac{1}{3} & \frac{2}{3}\\ \frac{2}{3} & -\frac{1}{3}\end{bmatrix}$ we get

$$\begin{bmatrix}-\frac{1}{3} & \frac{2}{3}\\ \frac{2}{3} & -\frac{1}{3}\end{bmatrix}\begin{bmatrix}1 & 2\\2 & 1\end{bmatrix}\begin{bmatrix}\alpha_1\\\alpha_2\end{bmatrix} = \begin{bmatrix}-\frac{1}{3} & \frac{2}{3}\\ \frac{2}{3} & -\frac{1}{3}\end{bmatrix}\begin{bmatrix}0\\0\end{bmatrix}$$

which upon simplifying yields

$$\begin{bmatrix}\alpha_1\\\alpha_2\end{bmatrix} = \begin{bmatrix}0\\0\end{bmatrix}$$

Thus, the equation

$$\alpha_1\begin{bmatrix}1\\2\end{bmatrix} + \alpha_2\begin{bmatrix}2\\1\end{bmatrix} = \begin{bmatrix}0\\0\end{bmatrix}$$

has the unique solution $\alpha_1 = \alpha_2 = 0$.

In this case we say that the set of vectors $\left\{\begin{bmatrix}1\\2\end{bmatrix}, \begin{bmatrix}2\\1\end{bmatrix}\right\}$ is linearly independent.

Example 7 Consider the elements e^x, e^{2x}, and e^{3x} of the vector space F of all real-valued functions defined on the real line. Show that if

$$\alpha_1 e^x + \alpha_2 e^{2x} + \alpha_3 e^{3x} = 0$$

for all x then we must have $\alpha_1 = \alpha_2 = \alpha_3 = 0$.

Solution Since the equation

$$\alpha_1 e^x + \alpha_2 e^{2x} + \alpha_3 e^{3x} = 0$$

holds for all x, it is satisfied in particular for $x = 0$, $x = \ln 2$ and $x = \ln 3$. We choose these values for x because substituting them into the equation above leads to a fairly simple system of linear equations for the unknowns α_1, α_2, and α_3. Substituting, we get the system

$$\begin{aligned} \alpha_1 e^0 + \alpha_2 e^{2\cdot 0} + \alpha_3 e^{3\cdot 0} &= 0 \\ \alpha_1 e^{\ln 2} + \alpha_2 e^{2(\ln 2)} + \alpha_3 e^{3(\ln 2)} &= 0 \\ \alpha_1 e^{\ln 3} + \alpha_3 e^{2(\ln 3)} + \alpha_3 e^{3(\ln 3)} &= 0 \end{aligned}$$

Remember:
$b(\ln b) = \ln (b^b)$
and
$e^{\ln b} = b$
Thus
$e^{\ln 2} = 2$
$e^{3(\ln 3)} = e^{\ln (3^3)} = 27$
and so on.

Simplification yields

$$\begin{aligned} \alpha_1 + \alpha_2 + \alpha_3 &= 0 \\ 2\alpha_1 + 4\alpha_2 + 8\alpha_3 &= 0 \\ 3\alpha_1 + 9\alpha_2 + 27\alpha_3 &= 0 \end{aligned}$$

Using matrix notation we may write

$$\begin{bmatrix}1 & 1 & 1\\2 & 4 & 8\\3 & 9 & 27\end{bmatrix}\begin{bmatrix}\alpha_1\\\alpha_2\\\alpha_3\end{bmatrix} = \begin{bmatrix}0\\0\\0\end{bmatrix}$$

$$\begin{bmatrix}\alpha_1\\\alpha_2\\\alpha_3\end{bmatrix} = \begin{bmatrix}1 & 1 & 1\\2 & 4 & 8\\3 & 9 & 27\end{bmatrix}^{-1}\begin{bmatrix}0\\0\\0\end{bmatrix} = \begin{bmatrix}0\\0\\0\end{bmatrix}$$

The reader can easily check that the matrix of coefficients of the linear system is invertible; thus, the only solution of the homogeneous system is $\alpha_1 = \alpha_2 = \alpha_3 = 0$.

We have therefore shown that the only possible way to produce the zero function of F as a linear combination of the three elements e^x, e^{2x}, and e^{3x} of F is by constructing their **trivial** linear combination $0 \cdot e^x + 0 \cdot e^{2x} + 0 \cdot e^{3x}$.

The reader might wonder why we chose only three different values of x from the infinitely many possible ones. In order to see the reason for that let us take one more value of x, say, $x = \ln 4$. In this case we will end up with a system of four linear equations

$$\begin{aligned} \alpha_1 + \alpha_2 + \alpha_3 &= 0 \\ 2\alpha_1 + 4\alpha_2 + 8\alpha_3 &= 0 \\ 3\alpha_1 + 9\alpha_2 + 27\alpha_3 &= 0 \\ 4\alpha_1 + 16\alpha_2 + 64\alpha_3 &= 0 \end{aligned}$$

Since the system is homogeneous, the last equation is certainly satisfied by the unique solution of the system made out of the first three equations only. Thus adding any further equations to the original three will always result in a homogeneous system which still has the unique solution $\alpha_1 = \alpha_2 = \alpha_3 = 0$.

The elements e^x, e^{2x}, and e^{3x} of the vector space F are said to be a linearly independent set of vectors because the only way to generate the zero vector of F in terms of a linear combination of them is to take the trivial one, namely $0 \cdot e^x + 0 \cdot e^{2x} + 0 \cdot e^{3x}$.

We now define the concept of linear independence in a general vector space.

Definition 3-2 Let $S = \{\mathbf{v}_1, \mathbf{v}_2, \ldots, \mathbf{v}_n\}$ be a set of vectors in a vector space V. The set S is called **linearly independent** if and only if the vector equation $\alpha_1\mathbf{v}_1 + \cdots + \alpha_n\mathbf{v}_n = \mathbf{0}$ has the unique solution $\alpha_1 = 0$, $\alpha_2 = 0, \ldots, \alpha_n = 0$.

Linear independence.

It follows from this definition that every nonzero vector in a vector space V constitutes a linearly independent set. To see this, assume $\mathbf{v} \in V$ and $\mathbf{v} \neq \mathbf{0}$. We construct a linear combination of $\mathbf{v}$, namely $\alpha\mathbf{v}$, and seek all possible solutions to the equation $\alpha\mathbf{v} = \mathbf{0}$. If $\alpha \neq 0$ then we may multiply the last equation by the scalar $\frac{1}{\alpha}$, obtaining $\frac{1}{\alpha}(\alpha\mathbf{v}) = \frac{1}{\alpha}\mathbf{0}$ which leads to $\mathbf{v} = \mathbf{0}$. Since we assumed that $\mathbf{v} \neq \mathbf{0}$ we must have $\alpha = 0$. Thus, the only solution of $\alpha\mathbf{v} = \mathbf{0}$ when $\mathbf{v} \neq \mathbf{0}$ is $\alpha = 0$. Therefore every nonzero vector in a vector space constitutes a linearly independent set.

Note:

$$\frac{1}{\alpha}(\alpha\mathbf{v}) = \left(\frac{1}{\alpha}\alpha\right)\mathbf{v} = 1\mathbf{v} = \mathbf{v}$$

and

$$\frac{1}{\alpha}\mathbf{0} = \mathbf{0}$$

See Theorem 2-1.

It is obvious from this argument that the zero vector does not satisfy the condition stated in Definition 3-2. As a matter of fact, the equation $\alpha\mathbf{0} = \mathbf{0}$ has infinitely many solutions for α in addition to the solution $\alpha = 0$. Therefore the set consisting of the zero vector of any vector space is *not* linearly independent. We can go one step further with the zero vector. Suppose we add the zero vector to a given set of nonzero vectors $\{\mathbf{v}_1, \mathbf{v}_2, \ldots, \mathbf{v}_n\}$ from a vector space V. Thus, we now look at the expanded set $T = \{\mathbf{v}_1, \mathbf{v}_2, \ldots, \mathbf{v}_n, \mathbf{0}\}$, consisting of $n + 1$ vectors. We assert again that the set T is *not* linearly independent. To see this, we examine the vector equation

$$\alpha_1\mathbf{v}_1 + \alpha_2\mathbf{v}_2 + \cdots + \alpha_n\mathbf{v}_n + \alpha_{n+1}\mathbf{0} = \mathbf{0}$$

for all possible solutions for $\alpha_1, \alpha_2, \ldots, \alpha_{n+1}$. One possible solution is obviously

$$\alpha_1 = \alpha_2 = \cdots = \alpha_n = \alpha_{n+1} = 0$$

But this is not the only solution. If we choose $\alpha_1 = \alpha_2 = \cdots = \alpha_n = 0$ and $\alpha_{n+1} = 1$ (or any number different from 0), then the equation above is satisfied because $0\mathbf{v}_i = \mathbf{0}$ for $i = 1, 2, \ldots, n$ and $1\mathbf{0} = \mathbf{0}$. Therefore we may state that any set of vectors from a vector space V which contains the zero vector of V is *not* linearly independent.

There exist also sets of vectors which do not contain the zero vector and still are *not* linearly independent. The following example exhibits such a set.

Example 8 Consider the vectors $\begin{bmatrix}1\\2\end{bmatrix}$, $\begin{bmatrix}2\\1\end{bmatrix}$, and $\begin{bmatrix}8\\7\end{bmatrix}$ in $\mathbf{R}^2$. Show that the set $\left\{\begin{bmatrix}1\\2\end{bmatrix}, \begin{bmatrix}2\\1\end{bmatrix}, \begin{bmatrix}8\\7\end{bmatrix}\right\}$ is *not* linearly independent.

Solution We examine the vector equation

$$\alpha_1\begin{bmatrix}1\\2\end{bmatrix} + \alpha_2\begin{bmatrix}2\\1\end{bmatrix} + \alpha_3\begin{bmatrix}8\\7\end{bmatrix} = \begin{bmatrix}0\\0\end{bmatrix}$$

for all possible solutions α_1, α_2, and α_3. Performing the operations of addition and scalar multiplication indicated on the left-hand side of the equation, we get

$$\begin{bmatrix}\alpha_1 + 2\alpha_2 + 8\alpha_3\\2\alpha_1 + \alpha_2 + 7\alpha_3\end{bmatrix} = \begin{bmatrix}0\\0\end{bmatrix}$$

The vector equation above can be written as a system:

$$\begin{aligned}\alpha_1 + 2\alpha_2 + 8\alpha_3 &= 0\\2\alpha_1 + \alpha_2 + 7\alpha_3 &= 0\end{aligned}$$

This is a homogeneous system of linear equations with more unknowns than equations. By Theorem 2-15, there exists a nontrivial solution for this system. The reader can check that the augmented matrix of the system of equations can be transformed into the row-reduced echelon matrix

$$\left[\begin{array}{ccc:c}1 & 0 & 2 & 0\\0 & 1 & 3 & 0\end{array}\right]$$

The corresponding system of equations is given by

$$\begin{aligned}\alpha_1 \quad + 2\alpha_3 &= 0\\\alpha_2 + 3\alpha_3 &= 0\end{aligned}$$

The solution is given by

$\alpha_1 = -2\alpha_3$
$\alpha_2 = -3\alpha_3$

Thus, for $\alpha_3 = -1$ we get $\alpha_1 = 2$ and $\alpha_2 = 3$. Therefore, we have shown that the vectors $\begin{bmatrix}1\\2\end{bmatrix}$, $\begin{bmatrix}2\\1\end{bmatrix}$, and $\begin{bmatrix}8\\7\end{bmatrix}$ are *not* linearly independent because we have

$$2\begin{bmatrix}1\\2\end{bmatrix} + 3\begin{bmatrix}2\\1\end{bmatrix} + (-1)\begin{bmatrix}8\\7\end{bmatrix} = \begin{bmatrix}0\\0\end{bmatrix}$$

It is customary to call any set of vectors which is *not* linearly independent a *linearly dependent* set. We give here the following "working" definition for a linearly dependent set of vectors.

Definition 3-3 Let $S = \{\mathbf{v}_1, \mathbf{v}_2, \ldots, \mathbf{v}_n\}$ be a set of vectors in a vector space V. The set S is called **linearly dependent** if and only if the vector equation $\alpha_1\mathbf{v}_1 + \alpha_2\mathbf{v}_2 + \cdots + \alpha_n\mathbf{v}_n = \mathbf{0}$ has at least two different solutions for $\alpha_1, \alpha_2, \ldots, \alpha_n$.

Linearly dependent set.

Note: One solution is always $\alpha_1 = \alpha_2 = \cdots = \alpha_n = 0$

Thus, in order to show that a given set of vectors $\{\mathbf{v}_1, \mathbf{v}_2, \ldots, \mathbf{v}_n\}$ is linearly dependent, all we have to do is show that we can find $\alpha_1, \alpha_2, \ldots, \alpha_n$, *not all zero,* such that the linear combination $\alpha_1\mathbf{v}_1 + \alpha_2\mathbf{v}_2 + \cdots + \alpha_n\mathbf{v}_n$ produces the zero vector. The vectors $\begin{bmatrix}1\\2\end{bmatrix}$, $\begin{bmatrix}2\\1\end{bmatrix}$, and $\begin{bmatrix}8\\7\end{bmatrix}$ in Example 8 form a linearly dependent set because in addition to the solution $\alpha_1 = \alpha_2 = \alpha_3 = 0$ for the vector equation

$$\alpha_1\begin{bmatrix}1\\2\end{bmatrix} + \alpha_2\begin{bmatrix}2\\1\end{bmatrix} + \alpha_3\begin{bmatrix}8\\7\end{bmatrix} = \begin{bmatrix}0\\0\end{bmatrix}$$

we also have the solution $\alpha_1 = 2$, $\alpha_2 = 3$, and $\alpha_3 = -1$.

Example 9 Consider the polynomials $2x^2 - x - 1$, $x^2 + x + 1$, and $8x^2 - x - 1$ in the vector space P_2. Show that the set $\{2x^2 - x - 1, x^2 + x + 1, 8x^2 - x - 1\}$ is linearly dependent.

Solution We have to find α_1, α_2, and α_3 not all zero such that

$$\alpha_1(2x^2 - x - 1) + \alpha_2(x^2 + x + 1) + \alpha_3(8x^2 - x - 1) = 0$$

Simplifying the left-hand side of the equation gives us

$$(2\alpha_1 + \alpha_2 + 8\alpha_3)x^2 + (-\alpha_1 + \alpha_2 - \alpha_3)x + (-\alpha_1 + \alpha_2 - \alpha_3) = 0$$

which must be satisfied for all x. This is possible only if the coefficients of x^2, x, and the constant term are all 0. We get the following system of equations

$$\begin{aligned} 2\alpha_1 + \alpha_2 + 8\alpha_3 &= 0 \\ -\ \alpha_1 + \alpha_2 - \ \ \alpha_3 &= 0 \\ -\ \alpha_1 + \alpha_2 - \ \ \alpha_3 &= 0 \end{aligned}$$

We may omit the third equation because it is identical to the second equation. Now the system

$$\begin{aligned} 2\alpha_1 + \alpha_2 + 8\alpha_3 &= 0 \\ -\ \alpha_1 + \alpha_2 - \ \ \alpha_3 &= 0 \end{aligned}$$

See Theorem 1-15.

has more unknowns than equations; therefore, it has a nontrivial solution (in addition to the trivial solution $\alpha_1 = \alpha_2 = \alpha_3 = 0$).

The reader should check that $\alpha_1 = 3$, $\alpha_2 = 2$, and $\alpha_3 = -1$ is a solution of the system above. Thus,

$3(2x^2 - x - 1) + 2(x^2 + x + 1) + (-1)(8x^2 - x - 1) = 0$ for all x

and the linear dependence of the set $\{2x^2 - x - 1,\ x^2 + x + 1,\ 8x^2 - x - 1\}$ is established.

In an effort to shed some more light on the concepts of linear independence and linear dependence, we shall find an answer to the following question: Let $S = \{\mathbf{v}_1, \mathbf{v}_2, \ldots, \mathbf{v}_n\}$ be a set of vectors in a vector space V. Suppose that one of the vectors, say $\mathbf{v}_1$, can be expressed as a linear combination of the remaining vectors $\mathbf{v}_2$, $\mathbf{v}_3, \ldots, \mathbf{v}_n$. Can we conclude anything about the set S with regard to its being linearly independent or linearly dependent?

In order to answer this question, let us make use of the assumption that $\mathbf{v}_1$ can be expressed as a linear combination of $\mathbf{v}_2, \mathbf{v}_3, \ldots, \mathbf{v}_n$. Thus, there exist scalars $\beta_2, \beta_3, \ldots, \beta_n$ such that $\mathbf{v}_1 = \beta_2\mathbf{v}_2 + \beta_3\mathbf{v}_3 + \cdots + \beta_n\mathbf{v}_n$. By adding $-\mathbf{v}_1$ to both sides of this equation we get

$$\mathbf{0} = -\mathbf{v}_1 + \beta_2\mathbf{v}_2 + \beta_3\mathbf{v}_3 + \cdots + \beta_n\mathbf{v}_n$$

Since we know that $-\mathbf{v}_1 = (-1)\mathbf{v}_1$ holds in every vector space, we may rewrite the equation as

$$\mathbf{0} = (-1)\mathbf{v}_1 + \beta_2\mathbf{v}_2 + \beta_3\mathbf{v}_3 + \cdots + \beta_n\mathbf{v}_n$$

This equation tells us that the set $S = \{\mathbf{v}_1, \mathbf{v}_2, \ldots, \mathbf{v}_n\}$ is linearly dependent because the equation $\alpha_1\mathbf{v}_1 + \alpha_2\mathbf{v}_2 + \cdots + \alpha_n\mathbf{v}_n = \mathbf{0}$ has a nontrivial solution given by $\alpha_1 = -1$, $\alpha_2 = \beta_2, \ldots, \alpha_n = \beta_n$. (The trivial solution given by $\alpha_1 = 0$, $\alpha_2 = 0, \ldots, \alpha_n = 0$ is obvious.) Notice that the values of $\beta_2, \beta_3, \ldots, \beta_n$ do not matter here because $\alpha_1 = -1 \neq 0$. We have thus proved the following theorem.

Theorem 3-1 Let $S = \{\mathbf{v}_1, \mathbf{v}_2, \ldots, \mathbf{v}_n\}$ be a set of vectors in a vector space V. If S contains a vector which is a linear combination of the remaining vectors in S, then S is linearly dependent.

The converse of this theorem is also true. To see this, let us assume that the set $S = \{\mathbf{v}_1, \mathbf{v}_2, \ldots, \mathbf{v}_n\}$ is linearly dependent. This implies that there exist scalars $\alpha_1, \alpha_2, \ldots, \alpha_n$ *not* all zero such that $\alpha_1\mathbf{v}_1 + \alpha_2\mathbf{v}_2 + \cdots + \alpha_n\mathbf{v}_n = \mathbf{0}$. Without loss of generality we may assume that $\alpha_1 \neq 0$ (we can always relabel the α's and the $\mathbf{v}$'s in such a way that we guarantee $\alpha_1 \neq 0$, because *not* all α's are zero). We now solve for $\alpha_1\mathbf{v}_1$ by writing

$$\alpha_1\mathbf{v}_1 = -\alpha_2\mathbf{v}_2 - \alpha_3\mathbf{v}_3 + \cdots - \alpha_n\mathbf{v}_n$$

Since $\alpha_1 \neq 0$ we can multiply both sides of the equation by $\dfrac{1}{\alpha_1}$ and thus solve for $\mathbf{v}_1$. We get the equation

$$\mathbf{v}_1 = -\frac{\alpha_2}{\alpha_1}\mathbf{v}_2 - \frac{\alpha_3}{\alpha_1}\mathbf{v}_3 - \cdots - \frac{\alpha_n}{\alpha_1}\mathbf{v}_n$$

Hence $\mathbf{v}_1$ is a linear combination of $\mathbf{v}_2, \mathbf{v}_3, \ldots, \mathbf{v}_n$ with the scalars $-\dfrac{\alpha_2}{\alpha_1}, -\dfrac{\alpha_3}{\alpha_1}, \ldots, -\dfrac{\alpha_n}{\alpha_1}$. Thus we have proved the following theorem.

Theorem 3-2 Let $S = \{\mathbf{v}_1, \mathbf{v}_2, \ldots, \mathbf{v}_n\}$ be a set of vectors in a vector space V. If S is linearly dependent then S contains at least one vector which is a linear combination of the remaining vectors of S.

From the theorems above we find that if $T = \{\mathbf{u}_1, \mathbf{u}_2, \ldots, \mathbf{u}_n\}$ is a linearly independent set of vectors in a vector space U, then T does *not* contain any vector which is a linear combination of the remaining vectors of T.

Recall Example 7.

We have shown that the functions e^x, e^{2x}, and e^{3x} are linearly independent. Thus we can also say that e^x is *not* a linear combination of e^{2x} and e^{3x}, e^{2x} is *not* a linear combination of e^x and e^{3x}, and e^{3x} is *not* a linear combination of e^x and e^{2x}.

Originally one might have tried to set $e^x = \alpha_1 e^{2x} + \alpha_2 e^{3x}$ and then determine values for α_1 and α_2. Since the equation must hold for all x we may choose the values $x = 0$, $x = \ln 2$, and $x = \ln 3$, obtaining the following system of equations:

$$\begin{aligned} \alpha_1 + \alpha_2 &= 1 \\ 4\alpha_1 + 8\alpha_2 &= 2 \\ 9\alpha_1 + 27\alpha_2 &= 3 \end{aligned}$$

A common mistake made by some students is to choose only two values for x, say $x = 0$ and $x = \ln 2$, which lead to the first two equations of the system above, namely.

$$\alpha_1 + \alpha_2 = 1$$
$$4\alpha_1 + 8\alpha_2 = 2$$

This last system has the solution $\alpha_1 = \frac{3}{2}$ and $\alpha_2 = -\frac{1}{2}$. Unfortunately the equation $e^x = \frac{3}{2}e^{2x} - \frac{1}{2}e^{3x}$ does not hold for all x. It is satisfied for $x = 0$ and $x = \ln 2$ of course, but it is not satisfied for $x = \ln 3$.

The augmented matrix of this system is reduced to

$$\begin{bmatrix} 1 & 0 & \vdots & \frac{3}{2} \\ 0 & 1 & \vdots & -\frac{1}{2} \\ 0 & 0 & \vdots & \frac{1}{2} \end{bmatrix}$$

From the third row of the row-reduced echelon matrix we conclude that the system of linear equations does not have a solution. Thus, it is *impossible* to find numbers α_1, α_2 satisfying the equation $e^x = \alpha_1 e^{2x} + \alpha_2 e^{3x}$ for *all* x.

We conclude this section with a theorem which will be found to be very useful in later sections.

Theorem 3-3 Let $S = \{\mathbf{v}_1, \mathbf{v}_2, \ldots, \mathbf{v}_k\}$ be a linearly independent set of vectors in a vector space V. Let $\mathbf{v}_{k+1}$ be an element of V such that $\mathbf{v}_{k+1}$ is *not* a linear combination of the vectors in S. Then the expanded set $\{\mathbf{v}_1, \mathbf{v}_2, \ldots, \mathbf{v}_k, \mathbf{v}_{k+1}\}$ is also linearly independent.

See problem 11.

Proof We prove the theorem for the particular case $k = 2$, leaving the general case as an exercise. We must therefore prove that if the set $\{\mathbf{v}_1, \mathbf{v}_2\}$ is linearly independent and $\mathbf{v}_3$ is *not* a linear combination of $\mathbf{v}_1$ and $\mathbf{v}_2$, then the set $\{\mathbf{v}_1, \mathbf{v}_2, \mathbf{v}_3\}$ is linearly independent. Consider the vector equation

$$\alpha_1 \mathbf{v}_1 + \alpha_2 \mathbf{v}_2 + \alpha_3 \mathbf{v}_3 = \mathbf{0} \quad \textbf{[1]}$$

We must show that equation 1 has the unique solution

$$\alpha_1 = \alpha_2 = \alpha_3 = 0$$

The proof will follow easily if we focus our attention on the scalar α_3. We claim that any solution α_1, α_2, and α_3 of equation 1 must satisfy the condition $\alpha_3 = 0$. To see that, assume $\alpha_3 \neq 0$. Then we can solve for $\mathbf{v}_3$ in terms of $\mathbf{v}_1$ and $\mathbf{v}_2$ from equation 1, obtaining

$$\mathbf{v}_3 = -\frac{\alpha_1}{\alpha_3}\mathbf{v}_1 - \frac{\alpha_2}{\alpha_3}\mathbf{v}_2$$

But this is a contradiction, because it is given that $\mathbf{v}_3$ is *not* a linear combination of $\mathbf{v}_1$ and $\mathbf{v}_2$. Thus, we must have $\alpha_3 = 0$. Equation 1 reduces therefore to

$$\alpha_1 \mathbf{v}_1 + \alpha_2 \mathbf{v}_2 = \mathbf{0}$$

Now, since the set $\{\mathbf{v}_1, \mathbf{v}_2\}$ is linearly independent, it follows that $\alpha_1 = \alpha_2 = 0$. Thus, equation 1 has the unique solution $\alpha_1 = \alpha_2 = \alpha_3 = 0$. Therefore, the set $\{\mathbf{v}_1, \mathbf{v}_2, \mathbf{v}_3\}$ is linearly independent.

Biographical Sketch

JOSEPH LOUIS LAGRANGE (1736–1813) *was born in Turin, Italy, on January 25, 1736, though his family originated in Touraine, France. His father was the son of a cavalry officer who settled in Turin when he married into the famous Conti family. His mother was the daughter of a wealthy physician of French descent.*

Lagrange was educated at Turin College, where he excelled in the classics. His interest in mathematics arose suddenly after a chance reading of a memoir by Halley, the famous English astronomer, on the superiority of the calculus over the synthetic methods. He developed a passion for the new concepts of mathematics which led him to master modern analysis rapidly. He became the greatest analytical mathematician of his day.

Lagrange's important mathematical work was in partial differential equations. He made his greatest contributions to the development of the calculus of variations, which became a new branch of mathematics. He also made contributions to the theory of numbers, and of probability. His interests ranged over a variety of mathematical, physical, and astronomical research. In physics he made contributions to wave theory, study of the propagation of sound, and hydrodynamics. In astronomy he offered solutions for the libration of the moon, the problems of the satellites of Jupiter, the shape of the earth, and the orbits of comets.

While Lagrange was still in college, his family's fortune met with disaster, and he had to work for a living. At 18, he became professor of mathematics at the Royal Artillery School of Turin, where he taught students much older than himself. Less than a year later he sent Euler his famous communication on his method of solving isoperimetrical problems. Euler recognized its worth and encouraged him even delaying publication of some of his completed work to allow the younger mathematician time to publish his own discovery first and receive the full credit. Lagrange's paper immediately established his reputation among the great mathematicians of the time, before he was 20.

In 1758 with a group of friends and students he founded a private scientific society which later became the Academy of Sciences of Turin. In 1759 this society published the first volume of his papers in "Miscellanea Taurinensia." At 23, Lagrange was elected a foreign member of the Academy of Berlin. At 28, he won the coveted prize of mathematics of the French Academy of Sciences, then considered the scientific center of the world. He presented his theory on the libration of the moon, giving the reason why it always shows the same face to the earth. In the succeeding years, he was to win this prize several times.

In 1766, on the recommendation of his friends Euler and d'Alembert, he was called upon by Frederick II of Prussia to replace Euler as director of the Mathematics Department of the Academy of Berlin. He sent for a young relative in Turin and married her. Unfortunately, his young wife died two years later after a long illness. He then devoted himself completely to his work. He stayed in Berlin twenty years. During this period his output was enormous: He produced an average of one scientific paper a month for the various academies of Berlin, Turin, and Paris, while writing his monumental treatise, the Mécanique analytique *(Analytical mechanics) which was to be published in Paris in 1787.*

As he reached middle age, Lagrange suffered a nervous depression, probably due to sheer exhaustion. He began to lose interest in his mathematical work and to be more attracted by philosophy, metaphysics, the history of religions, and the evolution of thought. However, he never allowed any of his ideas on those subjects to be published.

In 1787, after the death of Frederick II, Lagrange accepted an invitation from Louis XVI to come to Paris to continue his work at the French Academy. He was given an apartment in the Louvre. In 1792, at the age of 56, Lagrange married the young daughter of his friend, the astronomer Lemonnier. She was only 16. Contrary to what may be surmised, it was she who insisted on the marriage. She admired him immensely and was determined to bring him happiness and to restore his health. He gradually regained all his intellectual powers and his interest in his work. In 1795 he was appointed professor of mathematics at the newly established Ecole Normale, and when the Ecole Polytechnique was created in 1797, he became its first professor.

Napoleon had the greatest admiration for Lagrange: he treated him as a friend, gave him the title of count, and made him a senator and a high officer of the Legion of Honor.

In 1810, Lagrange started the revision of his "Mécanique analytique," but he was not able to achieve this last task. If his mind had regained all its power, his health never returned completely. He died April 10, 1813, at the age of 76, and was buried in the Pantheon with all honors due to the great.

Exercises

1 Show that the following sets of vectors in $\mathbf{R}^2$ are linearly dependent.

a $\left\{\begin{bmatrix}2\\3\end{bmatrix}, \begin{bmatrix}1\\4\end{bmatrix}, \begin{bmatrix}-1\\2\end{bmatrix}\right\}$ **b** $\left\{\begin{bmatrix}5\\6\end{bmatrix}, \begin{bmatrix}10\\12\end{bmatrix}\right\}$

2 Show that the following sets of vectors in $\mathbf{R}^2$ are linearly independent.

a $\left\{\begin{bmatrix} 1 \\ -1 \end{bmatrix}, \begin{bmatrix} 3 \\ 4 \end{bmatrix}\right\}$ **b** $\left\{\begin{bmatrix} 3 \\ -1 \end{bmatrix}, \begin{bmatrix} 1 \\ 3 \end{bmatrix}\right\}$

3 Determine which of the following sets of vectors in $\mathbf{R}^3$ are linearly dependent.

a $\left\{\begin{bmatrix} 1 \\ 2 \\ 3 \end{bmatrix}, \begin{bmatrix} 2 \\ 1 \\ -1 \end{bmatrix}, \begin{bmatrix} 0 \\ 1 \\ 3 \end{bmatrix}, \begin{bmatrix} -2 \\ -1 \\ 0 \end{bmatrix}\right\}$

b $\left\{\begin{bmatrix} -1 \\ 2 \\ 3 \end{bmatrix}, \begin{bmatrix} 5 \\ 0 \\ 2 \end{bmatrix}, \begin{bmatrix} 3 \\ 4 \\ 8 \end{bmatrix}\right\}$

c $\left\{\begin{bmatrix} 2 \\ -1 \\ 1 \end{bmatrix}, \begin{bmatrix} 3 \\ 0 \\ 4 \end{bmatrix}\right\}$

d $\left\{\begin{bmatrix} 3 \\ 1 \\ 2 \end{bmatrix}, \begin{bmatrix} 9 \\ 3 \\ 6 \end{bmatrix}\right\}$

4 Show that the set of vectors $\{\sin^2 x, \cos^2 x, \cos 2x\}$ in the vector space F of all real-valued functions defined on the real line is linearly dependent.

5 Show that the set of vectors $\{e^x, e^{-x}, e^{2x}, e^{-2x}\}$ in the vector space F of all real-valued functions is linearly independent.

6 Determine which of the following sets of vectors in P_2 are linearly dependent.

a $\{2x^2 - x - 5,\ 3x + 1,\ x^2 + 2x - 3\}$
b $\{x^2 - 1,\ x^2 + 5,\ 5x^2 + 7\}$
c $\{x^2 + 4x,\ 3x^2 + 12x\}$

7 Show that the set of matrices $\left\{\begin{bmatrix} 1 & 0 \\ 0 & 0 \end{bmatrix}, \begin{bmatrix} 0 & 1 \\ 0 & 0 \end{bmatrix}, \begin{bmatrix} 0 & 0 \\ 1 & 0 \end{bmatrix}, \begin{bmatrix} 0 & 0 \\ 0 & 1 \end{bmatrix}\right\}$ in the vector space M_2 is linearly independent.

8 Prove that each of the following is a linearly dependent set.

a any set of three vectors in $\mathbf{R}^2$
b any set of four vectors in $\mathbf{R}^3$
c any set of five vectors in $\mathbf{R}^4$
d any set of five 2×2 matrices in M_2

9 Let $S = \{\mathbf{v}_1, \mathbf{v}_2, \ldots, \mathbf{v}_m\}$ be a set of vectors in $\mathbf{R}^n$. Show that if $m > n$ then S is linearly dependent.

10 Let $\{\mathbf{v}_1,\mathbf{v}_2,\mathbf{v}_3,\mathbf{v}_4\}$ be a linearly independent set of vectors. Show that the subsets $\{\mathbf{v}_1,\mathbf{v}_2,\mathbf{v}_3\}$, $\{\mathbf{v}_1,\mathbf{v}_2\}$, and $\{\mathbf{v}_1\}$ are also linearly independent.

11 Prove Theorem 3-3 in its complete generality.

12 Let $\{\mathbf{v}_1,\mathbf{v}_2,\mathbf{v}_3\}$ be a linearly dependent set of vectors in a vector space V. Let $\mathbf{v}_4$ be a vector in V which is different from $\mathbf{v}_1$, $\mathbf{v}_2$, and $\mathbf{v}_3$. Show that the set $\{\mathbf{v}_1,\mathbf{v}_2,\mathbf{v}_3,\mathbf{v}_4\}$ is also linearly dependent.

13 Show that the set of vectors $\{1, x, x^2, \ldots, x^n\}$ in the vector space F is linearly independent.

3-2 Spanning Sets for Vector Spaces

Objectives

1 Introduce the concept of a spanning set for a vector space.
2 Introduce the concept of a finite-dimensional vector space.

Consider the vectors $\begin{bmatrix}1\\0\\0\end{bmatrix}$ and $\begin{bmatrix}0\\1\\0\end{bmatrix}$ in $\mathbf{R}^3$. Suppose we construct a set S which contains all possible linear combinations of $\begin{bmatrix}1\\0\\0\end{bmatrix}$ and $\begin{bmatrix}0\\1\\0\end{bmatrix}$. Thus,

$$S = \left\{\alpha_1\begin{bmatrix}1\\0\\0\end{bmatrix} + \alpha_2\begin{bmatrix}0\\1\\0\end{bmatrix} : \alpha_1, \alpha_2 \in \mathbf{R}\right\}$$

The set S is obviously a subset of the vector space $\mathbf{R}^3$. A natural question arises: Is the subset S also a subspace of $\mathbf{R}^3$? To answer this question we use our previously established criteria for a subspace. We must therefore check whether S is closed under addition and scalar multiplication defined on $\mathbf{R}^3$. Let

$$\alpha_1\begin{bmatrix}1\\0\\0\end{bmatrix} + \alpha_2\begin{bmatrix}0\\1\\0\end{bmatrix} \quad \text{and} \quad \beta_1\begin{bmatrix}1\\0\\0\end{bmatrix} + \beta_2\begin{bmatrix}0\\1\\0\end{bmatrix}$$

be any two elements in S. Then

$$\left(\alpha_1\begin{bmatrix}1\\0\\0\end{bmatrix} + \alpha_2\begin{bmatrix}0\\1\\0\end{bmatrix}\right) + \left(\beta_1\begin{bmatrix}1\\0\\0\end{bmatrix} + \beta_2\begin{bmatrix}0\\1\\0\end{bmatrix}\right)$$
$$= (\alpha_1 + \beta_1)\begin{bmatrix}1\\0\\0\end{bmatrix} + (\alpha_2 + \beta_2)\begin{bmatrix}0\\1\\0\end{bmatrix}$$

Since

$$(\alpha_1 + \beta_1)\begin{bmatrix}1\\0\\0\end{bmatrix} + (\alpha_2 + \beta_2)\begin{bmatrix}0\\1\\0\end{bmatrix}$$

is obviously a linear combination of $\begin{bmatrix}1\\0\\0\end{bmatrix}$ and $\begin{bmatrix}0\\1\\0\end{bmatrix}$, it must be an element of S. Thus, S is closed under addition.

We now check whether S is closed under scalar multiplication. Let

$$\alpha_1\begin{bmatrix}1\\0\\0\end{bmatrix} + \alpha_2\begin{bmatrix}0\\1\\0\end{bmatrix}$$

be an element of S and let γ be an arbitrary scalar. Then

$$\gamma\left\{\alpha_1\begin{bmatrix}1\\0\\0\end{bmatrix} + \alpha_2\begin{bmatrix}0\\1\\0\end{bmatrix}\right\} = (\gamma\alpha_1)\begin{bmatrix}1\\0\\0\end{bmatrix} + (\gamma\alpha_2)\begin{bmatrix}0\\1\\0\end{bmatrix}$$

Since

$$(\gamma\alpha_1)\begin{bmatrix}1\\0\\0\end{bmatrix} + (\gamma\alpha_2)\begin{bmatrix}0\\1\\0\end{bmatrix}$$

is an element of S, it follows that S is closed under scalar multiplication.

We conclude that S is a subspace of $\mathbf{R}^3$. We describe this fact also by saying that S is the subspace of $\mathbf{R}^3$ **spanned** by the vectors $\begin{bmatrix}1\\0\\0\end{bmatrix}$ and $\begin{bmatrix}0\\1\\0\end{bmatrix}$. We have the following general result:

Theorem 3-4 Let $\{\mathbf{v}_1, \mathbf{v}_2, \ldots, \mathbf{v}_n\}$ be a subset of a vector space V. Then the set S of all linear combinations of $\mathbf{v}_1, \mathbf{v}_2, \ldots, \mathbf{v}_n$ is a subspace of V.

Proof Let

$$\alpha_1\mathbf{v}_1 + \alpha_2\mathbf{v}_2 + \cdots + \alpha_n\mathbf{v}_n$$

and

$$\beta_1\mathbf{v}_1 + \beta_2\mathbf{v}_2 + \cdots + \beta_n\mathbf{v}_n$$

be any two elements of S. Then

$$(\alpha_1\mathbf{v}_1 + \alpha_2\mathbf{v}_2 + \cdots + \alpha_n\mathbf{v}_n) + (\beta_1\mathbf{v}_1 + \alpha_2\mathbf{v}_2 + \cdots + \beta_n\mathbf{v}_n)$$
$$= (\alpha_1 + \beta_1)\mathbf{v}_1 + (\alpha_2 + \beta_2)\mathbf{v}_2 + \cdots + (\alpha_n + \beta_n)\mathbf{v}_n$$

which implies that S is closed under addition. We also have

$$\gamma(\alpha_1\mathbf{v}_1 + \alpha_2\mathbf{v}_2 + \cdots + \alpha_n\mathbf{v}_n) = (\gamma\alpha_1)\mathbf{v}_1 + (\gamma\alpha_2)\mathbf{v}_2 + \cdots + (\gamma\alpha_n)\mathbf{v}_n$$

for any scalar γ, which implies that S is closed under scalar multiplication. Hence, S is a subspace of the vector space V.

Definition 3-4 Let $\{\mathbf{v}_1, \mathbf{v}_2, \ldots, \mathbf{v}_n\}$ be a subset of a vector space V. Let S be the subspace of V consisting of all linear combinations of $\mathbf{v}_1, \mathbf{v}_2, \ldots, \mathbf{v}_n$. We say that **$S$ is the subspace of V spanned by** $\mathbf{v}_1, \mathbf{v}_2, \ldots, \mathbf{v}_n$.

We also say that $\{\mathbf{v}_1, \mathbf{v}_2, \ldots, \mathbf{v}_n\}$ is a spanning set for the subspace S.

We have seen that the set of all linear combinations of the vectors $\begin{bmatrix}1\\0\\0\end{bmatrix}$ and $\begin{bmatrix}0\\1\\0\end{bmatrix}$ is a subspace of $\mathbf{R}^3$. Can this subspace be identical to $\mathbf{R}^3$? In other words, does the set $\left\{\begin{bmatrix}1\\0\\0\end{bmatrix}, \begin{bmatrix}0\\1\\0\end{bmatrix}\right\}$ span the vector space $\mathbf{R}^3$? In order to answer this question, we take an arbitrary vector $\begin{bmatrix}a\\b\\c\end{bmatrix}$ in $\mathbf{R}^3$ and try to determine if some linear combination of $\begin{bmatrix}1\\0\\0\end{bmatrix}$ and $\begin{bmatrix}0\\1\\0\end{bmatrix}$ will generate the vector $\begin{bmatrix}a\\b\\c\end{bmatrix}$. Thus, we write

$$\alpha_1\begin{bmatrix}1\\0\\0\end{bmatrix} + \alpha_2\begin{bmatrix}0\\1\\0\end{bmatrix} = \begin{bmatrix}a\\b\\c\end{bmatrix}$$

and try to determine α_1 and α_2 (in terms of a, b, c) such that the equation above is satisfied. Performing the operations on the left-hand side of the equation yields

$$\begin{bmatrix}\alpha_1\\\alpha_2\\0\end{bmatrix} = \begin{bmatrix}a\\b\\c\end{bmatrix}$$

The conclusion is therefore that if $c \neq 0$ then the equation

$$\alpha_1 \begin{bmatrix} 1 \\ 0 \\ 0 \end{bmatrix} + \alpha_2 \begin{bmatrix} 0 \\ 1 \\ 0 \end{bmatrix} = \begin{bmatrix} a \\ b \\ c \end{bmatrix}$$

cannot be satisfied by any scalars α_1 and α_2. Thus, none of the vectors $\begin{bmatrix} a \\ b \\ c \end{bmatrix}$ in $\mathbf{R}^3$ with $c \neq 0$ can be generated by linear combinations of $\begin{bmatrix} 1 \\ 0 \\ 0 \end{bmatrix}$ and $\begin{bmatrix} 0 \\ 1 \\ 0 \end{bmatrix}$. We conclude that $\mathbf{R}^3$ is not spanned by the vectors $\begin{bmatrix} 1 \\ 0 \\ 0 \end{bmatrix}$ and $\begin{bmatrix} 0 \\ 1 \\ 0 \end{bmatrix}$.

If we enlarge the set $\left\{\begin{bmatrix} 1 \\ 0 \\ 0 \end{bmatrix}, \begin{bmatrix} 0 \\ 0 \\ 1 \end{bmatrix}\right\}$ by adding the vector $\begin{bmatrix} 0 \\ 0 \\ 1 \end{bmatrix}$ we get the set $\left\{\begin{bmatrix} 1 \\ 0 \\ 0 \end{bmatrix}, \begin{bmatrix} 0 \\ 1 \\ 0 \end{bmatrix}, \begin{bmatrix} 0 \\ 0 \\ 1 \end{bmatrix},\right\}$ which spans $\mathbf{R}^3$. To see this, we must show that an arbitrary vector $\begin{bmatrix} a \\ b \\ c \end{bmatrix}$ of $\mathbf{R}^3$ is generated by $\begin{bmatrix} 1 \\ 0 \\ 0 \end{bmatrix}$, $\begin{bmatrix} 0 \\ 1 \\ 0 \end{bmatrix}$, and $\begin{bmatrix} 0 \\ 0 \\ 1 \end{bmatrix}$. Let us find α_1, α_2, and α_3 such that

$$\alpha_1 \begin{bmatrix} 1 \\ 0 \\ 0 \end{bmatrix} + \alpha_2 \begin{bmatrix} 0 \\ 1 \\ 0 \end{bmatrix} + \alpha_3 \begin{bmatrix} 0 \\ 0 \\ 1 \end{bmatrix} = \begin{bmatrix} a \\ b \\ c \end{bmatrix}$$

Performing the operations indicated on the left-hand side of the equation yields

$$\begin{bmatrix} \alpha_1 \\ \alpha_2 \\ \alpha_3 \end{bmatrix} = \begin{bmatrix} a \\ b \\ c \end{bmatrix}$$

Therefore, every vector $\begin{bmatrix} a \\ b \\ c \end{bmatrix}$ of $\mathbf{R}^3$ is a linear combination of the vectors $\begin{bmatrix} 1 \\ 0 \\ 0 \end{bmatrix}$, $\begin{bmatrix} 0 \\ 1 \\ 0 \end{bmatrix}$, and $\begin{bmatrix} 0 \\ 0 \\ 1 \end{bmatrix}$ because we have

$$\begin{bmatrix} a \\ b \\ c \end{bmatrix} = a\begin{bmatrix} 1 \\ 0 \\ 0 \end{bmatrix} + b\begin{bmatrix} 0 \\ 1 \\ 0 \end{bmatrix} + c\begin{bmatrix} 0 \\ 0 \\ 1 \end{bmatrix}$$

The problem of determining whether a given subset of a vector space V is a spanning set of V plays a significant role in studying the structure of vector spaces. To provide more insight into this problem, we discuss some examples.

Example 1 Consider the subset $S = \left\{\begin{bmatrix} 1 \\ 2 \end{bmatrix}, \begin{bmatrix} 2 \\ 1 \end{bmatrix}, \begin{bmatrix} 3 \\ 4 \end{bmatrix}\right\}$ of $\mathbf{R}^2$. Determine whether S spans the vector space $\mathbf{R}^2$.

Solution Let $\begin{bmatrix} a \\ b \end{bmatrix}$ be an arbitrary element of $\mathbf{R}^2$. We will try to determine if $\begin{bmatrix} a \\ b \end{bmatrix}$ can be expressed as a linear combination of $\begin{bmatrix} 1 \\ 2 \end{bmatrix}$, $\begin{bmatrix} 2 \\ 1 \end{bmatrix}$, and $\begin{bmatrix} 3 \\ 4 \end{bmatrix}$. We write the equation

$$\alpha_1\begin{bmatrix} 1 \\ 2 \end{bmatrix} + \alpha_2\begin{bmatrix} 2 \\ 1 \end{bmatrix} + \alpha_3\begin{bmatrix} 3 \\ 4 \end{bmatrix} = \begin{bmatrix} a \\ b \end{bmatrix}$$

and try to find a solution for α_1, α_2, and α_3 (in terms of a and b). The vector equation can be written as a system

$$\begin{aligned} \alpha_1 + 2\alpha_2 + 3\alpha_3 &= a \\ 2\alpha_1 + \alpha_2 + 4\alpha_4 &= b \end{aligned}$$

The augmented matrix of the system is given by

$$\left[\begin{array}{ccc:c} 1 & 2 & 3 & a \\ 2 & 1 & 4 & b \end{array}\right]$$

The corresponding row-reduced echelon matrix is given by

$$\left[\begin{array}{cc|c|c} 1 & 0 & \frac{5}{3} & \frac{2b-a}{3} \\ 0 & 1 & \frac{2}{3} & \frac{2a-b}{3} \end{array}\right]$$

Thus, the solution is given by

$$\alpha_1 = \frac{2b-a}{3} - \frac{5}{3}\alpha_3$$

$$\alpha_2 = \frac{2a-b}{3} - \frac{2}{3}\alpha_3$$

Since we need just *one* solution for α_1, α_2, and α_3, let us choose $\alpha_3 = 0$. This yields $\alpha_1 = \frac{2b-a}{3}$ and $\alpha_2 = \frac{2a-b}{3}$. Thus, the vector $\begin{bmatrix} a \\ b \end{bmatrix}$ may be expressed as a linear combination of the vectors $\begin{bmatrix} 1 \\ 2 \end{bmatrix}$, $\begin{bmatrix} 2 \\ 1 \end{bmatrix}$, and $\begin{bmatrix} 3 \\ 4 \end{bmatrix}$ by the equation

$$\frac{2b-a}{3}\begin{bmatrix} 1 \\ 2 \end{bmatrix} + \frac{2a-b}{3}\begin{bmatrix} 2 \\ 1 \end{bmatrix} + 0\begin{bmatrix} 3 \\ 4 \end{bmatrix} = \begin{bmatrix} a \\ b \end{bmatrix}$$

We conclude that the set $S = \left\{\begin{bmatrix} 1 \\ 2 \end{bmatrix}, \begin{bmatrix} 2 \\ 1 \end{bmatrix}, \begin{bmatrix} 3 \\ 4 \end{bmatrix}\right\}$ spans $\mathbf{R}^2$.

Example 2 Consider the subset $S = \left\{\begin{bmatrix} 1 \\ 1 \\ 1 \end{bmatrix}, \begin{bmatrix} 1 \\ 2 \\ 2 \end{bmatrix}, \begin{bmatrix} 1 \\ 3 \\ 3 \end{bmatrix}, \begin{bmatrix} 1 \\ 4 \\ 4 \end{bmatrix}, \begin{bmatrix} 1 \\ 5 \\ 6 \end{bmatrix}\right\}$ of $\mathbf{R}^3$. Determine whether S spans the vector space $\mathbf{R}^3$.

Solution Let $\begin{bmatrix} a \\ b \\ c \end{bmatrix}$ be an arbitrary vector in $\mathbf{R}^3$. We must determine if the vector equation

$$\alpha_1\begin{bmatrix} 1 \\ 1 \\ 1 \end{bmatrix} + \alpha_2\begin{bmatrix} 1 \\ 2 \\ 2 \end{bmatrix} + \alpha_3\begin{bmatrix} 1 \\ 3 \\ 3 \end{bmatrix} + \alpha_4\begin{bmatrix} 1 \\ 4 \\ 4 \end{bmatrix} + \alpha_5\begin{bmatrix} 1 \\ 5 \\ 6 \end{bmatrix} = \begin{bmatrix} a \\ b \\ c \end{bmatrix}$$

has a solution for α_1, α_2, α_3, α_4, and α_5 in terms of a, b, and c. The vector equation can be written as a system:

$$\begin{bmatrix} 1 & 1 & 1 & 1 & 1 \\ 1 & 2 & 3 & 4 & 5 \\ 1 & 2 & 3 & 4 & 6 \end{bmatrix} \begin{bmatrix} \alpha_1 \\ \alpha_2 \\ \alpha_3 \\ \alpha_4 \\ \alpha_5 \end{bmatrix} = \begin{bmatrix} a \\ b \\ c \end{bmatrix}$$

Row-reducing the augmented matrix of the system, we get

$$\left[\begin{array}{ccccc|c} 1 & 0 & -1 & -2 & 0 & 2a - 4b + 3c \\ 0 & 1 & 2 & 3 & 0 & -a + 5b - 4c \\ 0 & 0 & 0 & 0 & 1 & -b \quad + c \end{array}\right]$$

The solution of the corresponding system is given by

$$\begin{aligned} \alpha_1 &= \quad 2a - 4b + 3c + \alpha_3 + 2\alpha_4 \\ \alpha_2 &= -\ a + 5b - 4c - 2\alpha_3 - 3\alpha_4 \\ \alpha_5 &= -\ b + c \end{aligned}$$

The choice $\alpha_3 = 0$ and $\alpha_4 = 0$ leads to $\alpha_1 = 2a - 4b + 3c$, $\alpha_2 = -a + 5b - 4c$, and $\alpha_5 = -b + c$. We conclude that the set

$$S = \left\{ \begin{bmatrix} 1 \\ 1 \\ 1 \end{bmatrix}, \begin{bmatrix} 1 \\ 2 \\ 2 \end{bmatrix}, \begin{bmatrix} 1 \\ 3 \\ 3 \end{bmatrix}, \begin{bmatrix} 1 \\ 4 \\ 4 \end{bmatrix}, \begin{bmatrix} 1 \\ 5 \\ 6 \end{bmatrix} \right\} \text{ spans } \mathbf{R}^3$$

The reader can easily check that indeed we have

$$[2a - 4b + 3c] \begin{bmatrix} 1 \\ 1 \\ 1 \end{bmatrix} +$$

$$[-a + 5b - 4c] \begin{bmatrix} 1 \\ 2 \\ 2 \end{bmatrix} + 0 \begin{bmatrix} 1 \\ 3 \\ 3 \end{bmatrix} +$$

$$0 \begin{bmatrix} 1 \\ 4 \\ 4 \end{bmatrix} + [-b + c] \begin{bmatrix} 1 \\ 5 \\ 6 \end{bmatrix} = \begin{bmatrix} a \\ b \\ c \end{bmatrix}$$

Example 3 Let $S = \{x^2 - x + 1, 2x + 1\}$ be a subset of P_2. Determine whether S spans P_2.

Solution Let $ax^2 + bx + c$ be an arbitrary vector in P_2. We must determine if the equation

$$\alpha_1(x^2 - x + 1) + \alpha_2(2x + 1) = ax^2 + bx + c$$

has a solution for α_1 and α_2.

Note that there is no condition on a, b, c except the fact that they are real numbers.

Comparing coefficients of the same powers of x, we get

$$\begin{aligned} \alpha_1 \qquad &= a \\ -\alpha_1 + 2\alpha_2 &= b \\ \alpha_1 + \alpha_2 &= c \end{aligned}$$

or

$$\begin{bmatrix} 1 & 0 \\ -1 & 2 \\ 1 & 1 \end{bmatrix} \begin{bmatrix} \alpha_1 \\ \alpha_2 \end{bmatrix} = \begin{bmatrix} a \\ b \\ c \end{bmatrix}$$

The augmented matrix of the system reduces to

$$\begin{bmatrix} 1 & 0 & \vdots & a \\ 0 & 1 & \vdots & c - a \\ 0 & 0 & \vdots & 3a + b - 2c \end{bmatrix}$$

Therefore, the corresponding system of equations does not have a solution unless $3a + b - 2c = 0$. This means that when $3a + b - 2c \neq 0$, it is impossible to generate $ax^2 + bx + c$ by any linear combination of $x^2 - x + 1$ and $2x + 1$. We conclude that the set $S = \{x^2 - x + 1, 2x + 1\}$ does *not* span the vector space P_2. In other words, the subspace of P_2 spanned by $S = \{x^2 - x + 1, 2x + 1\}$ does *not* coincide with P_2.

Example 4 Let P_n be the vector space consisting of the zero polynomial and all polynomials of degree not exceeding n. Since an arbitrary element of P_n has the form $a_n x^n + \cdots + a_1 x + a_0$ it is obvious that the set $\{1, x, \ldots, x^n\}$ spans P_n.

We omit the proof of these facts at this point, but will deal with them in a later section after we have acquired some more knowledge about properties of the so-called *finite-dimensional* vector spaces.

It is important to note that there exist vector spaces which are *not* spanned by any of their *finite* subsets. Such an example is given by the vector space F of all real-valued functions defined on the real line, or the vector space P_∞ of *all* polynomials, which is a subspace of F.

Definition 3-5

Definition of finite-dimensional vector space.

Let V be a vector space. If V has a finite subset $\{\mathbf{v}_1, \mathbf{v}_2, \ldots, \mathbf{v}_n\}$ which spans V, then V is called a **finite-dimensional vector space.** If *every finite* subset of V *fails* to span V, then V is called an **infinite-dimensional vector space.**

The vector spaces $\mathbf{R}^2$, $\mathbf{R}^3$, P_2, P_n discussed in Examples 1, 2, 3, and 4 are all finite-dimensional vector spaces.

The vector space F of all real-valued functions defined on the real line, and the vector space P_∞ of all polynomials, are infinite-dimensional vector spaces.

Unless otherwise specified, the term vector space shall always mean a finite-dimensional vector space.

Exercises

1 Consider the vectors $\mathbf{v}_1 = \begin{bmatrix} 1 \\ -1 \\ 2 \end{bmatrix}$ and $\mathbf{v}_2 = \begin{bmatrix} 2 \\ 1 \\ -1 \end{bmatrix}$ in $\mathbf{R}^3$. Determine which of the following are linear combinations of $\mathbf{v}_1$ and $\mathbf{v}_2$:

a $\begin{bmatrix} 3 \\ 0 \\ 2 \end{bmatrix}$ **b** $\begin{bmatrix} 5 \\ -2 \\ 5 \end{bmatrix}$ **c** $\begin{bmatrix} -3 \\ 0 \\ 1 \end{bmatrix}$ **d** $\begin{bmatrix} 0 \\ 0 \\ 0 \end{bmatrix}$ **e** $\begin{bmatrix} -6 \\ 0 \\ -3 \end{bmatrix}$

2 Consider the vectors $\mathbf{v}_1 = \begin{bmatrix}1\\1\\1\end{bmatrix}$, $\mathbf{v}_2 = \begin{bmatrix}1\\0\\1\end{bmatrix}$, and $\mathbf{v}_3 = \begin{bmatrix}1\\2\\0\end{bmatrix}$ in $\mathbf{R}^3$.

Express each of the following vectors as a linear combination of $\mathbf{v}_1$, $\mathbf{v}_2$, and $\mathbf{v}_3$:

a $\begin{bmatrix}1\\0\\0\end{bmatrix}$ **b** $\begin{bmatrix}0\\1\\0\end{bmatrix}$ **c** $\begin{bmatrix}0\\0\\1\end{bmatrix}$ **d** $\begin{bmatrix}0\\0\\0\end{bmatrix}$ **e** $\begin{bmatrix}1\\2\\3\end{bmatrix}$

3 Show that the vectors $\mathbf{v}_1$, $\mathbf{v}_2$, and $\mathbf{v}_3$ in problem 2 span the vector space $\mathbf{R}^3$.

4 Consider the vectors $\mathbf{p}_1 = 2x^2 - x + 1$, $\mathbf{p}_2 = x^2 + x + 1$, and $\mathbf{p}_3 = x^2 + 2x + 3$ in P_2. Express each of the following vectors as a linear combination of $\mathbf{p}_1$, $\mathbf{p}_2$, and $\mathbf{p}_3$:

a $-x^2 - 2x - 1$ **b** $3x^2 + 2x - 4$

c $5x^2 - x + 3$ **d** $2x^2 + 3x$

5 Show that the vectors $\mathbf{p}_1$, $\mathbf{p}_2$, and $\mathbf{p}_3$ in problem 4 span the vector space P_2.

6 Show that $\mathbf{p}_1 = 2x^2 - x + 1$ and $\mathbf{p}_2 = x^2 + x + 1$ do *not* span P_2.

7 Let S be the subspace of P_2 spanned by $\mathbf{p}_1$ and $\mathbf{p}_2$ in problem 6. Find a vector $\mathbf{p}$ such that $\mathbf{p} \in P_2$ and $\mathbf{p} \notin S$.

8 Consider the elements $\mathbf{A}_1 = \begin{bmatrix}1 & 1\\1 & 1\end{bmatrix}$, $\mathbf{A}_2 = \begin{bmatrix}1 & -1\\1 & 0\end{bmatrix}$, $\mathbf{A}_3 = \begin{bmatrix}1 & 0\\0 & 1\end{bmatrix}$, $\mathbf{A}_4 = \begin{bmatrix}0 & 1\\-1 & 0\end{bmatrix}$, and $\mathbf{A}_5 = \begin{bmatrix}1 & 2\\3 & 4\end{bmatrix}$ of the vector space M_2 of all 2×2 matrices.

a Show that the set $\{\mathbf{A}_1,\mathbf{A}_2,\mathbf{A}_3,\mathbf{A}_4,\mathbf{A}_5\}$ spans M_2.

b Show that $\mathbf{A}_5$ is a linear combination of $\mathbf{A}_1$, $\mathbf{A}_2$, $\mathbf{A}_3$, and $\mathbf{A}_4$.

c Show that M_2 is also spanned by the set $\{\mathbf{A}_1,\mathbf{A}_2,\mathbf{A}_3,\mathbf{A}_4\}$.

9 Show that if a vector space V is spanned by a set of vectors $\{\mathbf{v}_1, \mathbf{v}_2, \ldots, \mathbf{v}_k, \mathbf{v}_{k+1}\}$ and $\mathbf{v}_{k+1}$ is a linear combination of $\{\mathbf{v}_1, \mathbf{v}_2, \ldots, \mathbf{v}_k\}$, then V is spanned by $\{\mathbf{v}_1, \mathbf{v}_2, \ldots, \mathbf{v}_k\}$.

10 **a** Show that the set $S = \left\{\begin{bmatrix}1\\2\end{bmatrix}, \begin{bmatrix}3\\2\end{bmatrix}\right\}$ spans the vector space $\mathbf{R}^2$.

b Construct a set containing three vectors that spans $\mathbf{R}^2$.

c Construct a set containing five vectors that spans $\mathbf{R}^2$.

11 Let $\{\mathbf{v}_1, \mathbf{v}_2, \ldots, \mathbf{v}_n\}$ span the vector space V and let $\mathbf{v}_{n+1} \in V$. Show that the expanded set $\{\mathbf{v}_1, \mathbf{v}_2, \ldots, \mathbf{v}_{n+1}\}$ also spans V.

12 **a** Construct an example of a set S that spans $\mathbf{R}^2$ such that the deletion of any one vector from S results in a set which does not span R^2.

b Construct an example of a set S that spans P_2 such that the deletion of any one vector from S results in a set which does not span P_2.

3-3 Basis for a Vector Space

Objectives
1. Introduce the concept of a basis for a vector space.
2. Show that the number of vectors in a linearly independent set can not exceed the number of vectors in a basis for a finite dimensional vector space.

A vector space V which contains more than one vector will have many subsets which are spanning sets for V. For example, the vector space $\mathbf{R}^2$ may have spanning sets containing two vectors, three vectors, or five vectors. Actually, one can show that if n is any integer such that $n \geq 2$, then there exists a set containing exactly n vectors which spans $\mathbf{R}^2$. The reader can easily check that the sets

See problem 10 of Section 3–2.

$$\left\{\begin{bmatrix} 1 \\ -1 \end{bmatrix}, \begin{bmatrix} 3 \\ 2 \end{bmatrix}\right\}, \left\{\begin{bmatrix} 1 \\ 2 \end{bmatrix}, \begin{bmatrix} 2 \\ 1 \end{bmatrix}, \begin{bmatrix} 3 \\ 3 \end{bmatrix}\right\}, \left\{\begin{bmatrix} -1 \\ 1 \end{bmatrix}, \begin{bmatrix} 5 \\ 3 \end{bmatrix}, \begin{bmatrix} 1 \\ 4 \end{bmatrix}, \begin{bmatrix} -2 \\ 1 \end{bmatrix}\right\}$$

are all spanning sets for $\mathbf{R}^2$. But while the set $\left\{\begin{bmatrix} 1 \\ -1 \end{bmatrix}, \begin{bmatrix} 3 \\ 2 \end{bmatrix}\right\}$ is linearly independent, the other two sets are linearly dependent. This fact makes the set $\left\{\begin{bmatrix} 1 \\ -1 \end{bmatrix}, \begin{bmatrix} 3 \\ 2 \end{bmatrix}\right\}$ something special, a spanning set and a linearly independent set. We refer to the set $\left\{\begin{bmatrix} 1 \\ -1 \end{bmatrix}, \begin{bmatrix} 3 \\ 2 \end{bmatrix}\right\}$ as a *basis* for the vector space $\mathbf{R}^2$.

The following definition deals with the concept of basis for a vector space V.

Definition 3-6 Let V be a vector space. Let $S = \{\mathbf{v}_1, \mathbf{v}_2, \ldots, \mathbf{v}_k\}$ be a finite subset of V. Then S is called a **basis** for V, provided the following conditions are satisfied:

i S is a linearly independent set

ii S spans the vector space V

Remember: Any finite subset of a vector space V which contains the zero vector is linearly dependent. In particular, a vector space consisting of *one vector only* (namely the zero vector) does not have a basis.

Since any basis for a vector space V must be a linearly independent set, it will *never* contain the zero vector V.

We wish to emphasize that a basis (when it exists) is never unique. In the following examples, we show different bases for every vector space given.

Example 1 Let $\mathbf{R}^1$ be the vector space of all real numbers under standard addition and scalar multiplication. In this case real numbers play the dual roles of vectors and scalars. The real number 1 can serve as a basis for $\mathbf{R}^1$ because $\{1\}$ is a linearly independent set which also spans $\mathbf{R}^1$. The linear independence follows from the fact that

if $\alpha \cdot 1 = 0$ then $\alpha = 0$

The spanning property follows from the fact that if a is an arbitrary element of $\mathbf{R}^1$ then the equation

$\alpha \cdot 1 = a$ has a solution, namely, $\alpha = a$

In a similar fashion, if we choose any real number $b \neq 0$ then $\{b\}$ is also a linearly independent set which spans $\mathbf{R}^1$. Thus, $\mathbf{R}^1$ has infinitely many bases which consist of sets having one element each.

Example 2 The set $\left\{\begin{bmatrix}1\\0\end{bmatrix}, \begin{bmatrix}0\\1\end{bmatrix}\right\}$ is the so-called **standard basis** for the vector space $\mathbf{R}^2$. The set $\left\{\begin{bmatrix}1\\0\end{bmatrix}, \begin{bmatrix}0\\1\end{bmatrix}\right\}$ is linearly independent because the equation

$$\alpha_1\begin{bmatrix}1\\0\end{bmatrix} + \alpha_2\begin{bmatrix}0\\1\end{bmatrix} = \begin{bmatrix}0\\0\end{bmatrix} \text{ implies } \alpha_1 = 0 \quad \text{and} \quad \alpha_2 = 0$$

The set $\left\{\begin{bmatrix}1\\0\end{bmatrix}, \begin{bmatrix}0\\1\end{bmatrix}\right\}$ spans $\mathbf{R}^2$ because if $\begin{bmatrix}a\\b\end{bmatrix}$ is an arbitrary vector in $\mathbf{R}^2$ then the equation

$$\alpha_1\begin{bmatrix}1\\0\end{bmatrix} + \alpha_2\begin{bmatrix}0\\1\end{bmatrix} = \begin{bmatrix}a\\b\end{bmatrix}$$

has the solution $\alpha_1 = a$ and $\alpha_2 = b$.

The set $\left\{\begin{bmatrix}1\\1\end{bmatrix}, \begin{bmatrix}1\\2\end{bmatrix}\right\}$ is also a basis for $\mathbf{R}^2$. The linear independence follows from the fact that the vector equation

$$\alpha_1\begin{bmatrix}1\\1\end{bmatrix} + \alpha_2\begin{bmatrix}1\\2\end{bmatrix} = \begin{bmatrix}0\\0\end{bmatrix}$$

has the unique solution $\alpha_1 = \alpha_2 = 0$.

To prove that $\left\{\begin{bmatrix}1\\1\end{bmatrix}, \begin{bmatrix}1\\2\end{bmatrix}\right\}$ spans $\mathbf{R}^2$ we let $\begin{bmatrix}a\\b\end{bmatrix}$ be an arbitrary vector in $\mathbf{R}^2$ and write

$$\alpha_1\begin{bmatrix}1\\1\end{bmatrix} + \alpha_2\begin{bmatrix}1\\2\end{bmatrix} = \begin{bmatrix}a\\b\end{bmatrix}$$

Solving the system of two equations with the two unknowns, α_1 and

α_2, we get

$$\alpha_1 = 2a - b \qquad \text{and} \qquad \alpha_2 = b - a$$

It can be shown that any two vectors $\begin{bmatrix} c \\ d \end{bmatrix}$ and $\begin{bmatrix} e \\ f \end{bmatrix}$ in $\mathbf{R}^2$ will form a basis for $\mathbf{R}^2$ provided that neither of the two is a scalar multiple of the other vector.

Example 3 Let $M_{2\times 3}$ be the vector space of all 2×3 matrices under standard addition and scalar multiplication. The set

$$S = \left\{ \begin{bmatrix} 1 & 0 & 0 \\ 0 & 0 & 0 \end{bmatrix}, \begin{bmatrix} 0 & 1 & 0 \\ 0 & 0 & 0 \end{bmatrix}, \begin{bmatrix} 0 & 0 & 1 \\ 0 & 0 & 0 \end{bmatrix}, \begin{bmatrix} 0 & 0 & 0 \\ 1 & 0 & 0 \end{bmatrix}, \begin{bmatrix} 0 & 0 & 0 \\ 0 & 1 & 0 \end{bmatrix}, \begin{bmatrix} 0 & 0 & 0 \\ 0 & 0 & 1 \end{bmatrix} \right\}$$

is a basis for $M_{2\times 3}$. To prove linear independence of S we write

$$\alpha_1 \begin{bmatrix} 1 & 0 & 0 \\ 0 & 0 & 0 \end{bmatrix} + \alpha_2 \begin{bmatrix} 0 & 1 & 0 \\ 0 & 0 & 0 \end{bmatrix} + \alpha_3 \begin{bmatrix} 0 & 0 & 1 \\ 0 & 0 & 0 \end{bmatrix} + \alpha_4 \begin{bmatrix} 0 & 0 & 0 \\ 1 & 0 & 0 \end{bmatrix} + \alpha_5 \begin{bmatrix} 0 & 0 & 0 \\ 0 & 1 & 0 \end{bmatrix} + \alpha_6 \begin{bmatrix} 0 & 0 & 0 \\ 0 & 0 & 1 \end{bmatrix} = \begin{bmatrix} 0 & 0 & 0 \\ 0 & 0 & 0 \end{bmatrix}$$

Performing the operations on the left-hand side of the equation yields

$$\begin{bmatrix} \alpha_1 & \alpha_2 & \alpha_3 \\ \alpha_4 & \alpha_5 & \alpha_6 \end{bmatrix} = \begin{bmatrix} 0 & 0 & 0 \\ 0 & 0 & 0 \end{bmatrix}$$

Hence the unique solution is given by

$$\alpha_1 = \alpha_2 = \alpha_3 = \alpha_4 = \alpha_5 = \alpha_6 = 0$$

To prove that S spans $M_{2\times 3}$ we take an arbitrary element $\begin{bmatrix} a_{11} & a_{12} & a_{13} \\ a_{21} & a_{22} & a_{23} \end{bmatrix}$ in $M_{2\times 3}$ and write

$$\alpha_1 \begin{bmatrix} 1 & 0 & 0 \\ 0 & 0 & 0 \end{bmatrix} + \alpha_2 \begin{bmatrix} 0 & 1 & 0 \\ 0 & 0 & 0 \end{bmatrix} + \alpha_3 \begin{bmatrix} 0 & 0 & 1 \\ 0 & 0 & 0 \end{bmatrix} + \alpha_4 \begin{bmatrix} 0 & 0 & 0 \\ 1 & 0 & 0 \end{bmatrix} + \alpha_5 \begin{bmatrix} 0 & 0 & 0 \\ 0 & 1 & 0 \end{bmatrix} + \alpha_6 \begin{bmatrix} 0 & 0 & 0 \\ 0 & 0 & 1 \end{bmatrix} = \begin{bmatrix} a_{11} & a_{12} & a_{13} \\ a_{21} & a_{22} & a_{23} \end{bmatrix}$$

The solution of the equation is given by

$$\alpha_1 = a_{11},\ \alpha_2 = a_{12},\ \alpha_3 = a_{13},\ \alpha_4 = a_{21},\ \alpha_5 = a_{22}, \quad \text{and} \quad \alpha_6 = a_{23}$$

This completes the proof that S is a basis for $M_{2\times 3}$.

One can construct infinitely many more bases for $M_{2\times 3}$. This can be done by simply changing zero entries into nonzero entries in one or more of the elements of S. Thus, the set T defined by

The student is encouraged to show that T is a basis for $M_{2\times 3}$.

$$T = \left\{\begin{bmatrix}1 & 0 & 0\\ 2 & 0 & 0\end{bmatrix}, \begin{bmatrix}0 & 1 & 0\\ 0 & 0 & 0\end{bmatrix}, \begin{bmatrix}0 & 0 & 1\\ 0 & 3 & 0\end{bmatrix}, \begin{bmatrix}0 & 0 & 0\\ 1 & 0 & 0\end{bmatrix}, \begin{bmatrix}0 & 0 & 0\\ 0 & 1 & 0\end{bmatrix}, \begin{bmatrix}0 & 0 & 0\\ 0 & 0 & 1\end{bmatrix}\right\}$$

is also a basis for $M_{2\times 3}$.

Example 4 The set $\{1,x,x^2\}$ is a basis for the vector space P_2. Another basis for P_2 is given by $\{x - 1,\ x + 1,\ x^2 - 2x\}$. First we prove linear independence. We write

$$\alpha_1(x - 1) + \alpha_2(x + 1) + \alpha_3(x^2 - 2x) = 0$$

which is equivalent to

$$\alpha_3 x^2 + (\alpha_1 + \alpha_2 - 2\alpha_3)x + (\alpha_2 - \alpha_1) = 0$$

Since the last equation must be satisfied for all x we have

$$\begin{aligned} \alpha_3 &= 0\\ \alpha_1 + \alpha_2 - 2\alpha_3 &= 0\\ -\alpha_1 + \alpha_2 \qquad &= 0 \end{aligned}$$

Why?

The system of equations possesses the unique solution $\alpha_1 = \alpha_2 = \alpha_3 = 0$. This proves that $\{x - 1,\ x + 1,\ x^2 - 2x\}$ is linearly independent.

To prove the spanning property we let $ax^2 + bx + c$ be an arbitrary element of P_2. We write the equation

$$\alpha_1(x - 1) + \alpha_2(x + 1) + \alpha_3(x^2 - 2x) = ax^2 + bx + c$$

and seek a solution α_1, α_2, and α_3 in terms of a, b, and c. Simplifying the left-hand side of the above equation we get

$$\alpha_3 x^2 + (\alpha_1 + \alpha_2 - 2\alpha_3)x + (-\alpha_1 + \alpha_2) = ax^2 + bx + c$$

Since the equation must be satisfied for all x we must have

$$\begin{aligned} \alpha_3 &= a\\ \alpha_1 + \alpha_2 - 2\alpha_3 &= b\\ -\alpha_1 + \alpha_2 \qquad &= c \end{aligned}$$

The system yields the solution

$$\alpha_1 = \frac{2a + b - c}{2},\ \alpha_2 = \frac{2a + b + c}{2},\ \alpha_3 = a$$

This shows that $\{x - 1,\ x + 1,\ x^2 - 2x\}$ spans P_2. Thus, the set $\{x - 1,\ x + 1,\ x^2 - 2x\}$ is a basis for P_2.

The following theorem will indicate the importance of the concept of a basis in investigating the structure of vector spaces. In fact, a basis for a vector space V determines the representation of each vector in V in terms of the vectors in that basis.

Theorem 3-5 Let $S = \{\mathbf{v}_1, \mathbf{v}_2, \ldots, \mathbf{v}_n\}$ be a basis for the vector space V. Then *every* vector in V has a *unique* representation as a linear combination of the vectors in S.

Proof Since S is a basis for V it must span V. This implies that if $\mathbf{v} \in V$, then there exist scalars $\alpha_1, \alpha_2, \ldots, \alpha_n$ such that

$$\mathbf{v} = \alpha_1\mathbf{v}_1 + \alpha_2\mathbf{v}_2 + \cdots + \alpha_n\mathbf{v}_n$$

To prove that this representation is unique we must show that if $\mathbf{v}$ also has the representation

$$\mathbf{v} = \beta_1\mathbf{v}_1 + \beta_2\mathbf{v}_2 + \cdots + \beta_n\mathbf{v}_n$$

then

$$\alpha_i = \beta_i \text{ for } i = 1, 2, \ldots, n$$

We set the two representations of $\mathbf{v}$ equal to each other, obtaining the equation

$$\alpha_1\mathbf{v}_1 + \alpha_2\mathbf{v}_2 + \cdots + \alpha_n\mathbf{v}_n = \beta_1\mathbf{v}_1 + \beta_2\mathbf{v}_2 + \cdots + \beta_n\mathbf{v}_n$$

Adding $-(\beta_1\mathbf{v}_1 + \beta_2\mathbf{v}_2 + \cdots + \beta_n\mathbf{v}_n)$ to both sides of the equation and rearranging terms leads to the equation

$$(\alpha_1 - \beta_1)\mathbf{v}_1 + (\alpha_2 - \beta_2)\mathbf{v}_2 + \cdots + (\alpha_n - \beta_n)\mathbf{v}_n = \mathbf{0}$$

Since the set $S = \{\mathbf{v}_1, \mathbf{v}_2, \ldots, \mathbf{v}_n\}$ is linearly independent, it follows that

This is because the set S is a basis.

$$\alpha_1 - \beta_1 = 0,\ \alpha_2 - \beta_2 = 0,\ \ldots,\ \alpha_n - \beta_n = 0$$

This shows that

$$\alpha_i = \beta_i \text{ for } i = 1, 2, \ldots, n$$

and the uniqueness of the representation is proved.

We wish to emphasize that if a spanning set for a vector space is not a basis (because it is linearly dependent) then a vector may have different representations in terms of the vectors of the spanning set. The following is such an example.

Example 5 The set $S = \left\{\begin{bmatrix}1\\2\end{bmatrix}, \begin{bmatrix}2\\1\end{bmatrix}, \begin{bmatrix}1\\3\end{bmatrix}\right\}$ spans $\mathbf{R}^2$. The vector $\begin{bmatrix}1\\1\end{bmatrix}$ has the following two different representations

$$\begin{bmatrix}1\\1\end{bmatrix} = \frac{1}{3}\begin{bmatrix}1\\2\end{bmatrix} + \frac{1}{3}\begin{bmatrix}2\\1\end{bmatrix} + 0\begin{bmatrix}1\\3\end{bmatrix}$$

and

$$\begin{bmatrix}1\\1\end{bmatrix} = -\frac{4}{3}\begin{bmatrix}1\\2\end{bmatrix} + \frac{2}{3}\begin{bmatrix}2\\1\end{bmatrix} + 1\begin{bmatrix}1\\3\end{bmatrix}$$

Actually, the vector $\begin{bmatrix}1\\1\end{bmatrix}$ has infinitely many different representations in terms of linear combinations of the vectors $\begin{bmatrix}1\\2\end{bmatrix}$, $\begin{bmatrix}2\\1\end{bmatrix}$, and $\begin{bmatrix}1\\3\end{bmatrix}$. This follows from the fact that the vector equation

$$\alpha_1\begin{bmatrix}1\\2\end{bmatrix} + \alpha_2\begin{bmatrix}2\\1\end{bmatrix} + \alpha_3\begin{bmatrix}1\\3\end{bmatrix} = \begin{bmatrix}1\\1\end{bmatrix}$$

has infinitely many solutions for α_1, α_2, and α_3.

We wish to conclude this section with a theorem that will indicate how large a linearly independent set can be in a finite-dimensional vector space V. This theorem will be very useful in treating the concept of *dimension* in the next section.

Theorem 3-6 Let $S = \{\mathbf{v}_1, \mathbf{v}_2, \ldots, \mathbf{v}_n\}$ be a basis for a vector space V. Then every finite subset of V containing more vectors that n is linearly dependent.

Proof Let $T = \{\mathbf{u}_1, \mathbf{u}_2, \ldots, \mathbf{u}_s\}$ be a subset of V such that $s > n$. We must show that T is linearly dependent. This will follow if the vector equation

Remark: The conclusion of this theorem could also have stated that *every* linearly independent subset of V contains *at least* n vectors.

$$\alpha_1\mathbf{u}_1 + \alpha_2\mathbf{u}_2 + \cdots + \alpha_s\mathbf{u}_s = \mathbf{0} \quad \textbf{[1]}$$

Note that S is a basis for V; hence, S is a spanning set for V.

has a nontrivial solution $\alpha_1, \alpha_2, \ldots, \alpha_s$. Since $S = \{\mathbf{v}_1, \mathbf{v}_2, \ldots, \mathbf{v}_n\}$ is a spanning set for the vector space V, each vector $\mathbf{u}_i$ ($i = 1, 2, \ldots, s$) is a linear combination of $\mathbf{v}_1, \mathbf{v}_2, \ldots, \mathbf{v}_n$. To express this fact we choose a notation which will lead to our familiar matrix notation. We may write

$$\begin{aligned}\mathbf{u}_1 &= a_{11}\mathbf{v}_1 + a_{21}\mathbf{v}_2 + \cdots + a_{n1}\mathbf{v}_n\\ \mathbf{u}_2 &= a_{12}\mathbf{v}_1 + a_{22}\mathbf{v}_2 + \cdots + a_{n2}\mathbf{v}_n\\ &\vdots \\ \mathbf{u}_s &= a_{1s}\mathbf{v}_1 + a_{2s}\mathbf{v}_2 + \cdots + a_{ns}\mathbf{v}_n\end{aligned} \quad \textbf{[2]}$$

where a_{ij} are scalars ($i = 1, 2, \ldots, n$ and $j = 1, 2, \ldots, s$). Substituting system 2 into equation 1 and rearranging to find coefficients of $\mathbf{v}_1, \mathbf{v}_2, \ldots, \mathbf{v}_n$, we get

$$\begin{aligned}&(a_{11}\alpha_1 + a_{12}\alpha_2 + \cdots + a_{1s}\alpha_s)\mathbf{v}_1\\ &+ (a_{21}\alpha_1 + a_{22}\alpha_2 + \cdots + a_{2s}\alpha_s)\mathbf{v}_2 + \cdots\\ \cdots &+ (a_{n1}\alpha_1 + a_{n2}\alpha_2 + \cdots + a_{ns}\alpha_s)\mathbf{v}_n = \mathbf{0}\end{aligned}$$

Note that S is a basis for V; hence, S is a linearly independent set.

Since $S = \{\mathbf{v}_1, \mathbf{v}_2, \ldots, \mathbf{v}_n\}$ is a linearly independent set, all the coefficients of the $\mathbf{v}_i$'s are zero. Therefore, we obtain the following system of equations for $\alpha_1, \alpha_2, \ldots, \alpha_s$:

$$\begin{aligned}a_{11}\alpha_1 + a_{12}\alpha_2 + \cdots + a_{1s}\alpha_s &= 0\\ a_{21}\alpha_1 + a_{22}\alpha_2 + \cdots + a_{2s}\alpha_s &= 0\\ \vdots \qquad \vdots \qquad\quad \vdots \quad &\;\; \vdots\\ a_{n1}\alpha_1 + a_{n2}\alpha_2 + \cdots + a_{ns}\alpha_s &= 0\end{aligned} \qquad [3]$$

But system 3 is a linear homogeneous system with more unknowns than equations, because by hypothesis $s > n$. Such a system always has a nontrivial solution $\alpha_1, \alpha_2, \ldots, \alpha_s$. This proves that there exist $\alpha_1, \alpha_2, \ldots, \alpha_s$, not all zero, such that $\alpha_1\mathbf{u}_1 + \alpha_2\mathbf{u}_2 + \cdots + \alpha_s\mathbf{u}_s = \mathbf{0}$. Thus, the set $T = \{\mathbf{u}_1, \mathbf{u}_2, \ldots, \mathbf{u}_s\}$ is linearly dependent.

An equivalent form of stating Theorem 3-6 is the following:

Theorem 3-7 Let $S = \{\mathbf{v}_1, \mathbf{v}_2, \ldots, \mathbf{v}_n\}$ be a basis for a vector space V. Then every linearly independent subset of V contains at most n vectors.

The following example is a simple consequence of Theorem 3-6.

Example 6 Consider the subset $S = \{3x^2 - x,\ x^2 + 5x - 1,\ -4x^2 + 1,\ x - 1\}$ of P_2. Determine whether S is linearly dependent or not.

Solution We have shown in Example 4 that P_2 has a basis containing three vectors. Therefore the set S which contains four vectors must be linearly dependent.

This follows from Theorem 3-6.

Exercises

1 Determine which of the following sets are bases for $\mathbf{R}^2$.

a $\left\{\begin{bmatrix}1\\3\end{bmatrix}, \begin{bmatrix}2\\1\end{bmatrix}\right\}$ **b** $\left\{\begin{bmatrix}3\\2\end{bmatrix}, \begin{bmatrix}-6\\-4\end{bmatrix}\right\}$

c $\left\{\begin{bmatrix}-2\\1\end{bmatrix}, \begin{bmatrix}3\\1\end{bmatrix}\right\}$ **d** $\left\{\begin{bmatrix}1\\1\end{bmatrix}, \begin{bmatrix}-5\\3\end{bmatrix}\right\}$

2 Show that the following sets are *not* bases for $\mathbf{R}^2$.

a $\left\{\begin{bmatrix}1\\2\end{bmatrix}, \begin{bmatrix}1\\3\end{bmatrix}, \begin{bmatrix}1\\4\end{bmatrix}\right\}$ **b** $\left\{\begin{bmatrix}2\\3\end{bmatrix}, \begin{bmatrix}4\\6\end{bmatrix}, \begin{bmatrix}6\\9\end{bmatrix}\right\}$

c $\left\{\begin{bmatrix}1\\0\end{bmatrix}\right\}$ **d** $\left\{\begin{bmatrix}1\\0\end{bmatrix}, \begin{bmatrix}0\\0\end{bmatrix}\right\}$

3 Determine which of the following sets are bases for $\mathbf{R}^3$.

a $\left\{\begin{bmatrix}1\\1\\1\end{bmatrix}, \begin{bmatrix}1\\2\\1\end{bmatrix}, \begin{bmatrix}2\\3\\1\end{bmatrix}\right\}$ **b** $\left\{\begin{bmatrix}1\\0\\2\end{bmatrix}, \begin{bmatrix}2\\0\\1\end{bmatrix}, \begin{bmatrix}3\\0\\5\end{bmatrix}\right\}$

c $\left\{\begin{bmatrix}-1\\2\\1\end{bmatrix}, \begin{bmatrix}0\\1\\1\end{bmatrix}, \begin{bmatrix}1\\-2\\2\end{bmatrix}\right\}$

4 Show that the following sets are *not* bases for $\mathbf{R}^3$.

a $\left\{\begin{bmatrix}1\\2\\3\end{bmatrix}\right\}$ **b** $\left\{\begin{bmatrix}1\\2\\3\end{bmatrix}, \begin{bmatrix}1\\1\\1\end{bmatrix}\right\}$

c $\left\{\begin{bmatrix}1\\2\\3\end{bmatrix}, \begin{bmatrix}1\\1\\1\end{bmatrix}, \begin{bmatrix}0\\0\\0\end{bmatrix}\right\}$ **d** $\left\{\begin{bmatrix}1\\2\\3\end{bmatrix}, \begin{bmatrix}1\\1\\1\end{bmatrix}, \begin{bmatrix}1\\4\\1\end{bmatrix}, \begin{bmatrix}-1\\-1\\0\end{bmatrix}\right\}$

5 Determine which of the following sets are bases for P_2.
a $\{1 + x, 3x + 3, x^2 + x\}$ **b** $\{2x, 2x + 1, 3x^2 - 5\}$
c $\{x^2 + x + 1, x + 1, x^2 + 1\}$

6 Show that the following sets are *not* bases for P_2.
a $\{x^2, x\}$ **b** $\{x^2, 3\}$ **c** $\{1, x, x^2, 3x - x - 5\}$
d $\{x, x^2 + 1, x^2 - x + 2, -2x^2 + x\}$

7 Consider the vectors $\mathbf{e}_1 = \begin{bmatrix}1\\0\\0\\\vdots\\0\end{bmatrix}$, $\mathbf{e}_2 = \begin{bmatrix}0\\1\\0\\\vdots\\0\end{bmatrix}$, $\dots$, $\mathbf{e}_{n-1} = \begin{bmatrix}0\\\vdots\\0\\1\\0\end{bmatrix}$

and $\mathbf{e}_n = \begin{bmatrix}0\\\vdots\\0\\0\\1\end{bmatrix}$ in $\mathbf{R}^n$. Show that the set $\{\mathbf{e}_1, \mathbf{e}_2, \dots, \mathbf{e}_{n-1}, \mathbf{e}_n\}$ is a basis for $\mathbf{R}^n$.

8 Let $S = \{\mathbf{v}_1, \mathbf{v}_2, \dots, \mathbf{v}_m\}$ be a subset of $\mathbf{R}^n$ where $m > n$. Explain why S must be a linearly dependent set.

Hint: See Example 3.

9 Show that any set of 2×3 matrices containing seven elements or more of the vector space $M_{2\times 3}$ is linearly dependent.

10 Let M_5 be the vector space of all 5×5 matrices under the standard addition and scalar multiplication of matrices.

a Find a basis for M_5.

b Let S be a subset of M_5 such that S contains 26 elements. Determine whether S is linearly dependent or not.

11 Let $\begin{bmatrix} a \\ b \end{bmatrix}$ and $\begin{bmatrix} c \\ d \end{bmatrix}$ be two vectors in $\mathbf{R}^2$. Show that if neither of the two vectors is a scalar multiple of the other, then the set $\left\{\begin{bmatrix} a \\ b \end{bmatrix}, \begin{bmatrix} c \\ d \end{bmatrix}\right\}$ is a basis for $\mathbf{R}^2$.

3-4 Finite-dimensional Vector Spaces

In this section we shall investigate a property concerning the number of vectors in a basis for a given vector space. The question we now ask concerns the connection between the number of vectors in a basis for a given vector space and the structure of the vector space.

We prove the following theorem:

Theorem 3-8 Let $S = \{\mathbf{v}_1, \mathbf{v}_2, \ldots, \mathbf{v}_k\}$ and $T = \{\mathbf{u}_1, \mathbf{u}_2, \ldots, \mathbf{u}_m\}$ be two bases for a vector space V. Then we must have $m = k$.

Objectives

1 Show that every two bases for a finite-dimensional vector space must contain the same number of vectors.
2 Introduce the concept of dimension of a vector space.
3 Construct a basis from a given spanning set for a vector space.
4 Construct a basis containing a given linearly independent set of vectors.

Motivation: The reader has seen in the previous section that every basis for $\mathbf{R}^2$ is a subset of $\mathbf{R}^2$ containing exactly two vectors; and every basis for P_2 is a subset of P_2 containing exactly three vectors. Other such examples are given in Examples 1–4 in Section 3-3.

Proof We know from Theorem 3-7 that the number of elements in any linearly independent subset of a vector space V cannot exceed the number of elements in a basis for V. If we let S play the role of a basis and T the role of a linearly independent subset of V, then we conclude that $m \leq k$. But we can also let T play the role of a basis while S plays the role of a linearly independent subset of V. This implies that $k \leq m$. Since $m \leq k$ and $k \leq m$ can hold simultaneously only if $m = k$, the proof is complete.

It follows from Theorem 3-8 that every basis for $\mathbf{R}^2$ must contain two vectors, every basis for P_2 must contain three vectors, and every basis for $\mathbf{R}^3$ must contain three vectors.

At this point we consider the following important problem: Suppose a vector space V is finite-dimensional and $S = \{\mathbf{v}_1, \mathbf{v}_2, \ldots, \mathbf{v}_m\}$ is a spanning set for V. If S is linearly dependent, it will certainly not be a basis for V. Is it possible to construct a basis for V by using any part of the set S? In other words, is it possible to discard some of the elements of S so that the remaining vectors will form a subset $\bar{S}$ of S with the property that $\bar{S}$ still spans V, and in addition will be a linearly independent set?

The answer to this problem is given in the following theorem.

Theorem 3-9 Let $S = \{\mathbf{v}_1, \mathbf{v}_2, \ldots, \mathbf{v}_m\}$ be a set of nonzero vectors spanning a vector space V. Then there exists a subset $\bar{S}$ of S such that $\bar{S}$ is a basis for V.

Proof

The main idea of the proof is that if S contains a vector $\mathbf{v}_i$ which is a linear combination of the remaining vectors in S, then we can delete $\mathbf{v}_i$ from S, obtaining a new set S^* which still spans the vector space V (as S does) (see problem 9 of Section 3-2).

If $S = \{\mathbf{v}_1, \mathbf{v}_2, \ldots, \mathbf{v}_m\}$ is linearly independent then S is a basis for V (because it is also a spanning set for V). In this case we let $\bar{S} = S$ and the proof is complete. If on the other hand S is linearly dependent, then one of its elements, say $\mathbf{v}_i$, is a linear combination of the remaining vectors in S. Thus, we delete $\mathbf{v}_i$ from S, obtaining a set $S^* = \{\mathbf{u}_1, \mathbf{u}_2, \ldots, \mathbf{u}_{m-1}\}$, where the $\mathbf{u}$'s are a relabeling of the remaining $\mathbf{v}$'s. The set S^* still spans V. If S^* is linearly independent then it is a basis for V. If S^* is linearly dependent then one of its elements, say $\mathbf{u}_j$, is a linear combination of the remaining elements in S^* and can be deleted as before. Since the original set $S = \{\mathbf{v}_1, \mathbf{v}_2, \ldots, \mathbf{v}_m\}$ is a finite set, this process will terminate after a finite number of deletions, leaving us with a set $\bar{S}$ containing *at least* one nonzero vector that is linearly independent and spans V. Thus, $\bar{S}$ will be a basis for V.

Note: A set consisting of one nonzero vector is linearly independent.

The following example illustrates the process mentioned in Theorem 3-9.

Example 1 The set $S = \left\{\begin{bmatrix}1\\2\end{bmatrix}, \begin{bmatrix}2\\4\end{bmatrix}, \begin{bmatrix}2\\1\end{bmatrix}, \begin{bmatrix}3\\3\end{bmatrix}, \begin{bmatrix}4\\5\end{bmatrix}\right\}$ spans $\mathbf{R}^2$. Find a subset $\bar{S}$ of S such that $\bar{S}$ is a basis for $\mathbf{R}^2$.

Solution We observe that the vector $\begin{bmatrix}2\\4\end{bmatrix}$ is a multiple of the vector $\begin{bmatrix}1\\2\end{bmatrix}$ by the scalar 2. Therefore

$$\begin{bmatrix}2\\4\end{bmatrix} = 2\begin{bmatrix}1\\2\end{bmatrix} + 0\begin{bmatrix}2\\1\end{bmatrix} + 0\begin{bmatrix}3\\3\end{bmatrix} + 0\begin{bmatrix}4\\5\end{bmatrix}$$

and we can delete $\begin{bmatrix}2\\4\end{bmatrix}$ from S, obtaining the set

$$S^* = \left\{\begin{bmatrix}1\\2\end{bmatrix}, \begin{bmatrix}2\\1\end{bmatrix}, \begin{bmatrix}3\\3\end{bmatrix}, \begin{bmatrix}4\\5\end{bmatrix}\right\}$$

The vector $\begin{bmatrix}3\\3\end{bmatrix}$ may be expressed as

$$\begin{bmatrix}3\\3\end{bmatrix} = 1\begin{bmatrix}1\\2\end{bmatrix} + 1\begin{bmatrix}2\\1\end{bmatrix} + 0\begin{bmatrix}4\\5\end{bmatrix}$$

Thus we can delete the vector $\begin{bmatrix}3\\3\end{bmatrix}$ from S^*, obtaining the set $S^{**} = \left\{\begin{bmatrix}1\\2\end{bmatrix}, \begin{bmatrix}2\\1\end{bmatrix}, \begin{bmatrix}4\\5\end{bmatrix}\right\}$, which also spans $\mathbf{R}^2$, Finally, the vector $\begin{bmatrix}4\\5\end{bmatrix}$ may be expressed as a linear combination of $\begin{bmatrix}1\\2\end{bmatrix}$ and $\begin{bmatrix}2\\1\end{bmatrix}$. We have

$$\begin{bmatrix}4\\5\end{bmatrix} = 2\begin{bmatrix}1\\2\end{bmatrix} + 1\begin{bmatrix}2\\1\end{bmatrix}$$

Note that there is no unique result here because the sets $\left\{\begin{bmatrix}1\\2\end{bmatrix}, \begin{bmatrix}4\\5\end{bmatrix}\right\}$, $\left\{\begin{bmatrix}1\\2\end{bmatrix}, \begin{bmatrix}3\\3\end{bmatrix}\right\}$, $\left\{\begin{bmatrix}2\\1\end{bmatrix}, \begin{bmatrix}4\\5\end{bmatrix}\right\}$, and $\left\{\begin{bmatrix}2\\1\end{bmatrix}, \begin{bmatrix}3\\3\end{bmatrix}\right\}$ are also subsets of S, having the property that each one is a basis for $\mathbf{R}^2$.

Therefore we also delete the vector $\begin{bmatrix}4\\5\end{bmatrix}$, obtaining the set $\bar{S} = \left\{\begin{bmatrix}1\\2\end{bmatrix}, \begin{bmatrix}2\\1\end{bmatrix}\right\}$. Since $\bar{S}$ is linearly independent and spans $\mathbf{R}^2$, it is a basis for $\mathbf{R}^2$

We wish to emphasize again that a finite-dimensional nonzero vector space always has a finite basis, and the number of vectors in such a basis is uniquely determined (see Theorem 3-8). This number is given a special name.

Definition 3-7
The dimension of a vector space.

Let V be a nonzero finite-dimensional vector space. The **dimension** of V, denoted by $\dim \mathrm{V}$, is the number of vectors in a basis for V. If $\dim V = n$ we say that V is an ***n*-dimensional vector space**.

We define the dimension of the zero vector space to be 0. This definition is an agreement with the fact that the vector space consisting of one vector only, the zero vector, does not have a basis. (Why?)

In order to determine the dimension of a nonzero vector space V we need to find only one basis for V and then count the number of vectors in it. Since the set $\left\{\begin{bmatrix}1\\0\end{bmatrix}, \begin{bmatrix}0\\1\end{bmatrix}\right\}$ is a basis for $\mathbf{R}^2$, we have $\dim \mathbf{R}^2 = 2$. The set $\left\{\begin{bmatrix}1\\0\\0\end{bmatrix}, \begin{bmatrix}0\\1\\0\end{bmatrix}, \begin{bmatrix}0\\0\\1\end{bmatrix}\right\}$ is a basis for $\mathbf{R}^3$, therefore $\dim \mathbf{R}^3 = 3$. The set $\{1,x,x^2\}$ is a basis for P_2, therefore $\dim P_2 = 3$.

Do not be misled by the subscript 2.

The number $\dim V$ is closely associated with the size of any linearly independent set in V and the size of any spanning set of V. It follows from Theorem 3-6 that if $S = \{\mathbf{v}_1, \mathbf{v}_2, \ldots, \mathbf{v}_k\}$ is a linearly independent subset of V and $\dim V = n$, then $k \leq n$. Thus, the number $\dim V$ gives us the maximum number of elements in any linearly independent subset of V. It also follows from Theorem 3-9 that if $T = \{\mathbf{u}_1, \mathbf{u}_2, \ldots, \mathbf{u}_m\}$ is a spanning set for V with $\mathbf{u}_i \neq 0$

for $i = 1, 2, \ldots, m$ and $\dim V = n$, then $n \leq m$. Thus, the number $\dim V$ gives us the minimum number of elements in any spanning set for V. Hence, no two vectors can span $\mathbf{R}^3$ because $\dim \mathbf{R}^3 = 3$ and a spanning set for $\mathbf{R}^3$ must contain at least three vectors from $\mathbf{R}^3$.

Consider the following problems:

Problem 1 Suppose we are given a linearly independent subset $S = \{\mathbf{v}_1, \mathbf{v}_2, \ldots, \mathbf{v}_n\}$ of a vector space V where $\dim V = n$. Is S a spanning set for V?

Problem 2 Suppose we are given a spanning set of nonzero vectors $T = \{\mathbf{u}_1, \mathbf{u}_2, \ldots, \mathbf{u}_n\}$ for a vector space V where $\dim V = n$. Is T a linearly independent set?

The answer to both problems is given in the following two theorems.

Theorem 3-10 Let $S = \{\mathbf{v}_1, \mathbf{v}_2, \ldots, \mathbf{v}_n\}$ be a linearly independent set of n vectors in an n-dimensional vector space V. Then S is a basis for V.

Theorem 3-11 Let $T = \{\mathbf{u}_1, \mathbf{u}_2, \ldots, \mathbf{u}_n\}$ be a set of n nonzero vectors that spans an n-dimensional vector space V. Then T is a basis for V.

Note:
a Theorem 3-10 implies that S spans V.
b Theorem 3-11 implies that T is linearly independent.
See problem 22.

Here we give the proof of Theorem 3-10 and leave the proof of Theorem 3-11 as an exercise for the student.

Proof of Theorem 3-10 In order to prove that the linearly independent set $S = \{\mathbf{v}_1, \mathbf{v}_2, \ldots, \mathbf{v}_n\}$ is a basis for V, we must show that S spans V. We do this by contradiction.

See Theorem 3-3.

Suppose S does not span the vector space V. Then V must contain at least one vector, say $\mathbf{v}_{n+1}$, which is not a linear combination of the vectors $\mathbf{v}_1, \mathbf{v}_2, \ldots, \mathbf{v}_n$ in S. Therefore, the set $S^* = \{\mathbf{v}_1, \mathbf{v}_2, \ldots, \mathbf{v}_n, \mathbf{v}_{n+1}\}$ obtained from S by adding to it the vector $\mathbf{v}_{n+1}$ is a linearly independent set. Since V is an n-dimensional vector space it has some basis consisting of n vectors. From Theorem 3-6 we know that any subset of such a vector space which contains more vectors than n must be linearly dependent. But we have constructed a set S^* which contains $n + 1$ vectors and is linearly independent. This is a contradiction which stems from the assumption that S *does not span* the vector space V. Thus, S must span V and therefore S is a basis for V. This completes the proof of Theorem 3-10.

The following two examples will illustrate the above theorems.

Example 2 Show that the set of vectors $S = \left\{ \begin{bmatrix} 1 \\ 1 \\ 2 \end{bmatrix}, \begin{bmatrix} 0 \\ 1 \\ 3 \end{bmatrix}, \begin{bmatrix} 2 \\ 1 \\ 5 \end{bmatrix} \right\}$ is a basis for $\mathbf{R}^3$.

Solution Since we already know that $\dim \mathbf{R}^3 = 3$, it is sufficient to show that the set S is linearly independent, because Theorem 3-10 will then guarantee that S is also a basis for $\mathbf{R}^3$. Thus, we have to prove that the vector equation

$$\alpha_1 \begin{bmatrix} 1 \\ 1 \\ 2 \end{bmatrix} + \alpha_2 \begin{bmatrix} 0 \\ 1 \\ 3 \end{bmatrix} + \alpha_3 \begin{bmatrix} 2 \\ 1 \\ 5 \end{bmatrix} = \begin{bmatrix} 0 \\ 0 \\ 0 \end{bmatrix}$$

has the unique solution $\alpha_1 = \alpha_2 = \alpha_3 = 0$.

The vector equation may be written as the system

$$\begin{bmatrix} 1 & 0 & 2 \\ 1 & 1 & 1 \\ 2 & 3 & 5 \end{bmatrix} \begin{bmatrix} \alpha_1 \\ \alpha_2 \\ \alpha_3 \end{bmatrix} = \begin{bmatrix} 0 \\ 0 \\ 0 \end{bmatrix}$$

The augmented matrix of the system of equations reduces to

$$\left[\begin{array}{ccc:c} 1 & 0 & 0 & 0 \\ 0 & 1 & 0 & 0 \\ 0 & 0 & 1 & 0 \end{array}\right]$$

and therefore we have $\alpha_1 = \alpha_2 = \alpha_3 = 0$.

Thus, $\begin{bmatrix} 1 \\ 1 \\ 2 \end{bmatrix}$, $\begin{bmatrix} 0 \\ 1 \\ 3 \end{bmatrix}$, and $\begin{bmatrix} 2 \\ 1 \\ 5 \end{bmatrix}$ are linearly independent and therefore the set S is a basis for $\mathbf{R}^3$.

Example 3 Show that the set $S = \{x + 1, x - 1, x^2 + 1\}$ is a basis for P_2.

Solution We have previously shown that $\dim P_2 = 3$. Thus it is sufficient to prove that S spans P_2, for Theorem 3-11 will then guarantee that S is a basis for P_2.

Let $ax^2 + bx + c$ be an arbitrary element of P_2. We must show that α_1, α_2, and α_3 can be found such that

$$\alpha_1(x + 1) + \alpha_2(x - 1) + \alpha_3(x^2 + 1) = ax^2 + bx + c \text{ for all } x$$

Simplifying the left-hand side of the equation we get

$$\alpha_3 x^2 + (\alpha_1 + \alpha_2)x + (\alpha_1 - \alpha_2 + \alpha_3) = ax^2 + bx + c \text{ for all } x$$

The last equation is satisfied for all x if and only if

$$\begin{aligned} \alpha_3 &= a \\ \alpha_1 + \alpha_2 \quad &= b \\ \alpha_1 - \alpha_2 + \alpha_3 &= c \end{aligned}$$

The solution of this system is given by

$$\alpha_1 = \frac{-a+b+c}{2}, \qquad \alpha_2 = \frac{a+b-c}{2} \qquad \text{and} \qquad \alpha_3 = a$$

This shows that S spans P_2. We conclude from Theorem 3-11 that S is a basis for P_2.

See Theorem 3-9.

Recall that every finite spanning set for a vector space V (different from the zero vector space) has a subset which is a basis for V. We now pose the following problem:

Problem 3 Suppose $S = \{\mathbf{v}_1, \mathbf{v}_2, \ldots, \mathbf{v}_k\}$ is a linearly independent subset of a vector space V where $\dim V = n$. Is it possible to find a basis for V which invludes all the vectors $\mathbf{v}_1, \mathbf{v}_2, \ldots, \mathbf{v}_k$? The following theorem provides an answer to this problem and provides a constructive way of finding such a basis.

Theorem 3-12 Let $S = \{\mathbf{v}_1, \mathbf{v}_2, \ldots, \mathbf{v}_k\}$ be a linearly independent set in an n-dimensional vector space V, where $k < n$. Then there exist $n - k$ vectors $\mathbf{v}_{k+1}, \ldots, \mathbf{v}_n$ in V such that the enlarged set $\bar{S} = \{\mathbf{v}_1, \mathbf{v}_2, \ldots, \mathbf{v}_k, \mathbf{v}_{k+1}, \ldots, \mathbf{v}_n\}$ is a basis for V.

Proof We present here a constructive proof indicating how to enlarge the given set $S = \{\mathbf{v}_1, \mathbf{v}_2, \ldots, \mathbf{v}_k\}$ to a basis $\bar{S}$ for V.

Since $\dim V = n$, V must have some basis $\{\mathbf{u}_1, \mathbf{u}_2, \ldots, \mathbf{u}_n\}$. Let

$$S^* = \{\mathbf{v}_1, \mathbf{v}_2, \ldots, \mathbf{v}_k\} \cup \{\mathbf{u}_1, \mathbf{u}_2, \ldots, \mathbf{u}_n\}$$

The set S^* certainly spans V because it contains a subset $\{\mathbf{u}_1, \mathbf{u}_2, \ldots, \mathbf{u}_n\}$ which spans V. We can get a basis for V by deleting suitable vectors from S^* without destroying the spanning property of the set. We delete the vectors in such a way that $\mathbf{v}_1, \mathbf{v}_2, \ldots, \mathbf{v}_k$ are retained. This is done in a systematic way by starting with a check on the vector $\mathbf{u}_1$. If $\mathbf{u}_1$ is a linear combination of $\mathbf{v}_1, \mathbf{v}_2, \ldots, \mathbf{v}_k$ then we delete it, otherwise we retain it. We proceed with the remaining $\mathbf{u}$'s in a similar fashion until we have retained $n - k$ of them. At each stage of this procedure we still have a spanning set which contains $\mathbf{v}_1, \mathbf{v}_2, \ldots, \mathbf{v}_k$ and some $\mathbf{u}$'s. Since S^* consists of nonzero vectors and spans V, it follows from Theorem 3-9 that the procedure indicated above will yield a basis for V consisting of $\mathbf{v}_1, \mathbf{v}_2, \ldots, \mathbf{v}_k$ and $n - k$ vectors from the set $\{\mathbf{u}_1, \mathbf{u}_2, \ldots, \mathbf{u}_n\}$.

Remember that every basis for V contains the same number of vectors.

The following examples will illustrate the procedure mentioned in Theorem 3–12.

Example 4 Construct a basis for $\mathbf{R}^3$ which includes the vector $\begin{bmatrix}1\\2\\3\end{bmatrix}$.

Solution Let

$$S^* = \left\{\begin{bmatrix}1\\2\\3\end{bmatrix}, \begin{bmatrix}1\\0\\0\end{bmatrix}, \begin{bmatrix}0\\1\\0\end{bmatrix}, \begin{bmatrix}0\\0\\1\end{bmatrix}\right\}$$

Then S^* spans $\mathbf{R}^3$ because it contains a subset $\left\{\begin{bmatrix}1\\0\\0\end{bmatrix}, \begin{bmatrix}0\\1\\0\end{bmatrix}, \begin{bmatrix}0\\0\\1\end{bmatrix}\right\}$ which spans $\mathbf{R}^3$. We wish to retain the vector $\begin{bmatrix}1\\2\\3\end{bmatrix}$. Since the vector $\begin{bmatrix}1\\0\\0\end{bmatrix}$ is not a linear combination of the vector $\begin{bmatrix}1\\2\\3\end{bmatrix}$ it is retained too. We now examine whether the vector $\begin{bmatrix}0\\1\\0\end{bmatrix}$ is a linear combination of the vectors which have been retained so far. We write

$$\alpha_1\begin{bmatrix}1\\2\\3\end{bmatrix} + \alpha_2\begin{bmatrix}1\\0\\0\end{bmatrix} = \begin{bmatrix}0\\1\\0\end{bmatrix}$$

and look for α_1 and α_2 which could possibly satisfy the vector equation. The corresponding linear system is

$$\begin{aligned}\alpha_1 + \alpha_2 &= 0\\ 2\alpha_1 \quad &= 1\\ 3\alpha_1 \quad &= 0\end{aligned}$$

It is obvious that this system is inconsistent and therefore has no solution for α_1 and α_2. Thus, $\begin{bmatrix}0\\1\\0\end{bmatrix}$ is not a linear combination of the vectors $\begin{bmatrix}1\\2\\3\end{bmatrix}, \begin{bmatrix}1\\0\\0\end{bmatrix}$ and therefore will be retained. We have con-

We obviously do not retain $\begin{bmatrix} 0 \\ 0 \\ 1 \end{bmatrix}$ because this vector must be a linear combination of the vectors in $\bar{S}$.

structed a set

$$\bar{S} = \left\{ \begin{bmatrix} 1 \\ 2 \\ 3 \end{bmatrix}, \begin{bmatrix} 1 \\ 0 \\ 0 \end{bmatrix}, \begin{bmatrix} 0 \\ 1 \\ 0 \end{bmatrix} \right\}$$

which is linearly independent. Since $\dim \mathbf{R}^3 = 3$, it follows from Theorem 3-10 that $\bar{S}$ is a basis for $\mathbf{R}^3$.

Example 5 Construct a basis for P_2 which includes the set $\{x + 1, x - 1\}$.

Solution Let $S^* = \{x + 1, x - 1, 1, x, x^2\}$. The set S^* spans P_2 because it contains the subset $\{1,x,x^2\}$ which spans P_2. The vectors $x + 1$ and $x - 1$ are linearly independent because neither is a scalar multiple of the other. Thus we may retain both and the problem as posed is solvable. Since the vector 1 is a linear combination of $x + 1$ and $x - 1$ it is *not* retained. Also, since the vector x is a linear combination of $x + 1$ and $x - 1$ it is not retained either. We retain the vector x^2. Thus the set $\bar{S} = \{x + 1, x - 1, x^2\}$ is a basis for P_2 because $\bar{S}$ is linearly independent and $\dim P_2 = 3$.

Note:

$\frac{1}{2}(x + 1) - \frac{1}{2}(x - 1) = 1$
$\frac{1}{2}(x + 1) + \frac{1}{2}(x - 1) = x$

Every subspace S of a given vector space V is a vector space in its own right. Therefore it makes sense to discuss the dimension of any such subspace. We now wish to find an answer to the following problem.

Problem 4 Let V be a finite-dimensional vector space and let S be a subspace of V. What is the relation between the dimension of S and the dimension of V?

The answer to this problem is given in the following:

Theorem 3-13 Let V be a finite-dimensional vector space. Then every subspace S of V is also finite-dimensional and dim $S \leq \dim V$.

See problem 25.

We leave the proof for the student as an exercise and proceed with two examples concerned with the determination of the dimension of some subspaces of given vector spaces.

Example 6 Given that the set S of all vectors $\begin{bmatrix} a \\ b \\ c \end{bmatrix}$, where $a + b = 0$, is a subspace of $\mathbf{R}^3$, find its dimension.

Note: We must find a basis for S and then count the number of vectors in it to find the dimension of S.

Solution The condition $a + b = 0$ implies that $b = -a$. Therefore a typical element of S is expressed by $\begin{bmatrix} a \\ -a \\ c \end{bmatrix}$ where a and c are arbitrary real numbers. Notice that if we choose a value for a then the value of the second component, $-a$, is determined. Thus, one should not expect the standard basis of $\mathbf{R}^3$ to be a basis for the subspace S. A linear combination of $\begin{bmatrix} 1 \\ 0 \\ 0 \end{bmatrix}$, $\begin{bmatrix} 0 \\ 1 \\ 0 \end{bmatrix}$, and $\begin{bmatrix} 0 \\ 0 \\ 1 \end{bmatrix}$ can produce $\begin{bmatrix} 1 \\ 2 \\ 3 \end{bmatrix}$, which does not belong to S because if $a = 1$ then $-a \neq 2$. Here is what we do. We write

$$\begin{bmatrix} a \\ -a \\ c \end{bmatrix} = \begin{bmatrix} a \\ -a \\ 0 \end{bmatrix} + \begin{bmatrix} 0 \\ 0 \\ c \end{bmatrix}$$

where each of the vectors on the right-hand side depends on only one arbitrary real number. Since

$$\begin{bmatrix} a \\ -a \\ 0 \end{bmatrix} = a\begin{bmatrix} 1 \\ -1 \\ 0 \end{bmatrix} \quad \text{and} \quad \begin{bmatrix} 0 \\ 0 \\ c \end{bmatrix} = c\begin{bmatrix} 0 \\ 0 \\ 1 \end{bmatrix}$$

we may write

$$\begin{bmatrix} a \\ -a \\ c \end{bmatrix} = a\begin{bmatrix} 1 \\ -1 \\ 0 \end{bmatrix} + c\begin{bmatrix} 0 \\ 0 \\ 1 \end{bmatrix}$$

Thus, every vector $\begin{bmatrix} a \\ -a \\ c \end{bmatrix}$ in S is a linear combination of the two vectors $\begin{bmatrix} 1 \\ -1 \\ 0 \end{bmatrix}$ and $\begin{bmatrix} 0 \\ 0 \\ 1 \end{bmatrix}$ where a and c are the corresponding scalars. We have therefore constructed a spanning set $\left\{\begin{bmatrix} 1 \\ -1 \\ 0 \end{bmatrix}, \begin{bmatrix} 0 \\ 0 \\ 1 \end{bmatrix}\right\}$ for S.

Since neither of the vectors $\begin{bmatrix} 1 \\ -1 \\ 0 \end{bmatrix}$, $\begin{bmatrix} 0 \\ 0 \\ 1 \end{bmatrix}$ is a scalar multiple of the other, they are linearly independent. Thus, the set $\left\{\begin{bmatrix}1\\-1\\0\end{bmatrix}, \begin{bmatrix}0\\0\\1\end{bmatrix}\right\}$ is a basis for S. We conclude that $\dim S = 2$.

Example 7 The set S of all 2×2 matrices $\begin{bmatrix}a & b\\c & d\end{bmatrix}$ where $a + b = c$ is a subspace of M_2. Find its dimension.

Show that S is a subspace M_2.

Solution We proceed to find a basis for S. Since $a + b = c$ we may write

$$\begin{bmatrix}a & b\\c & d\end{bmatrix} = \begin{bmatrix}a & b\\a+b & d\end{bmatrix}$$

But

$$\begin{bmatrix}a & b\\a+b & d\end{bmatrix} = \begin{bmatrix}a & 0\\a & 0\end{bmatrix} + \begin{bmatrix}0 & b\\b & 0\end{bmatrix} + \begin{bmatrix}0 & 0\\0 & d\end{bmatrix}$$

where again each of the vectors (matrices) depends on only one arbitrary real number. This leads to the form

$$\begin{bmatrix}a & b\\a+b & d\end{bmatrix} = a\begin{bmatrix}1 & 0\\1 & 0\end{bmatrix} + b\begin{bmatrix}0 & 1\\1 & 0\end{bmatrix} + d\begin{bmatrix}0 & 0\\0 & 1\end{bmatrix}$$

Thus, the set S is spanned by the three matrices $\begin{bmatrix}1 & 0\\1 & 0\end{bmatrix}$, $\begin{bmatrix}0 & 1\\1 & 0\end{bmatrix}$, and $\begin{bmatrix}0 & 0\\0 & 1\end{bmatrix}$. Since the set $\left\{\begin{bmatrix}1 & 0\\1 & 0\end{bmatrix}, \begin{bmatrix}0 & 1\\1 & 0\end{bmatrix}, \begin{bmatrix}0 & 0\\0 & 1\end{bmatrix}\right\}$ is also linearly independent it is a basis for S. Thus it follows that $\dim S = 3$.

Show that the set $\left\{\begin{bmatrix}1 & 0\\1 & 0\end{bmatrix}, \begin{bmatrix}0 & 1\\1 & 0\end{bmatrix}, \begin{bmatrix}0 & 0\\0 & 1\end{bmatrix}\right\}$ is linearly independent.

We conclude this section by showing that *not* all vector spaces are finite-dimensional. We have stated previously that the vector space F of all real-valued functions defined on the real line is an infinite-dimensional vector space. We are now in a position to prove this statement. We must show that every *finite* subset of F *fails* to be a spanning set for F. We prove this by contradiction.

Suppose there exists a finite subset $S = \{f_1, f_2, \ldots, f_n\}$ which spans F. Then F is a finite-dimensional vector space spanned by S. Without loss of generality we may assume that S does not contain the zero function. It follows from Theorem 3-9 that there exists a subset $\bar{S}$ of S such that $\bar{S}$ is a basis for F. The number of elements in $\bar{S}$ will be at most n. This means that $\dim F \leq n$. Let us recall that a vector space of dimension n cannot have a linearly independent subset with more vectors than n. But the subset of F given by $\{1,$

If S contains the zero function we can delete it without affecting the spanning property of S.

See Section 3-1, problem 13.

$x, x^2, \ldots, x^n\}$ is a linearly independent set containing $n + 1$ vectors. This is a contradiction which stems from the assumption we made that the vector space F could be spanned by a finite subset S. We conclude that every finite subset of F fails to span F and therefore that the vector space F is infinite-dimensional.

Exercises

1 Let $S = \left\{\begin{bmatrix} a \\ b \end{bmatrix}, \begin{bmatrix} c \\ d \end{bmatrix}, \begin{bmatrix} e \\ f \end{bmatrix}\right\}$ be a subset of the vector space $\mathbf{R}^2$. Explain why S is *not* a basis for $\mathbf{R}^2$.

2 Let $T = \{a_2x^2 + a_1x + a_0, b_2x^2 + b_1x + b_0\}$ be a subset of the vector space P_2. Is it possible for T to be a basis for P_2? Explain.

3 Consider the vectors $\mathbf{v}_1 = \begin{bmatrix} 1 \\ 2 \\ 3 \end{bmatrix}$ and $\mathbf{v}_2 = \begin{bmatrix} 0 \\ 1 \\ 1 \end{bmatrix}$ in $\mathbf{R}^3$. Let S be the set of *all* linear combinations of $\mathbf{v}_1$ and $\mathbf{v}_2$. Show that S is a subspace of $\mathbf{R}^3$ and find $\dim S$.

4 Consider the vectors $\mathbf{u}_1 = \begin{bmatrix} 1 \\ 1 \\ 2 \end{bmatrix}$ and $\mathbf{u}_2 = \begin{bmatrix} -2 \\ -2 \\ -4 \end{bmatrix}$ in $\mathbf{R}^3$. Let T be the set of *all* linear combinations of $\mathbf{u}_1$ and $\mathbf{u}_2$. Show that T is a subspace of $\mathbf{R}^3$ and find $\dim T$.

5 Let $\{\mathbf{v}_1, \mathbf{v}_2, \ldots, \mathbf{v}_k\}$ be a linearly independent subset of a vector space V. Let S be the set of *all* linear combinations of $\mathbf{v}_1, \mathbf{v}_2, \ldots, \mathbf{v}_k$. Show that S is a subspace of V and find $\dim S$.

6 Consider the subset $S = \left\{\begin{bmatrix} 1 \\ 2 \end{bmatrix}, \begin{bmatrix} -3 \\ -6 \end{bmatrix}, \begin{bmatrix} \frac{1}{3} \\ \frac{2}{3} \end{bmatrix}, \begin{bmatrix} \frac{2}{5} \\ \frac{4}{5} \end{bmatrix}, \begin{bmatrix} 2 \\ 3 \end{bmatrix}\right\}$ of $\mathbf{R}^2$. Find a subset $\bar{S}$ of S such that $\bar{S}$ is a basis for $\mathbf{R}^2$.

7 Consider the subset $T = \left\{\begin{bmatrix} 1 \\ -1 \\ 2 \end{bmatrix}, \begin{bmatrix} -2 \\ 2 \\ -4 \end{bmatrix}, \begin{bmatrix} 1 \\ -2 \\ 6 \end{bmatrix}, \begin{bmatrix} 2 \\ -3 \\ 8 \end{bmatrix}, \begin{bmatrix} 0 \\ 1 \\ 2 \end{bmatrix}\right\}$ of $\mathbf{R}^3$. Find a subset $\bar{T}$ of T such that $\bar{T}$ is a basis for $\mathbf{R}^3$.

8 Consider the subset $S = \{2x^2 - x + 1, x^2 + 2x, -3x^2 + 1, x - 1, 5x^2 - 2x + 3\}$ of P_2. Find a subset $\bar{S}$ of S such that $\bar{S}$ is a basis for P_2.

9 Consider the subset $S = \left\{\begin{bmatrix} 1 \\ 2 \\ 3 \end{bmatrix}, \begin{bmatrix} 0 \\ 1 \\ 1 \end{bmatrix}, \begin{bmatrix} 2 \\ 3 \\ 2 \end{bmatrix}\right\}$ of $\mathbf{R}^3$. Show that S is a basis for $\mathbf{R}^3$ by using Theorem 3-10.

10 Consider the subset $S = \left\{ \begin{bmatrix} 1 & 0 \\ 0 & 0 \end{bmatrix}, \begin{bmatrix} 0 & 1 \\ 0 & 0 \end{bmatrix}, \begin{bmatrix} 0 & 0 \\ 1 & 0 \end{bmatrix}, \begin{bmatrix} 0 & 0 \\ 0 & 1 \end{bmatrix} \right\}$ of the vector space M_2 of all 2×2 matrices. Show that S is a basis for M_2. What is $\dim M_2$?

11 Consider the subset $T = \left\{ \begin{bmatrix} 1 & 0 \\ 0 & 1 \end{bmatrix}, \begin{bmatrix} 0 & 1 \\ 1 & 1 \end{bmatrix}, \begin{bmatrix} 1 & 1 \\ 0 & 0 \end{bmatrix}, \begin{bmatrix} 0 & 0 \\ 1 & 1 \end{bmatrix} \right\}$ of M_2. Show that T is a basis for M_2 by using Theorem 3-10.

12 Construct a basis for $\mathbf{R}^2$ which includes the vector $\begin{bmatrix} -2 \\ 3 \end{bmatrix}$.

13 Construct a basis for $\mathbf{R}^3$ which includes the vectors $\begin{bmatrix} 1 \\ 1 \\ 1 \end{bmatrix}, \begin{bmatrix} 1 \\ 2 \\ 1 \end{bmatrix}$.

14 Construct a basis for P_2 which includes the vector $2x^2 + 3x - 4$.

15 Construct a basis for M_2 which includes the vector $\begin{bmatrix} 1 & 3 \\ 2 & 4 \end{bmatrix}$.

16 Construct a basis for P_3 which includes the vectors $2x^3 + x^2 - 3x$ and $2x^2 - 4x$.

17 Find a basis for the vector space M_3 of all 3×3 matrices. What is $\dim M_3$?

18 Show that each of the following sets is a subspace of $\mathbf{R}^2$. Find a basis for each subspace and determine its dimension.

a the set of all vectors $\begin{bmatrix} a \\ b \end{bmatrix}$ where $a = b$

b the set of all vectors $\begin{bmatrix} a \\ b \end{bmatrix}$ where $a + b = 0$

c the set of all vectors $\begin{bmatrix} a \\ b \end{bmatrix}$ where $a = 5b$

19 Show that each of the following sets is a subspace of $\mathbf{R}^3$. Find a basis for each subspace and determine its dimension.

a all vectors $\begin{bmatrix} a \\ b \\ c \end{bmatrix}$ where $a = b = c$

b all vectors $\begin{bmatrix} a \\ b \\ c \end{bmatrix}$ where $a + c = 0$

c all vectors $\begin{bmatrix} a \\ b \\ c \end{bmatrix}$ where $a + b - c = 0$

d all vectors of the form $\begin{bmatrix} a \\ 0 \\ c \end{bmatrix}$

e all vectors $\begin{bmatrix} a \\ b \\ c \end{bmatrix}$ where $a = 3b$ and $b = 5c$

20 Show that each of the following sets is a subspace of P_2. Find a basis for each subspace and determine its dimension.

a all vectors $ax^2 + bx + c$ where $c = 0$
b all vectors $ax^2 + bx + c$ where $a - b = c$
c all vectors $ax^2 + bx + c$ where $a = b = -c$
d all vectors $ax^2 + bx + c$ where $a + b = 3c$
e all vectors $ax^2 + bx + c$ where $2a = b = 3c$

21 Find the dimension of the subspace S of M_2 spanned by the set

$$\left\{\begin{bmatrix} 1 & 1 \\ 1 & 1 \end{bmatrix}, \begin{bmatrix} 3 & 0 \\ 2 & 1 \end{bmatrix}, \begin{bmatrix} 2 & -1 \\ 1 & 0 \end{bmatrix}, \begin{bmatrix} 2 & 2 \\ 2 & 2 \end{bmatrix}, \begin{bmatrix} 4 & 1 \\ 3 & 2 \end{bmatrix}\right\}.$$

22 Prove Theorem 3-11.

23 Show that each of the following sets is a subspace of M_2. Find a basis for each subspace and determine its dimension.

a the set of all vectors $\begin{bmatrix} a & b \\ c & d \end{bmatrix}$ where $c = b$

b the set of all vectors $\begin{bmatrix} a & b \\ c & d \end{bmatrix}$ where $b = -c$ and $a = d = 0$

24 Show that each of the following sets is a subspace of M_3. Find a basis for each subspace and determine its dimension.

The set of all vectors $\begin{bmatrix} a_{11} & a_{12} & a_{13} \\ a_{21} & a_{22} & a_{23} \\ a_{31} & a_{32} & a_{33} \end{bmatrix}$

a where $a_{12} = -a_{21}$, $a_{13} = a_{31}$, and $a_{23} = a_{32}$
b where $a_{12} = -a_{21}$, $a_{13} = -a_{31}$, $a_{23} = -a_{32}$, and $a_{11} = a_{22} = a_{33} = 0$

25 Prove Theorem 3-13.

26 a Show that $\mathbf{R}^2$ cannot be spanned by any of its subsets which contain only one vector.
b Show that P_2 cannot be spanned by any of its subsets which contain only one or only two vectors.

3-5 Rank of a Matrix

Objectives
1 Introduce the concept of a rank of a matrix.
2 Show that an $n \times n$ matrix is invertible if and only if its rank is n.

GEORG F. FROBENIUS (1848–1917), *a German mathematician, was among the early researchers in the theory of matrices. In 1879 he introduced the notion of the rank of a matrix in connection with his studies of determinants. He also introduced a method of finding series solutions of differential equations about a regular singular point. In 1896 he developed the theory of group characters for non-Abelian groups—a useful tool in quantum mechanics.*

Let A be a matrix of order $m \times n$. Let us denote the rows of A by $\mathbf{u}_1$, $\mathbf{u}_2, \ldots, \mathbf{u}_m$. Thus, we have

$$\begin{aligned}\mathbf{u}_1 &= (a_{11}, a_{12}, \ldots, a_{1n})\\ \mathbf{u}_2 &= (a_{21}, a_{22}, \ldots, a_{2n})\\ &\vdots\\ \mathbf{u}_m &= (a_{m1}, a_{m2}, \ldots, a_{mn})\end{aligned}$$

Since each $\mathbf{u}_i$ has n entries it may be viewed as a vector in $\mathbf{R}^n$. We refer to the $\mathbf{u}_i$'s as **row vectors** of the matrix A. Thus

$$\begin{aligned}\mathbf{u}_i + \mathbf{u}_j &= (a_{i1}, a_{i2}, \ldots, a_{in}) + (a_{j1}, a_{j2}, \ldots, a_{jn})\\ &= (a_{i1} + a_{j1}, a_{i2} + a_{j2}, \ldots, a_{in} + a_{jn})\end{aligned}$$

and

$$\alpha\mathbf{u}_i = \alpha(a_{i1}, a_{i2}, \ldots, a_{in}) = (\alpha a_{i1}, \alpha a_{i2}, \ldots, \alpha a_{in})$$

Let S be the set of all linear combinations of $\mathbf{u}_1, \mathbf{u}_2, \ldots, \mathbf{u}_m$. Then we know that S is a subspace of $\mathbf{R}^n$ spanned by the vectors $\mathbf{u}_1, \mathbf{u}_2, \ldots, \mathbf{u}_m$ (see Theorem 3-4). This leads to the following definition.

Definition 3-8 Let A be an $m \times n$ matrix. The subspace of $\mathbf{R}^n$ spanned by the row vectors of A is called the **row space** of A.

The row space of a matrix.

Example 1 Let $A = \begin{bmatrix}1 & 0 & 2 & -1 & 3\\ 2 & 4 & 0 & 1 & 0\\ 3 & 4 & 2 & 0 & 3\end{bmatrix}$. Find a basis for the row space of A.

Solution The row vectors of A are $\mathbf{u}_1 = (1,0,2,-1,3)$, $\mathbf{u}_2 = (2,4,0,1,0)$, and $\mathbf{u}_3 = (3,4,2,0,3)$. By inspection we find that $\mathbf{u}_3 = \mathbf{u}_1 + \mathbf{u}_2$. Thus, the row space of A is also spanned by $\mathbf{u}_1$ and $\mathbf{u}_2$. Since $\mathbf{u}_1$ and $\mathbf{u}_2$ are not scalar multiples of each other, it follows that $\{\mathbf{u}_1,\mathbf{u}_2\}$ is a linearly independent set. Therefore, the row vectors $\mathbf{u}_1$ and $\mathbf{u}_2$ form a basis for the row space of A. The dimension of the row space of A is equal to 2 because a basis for it consists of the two row vectors $\mathbf{u}_1$ and $\mathbf{u}_2$.

If A is an $m \times n$ matrix, then its row space is spanned by a finite set of m vectors, namely, the row vectors $\mathbf{u}_1, \mathbf{u}_2, \ldots, \mathbf{u}_m$. The following procedure is used in constructing a basis for the row space of A (if it exists).

Discard a row vector $\mathbf{u}_i$ if it is a linear combination of the remaining $\mathbf{u}$'s. After a finite number of such discards, the procedure terminates with one of the following two possibilities:

1 We have discarded all the $\mathbf{u}$'s. This can happen if and only if all the rows of A consist of 0 entries. The row space of A consists of the zero vector only and therefore does not have a basis.

2 We have discarded some (but not all) of the **u**'s and the remaining k row vectors form a linearly independent set of vectors. In this case the dimension of the row space of the matrix A is $k > 0$.

This leads to the following definition.

Definition 3-9

The row rank of a matrix.

The dimension of the row space of a matrix A is called the **row rank** of A.

Thus, if we say that the row rank of a matrix A is a positive integer k, it means that we can find k row vectors in A which form a linearly independent set, while every collection of rows of A with more than k rows contains a row vector of the matrix A that is a linear combination of the remaining ones.

In order to establish an efficient procedure for finding a basis for the row space of a given matrix A and thereby also determine the row rank of A, we state the following important theorem.

Theorem 3-14

The proof of this theorem is left for the student as an exercise.

Let E be an elementary matrix. Let A and B be matrices such that $B = EA$. Then the row space of A is equal to the row space of B.

This theorem states that elementary row operations performed on a given matrix A do not change the row space of A. This suggests the following procedure for finding a basis for the row space of a given matrix A.

Apply elementary row operations to the given matrix until it is in the form of a row-reduced echelon matrix B. It turns out that the set of all nonzero row vectors in B forms a basis for the row space of A. The following example illustrates the procedure.

Example 2

Determine the row rank of the matrix A given by

$$A = \begin{bmatrix} 1 & 2 & 0 & 0 & 0 & 2 \\ -1 & -2 & 1 & 3 & 0 & -1 \\ 2 & 4 & 1 & 3 & 1 & 9 \\ 1 & 2 & 1 & 3 & 0 & 3 \end{bmatrix}$$

Solution

The student is encouraged to obtain B from A.

We perform elementary row operations on the matrix A until we have transformed it to row-reduced echelon form. The reader can easily show that the matrix B thus obtained is given by

$$B = \begin{bmatrix} 1 & 2 & 0 & 0 & 0 & 2 \\ 0 & 0 & 1 & 3 & 0 & 1 \\ 0 & 0 & 0 & 0 & 1 & 4 \\ 0 & 0 & 0 & 0 & 0 & 0 \end{bmatrix}$$

The nonzero row vectors in B are linearly independent. To show this, we write

$$\alpha(1,2,0,0,0,2) + \beta(0,0,1,3,0,1) + \gamma(0,0,0,0,1,4) = (0,0,0,0,0,0)$$

Performing the operations of scalar multiplication and addition on the left-hand side of the vector equation, we obtain

$$(\alpha, 2\alpha, \beta, 3\beta, \gamma, 2\alpha + \beta + 4\gamma) = (0,0,0,0,0,0)$$

The last vector equation is equivalent to the system

$$\begin{aligned} \alpha \qquad\qquad &= 0 \\ 2\alpha \qquad\qquad &= 0 \\ \beta \qquad &= 0 \\ 3\beta \qquad &= 0 \\ \gamma &= 0 \\ 2\alpha + \beta + 4\gamma &= 0 \end{aligned}$$

which obviously implies $\alpha = \beta = \gamma = 0$.

Hence, the nonzero row vectors $(1,2,0,0,0,2)$, $(0,0,1,3,0,1)$, and $(0,0,0,0,1,4)$ of the matrix B are linearly independent and form a basis for the row space of the matrix A. It follows that the row rank of A is equal to 3.

That the nonzero row vectors of the row-reduced echelon matrix B are linearly independent is not accidental, as the following theorem indicates.

Theorem 3-15 Let B be a row-reduced echelon matrix. Then the nonzero row vectors of B are linearly independent.

The proof of this theorem follows the pattern exhibited in Example 2 above and is left as an exercise.

So far we have dealt only with the rows of a matrix A. Now we turn our attention to the columns of the matrix.

Let A be an $m \times n$ matrix. Let us denote the columns of A by $\mathbf{v}_1, \mathbf{v}_2, \ldots, \mathbf{v}_n$. Thus, we have

$$\mathbf{v}_1 = \begin{bmatrix} a_{11} \\ a_{21} \\ \vdots \\ a_{m1} \end{bmatrix}, \mathbf{v}_2 = \begin{bmatrix} a_{12} \\ a_{22} \\ \vdots \\ a_{m2} \end{bmatrix}, \ldots, \mathbf{v}_n = \begin{bmatrix} a_{1n} \\ a_{2n} \\ \vdots \\ a_{mn} \end{bmatrix}$$

Since each $\mathbf{v}_i$ has m entries we may look upon $\mathbf{v}_i$ as an element of $\mathbf{R}^m$. We call each $\mathbf{v}_i$ a **column vector** of the matrix A.

Let T be the set of all linear combinations of $\mathbf{v}_1, \mathbf{v}_2, \ldots, \mathbf{v}_n$. Then we know that T is a subspace of $\mathbf{R}^m$ spanned by the vectors $\mathbf{v}_1, \mathbf{v}_2, \ldots, \mathbf{v}_n$. We are led to the following definition.

Definition 3-10 Let A be an $m \times n$ matrix. The subspace of $\mathbf{R}^m$ spanned by the column vectors of A is called the **column space** of A.

The column space of a matrix.

Example 3 Let $A = \begin{bmatrix} 1 & 0 & 2 & -1 & 3 \\ 2 & 4 & 0 & 1 & 0 \\ 3 & 4 & 2 & 0 & 3 \end{bmatrix}$. Find a basis for the column space of A.

Solution Let

$$\mathbf{v}_1 = \begin{bmatrix} 1 \\ 2 \\ 3 \end{bmatrix}, \mathbf{v}_2 = \begin{bmatrix} 0 \\ 4 \\ 4 \end{bmatrix}, \mathbf{v}_3 = \begin{bmatrix} 2 \\ 0 \\ 2 \end{bmatrix}, \mathbf{v}_4 = \begin{bmatrix} -1 \\ 1 \\ 0 \end{bmatrix}, \text{ and } \mathbf{v}_5 = \begin{bmatrix} 3 \\ 0 \\ 3 \end{bmatrix}$$

Note: $\mathbf{v}_1$ and $\mathbf{v}_2$ are not proportional to each other.

be the column vectors of A. The vectors $\mathbf{v}_1$ and $\mathbf{v}_2$ are obviously linearly independent. But the vectors $\mathbf{v}_3$, $\mathbf{v}_4$, and $\mathbf{v}_5$ are linear combinations of $\mathbf{v}_1$ and $\mathbf{v}_2$, as indicated by the following equations:

$$\mathbf{v}_3 = 2\mathbf{v}_1 + (-1)\mathbf{v}_2, \mathbf{v}_4 = (-1)\mathbf{v}_1 + \tfrac{3}{4}\mathbf{v}_2, \quad \text{and} \quad \mathbf{v}_5 = 3\mathbf{v}_1 - \tfrac{3}{2}\mathbf{v}_2$$

Thus, in discarding the vectors $\mathbf{v}_3$, $\mathbf{v}_4$, and $\mathbf{v}_5$ we are left with the set $\{\mathbf{v}_1,\mathbf{v}_2\}$, which spans the column space of A and is also linearly independent. Therefore, the column vectors $\mathbf{v}_1$ and $\mathbf{v}_2$ form a basis for the column space of A. The dimension of the column space of the matrix A is 2, as indicated by the number of vectors in a basis for it.

This example shows how one may construct a basis for the column space of a given matrix A of order $m \times n$. The number of column vectors in such a basis is given a special name, as the following definition explains.

Definition 3-11 The dimension of the column space of a matrix A is called the **column rank** of A.

The column rank of a matrix.

Thus, if the column rank of a matrix A is a positive number k, one can find k column vectors in A which form a linearly independent set, while every collection of columns of A with more than k columns is a linearly dependent set.

We wish to establish an efficient method of finding a basis for the column space of a given matrix A. In establishing a procedure for finding a basis for the row space of a matrix A, we made use of the fact that elementary row operations performed on A did not change the row space of A. Unfortunately this is not true with respect to the column space of A. A quick look at Example 2 shows that the matrix B was obtained from the matrix A by elementary row operations, but the column space of A is different from the column space of B. For instance, the first column of A, given by

$$\begin{bmatrix} 1 \\ -1 \\ 2 \\ 1 \end{bmatrix}$$

cannot be generated by any linear combination of the columns of the matrix B. (Why?) In order to overcome this difficulty, we consider a matrix A^t that is obtained from A by interchanging the roles of rows and columns in A. The following definition clarifies the structure of A^t.

Definition 3-12 The transpose of a matrix. Let A be an $m \times n$ matrix. The **transpose** of A, denoted by A^t, is an $n \times m$ matrix whose first row is the first column (from the left) of A, whose second row is the second column of A, and so on up to the nth row of A^t, which is the nth column of A.

Example 4 Let the matrices A, B, and C be given by

$$A = \begin{bmatrix} 1 & 2 & 3 \\ 4 & 5 & 6 \end{bmatrix}, B = [1,2,3,4,5], C = \begin{bmatrix} c_{11} & c_{12} & c_{13} & c_{14} \\ c_{21} & c_{22} & c_{23} & c_{24} \\ c_{31} & c_{32} & c_{33} & c_{34} \end{bmatrix}$$

Find A^t, B^t, and C^t.

Solution $$A^t = \begin{bmatrix} 1 & 4 \\ 2 & 5 \\ 3 & 6 \end{bmatrix}, B^t = \begin{bmatrix} 1 \\ 2 \\ 3 \\ 4 \\ 5 \end{bmatrix} C^t = \begin{bmatrix} c_{11} & c_{21} & c_{31} \\ c_{12} & c_{22} & c_{32} \\ c_{13} & c_{23} & c_{33} \\ c_{14} & c_{24} & c_{34} \end{bmatrix}$$

We are now ready to establish a procedure for finding a basis for the column space of a matrix A.

From the given matrix A we construct the matrix A^t. Since the rows of A^t are columns of A, it follows that the row space of A^t and the column space of A are identical. Thus, a basis for the row space of A^t will also yield a basis for the column space of A.

We transform the matrix A^t by elementary row operations into a row-reduced echelon matrix B. The nonzero rows of B form a basis for the row space of A^t. If we now construct B^t, we find that the nonzero columns of B^t form a basis for the column space of the original matrix A.

Example 5 Determine the column rank of the matrix A where

$$A = \begin{bmatrix} 1 & 2 & 0 & 0 & 0 & 2 \\ -1 & -2 & 1 & 3 & 0 & -1 \\ 2 & 4 & 1 & 3 & 1 & 9 \\ 1 & 2 & 1 & 3 & 0 & 3 \end{bmatrix}$$

Solution We construct the transpose of A and obtain

$$A^t = \begin{bmatrix} 1 & -1 & 2 & 1 \\ 2 & -2 & 4 & 2 \\ 0 & 1 & 1 & 1 \\ 0 & 3 & 3 & 3 \\ 0 & 0 & 1 & 0 \\ 2 & -1 & 9 & 3 \end{bmatrix}$$

We transform A^t by elementary row operations into a row-reduced echelon matrix B, given by

$$B = \begin{bmatrix} 1 & 0 & 0 & 2 \\ 0 & 1 & 0 & 1 \\ 0 & 0 & 1 & 0 \\ 0 & 0 & 0 & 0 \\ 0 & 0 & 0 & 0 \\ 0 & 0 & 0 & 0 \end{bmatrix}$$

Now, construct the transpose of B, obtaining

$$B^t = \begin{bmatrix} 1 & 0 & 0 & 0 & 0 & 0 \\ 0 & 1 & 0 & 0 & 0 & 0 \\ 0 & 0 & 1 & 0 & 0 & 0 \\ 2 & 1 & 0 & 0 & 0 & 0 \end{bmatrix}$$

The nonzero columns of B^t, given by

$$\mathbf{v}_1 = \begin{bmatrix} 1 \\ 0 \\ 0 \\ 2 \end{bmatrix}, \quad \mathbf{v}_2 = \begin{bmatrix} 0 \\ 1 \\ 0 \\ 1 \end{bmatrix}, \quad \text{and} \quad \mathbf{v}_3 = \begin{bmatrix} 0 \\ 0 \\ 1 \\ 0 \end{bmatrix}$$

form a basis for the column space of the matrix A. Thus, the column rank of A is equal to 3.

It follows from Examples 2 and 5 that the row rank of the matrix A and its column rank are both equal to 3. Actually, this is no coincidence because it is so for every matrix, as indicated by the following theorem.

Theorem 3-16 Let A be an arbitrary $m \times n$ matrix. Then the row rank of A and the column rank of A are equal.

The proof of Theorem 3-16, which is somewhat involved, will be omitted.

For a proof of Theorem 3-16, see *Introduction to Linear Algebra* by F. Hohn (Macmillan, 1972), p. 164.

Based on Theorem 3-16, we now define the concept of the rank of a matrix.

Definition 3-13 Let A be an $m \times n$ matrix. The **rank** of A, denoted by $r(A)$, is equal to the row rank (or column rank) of A.

The rank of a matrix.

Since the identity matrix I_n has n linearly independent rows (and also n linearly independent columns), it follows from Definition 3-13 that its rank is given by $r(I_n) = n$.

Suppose a matrix A is $n \times n$. If A can be transformed by elementary row operations into the identity matrix I_n (which is obviously a row-reduced echelon matrix) then it follows that the ranks of A and I_n satisfy the equation

Remember that row operations do not change the rank of A.

$$r(A) = r(I_n) = n$$

Conversely, given that a matrix A of order $n \times n$ has rank equal to n, the row-reduced echelon matrix B, obtained from A by elementary row operations performed on A, also has rank equal to n. This means that B must have n nonzero rows each of which contains a leading entry 1. Since B is a row-reduced echelon matrix we must have $B = I_n$.

We have therefore proved the following theorem.

Theorem 3-17 Let A be a square $n \times n$ matrix. Then the rank of A equals n if and only if there exists a sequence of elementary matrices E_1, E_2, . . . , E_k such that

Remember that elementary matrices represent elementary row operations.

$$(E_k \dots E_1)A = I_n \qquad [1]$$

The reader should recall that equation 1 may be interpreted as saying that A is a nonsingular matrix whose inverse A^{-1} is given by

$$A^{-1} = E_k \dots E_1.$$

This leads immediately to the following result.

Theorem 3-18 Let A be a square $n \times n$ matrix. Then A is nonsingular if and only if the rank of A is equal to n.

Proof Suppose A is nonsingular. Then there exists a matrix A^{-1} such that

$$A^{-1}A = I_n \qquad [2]$$

Recall that every nonsingular matrix may be expressed as a product of a finite number of elementary matrices. Since A^{-1} is nonsingular (as A is), there exist k elementary matrices $E_1, E_2, \ldots, E_k$ such that

$$A^{-1} = E_k \ldots E_2E_1 \qquad [3]$$

It follows from equations 2 and 3 that

$$(E_k \ldots E_2E_1)A = I_n \qquad [4]$$

Since elementary row operations do not change the rank of A, it follows from equation 4 that

$$r(A) = r(I_n) = n$$

Thus, we have proved that if A is nonsingular, the rank of A is equal to n. We must now show that if A has rank n, A is nonsingular.

Assume that the rank of A is equal to n. Then, Theorem 3-17 tells us that there exists a sequence of elementary matrices $E_1, E_2, \ldots, E_k$ such that

$$(E_k \ldots E_2E_1)A = I_n \qquad [5]$$

But equation 5 expresses the fact that A is nonsingular and that its inverse A^{-1} is given by

$$A^{-1} = E_k \ldots E_2E_1$$

This completes the proof.

Exercises

1 Find a basis for the row space of each of the following matrices. State the dimension of the row space for each matrix.

a $\begin{bmatrix} 3 & 4 \\ 1 & -1 \\ 5 & 0 \end{bmatrix}$ **b** $\begin{bmatrix} 2 & 1 & 0 \\ -1 & 3 & 4 \\ 2 & 8 & 8 \end{bmatrix}$

c $\begin{bmatrix} 5 & -1 & 2 \\ 4 & 4 & -8 \\ 3 & 1 & -2 \\ 2 & -2 & 4 \end{bmatrix}$ **d** $\begin{bmatrix} 2 & -1 & 3 & 0 & -4 \\ 7 & -1 & 6 & 2 & -9 \\ 5 & 5 & -6 & 6 & 5 \\ 3 & 1 & 0 & 2 & -1 \end{bmatrix}$

2 Express the rows of each matrix in problem 1 as a linear combination of the vectors in the basis which you found for the row space of the matrix.

3 Find the transpose of each of the following matrices.

a $\begin{bmatrix} 2 & 3 \\ 1 & 5 \end{bmatrix}$ **b** $\begin{bmatrix} 1 \\ 2 \\ 3 \end{bmatrix}$ **c** $\begin{bmatrix} 1 & 2 & 3 & 4 \\ 5 & 6 & 7 & 8 \end{bmatrix}$

d $\begin{bmatrix} 0 & 1 & 2 & 3 \\ -1 & 0 & 4 & 5 \\ -2 & -4 & 0 & 6 \\ -3 & -5 & -6 & 0 \end{bmatrix}$ **e** $\begin{bmatrix} 1 & 5 & 6 & 7 \\ 5 & 2 & 8 & 9 \\ 6 & 8 & 3 & 0 \\ 7 & 9 & 0 & 4 \end{bmatrix}$

4 Find a basis for the column space of each of the following matrices. State the dimension of the column space for each matrix.

a $\begin{bmatrix} 1 & 2 & 3 \\ 4 & 5 & 6 \end{bmatrix}$. **b** $\begin{bmatrix} 3 & 1 & -2 \\ -1 & 2 & 0 \\ -3 & 4 & 1 \end{bmatrix}$

c $\begin{bmatrix} 2 & -3 & 1 & -1 \\ -4 & 6 & -2 & 2 \\ 6 & -9 & 3 & -3 \end{bmatrix}$ **d** $\begin{bmatrix} 1 & -2 & -3 & 5 & 2 \\ 2 & 6 & 4 & 0 & -6 \\ 3 & 5 & 2 & 4 & -5 \\ 4 & 3 & -1 & 9 & -3 \end{bmatrix}$

5 Express the columns of each matrix in problem 4 as a linear combination of the vectors in the basis which you found for the column space of the matrix.

6 **a** Construct a 2×3 matrix having rank equal to 0.
b Construct a 3×2 matrix having rank equal to 1.
c Construct a 3×4 matrix having rank equal to 2.
d Construct a 4×4 matrix having rank equal to 3.
e Construct a 5×5 matrix having rank equal to 5.

7 Find the rank of each of the following matrices. Determine which of the matrices is nonsingular.

a $\begin{bmatrix} 3 & 2 \\ -6 & -4 \end{bmatrix}$ **b** $\begin{bmatrix} 1 & 0 \\ 0 & 1 \\ 3 & 5 \end{bmatrix}$

c $\begin{bmatrix} 0 & 2 & 3 \\ 3 & 1 & 0 \\ -2 & 0 & 1 \end{bmatrix}$ **d** $\begin{bmatrix} 0 & 1 & 1 & 1 \\ -1 & 0 & 1 & 1 \\ -1 & -1 & 0 & 1 \\ -1 & -1 & -1 & 0 \end{bmatrix}$

8 Determine the rank of all elementary 3×3 matrices. Explain your answer.

9 What is the relationship between $r(A)$ and $r(A^t)$ where A is an arbitrary matrix? Explain your answer.

10 Let A be an $m \times n$ matrix and let P be an $m \times m$ nonsingular matrix. Show that $r(PA) = r(A)$.

11 Let A be an $m \times n$ matrix. Explain why $(A^t)^t = A$.

12 Let A be an $m \times n$ matrix and B an $n \times k$ matrix. Show that $(AB)^t = B^tA^t$.

Notice the reverse in the order of the matrices A^t and B^t.

13 Show that if A is a nonsingular $n \times n$ matrix, then A^t is also nonsingular and $(A^t)^{-1} = (A^{-1})^t$.

Hint: Consider the equation $A^{-1}A = I$. Take the transpose of each side and apply problem 12.

14 Let A be an $m \times n$ matrix and let Q be an $n \times n$ nonsingular matrix. Show that $r(AQ) = r(A)$.

15 Let A be an $m \times n$ matrix, B an $m \times m$ matrix, and C an $n \times n$ matrix. Show that if B and C are nonsingular matrices, then $r(BAC) = r(A)$.

3-6 Structure of Solutions to Linear Systems of Equations

Objective
Study the structure of solutions to linear systems of equations with the aid of concepts such as linear independence, rank of a matrix, and dimension of a vector space.

The technique of solving a system of linear equations has already been established in Chapter 1. We have seen that a homogeneous system can have a unique solution or infinitely many solutions. We have also encountered nonhomogeneous linear systems having a unique solution, infinitely many solutions, or no solution at all. We did not have the mathematical tools to investigate such phenomena at an early stage of our study. But now, having gained familiarity with such concepts as vector space, basis, and dimension, we are ready to have a fresh look at the nature of the solutions of linear systems and to investigate thoroughly their structure.

If the homogeneous system $A\mathbf{x} = \mathbf{0}$ has a unique solution, it must be the trivial solution $\mathbf{x} = \mathbf{0}$.

First we shall deal with the homogeneous linear system

$$A\mathbf{x} = \mathbf{0} \qquad [1]$$

where A is an $m \times n$ matrix and $\mathbf{x} = \begin{bmatrix} x_1 \\ x_2 \\ \vdots \\ x_n \end{bmatrix}$.

Obviously N is a subset of $\mathbf{R}^n$.

Let us denote by N the set of all solutions of system 1. The trivial solution $\mathbf{x} = \mathbf{0}$ certainly belongs to N, so N is not an empty set. We wish to show that N together with the standard operations defined on $\mathbf{R}^n$ is a subspace of $\mathbf{R}^n$.

Let $\mathbf{u} = \begin{bmatrix} u_1 \\ \vdots \\ u_n \end{bmatrix}$ and $\mathbf{v} = \begin{bmatrix} v_1 \\ \vdots \\ v_n \end{bmatrix}$ be any elements of N and let α be any real number. Then we have

$$A\mathbf{u} = \mathbf{0} \qquad \text{and} \qquad A\mathbf{v} = \mathbf{0}$$

We must show that $\mathbf{u} + \mathbf{v}$ and $\alpha\mathbf{u}$ are also elements of the set N

and that therefore they must satisfy the equation

$$A(\mathbf{u} + \mathbf{v}) = \mathbf{0} \qquad \text{and} \qquad A(\alpha\mathbf{u}) = \mathbf{0}$$

Using properties of matrix algebra, we get

See Theorem 1-9.

$$A(\mathbf{u} + \mathbf{v}) = A\mathbf{u} + A\mathbf{v} = \mathbf{0} + \mathbf{0} = \mathbf{0}$$

and

See Theorem 1-7.

$$A(\alpha\mathbf{u}) = \alpha(A\mathbf{u}) = \alpha\mathbf{0} = \mathbf{0}$$

This proves that $\mathbf{u} + \mathbf{v}$ and $\alpha\mathbf{u}$ are also solutions of system 1. We have therefore proved the following theorem.

Theorem 3-19 Let A be an $m \times n$ matrix. Then the set of all solutions of the homogeneous system $A\mathbf{x} = \mathbf{0}$ is a subspace of $\mathbf{R}^n$.

The null space of a matrix

We call the subspace of $\mathbf{R}^n$ that is mentioned in Theorem 3-19 the **null space** of the matrix A. Since the null space of any $m \times n$ matrix A is a subspace of $\mathbf{R}^n$, it follows from Theorem 3-13 that

$$\dim\,(\text{null space of } A) \leq \dim\,(\mathbf{R}^n) = n$$

If the null space of A consists of the zero vector only, then it has no basis (why?) and $\dim\,(\text{null space of } A) = 0$. But if the null space of A contains nonzero vectors, then it has a finite basis consisting of *at most n* vectors, each of which is a solution of the linear system $A\mathbf{x} = \mathbf{0}$. In this case, every solution of the last equation can be represented as a linear combination of the solution vectors contained in a basis for the null space of the matrix A. The following example illustrates such a representation.

Example 1 Consider the homogeneous linear system

$$\begin{aligned} 2x_1 - x_2 + 4x_3 + x_4 - x_5 &= 0 \\ 3x_1 + 2x_2 - 2x_3 - 5x_4 + 2x_5 &= 0 \\ x_1 + x_2 + 6x_3 + 2x_4 + x_5 &= 0 \end{aligned} \qquad [2]$$

Find a basis for the null space of the matrix of coefficients of system 2 and express every solution of the system as a linear combination of the "solution vectors" in that basis.

Solution The matrix of coefficients is given by

$$A = \begin{bmatrix} 2 & -1 & 4 & 1 & -1 \\ 3 & 2 & -2 & -5 & 2 \\ 1 & 1 & 6 & 2 & 1 \end{bmatrix}$$

Using elementary row operations we transform A into row-reduced echelon form given by

$$B = \begin{bmatrix} 1 & 0 & 0 & -\frac{12}{13} & 0 \\ 0 & 1 & 0 & -\frac{7}{13} & 1 \\ 0 & 0 & 1 & \frac{15}{26} & 0 \end{bmatrix}$$

The corresponding system of equations is given by

$$\begin{aligned} x_1 \qquad + (-\tfrac{12}{13})x_4 \qquad &= 0 \\ x_2 \quad + (-\tfrac{7}{13})x_4 + x_5 &= 0 \\ x_3 + \quad (\tfrac{15}{26})x_4 \qquad &= 0 \end{aligned}$$

Solving for the leading unknowns x_1, x_2, and x_3, we obtain

$$\begin{aligned} x_1 &= \tfrac{12}{13}x_4 \\ x_2 &= \tfrac{7}{13}x_4 - x_5 \\ x_3 &= -\tfrac{15}{26}x_4 \end{aligned}$$

where x_4 and x_5 are arbitrary real numbers. Thus, if $\begin{bmatrix} x_1 \\ x_2 \\ x_3 \\ x_4 \\ x_5 \end{bmatrix}$ is a solution vector for the system $A\mathbf{x} = \mathbf{0}$, then we have

$$\begin{aligned} x_1 &= \tfrac{12}{13}t \\ x_2 &= \tfrac{7}{13}t - s \\ x_3 &= -\tfrac{15}{26}t \\ x_4 &= t \\ x_5 &= s \end{aligned}$$

where we have substituted t and s for x_4 and x_5, respectively, to indicate that they are arbitrary real numbers.

We now rewrite each equation and obtain

$$\begin{aligned} x_1 &= \tfrac{12}{13}t + 0s \\ x_2 &= \tfrac{7}{13}t + (-1)s \\ x_3 &= -\tfrac{15}{26}t + 0s \\ x_4 &= 1t + 0s \\ x_5 &= 0t + 1s \end{aligned}$$

The last equations can be written in terms of a vector equation as follows:

$$\begin{bmatrix} x_1 \\ x_2 \\ x_3 \\ x_4 \\ x_5 \end{bmatrix} = t \begin{bmatrix} \frac{12}{13} \\ \frac{7}{13} \\ -\frac{15}{26} \\ 1 \\ 0 \end{bmatrix} + s \begin{bmatrix} 0 \\ -1 \\ 0 \\ 0 \\ 1 \end{bmatrix} \qquad [3]$$

Equation 3 tells us that in order to obtain *all* possible solutions of linear system 2 we must generate *all* possible linear combinations of the two vectors

$$\mathbf{u} = \begin{bmatrix} \frac{12}{13} \\ \frac{7}{13} \\ -\frac{15}{26} \\ 1 \\ 0 \end{bmatrix} \quad \text{and} \quad \mathbf{v} = \begin{bmatrix} 0 \\ -1 \\ 0 \\ 0 \\ 1 \end{bmatrix}$$

Notice that $\mathbf{u}$ is obtained as a solution if we choose $t = 1$ and $s = 0$ in equation 3, while $\mathbf{v}$ is obtained as a solution if we choose $t = 0$ and $s = 1$ in equation 3. Since $\mathbf{u}$ and $\mathbf{v}$ are obviously linearly independent, they form a basis for the null space of the matrix A.

In general, one can easily determine the dimension of the null space of any given matrix by using the following theorem.

Theorem 3-20 Let A be an $m \times n$ matrix and let $r(A)$ denote the rank of A. Then the dimension of the null space of A is equal to $n - r(A)$.

The proof of this theorem is tedious and is therefore omitted. For a proof of Theorem 3-20, see *Introduction to Linear Algebra* by F. Hohn, p. 175.

In Example 1 we have $n = 5$ and $r(A) = 3$. Thus the dimension of the null space of A is 2. This matches our findings in Example 1.

We now turn our attention to the case of a nonhomogeneous linear system

$$A\mathbf{x} = \mathbf{b}, \quad (\mathbf{b} \neq \mathbf{0}) \tag{4}$$

The situation here is quite different from the one we encountered in the case of a homogeneous system. The set of all solutions of system 4 is *no longer* a subspace of $\mathbf{R}^n$. In order to see that, let $\mathbf{u}$ and $\mathbf{v}$ be solutions of system 4. Then we have

$$A\mathbf{u} = \mathbf{b} \quad \text{and} \quad A\mathbf{v} = \mathbf{b}$$

For the sum $\mathbf{u} + \mathbf{v}$ we obtain

$$A(\mathbf{u} + \mathbf{v}) = A\mathbf{u} + A\mathbf{v} = \mathbf{b} + \mathbf{b} = 2\mathbf{b}$$

Thus, $\mathbf{u} + \mathbf{v}$ is *not* a solution of system 4, because $2\mathbf{b} \neq \mathbf{b}$. We can also show that if $\mathbf{u}$ is a solution of system 4 and α is any real number different from 1, then $\alpha\mathbf{u}$ is not a solution of system 4.

Note that $2\mathbf{b} = \mathbf{b}$ if and only if $\mathbf{b} = \mathbf{0}$.

This follows immediately from the equation $A(\alpha\mathbf{u}) = \alpha(A\mathbf{u}) = \alpha\mathbf{b}$ and the fact that $\alpha\mathbf{b} \neq \mathbf{b}$ for $\alpha \neq 1$.

It turns out that there is a very close connection between the solutions of system 4 and the solutions of the corresponding homogeneous system

$$A\mathbf{x} = \mathbf{0} \tag{5}$$

Instead of investigating the sum of two solutions $\mathbf{u}$ and $\mathbf{v}$ of system 4, let us consider the difference $\mathbf{u} - \mathbf{v}$.

Theorem 3-21 Let $\mathbf{u}$ and $\mathbf{v}$ be two solutions of $A\mathbf{x} = \mathbf{b}$, $b \neq 0$ (system 4). Then $\mathbf{u} - \mathbf{v}$ is a solution of $A\mathbf{x} = \mathbf{0}$ (system 5).

Proof Since $A\mathbf{u} = \mathbf{b}$ and $A\mathbf{v} = \mathbf{b}$, it follows immediately that

$$A(\mathbf{u} - \mathbf{v}) = A\mathbf{u} - A\mathbf{v} = \mathbf{b} - \mathbf{b} = \mathbf{0}$$

Thus, $\mathbf{u} - \mathbf{v}$ is a solution of system 5.

We have therefore proved that the difference of any two solutions of the nonhomogeneous system 4 is always a solution of the corresponding homogeneous system 5.

Let us label the vector $\mathbf{u} - \mathbf{v}$ by $\mathbf{h}$. Thus,

$$\mathbf{u} - \mathbf{v} = \mathbf{h} \qquad \text{or} \qquad \mathbf{u} = \mathbf{v} + \mathbf{h}$$

The last equation gives us the clue about the structure of solutions of the nonhomogeneous system 4. If we think of $\mathbf{v}$ as being a particular solution of system 4, then the equation $\mathbf{u} = \mathbf{v} + \mathbf{h}$ tells us that an arbitrary solution $\mathbf{u}$ of system 4 consists of the sum of the particular solution $\mathbf{v}$ and an appropriately chosen solution $\mathbf{h}$ of system 5. The following theorem deals with the structure of solutions of the nonhomogeneous system 4.

Theorem 3-22 Let $\mathbf{v}$ be a particular solution of the nonhomogeneous system $A\mathbf{x} = \mathbf{b}$. Then every solution $\mathbf{u}$ of $A\mathbf{x} = \mathbf{b}$ has the form $\mathbf{u} = \mathbf{v} + \mathbf{h}$ where $\mathbf{h}$ is a solution of the homogeneous system $A\mathbf{x} = \mathbf{0}$.

Proof We have to show that the set $S = \{\mathbf{v} + \mathbf{h}\colon \mathbf{h} \text{ is any solution of } A\mathbf{x} = \mathbf{0}\}$ is precisely the set of all solutions of $A\mathbf{x} = \mathbf{b}$. We have

$$A(\mathbf{v} + \mathbf{h}) = A\mathbf{v} + A\mathbf{h} = \mathbf{b} + \mathbf{0} = \mathbf{b}$$

which proves that every vector in S is a solution of the nonhomogeneous system $A\mathbf{x} = \mathbf{b}$. The converse now follows from Theorem 3-21. If $\mathbf{u}$ is any solution of $A\mathbf{x} = \mathbf{b}$, then $\mathbf{u} - \mathbf{v}$ is a solution of $A\mathbf{x} = \mathbf{0}$. Since $\mathbf{u}$ can be expressed by

$$\mathbf{u} = \mathbf{v} + (\mathbf{u} - \mathbf{v})$$

it follows that $\mathbf{u} \in S$ (because of its structure). This completes the proof.

We may describe the conclusion of Theorem 3-22 as follows: By adding to *each* solution of the homogeneous system $A\mathbf{x} = \mathbf{0}$ a particular solution $\mathbf{v}$ of $A\mathbf{x} = \mathbf{b}$, one obtains *all* solutions of the nonhomogeneous system $A\mathbf{x} = \mathbf{b}$.

Example 2 Consider the nonhomogeneous linear system

$$\begin{aligned} 2x_1 + 3x_2 - x_3 - x_4 &= -1 \\ 4x_1 + x_2 + x_3 - 2x_4 &= 9 \end{aligned} \tag{6}$$

Find all solutions of the linear system. Express the solutions in the form mentioned in Theorem 3-22.

Solution The augmented matrix of the system is given by

$$C = \begin{bmatrix} 2 & 3 & -1 & -1 & \vdots & -1 \\ 4 & 1 & 1 & -2 & \vdots & 9 \end{bmatrix}$$

Using elementary row operations we transform C into row-reduced echelon form given by

$$D = \begin{bmatrix} 1 & 0 & \frac{2}{5} & -\frac{1}{2} & \vdots & \frac{14}{5} \\ 0 & 1 & -\frac{3}{5} & 0 & \vdots & -\frac{11}{5} \end{bmatrix}$$

The corresponding system of equations is given by

$$\begin{aligned} x_1 \quad + \tfrac{2}{5}x_3 - \tfrac{1}{2}x_4 &= \tfrac{14}{5} \\ x_2 - \tfrac{3}{5}x_3 \qquad &= -\tfrac{11}{5} \end{aligned}$$

Solving for the leading unknowns x_1 and x_2, we obtain

$$\begin{aligned} x_1 &= \tfrac{14}{15} - \tfrac{2}{5}x_3 + \tfrac{1}{2}x_4 \\ x_2 &= -\tfrac{11}{5} + \tfrac{3}{5}x_3 \end{aligned}$$

where x_3 and x_4 are arbitrary real numbers. If $\begin{bmatrix} x_1 \\ x_2 \\ x_3 \\ x_4 \end{bmatrix}$ is a vector solution of system 6, then we have

$$\begin{aligned} x_1 &= \tfrac{14}{5} - \tfrac{2}{5}s + \tfrac{1}{2}t \\ x_2 &= -\tfrac{11}{5} + \tfrac{3}{5}s \\ x_3 &= s \\ x_4 &= t \end{aligned}$$

where we have substituted s and t for x_3 and x_4, respectively, to indicate that they are arbitrary real numbers.

Rewriting each of the equations above, we obtain

$$\begin{aligned} x_1 &= \tfrac{14}{5} - \tfrac{2}{5}s + \tfrac{1}{2}t \\ x_2 &= -\tfrac{11}{5} + \tfrac{3}{5}s + 0t \\ x_3 &= 0 + 1s + 0t \\ x_4 &= 0 + 0s + 1t \end{aligned}$$

Using vector algebra we may write the equation above as

$$\begin{bmatrix} x_1 \\ x_2 \\ x_3 \\ x_4 \end{bmatrix} = \begin{bmatrix} \frac{14}{5} \\ -\frac{11}{5} \\ 0 \\ 0 \end{bmatrix} + s\begin{bmatrix} -\frac{2}{5} \\ \frac{3}{5} \\ 1 \\ 0 \end{bmatrix} + t\begin{bmatrix} \frac{1}{2} \\ 0 \\ 0 \\ 1 \end{bmatrix} \qquad [7]$$

For the choice $s = t = 0$ we obtain the particular solution $\mathbf{v} = \begin{bmatrix} \frac{14}{5} \\ -\frac{11}{5} \\ 0 \\ 0 \end{bmatrix}$ of the nonhomogeneous system 6.

It follows from Theorem 3-22 that the remaining part of equation 7, which is given by

$$\mathbf{h} = s\begin{bmatrix} -\frac{2}{5} \\ \frac{3}{5} \\ 1 \\ 0 \end{bmatrix} + t\begin{bmatrix} \frac{1}{2} \\ 0 \\ 0 \\ 1 \end{bmatrix}$$

contains *all* solutions for the homogeneous system corresponding to system 6, namely,

$$\begin{aligned} 2x_1 + 3x_2 - x_3 - x_4 &= 0 \\ 4x_1 + x_2 + x_3 - 2x_4 &= 0 \end{aligned}$$

We devote the remaining part of this section to problems of existence and uniqueness of solutions of linear systems. The reader may recall that many problems involving vector equations led eventually to systems of linear equations. A simple problem of this nature is the following:

Problem 1 Show that the vectors $\begin{bmatrix} 2 \\ 1 \end{bmatrix}$, $\begin{bmatrix} 1 \\ 2 \end{bmatrix}$, and $\begin{bmatrix} -3 \\ 5 \end{bmatrix}$ in $\mathbf{R}^2$ are linearly dependent.

In order to solve this problem we must show that the vector equation

$$x_1\begin{bmatrix} 2 \\ 1 \end{bmatrix} + x_2\begin{bmatrix} 1 \\ 2 \end{bmatrix} + x_3\begin{bmatrix} -3 \\ 5 \end{bmatrix} = \begin{bmatrix} 0 \\ 0 \end{bmatrix} \qquad \textbf{[8]}$$

has a nontrivial solution x_1, x_2, x_3.

Performing the operations of addition and scalar multiplication on the left-hand side of equation 8 leads to the system

$$\begin{aligned} 2x_1 + x_2 - 3x_3 &= 0 \\ x_1 + 2x_2 + 5x_3 &= 0 \end{aligned} \qquad \textbf{[9]}$$

which can also be expressed in matrix form as

$$\begin{bmatrix} 2 & 1 & -3 \\ 1 & 2 & 5 \end{bmatrix}\begin{bmatrix} x_1 \\ x_2 \\ x_3 \end{bmatrix} = \begin{bmatrix} 0 \\ 0 \end{bmatrix} \qquad \textbf{[10]}$$

By transforming the matrix $\begin{bmatrix} 2 & 1 & -3 \\ 1 & 2 & 5 \end{bmatrix}$ into a row-reduced echelon form all the solutions of system 10 are readily available. A particular nontrivial solution of system 10 is given by

The student is encouraged to check this result.

$$x_1 = 11, \qquad x_2 = -13, \qquad \text{and} \qquad x_3 = 3$$

Thus, the vectors $\begin{bmatrix} 2 \\ 1 \end{bmatrix}$, $\begin{bmatrix} 1 \\ 2 \end{bmatrix}$, and $\begin{bmatrix} -3 \\ 5 \end{bmatrix}$ are linearly dependent.

The reason for introducing this simple problem here is to emphasize the equivalence between the vector equation 8 and the system of linear equations given by system 9 or system 10. *We are interested in interpreting any system of linear equations as a vector equation involving the columns of the matrix of coefficients of the given system.* The following examples illustrate this point.

Example 3 Let a system of linear equations be given by

$$\begin{aligned} 5x_1 - 3x_2 + x_3 &= 0 \\ 4x_1 + 2x_2 - x_3 &= 0 \end{aligned}$$

Express the system as a vector equation in $\mathbf{R}^2$.

Solution We write the system as

$$\begin{bmatrix} 5x_1 \\ 4x_1 \end{bmatrix} + \begin{bmatrix} -3x_2 \\ 2x_2 \end{bmatrix} + \begin{bmatrix} x_3 \\ -x_3 \end{bmatrix} = \begin{bmatrix} 0 \\ 0 \end{bmatrix}$$

Now, we factor x_1, x_2, and x_3, respectively, from each vector on the left-hand side of the equation, obtaining

$$x_1 \begin{bmatrix} 5 \\ 4 \end{bmatrix} + x_2 \begin{bmatrix} -3 \\ 2 \end{bmatrix} + x_3 \begin{bmatrix} 1 \\ -1 \end{bmatrix} = \begin{bmatrix} 0 \\ 0 \end{bmatrix}$$

Example 4 Let a system of linear equation be given by

$$\begin{bmatrix} a_{11} & a_{12} & a_{13} & a_{14} \\ a_{21} & a_{22} & a_{23} & a_{24} \\ a_{31} & a_{32} & a_{33} & a_{34} \\ a_{41} & a_{42} & a_{43} & a_{44} \end{bmatrix} \begin{bmatrix} x_1 \\ x_2 \\ x_3 \\ x_4 \end{bmatrix} = \begin{bmatrix} b_1 \\ b_2 \\ b_3 \\ b_4 \end{bmatrix} \qquad [11]$$

Express system 11 as a vector equation in $\mathbf{R}^4$.

Solution The linear system 11 can be written as

$$\begin{aligned} a_{11}x_1 + a_{12}x_2 + a_{13}x_3 + a_{14}x_4 &= b_1 \\ a_{21}x_1 + a_{22}x_2 + a_{23}x_3 + a_{24}x_4 &= b_2 \\ a_{31}x_1 + a_{32}x_2 + a_{33}x_3 + a_{34}x_4 &= b_3 \\ a_{41}x_1 + a_{42}x_2 + a_{43}x_3 + a_{44}x_4 &= b_4 \end{aligned} \qquad [12]$$

Viewing system 12 as a vector equation in $\mathbf{R}^4$, we get

$$\begin{bmatrix} a_{11}x_1 \\ a_{21}x_1 \\ a_{31}x_1 \\ a_{41}x_1 \end{bmatrix} + \begin{bmatrix} a_{12}x_2 \\ a_{22}x_2 \\ a_{32}x_2 \\ a_{42}x_2 \end{bmatrix} + \begin{bmatrix} a_{13}x_3 \\ a_{23}x_3 \\ a_{33}x_3 \\ a_{43}x_3 \end{bmatrix} + \begin{bmatrix} a_{14}x_4 \\ a_{24}x_4 \\ a_{34}x_4 \\ a_{44}x_4 \end{bmatrix} = \begin{bmatrix} b_1 \\ b_2 \\ b_3 \\ b_4 \end{bmatrix} \qquad [13]$$

Factoring x_1, x_2, x_3, and x_4, respectively, from each vector on the left-hand side of system 13, we obtain

$$x_1\begin{bmatrix} a_{11} \\ a_{21} \\ a_{31} \\ a_{41} \end{bmatrix} + x_2\begin{bmatrix} a_{12} \\ a_{22} \\ a_{32} \\ a_{42} \end{bmatrix} + x_3\begin{bmatrix} a_{13} \\ a_{23} \\ a_{33} \\ a_{43} \end{bmatrix} + x_4\begin{bmatrix} a_{14} \\ a_{24} \\ a_{34} \\ a_{44} \end{bmatrix} = \begin{bmatrix} b_1 \\ b_2 \\ b_3 \\ b_4 \end{bmatrix}$$

Thus, the left-hand side of equation 11 is expressed as a linear combination of the columns of the matrix of coefficients of system 11 and the scalars involved are the unknowns x_1, x_2, x_3, and x_4.

The generalization to any linear system of equations is now obvious. Let $A\mathbf{x} = \mathbf{b}$ be a system of m equations in n unknowns. We may represent the system as a vector equation in $\mathbf{R}^m$ by the following

$$x_1\begin{bmatrix} a_{11} \\ a_{21} \\ \vdots \\ a_{m1} \end{bmatrix} + x_2\begin{bmatrix} a_{12} \\ a_{22} \\ \vdots \\ a_{m2} \end{bmatrix} + \cdots + x_n\begin{bmatrix} a_{1n} \\ a_{2n} \\ \vdots \\ a_{mn} \end{bmatrix} = \begin{bmatrix} b_1 \\ b_2 \\ \vdots \\ b_m \end{bmatrix} \qquad [14]$$

Notice that the left-hand side of equation 14 represents a linear combination of the columns of the matrix A. Therefore, a solution of the system $A\mathbf{x} = \mathbf{b}$ exists if and only if the vector $\begin{bmatrix} b_1 \\ b_2 \\ \vdots \\ b_m \end{bmatrix}$ can be expressed as a linear combination of the column vectors of the matrix A. This interpretation for the system $A\mathbf{x} = \mathbf{b}$ leads to the following theorem.

Theorem 3-23 Let $A\mathbf{x} = \mathbf{b}$ be a nonhomogeneous system of m equations in n unknowns. A necessary and sufficient condition for the existence of a solution $\mathbf{x}$ is that the rank of the matrix of coefficients A be equal to the rank of the augmented matrix $[A \vdots \mathbf{b}]$.

Proof To prove sufficiency we assume that the matrices A and $[A \vdots \mathbf{b}]$ have the same rank. Recall that the rank of A is equal to the dimension of its column space. If the rank of the augmented matrix $[A \vdots \mathbf{b}]$ is equal to the rank of A, it means that the vector $\mathbf{b}$ belongs to the column space of the matrix A. Thus, there exist scalars $x_1, x_2, \ldots, x_n$ such that the linear combination of the columns of A generates the vector $\mathbf{b}$, namely,

$$x_1 \begin{bmatrix} a_{11} \\ a_{21} \\ \vdots \\ a_{m1} \end{bmatrix} + x_2 \begin{bmatrix} a_{12} \\ a_{22} \\ \vdots \\ a_{m2} \end{bmatrix} + \cdots + x_n \begin{bmatrix} a_{1n} \\ a_{2n} \\ \vdots \\ a_{mn} \end{bmatrix} = \begin{bmatrix} b_1 \\ b_2 \\ \vdots \\ b_m \end{bmatrix}$$

Proof of necessity follows easily and is left as an exercise for the reader (see problem 4).

Thus, the system $A\mathbf{x} = \mathbf{b}$ has a solution.

Let us now turn our attention to linear systems consisting of n equations and n unknowns. We may use our knowledge about vector spaces to deduce the following uniqueness theorem.

Theorem 3-24 Let A be a square matrix of order n . If the system $A\mathbf{x} = \mathbf{0}$ has a unique solution (which must be the trivial solution $\mathbf{x} = \mathbf{0}$) then for every $\mathbf{b}$ in $\mathbf{R}^n$, the system $A\mathbf{x} = \mathbf{b}$ has a unique solution.

Remark: We provide two proofs for Theorem 3-24 and encourage the reader to study both because of their instructive value.

Proof The condition that the system $A\mathbf{x} = \mathbf{0}$ has the unique solution $\mathbf{x} = \mathbf{0}$ may be interpreted as saying that the column vectors of A are linearly independent vectors in $\mathbf{R}^n$. Since $\dim(\mathbf{R}^n) = n$ it follows from theorem 3-10 that the column vectors of A also span $\mathbf{R}^n$ and hence form a basis for $\mathbf{R}^n$. Therefore, if $\mathbf{b}$ is any element of $\mathbf{R}^n$, then it follows from Theorem 3-5 that $\mathbf{b}$ has a unique representation as a linear combination of the column vectors of A. Hence, the system $A\mathbf{x} = \mathbf{b}$ has a unique solution.

Alternate Proof The condition that the system $A\mathbf{x} = \mathbf{0}$ has the unique solution $\mathbf{x} = \mathbf{0}$ implies that the column vectors of the matrix A are linearly independent. Thus, the column space of A has dimension n. This means that the rank of the matrix A is equal to n. It follows from Theorem 3-18 that A is a nonsingular matrix. We may now use Theorem 1-11 to deduce that the system $A\mathbf{x} = \mathbf{b}$ has the unique solution given by $\mathbf{x} = A^{-1}\mathbf{b}$.

An alternate way to state Theorem 3-24 is as follows:

Theorem 3-25 Let A be a square matrix of order n. If the rank of A is n, then for every $\mathbf{b}$ in $\mathbf{R}^n$, the system $A\mathbf{x} = \mathbf{b}$ has a unique solution given by $\mathbf{x} = A^{-1}\mathbf{b}$.

Summary

We conclude this section by providing a list of equivalent statements summarizing many of the theorems that have been given throughout Chapters 1–3.

Let A be a square matrix of order n. Each of the following statements implies all the others.

a A is a nonsingular matrix.

b The rank of A is equal to n.

c The columns of A are linearly independent vectors in $\mathbf{R}^n$.

d The system $A\mathbf{x} = \mathbf{0}$ has the unique solution $\mathbf{x} = \mathbf{0}$.

e The system $A\mathbf{x} = \mathbf{b}$ has a unique solution for every $\mathbf{b}$ in $\mathbf{R}^n$.

f A is a product of elementary matrices.

g A can be transformed by elementary row operations into the identity matrix I_n.

h The rows of A are linearly independent vectors in $\mathbf{R}^n$.

i The columns of A span $\mathbf{R}^n$.

j The rows of A span $\mathbf{R}^n$.

Exercises

1 Find a basis for the null space of each of the following matrices. State the dimension of the null space for each matrix.

a $[2 \quad 1 \quad -3]$ **b** $\begin{bmatrix} 3 & 0 & 1 \\ 1 & 2 & 0 \end{bmatrix}$

c $\begin{bmatrix} 2 & 4 \\ 1 & 2 \\ -3 & -6 \end{bmatrix}$ **d** $\begin{bmatrix} 1 & 2 & 3 & 4 & 5 \\ 1 & 1 & 1 & 1 & 1 \\ 2 & -1 & 2 & -1 & 2 \end{bmatrix}$

See Example 2.

2 Solve each of the following nonhomogeneous systems and express the solutions in vector form.

a $\begin{aligned} 2x_1 - 3x_2 - 4x_3 &= 5 \\ x_1 + 2x_2 + 3x_3 &= 1 \end{aligned}$ **b** $\begin{aligned} 3x_1 - x_2 - 4x_3 + x_4 &= 8 \\ x_1 + 2x_2 - 3x_3 - x_4 &= 6 \\ 5x_1 - 4x_2 - 5x_3 + 3x_4 &= 10 \end{aligned}$

3 For each of the following systems, compute the ranks of the matrix of coefficients and the augmented matrix. Determine which of the systems have solutions. Determine which of the systems have a unique solution.

a $\begin{aligned} 2x_1 + 3x_2 &= 1 \\ 4x_1 + 6x_2 &= 3 \end{aligned}$ **b** $\begin{aligned} 3x_1 - 4x_2 &= 0 \\ 2x_1 + 3x_2 &= 17 \end{aligned}$ **c** $\begin{aligned} x_1 + x_2 &= 2 \\ 2x_1 + x_2 &= 3 \\ 3x_1 + x_2 &= 5 \end{aligned}$

d $x_1 + 2x_2 + 3x_3 = 10$
$2x_1 + 3x_2 + 4x_3 = 16$
$3x_1 + 4x_2 + 5x_3 = 22$

e $5x_1 + 3x_2 - x_3 = 0$
$3x_1 - x_2 + 4x_3 = 12$
$2x_1 + x_2 - 3x_3 = -5$

4 Prove necessity in Theorem 3-23.

5 Let A be a square matrix of order n. Use any stated theorems from the text to prove the following statement:

The matrix A is nonsingular if and only if the rows of A span the vector space $\mathbf{R}^n$.

Review of Chapter 3

1 What do we mean by saying that a vector $\mathbf{v}$ is a linear combination of the vectors $\mathbf{v}_1, \mathbf{v}_2, \ldots, \mathbf{v}_n$?

2 Is the vector $\begin{bmatrix} 7 \\ 7 \end{bmatrix}$ a linear combination of the vectors $\begin{bmatrix} 2 \\ 3 \end{bmatrix}$ and $\begin{bmatrix} 3 \\ 1 \end{bmatrix}$?

3 Express the matrix $\begin{bmatrix} 2 & 3 \\ 5 & 6 \end{bmatrix}$ as a linear combination of the matrices $\begin{bmatrix} 1 & 0 \\ 0 & 0 \end{bmatrix}$, $\begin{bmatrix} 0 & 1 \\ 0 & 0 \end{bmatrix}$, $\begin{bmatrix} 0 & 0 \\ 1 & 0 \end{bmatrix}$, and $\begin{bmatrix} 0 & 0 \\ 0 & 1 \end{bmatrix}$.

4 Show that the element $2x^2 + 5x + 7$ of P_2 is a linear combination of $x^2 + 1$ and $x + 1$ (which also belong to P_2).

5 What do we mean by saying that a set of vectors $S = \{\mathbf{v}_1, \mathbf{v}_2, \ldots, \mathbf{v}_n\}$ is linearly independent?

6 What do we mean by saying that a set of vectors $T = \{\mathbf{u}_1, \mathbf{u}_2, \ldots, \mathbf{u}_m\}$ is linearly dependent?

7 Let S be the set consisting of the $\mathbf{0}$ vector of a vector space V. Determine whether S is a linearly dependent or independent set.

8 Let T be the set consisting of a given nonzero vector $\mathbf{v}$ of a vector space V. Determine whether T is a linearly dependent or independent set.

9 Let $T = \{\mathbf{u}_1, \mathbf{u}_2, \ldots, \mathbf{u}_n\}$ be a linearly independent set of vectors in a vector space U. Does T contain a vector which is a linear combination of the remaining vectors in T?

10 Let $S = \{\mathbf{v}_1, \mathbf{v}_2, \ldots, \mathbf{v}_k\}$ be a linearly independent set of vectors in a vector space V. Let V_{k+1} also be an element of V. State a condition under which the expanded set $\{\mathbf{v}_1, \mathbf{v}_2, \ldots, \mathbf{v}_k, \mathbf{v}_{k+1}\}$ is linearly dependent. Also, state a condition under which the expanded set $\{\mathbf{v}_1, \mathbf{v}_2, \ldots, \mathbf{v}_k, \mathbf{v}_{k+1}\}$ is linearly independent.

11 What do we mean by saying that a vector space V is spanned by the vectors $\mathbf{v}_1, \mathbf{v}_2, \ldots, \mathbf{v}_n$?

12 Does a spanning set for a vector space V have to be linearly independent?

13 Let S and T be two spanning sets for the vector space $\mathbf{R}^2$. Do S and T contain the same number of vectors?

14 What do we mean by saying that a vector space V is finite-dimensional?

15 Let S be a finite subset of a vector space V. Under what conditions is S a basis for V?

16 Can a basis for a vector space V contain the zero vector of V?

17 Show two different bases for each of the following vector spaces:
a $\mathbf{R}^1$ **b** $\mathbf{R}^2$ **c** $\mathbf{R}^3$ **d** P_3 **e** P_4
f M_2 (all 2×2 matrices under the operations of matrix addition and scalar multiplication)

18 Let $S = \{\mathbf{v}_1, \mathbf{v}_2, \ldots, \mathbf{v}_k\}$ be a spanning set for a vector space V and let $\mathbf{v} \in V$. Under what condition does $\mathbf{v}$ have a **unique** representation as a linear combination of the vectors in S?

19 Let $S = \{\mathbf{v}_1,\mathbf{v}_2,\mathbf{v}_3,\mathbf{v}_4,\mathbf{v}_5\}$ be a basis for a vector space V. Does there exist a linearly independent subset of V containing 4 vectors from V? Does there exist a linearly independent subset of V containing 6 vectors from V?

20 Let $S = \{\mathbf{v}_1, \mathbf{v}_2, \ldots, \mathbf{v}_n\}$ and $T = \{\mathbf{u}_1, \mathbf{u}_2, \ldots, \mathbf{u}_m\}$ be bases for a vector space V. What is the relation between n and m?

21 What do we mean by saying that the dimension of some vector space V is 5 ($\dim V = 5$)?

22 Let V be a vector space such that $\dim V = 4$. Does there exist a subset of V consisting of 3 vectors which spans V?

23 Let M_5 be the vector space of all 5×5 matrices with the operations of matrix addition and scalar multiplication. Does there exist a subset of M_5 consisting of 24 elements which spans M_5?

24 Let A be a 3×4 matrix. What is the meaning of the statement "The rank of A is 2"
a with respect to the rows of A?
b with respect to the columns of A?

25 Let A be a 3×3 matrix and let A^{-1} exist. What is the rank of A? What is the rank of A^{-1}?

26 What are the ranks of the following matrices?

a $\begin{bmatrix} 1 & 0 & 0 & 5 \\ 0 & 1 & 0 & 4 \\ 0 & 0 & 1 & 3 \end{bmatrix}$ **b** $\begin{bmatrix} 0 & 0 & 0 \\ 0 & 0 & 0 \\ 0 & 0 & 0 \end{bmatrix}$ **c** $\begin{bmatrix} 1 & 2 & 0 & 3 & 0 \\ 0 & 0 & 1 & 1 & 0 \\ 0 & 0 & 0 & 0 & 1 \\ 0 & 0 & 0 & 0 & 0 \end{bmatrix}$

27 What is the relation between the ranks of the matrix products AB and B^tA^t?

28 Let $A\mathbf{x} = \mathbf{b}$ be a nonhomogeneous system of m equations and n unknowns. Let the rank of the augmented matrix $[A \vdots \mathbf{b}]$ be greater than the rank of the matrix A. Does the system $A\mathbf{x} = \mathbf{b}$ have a solution?

29 Let $A\mathbf{x} = \mathbf{b}$ be a nonhomogeneous system of n equation and n unknowns. State conditions which will guarantee that the system $A\mathbf{x} = \mathbf{b}$ has at least two different solutions.

30 Let A be an $m \times n$ matrix whose rank is less than n. Show that the homogeneous system $A\mathbf{x} = \mathbf{0}$ has infinitely many solutions. (Hint: Use Theorem 3-20 or Theorem 1-15.)

Determinants

4

4-1 The Determinant Function

Objectives
1 Introduce the concept of a determinant function.
2 Establish techniques of computing the value of a determinant using cofactor expansion with respect to a row or a column.

Let a, b, c, and d be any real numbers satisfying the condition $ad - bc \neq 0$. It is a simple matter to verify that the matrix $\begin{bmatrix} a & b \\ c & d \end{bmatrix}$ has an inverse given by (See Problem 10 in Section 1-5.)

$$\begin{bmatrix} a & b \\ c & d \end{bmatrix}^{-1} = \frac{1}{ad - bc}\begin{bmatrix} d & -b \\ -c & a \end{bmatrix} \qquad [1]$$

It is obvious from equation 1 that the condition $ad - bc \neq 0$ guarantees the existence of the inverse to the given matrix $\begin{bmatrix} a & b \\ c & d \end{bmatrix}$.

Note: The use of vertical lines to denote the determinant of a matrix is due to the work of Cayley, who introduced them in 1841.

The number $ad - bc$, which is constructed from all the entries in the matrix $\begin{bmatrix} a & b \\ c & d \end{bmatrix}$, is called the **determinant** of the matrix and is denoted by $\det \begin{bmatrix} a & b \\ c & d \end{bmatrix}$ $\left(\text{or by } \begin{vmatrix} a & b \\ c & d \end{vmatrix}\right)$.

We may therefore think of $\det \begin{bmatrix} a & b \\ c & d \end{bmatrix}$ as a function whose domain is the set of all 2×2 matrices with real entries and whose range is the set $\mathbf{R}$ of all real numbers. We describe the determinant function by the equation

$$\det \begin{bmatrix} a & b \\ c & d \end{bmatrix} = ad - bc$$

A simple application of the determinant function is given in the following example.

Example 1 Let $A\mathbf{x} = \mathbf{b}$ be a linear system of two equations in two unknowns, x_1 and x_2. Let $A = \begin{bmatrix} 2 & 5 \\ 7 & 3 \end{bmatrix}$ and $\mathbf{b} = \begin{bmatrix} 1 \\ 2 \end{bmatrix}$. Find all solutions of the system $A\mathbf{x} = \mathbf{b}$.

Solution We know that if A is a nonsingular matrix then the system $A\mathbf{x} = \mathbf{b}$ has the unique solution $\mathbf{x} = A^{-1}\mathbf{b}$. In order to ascertain that A^{-1} exists, we find $\det A$. We have

See Theorem 1-11.

$$\det \begin{bmatrix} 2 & 5 \\ 7 & 3 \end{bmatrix} = 2 \cdot 3 - 5 \cdot 7 = 6 - 35 = -29$$

Since $\det \begin{bmatrix} 2 & 5 \\ 7 & 3 \end{bmatrix} \neq 0$ we may use equation 1 to obtain the unique solution of the system $A\mathbf{x} = \mathbf{b}$. We have

$$\mathbf{x} = \begin{bmatrix} 2 & 5 \\ 7 & 3 \end{bmatrix}^{-1} \begin{bmatrix} 1 \\ 2 \end{bmatrix} = 1/(-29) \begin{bmatrix} 3 & -5 \\ -7 & 2 \end{bmatrix} \begin{bmatrix} 1 \\ 2 \end{bmatrix}$$

$$= -1/29 \begin{bmatrix} -7 \\ -3 \end{bmatrix} = \begin{bmatrix} \frac{7}{29} \\ \frac{3}{29} \end{bmatrix}$$

Thus, the unique solution of the system $A\mathbf{x} = \mathbf{b}$ is given by $x_1 = \frac{7}{29}$ and $x_2 = \frac{3}{29}$.

Having discovered that the concept of the determinant function is a useful one, we intend to extend its definition to square matrices of any order $n \geq 1$ and demonstrate its theoretical as well as practical value in solving problems of linear algebra. Let us start by defining the determinant function for square matrices of order $1, 2$, and 3.

If $A = [a_{11}]$ then $\det A = a_{11}$

If $A = \begin{bmatrix} a_{11} & a_{12} \\ a_{21} & a_{22} \end{bmatrix}$ then $\det A = a_{11} \det [a_{22}] - a_{12} \det [a_{21}]$

If $A = \begin{bmatrix} a_{11} & a_{12} & a_{13} \\ a_{21} & a_{22} & a_{23} \\ a_{31} & a_{32} & a_{33} \end{bmatrix}$ then $\det A = a_{11} \det \begin{bmatrix} a_{22} & a_{23} \\ a_{32} & a_{33} \end{bmatrix}$

$$-a_{12} \det \begin{bmatrix} a_{21} & a_{23} \\ a_{31} & a_{33} \end{bmatrix} + a_{13} \det \begin{bmatrix} a_{21} & a_{22} \\ a_{31} & a_{32} \end{bmatrix}$$

A close inspection of these definitions reveals that an inductive approach has been employed. Let us examine the structure of the last definition. We find that if A is a 3×3 matrix then $\det A$ is expressed in terms of the entries a_{11}, a_{12}, a_{13} of the first row of A and some determinants of square matrices of order 2 whose entries are taken from the matrix A.

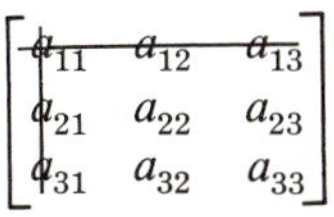

The factor $\det \begin{bmatrix} a_{22} & a_{23} \\ a_{32} & a_{33} \end{bmatrix}$ which multiplies the entry a_{11} is obtained as follows. We delete from the matrix A the row and the column in which the entry a_{11} is located (that is, the first row and the first column). We obtain a 2×2 matrix which is made of the remaining entries of A, namely, the matrix $\begin{bmatrix} a_{22} & a_{23} \\ a_{32} & a_{33} \end{bmatrix}$, and we find $\det \begin{bmatrix} a_{22} & a_{23} \\ a_{32} & a_{33} \end{bmatrix}$.

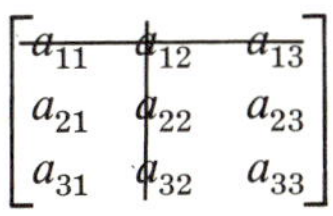

The factor $\det \begin{bmatrix} a_{21} & a_{23} \\ a_{31} & a_{33} \end{bmatrix}$ which multiplies the entry a_{12} is obtained in a similar fashion by deleting from the matrix A the first row and the second column (the row and column which contain a_{12}).

The reader will no doubt wonder about the minus sign which

precedes a_{12}. Each of the entries a_{11}, a_{12}, and a_{13} is multiplied not only by the corresponding determinant factor, but also by the factor $(-1)^{i+j}$ where i and j are the subscripts of a_{ij}. Thus, a_{12} is multiplied by $(-1)^{1+2} = -1$ while a_{11} and a_{13} are multiplied by $(-1)^{1+1} = 1$ and $(-1)^{1+3} = 1$, respectively.

After this short introduction, we now formulate a general definition for the determinant function of a matrix of order $n \times n$ using an inductive procedure. For that purpose we first introduce the concepts of the *minor* and the *cofactor* associated with an entry a_{ij} of the matrix A.

Definition 4-1 Let A be an $n \times n$ matrix with $n \geq 2$. Let S be the $(n-1) \times (n-1)$ matrix obtained from A by deleting the ith row and the jth column of A. Let $M_{ij} = \det S$. Then M_{ij} is called the **minor** belonging to the entry a_{ij} of the matrix A. The matrix S is called a **submatrix** of A.

Example 2 Let $A = \begin{bmatrix} 5 & 1 \\ 3 & 2 \end{bmatrix}$. Find the minor belonging to each entry of the matrix A.

Solution Constructing the proper submatrix for each entry of A and taking its determinant, we find immediately $M_{11} = \det [2] = 2$, $M_{12} = \det [3] = 3$, $M_{21} = \det [1] = 1$, $M_{22} = \det [5] = 5$.

The cofactor of a_{ij} is associated with the minor M_{ij}, as we see in the next definition.

Definition 4-2 Let A be an $n \times n$ matrix with $n \geq 2$. Let $A_{ij} = (-1)^{i+j} M_{ij}$ where M_{ij} is the minor belonging to a_{ij}. Then A_{ij} is called the **cofactor** of a_{ij}.

Example 3 Let $A = \begin{bmatrix} 1 & 2 & 3 \\ 4 & 5 & 6 \\ 7 & 8 & 9 \end{bmatrix}$. Find the cofactors of the entries 5 and 6 in A.

Solution Since $a_{22} = 5$ we delete from A the second row and the second column and obtain the submatrix $\begin{bmatrix} 1 & 3 \\ 7 & 9 \end{bmatrix}$. Therefore,

$$\begin{bmatrix} 1 & \not{2} & 3 \\ \not{4} & \not{5} & \not{6} \\ 7 & \not{8} & 9 \end{bmatrix}$$

$$M_{22} = \det \begin{bmatrix} 1 & 3 \\ 7 & 9 \end{bmatrix} = -12 \quad \text{and} \quad A_{22} = (-1)^{2+2} M_{22} = -12.$$

In order to find the cofactor of 6 we observe that we have $a_{23} = 6$. By deleting from A the second row and the third column we

$$\begin{bmatrix} 1 & 2 & 3 \\ 4 & 5 & 6 \\ 7 & 8 & 9 \end{bmatrix}$$

obtain the submatrix $\begin{bmatrix} 1 & 2 \\ 7 & 8 \end{bmatrix}$. Therefore, $M_{23} = \det\begin{bmatrix} 1 & 2 \\ 7 & 8 \end{bmatrix} = -6$ and $A_{23} = (-1)^{2+3}M_{23} = (-1)(-6) = 6$.

Equipped with the concept of cofactor, we are ready to formulate the general definition for the determinant function whose domain is the set of all $n \times n$ matrices and whose range is the set of all real numbers.

Definition 4-3 Let A be an $n \times n$ matrix.

If $n = 1$ then $\det A = a_{11}$.

If $n \geq 2$ then $\det A = a_{11}A_{11} + a_{12}A_{12} + \cdots + a_{1n}A_{1n}$ where A_{1j} is the cofactor of a_{1j} for $j = 1, 2, \ldots, n$.

Thus, if a matrix A is 1×1 then the value of $\det A$ is simply given by the single element a_{11} of the matrix A.

It is important for the reader to realize that in order to find the value of $\det A$ for a matrix A of order $n \times n$ with $n \geq 2$, one first computes the cofactors of a_{1j} ($j = 1, 2, \ldots, n$), which involves determinants of matrices that are $(n-1) \times (n-1)$. The following example illustrates the inductive nature of Definition 4-3.

Example 4 Let A be a 4×4 matrix. Find a formula for $\det A$ in terms of the cofactors of a_{1j} ($j = 1, 2, 3, 4$).

Solution First we find the minors M_{11}, M_{12}, M_{13}, and M_{14} belonging to the entries a_{11}, a_{12}, a_{13}, and a_{14}, respectively.

Deleting the first row and the first column of A, we obtain

$$\begin{bmatrix} a_{11} & a_{12} & a_{13} & a_{14} \\ a_{21} & a_{22} & a_{23} & a_{24} \\ a_{31} & a_{32} & a_{33} & a_{34} \\ a_{41} & a_{42} & a_{43} & a_{44} \end{bmatrix}$$

$$M_{11} = \begin{vmatrix} a_{22} & a_{23} & a_{24} \\ a_{32} & a_{33} & a_{34} \\ a_{42} & a_{43} & a_{44} \end{vmatrix}$$

Deleting the first row and the second column of A, we obtain

$$\begin{bmatrix} a_{11} & a_{12} & a_{13} & a_{14} \\ a_{21} & a_{22} & a_{23} & a_{24} \\ a_{31} & a_{32} & a_{33} & a_{34} \\ a_{41} & a_{42} & a_{43} & a_{44} \end{bmatrix}$$

$$M_{12} = \begin{vmatrix} a_{21} & a_{23} & a_{24} \\ a_{31} & a_{33} & a_{34} \\ a_{41} & a_{43} & a_{44} \end{vmatrix}$$

In a similar fashion, we obtain the remaining minors M_{13} and M_{14}. We have

$$M_{13} = \begin{vmatrix} a_{21} & a_{22} & a_{24} \\ a_{31} & a_{32} & a_{34} \\ a_{41} & a_{42} & a_{44} \end{vmatrix} \quad \text{and} \quad M_{14} = \begin{vmatrix} a_{21} & a_{22} & a_{23} \\ a_{31} & a_{32} & a_{33} \\ a_{41} & a_{42} & a_{43} \end{vmatrix}$$

See Definition 4-2

Since the cofactors A_{1j} satisfy the equation $A_{1j} = (-1)^{1+j}\, M_{1j}$ $(j = 1, 2, 3, 4)$ we may write

$$\det A = a_{11}(-1)^{1+1}M_{11} + a_{12}(-1)^{1+2}M_{12} + a_{13}(-1)^{1+3}M_{13} + a_{14}(-1)^{1+4}M_{14}$$

Substituting for M_{11}, M_{12}, M_{13}, and M_{14}, we obtain

$$\det A = a_{11}\begin{vmatrix} a_{22} & a_{23} & a_{24} \\ a_{32} & a_{33} & a_{34} \\ a_{42} & a_{43} & a_{44} \end{vmatrix} - a_{12}\begin{vmatrix} a_{21} & a_{23} & a_{24} \\ a_{31} & a_{33} & a_{34} \\ a_{41} & a_{43} & a_{44} \end{vmatrix} + a_{13}\begin{vmatrix} a_{21} & a_{22} & a_{24} \\ a_{31} & a_{32} & a_{34} \\ a_{41} & a_{42} & a_{44} \end{vmatrix} - a_{14}\begin{vmatrix} a_{21} & a_{22} & a_{23} \\ a_{31} & a_{32} & a_{33} \\ a_{41} & a_{42} & a_{43} \end{vmatrix}$$

Thus, in order to find the value of $\det A$ we must compute the values of M_{11}, M_{12}, M_{13}, and M_{14}. This can be quite cumbersome unless some simplifications occur in the expression of $\det A$. Let us examine the following numerical example.

Example 5 Let a 4×4 matrix A be given by

$$A = \begin{bmatrix} 0 & 2 & 0 & 0 \\ 5 & 1 & 0 & 0 \\ 0 & 3 & 0 & 6 \\ 0 & 4 & 7 & 0 \end{bmatrix}$$

Find the value of $\det A$.

Solution Making use of the expression for $\det A$ in Example 4, we obtain

$$\det A = 0\begin{vmatrix} 1 & 0 & 0 \\ 3 & 0 & 6 \\ 4 & 7 & 0 \end{vmatrix} - 2\begin{vmatrix} 5 & 0 & 0 \\ 0 & 0 & 6 \\ 0 & 7 & 0 \end{vmatrix} + 0\begin{vmatrix} 5 & 1 & 0 \\ 0 & 3 & 6 \\ 0 & 4 & 0 \end{vmatrix} - 0\begin{vmatrix} 5 & 1 & 0 \\ 0 & 3 & 0 \\ 0 & 4 & 7 \end{vmatrix}$$

Since $a_{11} = a_{13} = a_{14} = 0$ there is no need to compute the values of corresponding minors. The equation above reduces to

$$\det A = -2\begin{vmatrix} 5 & 0 & 0 \\ 0 & 0 & 6 \\ 0 & 7 & 0 \end{vmatrix}$$

We need therefore to find the value of $\begin{vmatrix} 5 & 0 & 0 \\ 0 & 0 & 6 \\ 0 & 7 & 0 \end{vmatrix}$. We have

$$\begin{vmatrix} 5 & 0 & 0 \\ 0 & 0 & 6 \\ 0 & 7 & 0 \end{vmatrix} = 5\begin{vmatrix} 0 & 6 \\ 7 & 0 \end{vmatrix} - 0\begin{vmatrix} 0 & 6 \\ 0 & 0 \end{vmatrix} + 0\begin{vmatrix} 0 & 0 \\ 0 & 7 \end{vmatrix} = 5\begin{vmatrix} 0 & 6 \\ 7 & 0 \end{vmatrix}$$

$$= 5(0 \cdot 0 - 6 \cdot 7) = -210$$

Thus, $\det A = (-2) \cdot (-210) = 420$

The last example shows that the presence of zero entries in the first row of A substantially simplifies the computation of $\det A$. As a matter of fact, the existence of zero entries in any row or column of the matrix A can be used to simplify the computation of $\det A$, as the following theorem indicates.

Theorem 4-1 Let A be a matrix of order $n \times n$ where $n \geq 2$. Then

a $\det A = a_{i1}A_{i1} + a_{i2}A_{i2} + \cdots + a_{in}A_{in}$
b $\det A = a_{1j}A_{1j} + a_{2j}A_{2j} + \cdots + a_{nj}A_{nj}$

Alexandre-Théophile Vandermonde *(1735–1796), a French mathematician, is considered the founder of the theory of determinants. He gave a rule for expanding determinants by using second-order minors. In 1772, Laplace proved some of Vandermonde's rules and presented a more general method for the expansion of determinants.*

It is important that the reader understand the meaning of Theorem 4-1 and its practical aspects. The first conclusion states that we may use any row (for example, the ith row) to obtain an expansion of $\det A$ in terms of the entries $a_{i1}, a_{i2}, \ldots, a_{in}$ in the ith row and their corresponding cofactors $A_{i1}, A_{i2}, \ldots, A_{in}$. We simply multiply each entry of the ith row by its cofactor and add all these products to obtain $\det A$. The second conclusion states that in order to obtain $\det A$ we may use any column of A. If we select the jth column of A then we simply multiply each of its entries by the corresponding cofactor and add all these products to obtain the value of $\det A$.

The application of Theorem 4-1 is exhibited in the following example.

For a proof of Theorem 4-1, see *Introduction to Linear Algebra* by F. Hohn, p. 196–97.

Example 6 Let a 4×4 matrix A be given by

$$A = \begin{bmatrix} 6 & 0 & 1 & 2 \\ 7 & 0 & 2 & 1 \\ 0 & 0 & 3 & 0 \\ 8 & 5 & 4 & 9 \end{bmatrix}$$

Find the value of $\det A$.

Solution We observe that the third row of A contains three zero entries, and therefore choose to expand $\det A$ in terms of the entries of the third row of A and their cofactors. We obtain

$$\det A = 0\begin{vmatrix} 0 & 1 & 2 \\ 0 & 2 & 1 \\ 5 & 4 & 9 \end{vmatrix} - 0\begin{vmatrix} 6 & 1 & 2 \\ 7 & 2 & 1 \\ 8 & 4 & 9 \end{vmatrix} + 3\begin{vmatrix} 6 & 0 & 2 \\ 7 & 0 & 1 \\ 8 & 5 & 9 \end{vmatrix} - 0\begin{vmatrix} 6 & 0 & 1 \\ 7 & 0 & 2 \\ 8 & 5 & 4 \end{vmatrix}$$

Thus, the value of $\det A$ reduces to

$$\det A = 3\begin{vmatrix} 6 & 0 & 2 \\ 7 & 0 & 1 \\ 8 & 5 & 9 \end{vmatrix}$$

In order to shorten the computation, we now expand the 3×3 determinant $\begin{vmatrix} 6 & 0 & 2 \\ 7 & 0 & 1 \\ 8 & 5 & 9 \end{vmatrix}$ with respect to the second column, which contains two zero entries. We obtain

$$\begin{vmatrix} 6 & 0 & 2 \\ 7 & 0 & 1 \\ 8 & 5 & 9 \end{vmatrix} = -0\begin{vmatrix} 7 & 1 \\ 8 & 9 \end{vmatrix} + 0\begin{vmatrix} 6 & 2 \\ 8 & 9 \end{vmatrix} - 5\begin{vmatrix} 6 & 2 \\ 7 & 1 \end{vmatrix}$$

$$= (-5)\cdot(6 - 14) = 40$$

Hence, $\det A = 3 \cdot 40 = 120$.

The last example obviously shows that the computation of $\det A$ is shortened provided we use a row or column that contains zero entries. In particular, it follows from Theorem 4-1 that if an $n \times n$ matrix A has a row (or column) consisting of zero entries only, then we must have $\det A = 0$. To see this, we suppose that the ith row of A consists entirely of zero entries. Expanding $\det A$ with respect to the ith row we obtain

We assume here that $n \geq 2$.

$$\det A = a_{i1}A_{i1} + a_{i2}A_{i2} + \cdots + a_{in}A_{in}$$

which reduces to

$$\det A = 0 \cdot A_{i1} + 0 \cdot A_{i2} + \cdots + 0 \cdot A_{in} = 0$$

Exercises

1 Let $A = \begin{bmatrix} 1 & 0 & 2 & 3 \\ 4 & 1 & 0 & -2 \\ 0 & -3 & 1 & 2 \\ -1 & -2 & 3 & 0 \end{bmatrix}$. Determine the minor M_{ij} and cofactor A_{ij} for each of the following entries:

a a_{12} **b** a_{23} **c** a_{34} **d** a_{42}

2 Evaluate each of the following determinants:

a $\begin{vmatrix} 1 & 2 \\ 3 & 4 \end{vmatrix}$ **b** $\begin{vmatrix} 5 & 0 \\ 0 & 6 \end{vmatrix}$ **c** $\begin{vmatrix} 0 & 8 \\ 0 & 9 \end{vmatrix}$

d $\begin{vmatrix} 7 & 5 \\ 0 & 0 \end{vmatrix}$ **e** $\begin{vmatrix} 3 & x \\ 2 & 1 \end{vmatrix}$

3 Solve for x in each of the following equations:

a $\begin{vmatrix} 5 & x \\ 3 & x \end{vmatrix} = 6$ **b** $\begin{vmatrix} 4x & 2x \\ 3x & 3x \end{vmatrix} = 12$ **c** $\begin{vmatrix} x^2 & x \\ 1 & x \end{vmatrix} = 0$

4 Evaluate each of the following determinants:

a $\begin{vmatrix} 1 & 2 & 3 \\ 0 & 0 & 4 \\ 5 & 6 & 7 \end{vmatrix}$ **b** $\begin{vmatrix} 2 & 0 & -1 \\ 3 & 4 & 1 \\ 5 & 0 & 6 \end{vmatrix}$ **c** $\begin{vmatrix} 1 & 2 & 3 \\ 1 & 3 & 6 \\ 1 & 4 & 10 \end{vmatrix}$

5 Solve for x in each of the following equations:

a $\begin{vmatrix} x^2 & x & 2 \\ 1 & -1 & 3 \\ 0 & 4 & 1 \end{vmatrix} = -6$ **b** $\begin{vmatrix} 1-x & 1 & -1 \\ 2 & 3-x & -4 \\ 4 & 1 & -4-x \end{vmatrix} = 0$

c $\begin{vmatrix} 1-x & -1 & -1 \\ 1 & 3-x & 1 \\ -3 & 1 & -1-x \end{vmatrix} = 0$

6 Use Theorem 4-1 to evaluate the following determinants:

a $\begin{vmatrix} 3 & 1 & 1 & -1 \\ 1 & 2 & 3 & 1 \\ 0 & 0 & 0 & 1 \\ 1 & -2 & 1 & 5 \end{vmatrix}$ **b** $\begin{vmatrix} 1 & -2 & 0 & 4 \\ 2 & 3 & 0 & -1 \\ 3 & 1 & 2 & 5 \\ 4 & 0 & 0 & 1 \end{vmatrix}$

c $\begin{vmatrix} 1 & 2 & 0 & 0 \\ 3 & 4 & 0 & 0 \\ 0 & 0 & 5 & 6 \\ 0 & 0 & 7 & 8 \end{vmatrix}$

7 Let A be a 5×5 matrix.

a Find a formula for $\det A$ in terms of the cofactors of a_{1j} ($j = 1, 2, 3, 4, 5$).

b Find a formula for $\det A$ in terms of the cofactors of a_{i3} ($i = 1, 2, 3, 4, 5$).

(Hint: Use Theorem 4-1 and the format of Example 4.)

4-2 Properties of Determinants

Objectives

1. Compute determinants associated with triangular matrices.
2. Establish rules of operation on determinants to simplify their computation.
3. Establish a necessary and sufficient condition for a matrix A to be nonsingular in terms of the value of $\det A$.

We have seen in the previous section that if an $n \times n$ matrix A has a row (or column) consisting of zero entries, then $\det A = 0$. Another class of matrices whose determinants are easy to compute consists of *triangular* matrices defined as follows.

Definition 4-4 Let A be a square matrix. A is called **upper triangular** if all the entries below its main diagonal are zeros. A is called **lower triangular** if all the entries above its main diagonal are zeros. A is called **triangular** if it is either upper triangular or lower triangular.

The following examples will clarify for the reader how simple it is to compute the determinant of a triangular matrix.

Example 1 Let $A = \begin{bmatrix} a_{11} & 0 & 0 \\ a_{21} & a_{22} & 0 \\ a_{31} & a_{32} & a_{33} \end{bmatrix}$. Find $\det A$.

Solution Since A is lower triangular, it is advantageous to expand $\det A$ with respect to the first row of A, which contains two zero entries. We obtain

Note: Another choice is to expand $\det A$ with respect to the third column of A.

$$\det A = a_{11} \begin{vmatrix} a_{22} & 0 \\ a_{32} & a_{33} \end{vmatrix} - 0 \begin{vmatrix} a_{21} & 0 \\ a_{31} & a_{33} \end{vmatrix} + 0 \begin{vmatrix} a_{21} & a_{22} \\ a_{31} & a_{32} \end{vmatrix}$$

which immediately simplifies to

$$\det A = a_{11}a_{22}a_{33}$$

Thus, the value of $\det A$ is the product of all the entries in the main diagonal whenever the 3×3 matrix A is lower triangular.

Example 2 Let $A = \begin{bmatrix} a_{11} & a_{12} & a_{13} & a_{14} \\ 0 & a_{22} & a_{23} & a_{24} \\ 0 & 0 & a_{33} & a_{34} \\ 0 & 0 & 0 & a_{44} \end{bmatrix}$. Find $\det A$.

Solution Since A is upper triangular, we expand $\det A$ with respect to the first column of A, which contains a maximum number of zero entries. We have

$$\det A = a_{11} \begin{vmatrix} a_{22} & a_{23} & a_{24} \\ 0 & a_{33} & a_{34} \\ 0 & 0 & a_{44} \end{vmatrix} - 0 \begin{vmatrix} a_{12} & a_{13} & a_{14} \\ 0 & a_{33} & a_{34} \\ 0 & 0 & a_{44} \end{vmatrix} + 0 \begin{vmatrix} a_{12} & a_{13} & a_{14} \\ a_{22} & a_{23} & a_{24} \\ 0 & 0 & a_{44} \end{vmatrix} - 0 \begin{vmatrix} a_{12} & a_{13} & a_{14} \\ a_{22} & a_{23} & a_{24} \\ 0 & a_{33} & a_{34} \end{vmatrix}$$

which reduces to

$$\det A = a_{11} \begin{vmatrix} a_{22} & a_{23} & a_{24} \\ 0 & a_{33} & a_{34} \\ 0 & 0 & a_{44} \end{vmatrix}$$

Expanding the 3×3 determinant above with respect to the first column, we obtain

$$\det A = a_{11}\left\{a_{22}\begin{vmatrix} a_{33} & a_{34} \\ 0 & a_{44} \end{vmatrix} - 0\begin{vmatrix} a_{23} & a_{24} \\ 0 & a_{44} \end{vmatrix} + 0\begin{vmatrix} a_{23} & a_{24} \\ a_{33} & a_{34} \end{vmatrix}\right\}$$

Simplifying yields the result

$$\det A = a_{11}a_{22}a_{33}a_{44}$$

Thus, the value of the determinant of the upper triangular matrix A of order 4×4 is equal to the product of all the entries in the main diagonal.

The general result with regard to triangular matrices is given by the following theorem.

Theorem 4-2 Let A be an $n \times n$ triangular matrix. Then $\det A$ is equal to the product of all the entries in the main diagonal of A; that is, $\det A = a_{11}a_{22} \ldots a_{nn}$.

Example 3 Find the value of the determinant for each of the following matrices:

a The identity matrix I_n

b $A = \begin{bmatrix} 3 & 8 & -1 & 5 & 6 \\ 0 & 1 & 2 & 9 & 1 \\ 0 & 0 & -1 & 7 & 2 \\ 0 & 0 & 0 & 2 & 3 \\ 0 & 0 & 0 & 0 & 4 \end{bmatrix}$

Solution **a** Since the identity matrix I_n is triangular, it follows from Theorem 4-2 that

$$\det(I_n) = 1 \cdot 1 \ldots \cdot 1 = 1$$

b Since A is (upper) triangular we get

$$\det A = (3)(1)(-1)(2)(4) = -24$$

The examples above show that the computation of determinants of triangular matrices is actually very simple. Obviously this is because each triangular matrix A has some zero entries that are "strategically" located and help to simplify the computation of $\det A$. In practice, however, we are often confronted with problems involving the computation of $\det A$ where A is not a triangular matrix. Although we can always use Definition 4-3 to compute the value of $\det A$, it becomes quite a long computation when the matrix A is of order 4 or higher. A natural question arises here: Is it possible to transform the matrix A into a matrix B containing additional zero entries located strategically to facilitate the computation of $\det B$, while keeping track of the relation between $\det A$ and $\det B$? A positive answer to this question will certainly simplify the procedure of computing the value of $\det A$.

See Section 1-6.

At this point the reader should recall that elementary row operations were used in finding solutions of linear systems of equations. Here again we will make use of these operations to transform a given square matrix A into a matrix B containing a maximal number of zero entries in some row, thereby simplifying the computation of $\det B$ as well as $\det A$. In order to determine the value of $\det A$ from the value of $\det B$ we shall need the following theorem.

Theorem 4-3 Let A be an $n \times n$ matrix.

a If $\bar{A}$ is a matrix obtained from A by interchanging two rows of A, then $\det \bar{A} = -\det A$.

b If $\bar{A}$ is a matrix obtained from A by multiplying a single row of A by α, then $\det \bar{A} = \alpha \det A$.

c If $\bar{A}$ is a matrix obtained from A by adding a multiple of one row of A to *another* row of A, then $\det \bar{A} = \det A$.

We shall show the validity of Theorem 4-3 for the case of a square matrix A of order 2.

a Let $A = \begin{bmatrix} a_{11} & a_{12} \\ a_{21} & a_{22} \end{bmatrix}$ and $\bar{A} = \begin{bmatrix} a_{21} & a_{22} \\ a_{11} & a_{12} \end{bmatrix}$. Thus, $\bar{A}$ is obtained from A by interchanging its two rows. We have $\det A = a_{11}a_{22} - a_{12}a_{21}$ and $\det \bar{A} = a_{21}a_{12} - a_{22}a_{11}$. It is therefore obvious that $\det \bar{A} = -\det A$.

b Let $A = \begin{bmatrix} a_{11} & a_{12} \\ a_{21} & a_{22} \end{bmatrix}$ and $\bar{A} = \begin{bmatrix} \alpha a_{11} & \alpha a_{12} \\ a_{21} & a_{22} \end{bmatrix}$. Thus, $\bar{A}$ is obtained from A by multiplying its first row by α. We have $\det A = a_{11}a_{22} - a_{12}a_{21}$ and $\det \bar{A} = \alpha a_{11}a_{22} - \alpha a_{12}a_{21}$. It follows that $\det \bar{A} = \alpha \det A$.

c Let $A = \begin{bmatrix} a_{11} & a_{12} \\ a_{21} & a_{22} \end{bmatrix}$ and $\bar{A} = \begin{bmatrix} a_{11} & a_{12} \\ a_{21} + \alpha a_{11} & a_{22} + \alpha a_{12} \end{bmatrix}$. Thus, $\bar{A}$ is obtained from A by adding a multiple of its first row by α to its second row. We have

$$\begin{aligned} \det \bar{A} &= a_{11}(a_{22} + \alpha a_{12}) - a_{12}(a_{21} + \alpha a_{11}) \\ &= a_{11}a_{22} + \alpha a_{11}a_{12} - a_{12}a_{21} - \alpha a_{12}a_{11} \\ &= a_{11}a_{22} - a_{12}a_{21} \end{aligned}$$

Thus, $\det \bar{A} = \det A$.

The following example illustrates the procedure of evaluating a determinant by making use of Theorem 4-3.

Example 4

Let $A = \begin{vmatrix} 1 & 4 & 6 & 8 \\ 2 & 4 & 8 & 10 \\ 3 & 9 & 12 & 16 \\ 4 & 4 & 4 & 4 \end{vmatrix}$. Find $\det A$.

Solution

$$\det A = \begin{vmatrix} 1 & 4 & 6 & 8 \\ 2 & 4 & 8 & 10 \\ 3 & 9 & 12 & 16 \\ 4 & 4 & 4 & 4 \end{vmatrix} = \begin{vmatrix} 1 & 4 & 6 & 8 \\ 0 & -4 & -4 & -6 \\ 3 & 9 & 12 & 16 \\ 4 & 4 & 4 & 4 \end{vmatrix}$$

$$= \begin{vmatrix} 1 & 4 & 6 & 8 \\ 0 & -4 & -4 & -6 \\ 0 & -3 & -6 & -8 \\ 4 & 4 & 4 & 4 \end{vmatrix} = \begin{vmatrix} 1 & 4 & 6 & 8 \\ 0 & -4 & -4 & -6 \\ 0 & -3 & -6 & -8 \\ 0 & -12 & -20 & -28 \end{vmatrix}$$

The last result was obtained by adding -2, -3, and -4 times the first row of A to the second, third, and fourth rows of A, respectively. Expanding the last determinant with respect to the first column, we get

$$\begin{vmatrix} 1 & 4 & 6 & 8 \\ 0 & -4 & -4 & -6 \\ 0 & -3 & -6 & -8 \\ 0 & -12 & -20 & -28 \end{vmatrix} = 1 \begin{vmatrix} -4 & -4 & -6 \\ -3 & -6 & -8 \\ -12 & -20 & -28 \end{vmatrix}$$

Thus, we have reduced the problem of evaluating $\det A$ to the problem of computing the determinant of a 3×3 matrix. Making repeated use of Theorem 4-3, we obtain

$$\begin{vmatrix} -4 & -4 & -6 \\ -3 & -6 & -8 \\ -12 & -20 & -28 \end{vmatrix} = -4 \begin{vmatrix} 1 & 1 & \frac{3}{2} \\ -3 & -6 & -8 \\ -12 & -20 & -28 \end{vmatrix}$$

$$= -4 \begin{vmatrix} 1 & 1 & \frac{3}{2} \\ 0 & -3 & -\frac{7}{2} \\ -12 & -20 & -28 \end{vmatrix} = -4 \begin{vmatrix} 1 & 1 & \frac{3}{2} \\ 0 & -3 & -\frac{7}{2} \\ 0 & -8 & -10 \end{vmatrix}$$

Expanding again with respect to the first column, we obtain

$$\begin{vmatrix} 1 & 1 & \frac{3}{2} \\ 0 & -3 & -\frac{7}{2} \\ 0 & -8 & -10 \end{vmatrix} = 1 \begin{vmatrix} -3 & -\frac{7}{2} \\ -8 & -10 \end{vmatrix}$$

But

$$\begin{vmatrix} -3 & -\frac{7}{2} \\ -8 & -10 \end{vmatrix} = (-3)(-10) - (-\tfrac{7}{2})(-8) = 30 - 28 = 2$$

Thus, we finally have $\det A = (-4)(2) = -8$.

We have seen in the previous section that if a square matrix A contains a row of zero entries, then $\det A = 0$. Making use of Theorem 4-3, we can go one step further and state that if a square matrix A has two proportional rows, then $\det A = 0$. This follows from the fact that we can get a row of zero entries by adding a proper multiple of one of the rows to the other. The following example illustrates this idea.

Example 5 Let $A = \begin{bmatrix} 1 & 2 & 3 \\ 4 & 5 & 6 \\ -3 & -6 & -9 \end{bmatrix}$. Find $\det A$.

Solution We observe that the third row of A is -3 times the first row. Therefore, adding 3 times the first row of A to the third row, we obtain

$$\det A = \begin{vmatrix} 1 & 2 & 3 \\ 4 & 5 & 6 \\ -3 & -6 & -9 \end{vmatrix}$$

$$= \begin{vmatrix} 1 & 2 & 3 \\ 4 & 5 & 6 \\ -3 + (1 \cdot 3) & -6 + (2 \cdot 3) & -9 + (3 \cdot 3) \end{vmatrix}$$

$$= \begin{vmatrix} 1 & 2 & 3 \\ 4 & 5 & 6 \\ 0 & 0 & 0 \end{vmatrix}$$

Thus, $\det A = 0$.

We now generalize the last result as follows. Let A be a square matrix having the property that one of its rows is a linear combination of all the remaining rows. Then, we must have $\det A = 0$. Let us exhibit this fact for a 3×3 matrix A.

Let $A = \begin{bmatrix} a_{11} & a_{12} & a_{13} \\ a_{21} & a_{22} & a_{23} \\ a_{31} & a_{32} & a_{33} \end{bmatrix}$ and assume that the first row of A is a linear combination of the second and third rows of A. Thus, we may write

$$[a_{11},a_{12},a_{13}] = \alpha[a_{21},a_{22},a_{23}] + \beta[a_{31},a_{32},a_{33}]$$

where α and β are some scalars.

To the first row of A we now add $-\alpha$ times the second row and $-\beta$ times the third row. We obtain a matrix B having the form

$$B = \begin{bmatrix} 0 & 0 & 0 \\ a_{21} & a_{22} & a_{23} \\ a_{31} & a_{32} & a_{33} \end{bmatrix}$$

Since $\det A = \det B$ and $\det B = 0$, it follows that $\det A = 0$.

We state the general result in the following theorem.

Theorem 4-4 Let A be an $n \times n$ matrix. If one row of A is a linear combination of the remaining rows of A, then $\det A = 0$.

The last theorem can also be stated in another equivalent form. The reader should recall the fact that if one row of the matrix A is a linear combination of the remaining rows of A then these rows, viewed as vectors in $\mathbf{R}^n$, form a linearly dependent set. This leads to the conclusion that the rank of A is less than n and that therefore the matrix A must be singular. We also know that singularity of the matrix A implies that the rank of A is less than n and thus the rows of A are linearly dependent. Hence, one row of A must be a linear combination of all the remaining rows of A. We may therefore state Theorem 4-4 in the following form:

See Theorem 3-1.

See Theorem 3-18.

Theorem 4-5 Let A be an $n \times n$ matrix. If A is a singular matrix, then $\det A = 0$.

A natural question which arises here is the following: Let A be an $n \times n$ matrix such that $\det A = 0$. Does it follow that A must be singular? The answer to this question will be readily available if we first resolve the following problem. Suppose A is a nonsingular matrix. Is it possible for A to satisfy the equation $\det A = 0$?

To answer this latter question we need the following result, the proof of which is beyond the scope of this text.

For a proof of Theorem 4-6, see *Introduction to Linear Algebra* by F. Hohn, p. 205.

Theorem 4-6 Let A and B be $n \times n$ matrices. Then we have $\det(AB) = (\det A)(\det B)$.

We are now able to answer the last question. If we assume that A is nonsingular then A^{-1} exists and we have the equation $AA^{-1} = I_n$.

It follows that $\det(AA^{-1}) = \det(I_n)$.

Since $\det(I_n) = 1$, we have from Theorem 4-6 that $(\det A)(\det(A^{-1})) = 1$ and therefore $\det A \neq 0$ whenever A is nonsingular. Thus, if $\det A = 0$, then A cannot possibly be a nonsingular matrix. Hence, it must be a singular matrix. We have therefore established the following result.

Theorem 4-7 Let A be an $n \times n$ matrix. Then A is a singular matrix if and only if $\det A = 0$.

We may conclude from Theorem 4-7 that the following theorem is valid.

Theorem 4-8 Let A be an $n \times n$ matrix. Then A is nonsingular if and only if $\det A \neq 0$.

Example 6 Let $A = \begin{bmatrix} 2 & 1 \\ 8 & 4 \end{bmatrix}$ and $B = \begin{bmatrix} 1 & 2 \\ 3 & 4 \end{bmatrix}$. Determine whether A and B are singular matrices or not.

Solution We have $\det A = (2)(4) - (1)(8) = 0$ and $\det B = (1)(4) - (2)(3) = -2$. Since $\det A = 0$, it follows from Theorem 4-7 that A is singular. Since $\det B \neq 0$, it follows from Theorem 4-8 that B is nonsingular.

We have seen that the computation of $\det A$ can be facilitated by performing "smart" row operations on the matrix A. Can we perform similar operations on the columns of A to simplify the computation of $\det A$?

The positive answer to this question is an immediate consequence of the following theorem.

See Section 3-5.

Theorem 4-9 Let A be an $n \times n$ matrix. Let A^t be the transpose of A. Then $\det A = \det(A^t)$.

Every row operation performed on A^t is equivalent to a similar operation on the corresponding column of A. Thus, combining Theorems 4-3 and 4-9, we can state the following.

Theorem 4-10 Let A be an $n \times n$ matrix.

a If $\bar{A}$ is a matrix obtained from A by interchanging two columns of A, then $\det \bar{A} = -\det A$.

b If $\bar{A}$ is a matrix obtained from A by multiplying a single column of A by α then $\det \bar{A} = \alpha \det A$.

c If $\bar{A}$ is a matrix obtained from A by adding a multiple of one column of A to *another* column of A, then $\det \bar{A} = \det A$.

Thus, whenever we have to compute $\det A$ we may combine row operations and column operations to obtain optimal simplification of the computation. The following example exhibits such a combination.

Example 7 Let $A = \begin{bmatrix} 1 & -4 & -4 & 3 \\ 2 & 0 & -2 & 0 \\ -4 & 2 & -5 & 2 \\ 3 & -4 & 1 & 0 \end{bmatrix}$. Find $\det A$.

Solution The second row of A already has two zero entries. By adding the first column of A (multiplied by 1) to the third column of A, we produce an additional zero entry in the second row of A. We have

$$\det A = \begin{vmatrix} 1 & -4 & -4 & 3 \\ 2 & 0 & -2 & 0 \\ -4 & 2 & -5 & 2 \\ 3 & -4 & 1 & 0 \end{vmatrix} = \begin{vmatrix} 1 & -4 & -3 & 3 \\ 2 & 0 & 0 & 0 \\ -4 & 2 & -9 & 2 \\ 3 & -4 & 4 & 0 \end{vmatrix}$$

Expanding the last determinant with respect to the second row, we obtain the equation

$$\det A = (-1)^{2+1}(2) \begin{vmatrix} -4 & -3 & 3 \\ 2 & -9 & 2 \\ -4 & 4 & 0 \end{vmatrix} \qquad [1]$$

We now produce an additional zero entry in the third row of the 3×3 determinant by adding the first column to the second column. We obtain

$$\begin{vmatrix} -4 & -3 & 3 \\ 2 & -9 & 2 \\ -4 & 4 & 0 \end{vmatrix} = \begin{vmatrix} -4 & -7 & 3 \\ 2 & -7 & 2 \\ -4 & 0 & 0 \end{vmatrix} \qquad [2]$$

Expanding the last determinant with respect to the third row, we obtain

$$\begin{vmatrix} -7 & 3 \\ -7 & 2 \end{vmatrix} = (-1) \begin{vmatrix} 7 & 3 \\ 7 & 2 \end{vmatrix}$$

$$\begin{vmatrix} -4 & -7 & 3 \\ 2 & -7 & 2 \\ -4 & 0 & 0 \end{vmatrix} = (-1)^{3+1}(-4) \begin{vmatrix} -7 & 3 \\ -7 & 2 \end{vmatrix} = (-4)(-1) \begin{vmatrix} 7 & 3 \\ 7 & 2 \end{vmatrix}$$
$$= 4(7 \cdot 2 - 3 \cdot 7) = -28$$

Combining our last result with equations 1 and 2, we have

$$\det A = (-1)^{2+1}(2)(-28) = 56$$

Viewing the columns of a matrix A as vectors in $\mathbf{R}^n$, we can prove (by repeated use of part **c** of Theorem 4-10) the following result:

Theorem 4-11 Let A be an $n \times n$ matrix. If one column of A is a linear combination of the remaining columns of A, then $\det A = 0$. In particular, it follows that if one column of A is proportional to another column of A, then $\det A = 0$.

Exercises

1 Evaluate the following determinants by inspection.

a $\begin{vmatrix} 2 & 1 & 5 \\ 0 & 3 & 6 \\ 0 & 0 & 4 \end{vmatrix}$ **b** $\begin{vmatrix} 3 & 0 & 4 \\ 2 & 0 & 5 \\ 1 & 0 & 6 \end{vmatrix}$ **c** $\begin{vmatrix} 3 & 5 & 7 \\ 4 & 9 & 1 \\ 3 & 5 & 7 \end{vmatrix}$

d $\begin{vmatrix} 2 & 1 & -6 \\ 5 & 7 & -15 \\ 3 & 2 & -9 \end{vmatrix}$ **e** $\begin{vmatrix} 1 & 0 & 0 & 0 \\ 3 & -1 & 0 & 0 \\ 5 & 9 & 2 & 0 \\ 7 & 1 & 5 & -2 \end{vmatrix}$

f $\begin{vmatrix} 3 & 5 & 1 & -7 \\ -1 & 7 & 2 & 6 \\ 0 & 0 & 0 & 0 \\ 2 & 1 & 4 & 9 \end{vmatrix}$ **g** $\begin{vmatrix} 3 & 1 & 5 & -2 \\ 2 & 0 & -3 & 1 \\ 7 & 9 & -8 & 5 \\ -4 & 0 & 6 & -2 \end{vmatrix}$

2 Evaluate the following determinants by using row and column operations.

a $\begin{vmatrix} 2 & 1 & 3 \\ -1 & 2 & 0 \\ 5 & 1 & 6 \end{vmatrix}$ **b** $\begin{vmatrix} 1 & 2 & 3 \\ 1 & 3 & 6 \\ 1 & 4 & 10 \end{vmatrix}$ **c** $\begin{vmatrix} 1 & 1 & 1 \\ 4 & 5 & 6 \\ 10 & 15 & 21 \end{vmatrix}$

d $\begin{vmatrix} 2 & 4 & -6 & 8 \\ 10 & 12 & 14 & 16 \\ 1 & 2 & -3 & -4 \\ 4 & 6 & 7 & 8 \end{vmatrix}$ **e** $\begin{vmatrix} 4 & 0 & 3 & 2 \\ 6 & -3 & 3 & 5 \\ 8 & 0 & -4 & -1 \\ -6 & -3 & 1 & 4 \end{vmatrix}$

f $\begin{vmatrix} 0 & 1 & 2 & 3 \\ -1 & 0 & 4 & 5 \\ -2 & -4 & 0 & 6 \\ -3 & -5 & -6 & 0 \end{vmatrix}$

3 Determine which of the following matrices are singular.

a $\begin{bmatrix} 3 & 1 \\ 6 & 2 \end{bmatrix}$ **b** $\begin{bmatrix} 1 & 3 \\ 2 & 4 \end{bmatrix}$

c $\begin{bmatrix} 2 & 3 & 1 \\ -1 & 5 & 1 \\ 4 & -7 & 1 \end{bmatrix}$ **d** $\begin{bmatrix} 3 & 1 & 8 \\ 0 & 4 & 8 \\ -2 & 1 & -2 \end{bmatrix}$

4 Find all values of x for which the following matrices fail to be invertible.

a $\begin{bmatrix} -x & 0 & 1 \\ 0 & -x-1 & 0 \\ 2 & 2 & 1-x \end{bmatrix}$ **b** $\begin{bmatrix} 1 & 1 & 1 \\ x & 2 & 3 \\ x^2 & 4 & 9 \end{bmatrix}$

5 Given that $\det A = 9$ for some 3×3 matrix A, find the values of the determinants of the following matrices:

a A^{-1} **b** A^t **c** $(A^t)^{-1}$

6 Let A and B be 2×2 matrices. Show that $\det(AB) = (\det A)(\det B)$.

7 Let A and B be $n \times n$ matrices and let A^{-1} exist. Show that $\det(ABA^{-1}) = \det B$.

8 Give an example to show that $\det(A + B) \neq \det A + \det B$.

9 a Let A be a 3×3 matrix and let α be a real number. Show that $\det(\alpha A) = \alpha^3 \det A$.

b Generalize the result of part **a** for a matrix A of order $n \times n$.

10 Find the value of each of the following:

a $\begin{vmatrix} 0 & 0 & 1 \\ 0 & 2 & 0 \\ 3 & 0 & 0 \end{vmatrix}$ **b** $\begin{bmatrix} 0 & 0 & 0 & 1 \\ 0 & 0 & 2 & 0 \\ 0 & 3 & 0 & 0 \\ 4 & 0 & 0 & 0 \end{bmatrix}$ **c** $\begin{vmatrix} 0 & 0 & \cdots & 0 & 1 \\ \vdots & \vdots & & 2 & 0 \\ 0 & & & \vdots & \vdots \\ n & 0 & \cdots & 0 & 0 \end{vmatrix}$

The determinants in exercise 11 are called Vandermonde determinants.

11 Prove the following

a $\begin{vmatrix} 1 & 1 & 1 \\ x_1 & x_2 & x_3 \\ (x_1)^2 & (x_2)^2 & (x_3)^2 \end{vmatrix} = (x_2 - x_1)(x_3 - x_1)(x_3 - x_2)$

b $\begin{vmatrix} 1 & 1 & 1 & 1 \\ x_1 & x_2 & x_3 & x_4 \\ (x_1)^2 & (x_2)^2 & (x_3)^2 & (x_4)^2 \\ (x_1)^3 & (x_2)^3 & (x_3)^3 & (x_4)^3 \end{vmatrix} = \begin{matrix} (x_2 - x_1)(x_3 - x_1)(x_3 - x_2) \\ (x_4 - x_1)(x_4 - x_2)(x_4 - x_3) \end{matrix}$

Note: The symbol $\prod_{i>j} (x_i - x_j)$ stands for the product of all numbers $x_i - x_j$ where $i > j$ for $i = 2, 3, \ldots, n$ and $j = 1, 2, \ldots, n - 1$.

c $\begin{vmatrix} 1 & 1 & \cdots & 1 \\ x_1 & x_2 & \cdots & x_n \\ (x_1)^2 & (x_2)^2 & \cdots & (x_n)^2 \\ \vdots & \vdots & & \vdots \\ (x_1)^{n-1} & (x_2)^{n-1} & \cdots & (x_n)^{n-1} \end{vmatrix} = \prod_{i>j} (x_i - x_j)$

4-3 Matrix Inversion by Cofactors and Cramer's Rule

Objectives

1 Introduce the concept of the adjoint of a matrix A.

The reader may recall that in Section 1-5 we established a method for computing the inverse of a given nonsingular matrix. We also introduced in Section 1-6 a procedure for finding the solutions of a linear system of equations. In both sections we used the technique

Objectives Continued

2 Establish a formula for the inverse of a matrix A by use of the adjoint of A.

3 Establish Cramer's Rule for solving the system of linear equations $A\mathbf{x} = \mathbf{b}$ when A is an $n \times n$ nonsingular matrix.

The method of using determinants in solving a linear system of n equations with n unknowns $(n \leq 4)$ was initiated by Colin Maclaurin (1698–1746), a professor at the University of Edinburgh, Scotland. This method was extended and published in 1750 by Gabriel Cramer (1704–1752). Cramer used this rule to solve for the coefficients of the general conic $x^2 + c_1y^2 + c_2xy + c_3x + c_4y + c_5 = 0$, passing through five given points, thus solving five equations in five unknowns.

of row reduction to achieve our goals. Making use of determinants, we will now establish another method for finding the inverse of a given matrix and still another method for solving a linear system of n equations with n unknowns whenever the matrix of coefficients of the system is nonsingular.

Let A be an $n \times n$ matrix. We know from Theorem 4-1 that $\det A$ can be expressed by

$$\det A = a_{i1}A_{i1} + a_{i2}A_{i2} + \cdots + a_{in}A_{in}$$

or by

$$\det A = a_{1j}A_{1j} + a_{2j}A_{2j} + \cdots + a_{nj}A_{nj}$$

The first formula is obtained when we multiply each of the entries in the ith row of A by its corresponding cofactor and add these products.

The second formula is obtained when we multiply each of the entries in the jth column of A by its corresponding cofactor and add these products.

We now need an answer to the following question: Suppose we multiply the entries of the ith row of A by the cofactors belonging to the entries of the jth row of A where $i \neq j$ and then add these products. What can we conclude about the resulting expression

$$a_{i1}A_{j1} + a_{i2}A_{j2} + \cdots + a_{in}A_{jn}?$$

In order to answer this question, let us examine the case where A is a 3×3 matrix.

Suppose we multiply the entries of the first row of A by the cofactors belonging to the entries of the third row of A and add these products. We obtain the expression

$$a_{11}A_{31} + a_{12}A_{32} + a_{13}A_{33}$$

which may be interpreted as the expansion of the determinant of a matrix $\bar{A}$ with respect to its third row where $\bar{A}$ is defined by

$$\bar{A} = \begin{bmatrix} a_{11} & a_{12} & a_{13} \\ a_{21} & a_{22} & a_{23} \\ a_{11} & a_{12} & a_{13} \end{bmatrix}$$

The reader should notice that the matrix $\bar{A}$ differs from A only in the third row. We obtain $\bar{A}$ by replacing the third row of A by a row which is identical to the first row of A. Since $\bar{A}$ has two identical rows we must have $\det \bar{A} = 0$. We conclude that if we multiply the entries a_{11}, a_{12}, a_{13} from the first row of A by the cofactors A_{31}, A_{32}, A_{33} belonging to the entries of the third row of A and add the products, we obtain zero.

The general result is stated in the following theorem.

Theorem 4-12 Let A be an $n \times n$ matrix. Then we have

$$a_{i1}A_{j1} + a_{i2}A_{j2} + \cdots + a_{in}A_{jn} = 0 \text{ for } i \neq j$$

Proof Construct a matrix $\bar{A}$ by replacing the jth row of A with a row identical to the ith row of A, keeping all the remaining rows of A intact. Thus, $\bar{A}$ contains two identical rows, its ith and jth rows. The jth row of $\bar{A}$ consists of the entries $a_{i1}, a_{i2}, \ldots, a_{in}$, and their corresponding cofactors are $A_{j1}, A_{j2}, \ldots, A_{jn}$. If we expand $\det \bar{A}$ with respect to its jth row and use the fact that $\bar{A}$ has two identical rows, we obtain $a_{i1}A_{j1} + a_{i2}A_{j2} + \cdots + a_{in}A_{jn} = 0$ whenever $i \neq j$.

A similar result holds also with respect to the columns of the matrix A. If we multiply the entries of one column of A by the cofactors belonging to another column and add the products, we obtain zero. We state this result in the following theorem.

Theorem 4-13 Let A be an $n \times n$ matrix. Then we have

$$a_{1i}A_{1j} + a_{2i}A_{2j} + \cdots + a_{ni}A_{nj} = 0 \text{ for } i \neq j$$

See problem 11.

The proof is left as an exercise for the reader.

Combining the results of Theorems 4-1, 4-12, and 4-13, we have the following equations:

$\delta_{ij} = \begin{cases} 1 & \text{if } i = j \\ 0 & \text{if } i \neq j \end{cases}$

$$a_{i1}A_{j1} + a_{i2}A_{j2} + \cdots + a_{in}A_{jn} = (\det A)\delta_{ij} \qquad [1]$$
$$a_{1i}A_{1j} + a_{2i}A_{2j} + \cdots + a_{ni}A_{nj} = (\det A)\delta_{ij} \qquad [2]$$

Let us concentrate on equation 1. We will show that if A is a nonsingular matrix (and therefore $\det A \neq 0$), then equation 1 can be used to establish a method for computing A^{-1}. A close inspection of equation 1 reveals that we may think of $(\det A)\delta_{ij}$ as representing an $n \times n$ matrix having all the entries of its main diagonal equal to $\det A$ while all the remaining entries are zeros. In order to obtain a useful interpretation of the expression $a_{i1}A_{j1} + a_{i2}A_{j2} + \cdots + a_{in}A_{jn}$ in equation 1, we need the following definition.

Definition 4-5 Let A be an $n \times n$ matrix. Let A_{ij} be the cofactor of a_{ij}. Then the matrix $\tilde{A}$ given by

Note: $\tilde{A}$ reads A tilde.

$$\tilde{A} = \begin{bmatrix} A_{11} & A_{12} & \cdots & A_{1n} \\ A_{21} & A_{22} & \cdots & A_{2n} \\ \vdots & \vdots & & \vdots \\ A_{n1} & A_{n2} & \cdots & A_{nn} \end{bmatrix}$$

is called the **matrix of cofactors** associated with A.

Definition of adj A.

The transpose of $\widetilde{A}$ is called the **adjoint** of A and is denoted by adj A.

Example 1 Let $A = \begin{bmatrix} 1 & 0 & -2 \\ 2 & 4 & 3 \\ 3 & 1 & -1 \end{bmatrix}$. Find adj A and (adj A)A.

Solution For any 3×3 matrix A we have

$$\operatorname{adj} A = \begin{bmatrix} A_{11} & A_{21} & A_{31} \\ A_{12} & A_{22} & A_{32} \\ A_{13} & A_{23} & A_{33} \end{bmatrix} = (\widetilde{A})^t$$

where A_{ij} is the cofactor of a_{ij}. Computation of the cofactors yields:

$$\begin{aligned} A_{11} &= -7, & A_{12} &= 11, & A_{13} &= -10 \\ A_{21} &= -2, & A_{22} &= 5, & A_{23} &= -1 \\ A_{31} &= 8, & A_{32} &= -7, & A_{33} &= 4 \end{aligned}$$

Thus, we obtain

$$\operatorname{adj} A = \begin{bmatrix} -7 & -2 & 8 \\ 11 & 5 & -7 \\ -10 & -1 & 4 \end{bmatrix}$$

Multiplying the matrices adj A and A yields

$$(\operatorname{adj} A)A = \begin{bmatrix} -7 & -2 & 8 \\ 11 & 5 & -7 \\ -10 & -1 & 4 \end{bmatrix} \begin{bmatrix} 1 & 0 & -2 \\ 2 & 4 & 3 \\ 3 & 1 & -1 \end{bmatrix} = \begin{bmatrix} 13 & 0 & 0 \\ 0 & 13 & 0 \\ 0 & 0 & 13 \end{bmatrix}$$

The reader can easily check that the matrices A and adj A commute. Thus, we also have

$$A(\operatorname{adj} A) = \begin{bmatrix} 13 & 0 & 0 \\ 0 & 13 & 0 \\ 0 & 0 & 13 \end{bmatrix}$$

Based on the definition of adj A we interpret the expression $a_{i1}A_{j1} + a_{i2}A_{j2} + \cdots + a_{in}A_{jn}$ in equation 1 as the (i,j) entry of the product $A(\operatorname{adj} A)$. This interpretation is valid because $a_{i1}, a_{i2}, \ldots, a_{in}$ are the entries of the ith row of A while $A_{j1}, A_{j2}, \ldots, A_{jn}$ are the entries of the jth column of adj A. We have therefore obtained the following result.

Theorem 4-14 Let A be an $n \times n$ matrix. Then we have

$$A(\text{adj}\, A) = (\det A)I_n$$

Making use of Theorem 4-14, we are ready to establish a formula for the inverse of a given nonsingular matrix.

Theorem 4-15 Let A be a nonsingular $n \times n$ matrix. Then we have

$$A^{-1} = \frac{1}{\det A}(\text{adj}\, A)$$

Proof Since A is nonsingular we have $\det A \neq 0$, and $\dfrac{1}{\det A}$ is a well-defined real number. Multiplying the matrices A and $\dfrac{1}{\det A}$ $(\text{adj}\, A)$ and making use of Theorem 4-14, we obtain

$$A\left(\frac{1}{\det A}(\text{adj}\, A)\right) = \frac{1}{\det A}A(\text{adj}\, A) = \frac{1}{\det A}(\det A)I_n = I_n$$

Thus, the matrix $\dfrac{1}{\det A}(\text{adj}\, A)$ is the inverse of A.

Example 2 Let $A = \begin{bmatrix} 1 & 0 & -2 \\ 2 & 4 & 3 \\ 3 & 1 & -1 \end{bmatrix}$. Show that A is nonsingular and find A^{-1} by using Theorem 4-15.

Solution We have shown in Example 1 that

$$A(\text{adj}\, A) = (13)(I_n) \qquad \text{where} \qquad \text{adj}\, A = \begin{bmatrix} -7 & -2 & 8 \\ 11 & 5 & -7 \\ -10 & -1 & 4 \end{bmatrix}$$

It is obvious from Theorem 4-14 that $\det A = 13 \neq 0$. Thus, A is a nonsingular matrix and its inverse is given by

$$A^{-1} = \tfrac{1}{13}\begin{bmatrix} -7 & -2 & 8 \\ 11 & 5 & -7 \\ -10 & -1 & 4 \end{bmatrix} = \begin{bmatrix} -\frac{7}{13} & -\frac{2}{13} & \frac{8}{13} \\ \frac{11}{13} & \frac{5}{13} & -\frac{7}{13} \\ -\frac{10}{13} & -\frac{1}{13} & \frac{4}{13} \end{bmatrix}$$

Now that we have the formula $A^{-1} = \dfrac{1}{\det A}(\text{adj}\, A)$ at our disposal, we are also in a position to establish a formula for the unique solution of a system $A\mathbf{x} = \mathbf{b}$ where A is a nonsingular $n \times n$ matrix. The formula is known as Cramer's Rule and is established in the following theorem.

Theorem 4-16 (Cramer's Rule) Let $A\mathbf{x} = \mathbf{b}$ be a system of n linear equations in n unknowns. Let the matrix of coefficients A be nonsingular. Then the system has a unique solution given by

$$x_1 = \frac{\det B_1}{\det A}, \qquad x_2 = \frac{\det B_2}{\det A}, \ldots, \qquad x_n = \frac{\det B_n}{\det A}$$

where B_i is the matrix obtained from A by replacing the entries of the ith column of A with the respective entries of the column $\mathbf{b}$.

$$b = \begin{bmatrix} b_1 \\ b_2 \\ \vdots \\ b_n \end{bmatrix}$$

Proof Since A is a nonsingular matrix it follows from Theorem 4-15 that A is invertible and its inverse A^{-1} is given by $A^{-1} = \frac{1}{\det A}$ $(\operatorname{adj} A)$. We know also from Theorem 1-11 that the system $A\mathbf{x} = \mathbf{b}$ has a unique solution $\mathbf{x} = A^{-1}\mathbf{b}$ whenever A is nonsingular. Combining these two results we may write

$$\mathbf{x} = \frac{1}{\det A}((\operatorname{adj} A)\mathbf{b})$$

Let us examine the product $(\operatorname{adj} A)\mathbf{b}$.

$$(\operatorname{adj} A)\mathbf{b} = \begin{bmatrix} A_{11} & A_{21} & \cdots & A_{n1} \\ A_{12} & A_{22} & \cdots & A_{n2} \\ \vdots & \vdots & & \vdots \\ A_{1n} & A_{2n} & \cdots & A_{nn} \end{bmatrix} \begin{bmatrix} b_1 \\ b_2 \\ \vdots \\ b_n \end{bmatrix}$$

$$= \begin{bmatrix} b_1A_{11} + b_2A_{21} + \cdots + b_nA_{n1} \\ b_1A_{12} + b_2A_{22} + \cdots + b_nA_{n2} \\ \vdots \qquad \vdots \qquad \vdots \\ b_1A_{1n} + b_2A_{2n} + \cdots + b_nA_{nn} \end{bmatrix}$$

It follows that we have

$$\begin{bmatrix} x_1 \\ x_2 \\ \vdots \\ x_n \end{bmatrix} = \frac{1}{\det A} \begin{bmatrix} b_1A_{11} + b_2A_{21} + \cdots + b_nA_{n1} \\ b_1A_{12} + b_2A_{22} + \cdots + b_nA_{n2} \\ \vdots \qquad \vdots \qquad \vdots \\ b_1A_{1n} + b_2A_{2n} + \cdots + b_nA_{nn} \end{bmatrix}$$

which we may also express by the equation

$$x_i = \frac{1}{\det A}(b_1A_{1i} + b_2A_{2i} + \cdots + b_nA_{ni}) \quad \text{for} \quad i = 1, 2, \ldots, n$$

Let us define a matrix B_i as follows:

$$B_i = \begin{bmatrix} a_{11} & a_{12} & \cdots & a_{1(i-1)} & b_1 & a_{1(i+1)} & \cdots & a_{1n} \\ a_{21} & a_{22} & \cdots & a_{2(i-1)} & b_2 & a_{2(i+1)} & \cdots & a_{2n} \\ \vdots & \vdots & & \vdots & \vdots & \vdots & & \vdots \\ a_{n1} & a_{n2} & \cdots & a_{n(i-1)} & b_n & a_{n(i+1)} & \cdots & a_{nn} \end{bmatrix}$$

The matrix B_i differs from A only in its ith column; therefore, the cofactors of the entries $b_1, b_2, \ldots, b_n$, which have replaced the entries $a_{1i}, a_{2i}, \ldots, a_{ni}$, are given by $A_{1i}, A_{2i}, \ldots, A_{ni}$, respectively. If we expand det B_i with respect to the ith column, we obtain

$$\det B_i = b_1A_{1i} + b_2A_{2i} + \cdots + b_nA_{ni} \quad \text{for } i = 1, 2, \ldots, n$$

Hence, it follows that $x_i = \dfrac{\det B_i}{\det A}$, that is,

$$x_1 = \frac{\det B_1}{\det A}, \qquad x_2 = \frac{\det B_2}{\det A}, \ldots, \qquad x_n = \frac{\det B_n}{\det A}$$

Example 3 Let a linear system be given by

$$\begin{aligned} 3x_1 - 2x_2 + x_3 &= -9 \\ x_1 + 2x_2 - x_3 &= 5 \\ 2x_1 - x_2 + 3x_3 &= -10 \end{aligned}$$

Use Cramer's Rule to solve the linear system.

Solution We have

$$\det A = \begin{vmatrix} 3 & -2 & 1 \\ 1 & 2 & -1 \\ 2 & -1 & 3 \end{vmatrix} = 20,$$

$$\det B_1 = \begin{vmatrix} -9 & -2 & 1 \\ 5 & 2 & -1 \\ 10 & -1 & 3 \end{vmatrix} = -20,$$

$$\det B_2 = \begin{vmatrix} 3 & -9 & 1 \\ 1 & 5 & -1 \\ 2 & -10 & 3 \end{vmatrix} = 40,$$

and $$\det B_3 = \begin{vmatrix} 3 & -2 & -9 \\ 1 & 2 & 5 \\ 2 & -1 & -10 \end{vmatrix} = -40$$

Thus, the solution is given by

$$x_1 = \frac{\det B_1}{\det A} = \frac{-20}{20} = -1, \qquad x_2 = \frac{\det B_2}{\det A} = \frac{40}{20} = 2,$$

$$x_3 = \frac{\det B_3}{\det A} = \frac{-40}{20} = -2$$

We now have two different methods for solving a linear system of n equations in n unknowns, namely, Cramer's Rule and Gauss-Jordan elimination.

A solution by Cramer's Rule requires the evaluation of $n+1$ determinants of $n \times n$ matrices, while the Gauss-Jordan elimination method requires the reduction of only one augmented matrix of order $n \times (n+1)$. Practice shows that for $n \geq 4$, Cramer's Rule is inferior to the Gauss-Jordan elimination method. However, Cramer's Rule, since it is expressed in terms of a formula, has a theoretical value lacking in the Gauss-Jordan elimination method because of the latter's purely numerical character. The same is true with respect to the two methods for finding A^{-1} whenever A is a nonsingular matrix. The sheer number of computations involved in finding $\frac{1}{\det A}(\operatorname{adj} A)$ for matrices of order 4×4 or larger should convince anyone to use the Gauss-Jordan elimination method for the purpose of obtaining numerical solutions for a linear system of n equations in n unknowns. As a matter of fact, even in finding A^{-1} with the aid of a computer the Gauss-Jordan elimination method is the faster and more efficient of the two methods.

Biographical Sketch

PIERRE-SIMON, Marquis de LAPLACE (1749–1827). *At his birth, Laplace seemed destined to live out his life as a poor farmer in a small village in Normandy, but a benefactor, recognizing his remarkable mind, made arrangements for him to attend the military school in his village. There he was introduced to mathematics, a subject which fascinated him. He managed to go to college at Caen and then to Paris.*

This second half of the eighteenth century was an era of enlightenment where new ideas, new trends, new discoveries abounded and where knowledge began to be recorded systematically and was widely shared. This circumstance helped Laplace pave the way to his famous scientific career.

Well acquainted with d'Alembert's works, Laplace obtained a first interview with this famous mathematician by sending him an outstanding paper on the general principles of "mécanique." Laplace not only got an interview the very same day his letter was received, but also was appointed professor of mathematics at the Ecole Militaire of Paris—he was 19 years old.

Laplace's scientific career never slowed in tempo, in spite of the tumultuous heavings of the government of France. In 1773 he was chosen an Associate of the Académie des Sciences (he was twenty-four!). Ten years later he became professor at the Ecole Normale Supérieure de Paris and in 1795 he became a member of the Institut de France. Napoleon gave him the title of Count in 1806. In 1816 he was elected member of the Académie Française. In 1817 Louis XVIII gave him the title of Marquis and Peer of France. Laplace has been called the Newton of France.

---

1 Let $A = \begin{bmatrix} 1 & 4 & 7 \\ 2 & 5 & 8 \\ 3 & 6 & 9 \end{bmatrix}$. Verify Theorem 4-12 for the matrix A by computing $a_{11}A_{21} + a_{12}A_{22} + a_{13}A_{23}$.

2 Let $A = \begin{bmatrix} 5 & 6 & 1 \\ 4 & 2 & 8 \\ 3 & 7 & 9 \end{bmatrix}$. Verify Theorem 4-13 for the matrix A by computing $a_{11}A_{13} + a_{21}A_{23} + a_{31}A_{33}$.

3 Let $A = \begin{bmatrix} 3 & 1 \\ 4 & 2 \end{bmatrix}$.

a Find $\operatorname{adj} A$.
b Compute $A(\operatorname{adj} A)$.
c Find $\det A$ from your result in part **b**.

4 Let $A = \begin{bmatrix} 2 & 1 & 0 \\ -1 & 3 & 1 \\ 0 & -1 & 4 \end{bmatrix}$

a Find $\operatorname{adj} A$.
b Compute $A(\operatorname{adj} A)$.
c Find $\det A$ from your result in part **b**.

5 Use Theorem 4-15 to find the inverse of each of the following matrices:

a $\begin{bmatrix} 5 & 2 \\ 6 & 3 \end{bmatrix}$ **b** $\begin{bmatrix} 5 & 1 & 2 \\ -1 & 3 & 0 \\ -2 & 0 & 4 \end{bmatrix}$ **c** $\begin{bmatrix} 1 & 1 & 1 & 1 \\ 1 & 2 & 3 & 4 \\ 1 & 3 & 6 & 10 \\ 1 & 4 & 10 & 20 \end{bmatrix}$

6 Let A be an $n \times n$ matrix such that all its entries a_{ij} are integers. Also let $\det A = 1$. Explain why all the entries in A^{-1} must be integers too.

7 Use Cramer's Rule (if it applies) to solve the following linear systems:

a $\begin{aligned} 2x_1 - 3x_2 &= 10 \\ 5x_1 + 4x_2 &= 48 \end{aligned}$

b $\begin{aligned} 2x_1 + x_2 - 3x_3 &= 1 \\ x_1 - 2x_2 - x_3 &= -1 \\ 5x_1 + 3x_2 + x_4 &= 20 \end{aligned}$

c
$$\begin{aligned} 2x_1 + 3x_2 \qquad\quad - x_4 &= 1 \\ -x_1 + 2x_2 + x_3 \qquad\quad &= -1 \\ 3x_1 - x_2 - x_3 \qquad\quad &= 2 \\ x_2 + x_3 + 2x_4 &= -3 \end{aligned}$$

8 Let A be an $n \times n$ matrix. What can you conclude about the rank of adj A if

a rank of A is n?

b rank of A is smaller than n?

9 Let A be an $n \times n$ matrix. Show that if $\det A = 0$ then $\det(\operatorname{adj} A) = 0$.

10 Let A be an $n \times n$ matrix. Show that $\det(\operatorname{adj} A) = (\det A)^{n-1}$.

11 Prove Theorem 4-13.

Hint: Let x and y be the number of male and female applicants interviewed, respectively, and set up a system of two linear equations in the two unknowns.

12 Betty and Frank are interviewing applicants for a performance of "Sleeping Beauty." Every applicant is to be interviewed by both Betty and Frank. Betty's interview with each female applicant takes 10 minutes and her interview with each male applicant takes 15 minutes. Frank's interview with each male applicant takes 10 minutes, while his interview with each female applicant takes 20 minutes. If Betty works for 4 hours and Frank works for 6 hours, find the maximum number of male and female applicants that can be interviewed by both.

Historical Note

The origin of determinants lies in the solutions of systems of linear equations. The discovery of determinants is credited to Leibnitz (in 1693). However, it is believed that the Japanese mathematician Seki Kowa had used determinants even prior to 1693. In 1750 Cramer, while working on the analysis of curves, rediscovered determinants. Since then, and especially in the nineteenth century, numerous papers have been written on the subject. Despite the attention paid to the concept of determinants, the subject does not constitute an outstanding innovation in mathematics. The concept provides an innovation in language. In fact, determinants are very useful tools that are indispensable in many branches of modern mathematics.

The early development of determinants is due to Cramer, Vandermonde, Lagrange, and Laplace, among others. It was Gauss who first used the word determinant; the use of square arrays of the elements is due to Cauchy. The modern theory of determinants is based on the works of Cauchy and Jacobi. In 1815, Cauchy proved the multiplication theorem for determinants in a general setting. In 1825, Heinrich Scherk (1789–1885) formulated rules concerning operations with determinants and demonstrated several important properties. Major contributions to determinant theory were made by Sylvester, Cayley, Weierstrass, and Kronecker.

Review of Chapter 4

1 Let $A = \begin{bmatrix} 1 & 2 & 3 & 4 \\ 0 & 0 & 0 & 0 \\ 2 & 3 & 4 & 5 \\ 3 & 4 & 5 & 6 \end{bmatrix}$. What is $\det A$?

2 Let $B = \begin{bmatrix} 9 & 1 & 0 & 3 \\ 8 & 2 & 0 & 4 \\ 7 & 3 & 0 & 5 \\ 6 & 4 & 0 & 6 \end{bmatrix}$. What is $\det B$?

3 Let $C = \begin{bmatrix} 1 & 0 & 0 & 0 \\ 2 & 3 & 0 & 0 \\ 4 & 5 & 6 & 0 \\ 1 & 2 & 3 & 2 \end{bmatrix}$. What is $\det C$?

4 Let $A = \begin{bmatrix} 1 & 3 & 4 \\ 3 & 2 & 1 \\ -2 & -6 & -8 \end{bmatrix}$. What is $\det A$?

5 Let $B = \begin{bmatrix} 2 & 5 & -6 \\ 1 & 7 & -3 \\ 3 & 8 & -9 \end{bmatrix}$. What is $\det B$?

6 Let A be an $n \times n$ matrix. How does one determine the singularity or nonsingularity of the matrix A in terms of $\det A$?

7 Let A and B be $n \times n$ matrices. If A and B are nonsingular, is it true that AB must be a nonsingular matrix too?

8 Let A be an $n \times n$ singular matrix and B an $n \times n$ nonsingular matrix. Is the matrix AB singular or nonsingular?

9 Let A be an $n \times n$ matrix such that $\det A = 3$. Does A^{-1} exist? If so, what is $\det(A^{-1})$?

10 Let A and B be $n \times n$ matrices. Is it necessarily true that $\det(A - B) = \det A - \det B$?

11 Let A be a 2×2 matrix. Is it necessarily true that $\det(5A) = 5 \det A$?

12 Let A be a 3×3 matrix such that $\det A = 2$. What is the form of the matrix $A(\operatorname{adj} A)$?

13 Let A and B be $n \times n$ nonsingular matrices. Express $(AB)^{-1}$ in terms of $\text{adj}\,(AB)$ and $\det\,(AB)$.

14 Show that if A and B are $n \times n$ nonsingular matrices then $\text{adj}\,(AB) = (\text{adj}\,B)(\text{adj}\,A)$.

15 Let A be an $n \times n$ matrix. Show that $\text{adj}\,(\alpha A) = \alpha^{n-1}(\text{adj}\,A)$ for any real number α.

Dot Products and Orthogonality

5

5-1 Dot Products in $\mathbf{R}^n$

Objectives

1 Introduce the concept of a dot product of two vectors $\mathbf{u}$ and $\mathbf{v}$ in $\mathbf{R}^n$.
2 Introduce the concept of length of a vector in $\mathbf{R}^n$.
3 Introduce the concept of an angle θ between two vectors $\mathbf{u}$ and $\mathbf{v}$ in $\mathbf{R}^n$.

The vector spaces $\mathbf{R}^2$ and $\mathbf{R}^3$ have been shown to have geometrical representations in terms of directed line segments. In this section we discuss geometrical concepts such as the length of a vector and the angle between two vectors. We shall see that one can easily extend the definitions of the concepts of length and angle in $\mathbf{R}^2$ and $\mathbf{R}^3$ to the vector space $\mathbf{R}^n$ with $n \geq 4$. To achieve this goal we need to introduce a new operation on vectors, called the *dot product* (or *inner product*) of two vectors.

Definition 5-1 Let $\mathbf{u} = [u_1, u_2, \ldots, u_n]$ and $\mathbf{v} = [v_1, v_2, \ldots, v_n]$ be vectors in $\mathbf{R}^n$. The **dot product** of $\mathbf{u}$ and $\mathbf{v}$, denoted by $\mathbf{u} \cdot \mathbf{v}$, is defined by

$$\mathbf{u} \cdot \mathbf{v} = u_1v_1 + u_2v_2 + \cdots + u_nv_n$$

Note: Expressions like $(\mathbf{u} \cdot \mathbf{v}) \cdot \mathbf{u}$ or $\mathbf{u} \cdot (\mathbf{v} \cdot \mathbf{u})$ are meaningless because we cannot perform a dot product between a scalar and a vector.

The reader should note that the dot product $\mathbf{u} \cdot \mathbf{v}$ is a scalar (rather than a vector). Thus, an expression like $(\mathbf{u} \cdot \mathbf{v})\mathbf{u}$ is simply a scalar multiple of $\mathbf{u}$ by the scalar $\mathbf{u} \cdot \mathbf{v}$.

The dot product has some properties which follow easily from Definition 5-1 and the properties of the real numbers.

For any vectors $\mathbf{u}$, $\mathbf{v}$, $\mathbf{w}$ in $\mathbf{R}^n$ and any real number α we have the following:

The reader is urged to show the validity of properties 1–4 of dot products.

1 $\mathbf{u} \cdot \mathbf{v} = \mathbf{v} \cdot \mathbf{u}$
2 $(\mathbf{u} + \mathbf{v}) \cdot \mathbf{w} = \mathbf{u} \cdot \mathbf{w} + \mathbf{v} \cdot \mathbf{w}$
3 $\alpha(\mathbf{u} \cdot \mathbf{v}) = (\alpha\mathbf{u}) \cdot \mathbf{v} = \mathbf{u} \cdot (\alpha\mathbf{v})$
4 $\mathbf{u} \cdot \mathbf{u} > 0$ for every $\mathbf{u} \neq \mathbf{0}$

Some computations involving dot products exhibit similarities to products involving real numbers. The following example will illustrate this fact.

Example 1 Prove the following identities:

a $(\alpha\mathbf{u} + \mathbf{v}) \cdot (\alpha\mathbf{u} + \mathbf{v}) = \alpha^2(\mathbf{u} \cdot \mathbf{u}) + 2\alpha(\mathbf{u} \cdot \mathbf{v}) + \mathbf{v} \cdot \mathbf{v}$
b $(\mathbf{u} - \mathbf{v}) \cdot (\mathbf{u} - \mathbf{v}) = \mathbf{u} \cdot \mathbf{u} - 2(\mathbf{u} \cdot \mathbf{v}) + \mathbf{v} \cdot \mathbf{v}$

Solution

The reader should note that we have made free use of properties 1–3 of dot products mentioned earlier.

a
$$\begin{aligned}(\alpha\mathbf{u} + \mathbf{v}) \cdot (\alpha\mathbf{u} + \mathbf{v}) &= (\alpha\mathbf{u}) \cdot (\alpha\mathbf{u} + \mathbf{v}) + \mathbf{v} \cdot (\alpha\mathbf{u} + \mathbf{v}) \\ &= (\alpha\mathbf{u}) \cdot (\alpha\mathbf{u}) + (\alpha\mathbf{u}) \cdot \mathbf{v} + \mathbf{v} \cdot (\alpha\mathbf{u}) + \mathbf{v} \cdot \mathbf{v} \\ &= \alpha^2(\mathbf{u} \cdot \mathbf{u}) + \alpha(\mathbf{u} \cdot \mathbf{v}) + \alpha(\mathbf{v} \cdot \mathbf{u}) + \mathbf{v} \cdot \mathbf{v} \\ &= \alpha^2(\mathbf{u} \cdot \mathbf{u}) + 2\alpha(\mathbf{u} \cdot \mathbf{v}) + \mathbf{v} \cdot \mathbf{v}\end{aligned}$$

b
$$\begin{aligned}(\mathbf{u} - \mathbf{v}) \cdot (\mathbf{u} - \mathbf{v}) &= \mathbf{u} \cdot (\mathbf{u} - \mathbf{v}) - \mathbf{v} \cdot (\mathbf{u} - \mathbf{v}) \\ &= \mathbf{u} \cdot \mathbf{u} - (\mathbf{u} \cdot \mathbf{v}) - (\mathbf{v} \cdot \mathbf{u}) + \mathbf{v} \cdot \mathbf{v} \\ &= \mathbf{u} \cdot \mathbf{u} - 2(\mathbf{u} \cdot \mathbf{v}) + \mathbf{v} \cdot \mathbf{v}\end{aligned}$$

Making use of the dot product, we now define the concepts of length and distance in $\mathbf{R}^n$.

Definition 5-2

Length of a vector.

Let $\mathbf{u} = [u_1, u_2, \ldots, u_n]$ be an element of $\mathbf{R}^n$. The **length** of $\mathbf{u}$, denoted by $|\mathbf{u}|$, is defined by

$$|\mathbf{u}| = \sqrt{\mathbf{u}\cdot\mathbf{u}} = [(u_1)^2 + (u_2)^2 + \cdots + (u_n)^2]^{1/2}$$

Distance between two elements of $\mathbf{R}^n$.

Note: $\mathbf{v} = [v_1, v_2, \ldots, v_n]$ and $|\mathbf{v}| = d(\mathbf{v},\mathbf{0})$.

Let $\mathbf{u}$ and $\mathbf{v}$ be elements of $\mathbf{R}^n$. The **distance** between $\mathbf{u}$ and $\mathbf{v}$, denoted by $d(\mathbf{u},\mathbf{v})$, is defined by

$$\begin{aligned} d(\mathbf{u},\mathbf{v}) &= |\mathbf{u} - \mathbf{v}| \\ &= [(u_1 - v_1)^2 + (u_2 - v_2)^2 + \cdots + (u_n - v_n)^2]^{1/2} \end{aligned}$$

The reader should note that for $n = 2$ or $n = 3$ this formula reduces to the familiar euclidean distance in the xy-plane or xyz-space, respectively.

If we confine ourselves to the xy-plane, then the algebraic definition of the dot product $\mathbf{u}\cdot\mathbf{v}$ can be given a geometric interpretation by making use of the law of cosines.

Law of cosines.

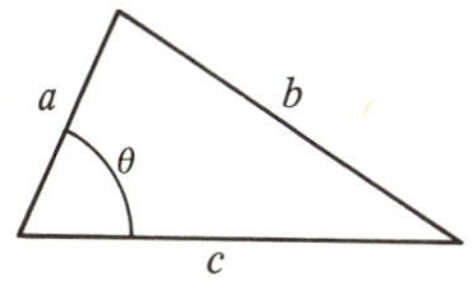

$$b^2 = a^2 + c^2 - 2ac\cos\theta$$

Let us therefore assume that $\mathbf{u}$ and $\mathbf{v}$ are elements of $\mathbf{R}^2$. Let $\mathbf{u}$ and $\mathbf{v}$ be represented by directed line segments leading from the origin to the points $P(x_1,y_1)$ and $Q(x_2,y_2)$, respectively. Then the vector $\mathbf{u} - \mathbf{v}$ may be represented by the directed line segment leading from Q to P. (See Figure 5-1.) Let θ denote the angle between $\mathbf{u}$ and $\mathbf{v}$; then by the law of cosines we have

$$|\mathbf{u} - \mathbf{v}|^2 = |\mathbf{u}|^2 + |\mathbf{v}|^2 - 2|\mathbf{u}|\,|\mathbf{v}|\cos\theta \qquad [1]$$

Since $|\mathbf{u} - \mathbf{v}|^2 = (\mathbf{u} - \mathbf{v})\cdot(\mathbf{u} - \mathbf{v})$, $|\mathbf{u}|^2 = \mathbf{u}\cdot\mathbf{u}$, and $|\mathbf{v}|^2 = \mathbf{v}\cdot\mathbf{v}$, we may write equation 1 as

$$(\mathbf{u} - \mathbf{v})\cdot(\mathbf{u} - \mathbf{v}) = \mathbf{u}\cdot\mathbf{u} + \mathbf{v}\cdot\mathbf{v} - 2|\mathbf{u}|\,|\mathbf{v}|\cos\theta \qquad [2]$$

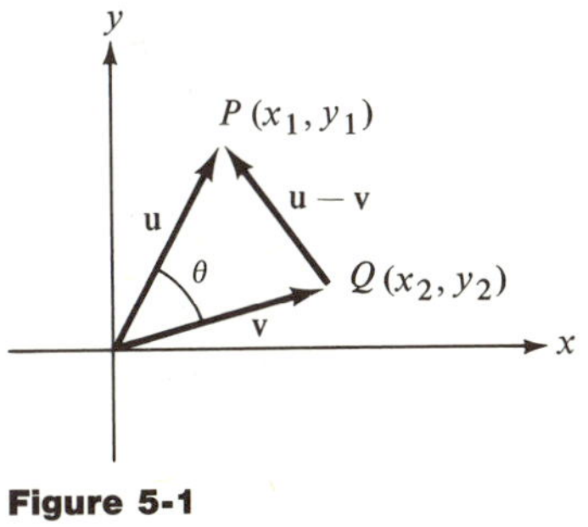

Figure 5-1

Using Example 1 part b, we obtain the equation

$$\mathbf{u}\cdot\mathbf{u} - 2(\mathbf{u}\cdot\mathbf{v}) + \mathbf{v}\cdot\mathbf{v} = \mathbf{u}\cdot\mathbf{u} + \mathbf{v}\cdot\mathbf{v} - 2|\mathbf{u}|\,|\mathbf{v}|\cos\theta \qquad [3]$$

Simple cancellations in equation 3 lead immediately to

$$\mathbf{u}\cdot\mathbf{v} = |\mathbf{u}|\,|\mathbf{v}|\cos\theta \qquad [4]$$

Thus, the dot product $\mathbf{u}\cdot\mathbf{v}$ is equal to the product of the lengths of $\mathbf{u}$ and $\mathbf{v}$ by the cosine of the angle θ between them.

Equation 4 has some interesting applications. We pose the following problem, the solution of which is an immediate consequence of equation 4.

Problem 1 Let $\mathbf{u}$ and $\mathbf{v}$ be two given nonzero vectors in $\mathbf{R}^2$. Let $\mathbf{u}$ and $\mathbf{v}$ be represented by directed line segments leading from the origin to the points P and Q, respectively. Determine if $\mathbf{u}$ and $\mathbf{v}$ are perpendicular.

Solution The vectors $\mathbf{u}$ and $\mathbf{v}$ are perpendicular if and only if the angle θ between them is equal to $90°$. It follows that $\mathbf{u}$ and $\mathbf{v}$ are perpen-

$0° \leq \theta \leq 180°$.

dicular if and only if $\cos\theta = 0$. Thus, we see from equation 4 that the nonzero vectors $\mathbf{u}$ and $\mathbf{v}$ are perpendicular if and only if $\mathbf{u} \cdot \mathbf{v} = 0$.

Remark Since the equation $\mathbf{u} \cdot \mathbf{v} = |\mathbf{u}|\,|\mathbf{v}| \cos\theta$ holds also for any vectors $\mathbf{u}$ and $\mathbf{v}$ in $\mathbf{R}^3$, it follows that a necessary and sufficient condition for two nonzero vectors $\mathbf{u}$ and $\mathbf{v}$ from $\mathbf{R}^3$ to be perpendicular is that $\mathbf{u} \cdot \mathbf{v} = 0$.

Another simple result follows from equation 4 if we rewrite it in the form

$$\cos\theta = \frac{\mathbf{u} \cdot \mathbf{v}}{|\mathbf{u}|\,|\mathbf{v}|} \tag{5}$$

which is of course valid only when $|\mathbf{u}| \neq 0$ and $|\mathbf{v}| \neq 0$. Equation 5 enables us to find the angle θ between two nonzero vectors $\mathbf{u}$ and $\mathbf{v}$ (in $\mathbf{R}^2$ or $\mathbf{R}^3$) by a simple computation. The following example illustrates this fact.

Example 2 Consider the vectors $\mathbf{u} = [-1,1,2]$ and $\mathbf{v} = [1,2,1]$ in $\mathbf{R}^3$. Find the angle θ between $\mathbf{u}$ and $\mathbf{v}$.

Solution We have $\mathbf{u} \cdot \mathbf{v} = u_1v_1 + u_2v_2 + u_3v_3 = (-1)(1) + (1)(2) \div (2)(1) = 3$, $|\mathbf{u}| = [(u_1)^2 + (u_2)^2 + (u_3)^2]^{1/2} = [(-1)^2 + (1)^2 + (2)^2]^{1/2} = \sqrt{6}$, and $|\mathbf{v}| = [(v_1)^2 + (v_2)^2 + (v_3)^2]^{1/2} = [(1)^2 + (2)^2 + (1)^2]^{1/2} = \sqrt{6}$. Thus,

$$\cos\theta = \frac{\mathbf{u} \cdot \mathbf{v}}{|\mathbf{u}|\,|\mathbf{v}|} = \frac{3}{\sqrt{6}\sqrt{6}} = \frac{3}{6} = \frac{1}{2}$$

It follows that $\theta = 60°$.

We have already introduced the ideas of length and distance in $\mathbf{R}^n$. We shall find it useful to define the concept of the angle between two nonzero vectors $\mathbf{u}$ and $\mathbf{v}$ in $\mathbf{R}^n$. This will be done by showing that equation 5 may be given a meaningful interpretation in $\mathbf{R}^n$. In particular, this will allow us to speak about perpendicular vectors $\mathbf{u}$ and $\mathbf{v}$ in $\mathbf{R}^n$. For that purpose we need to prove the Cauchy-Schwarz inequality, which is stated in the following theorem.

Theorem 5-1 Let $\mathbf{u}$ and $\mathbf{v}$ be any vectors in $\mathbf{R}^n$. Then

Cauchy-Schwarz inequality.

$$|\mathbf{u} \cdot \mathbf{v}| \leq |\mathbf{u}|\,|\mathbf{v}| \tag{6}$$

Proof If $\mathbf{u} = (0, \ldots, 0)$, then $\mathbf{u} \cdot \mathbf{v} = 0$, $|\mathbf{u}| = 0$, and the inequality 6 holds. (We actually have equality in this case.) If on the other hand $\mathbf{u} \neq (0, \ldots, 0)$, then $|\mathbf{u}| \neq 0$ and we proceed as follows: We consider the vector $\alpha\mathbf{u} + \mathbf{v}$ where α is any real number. It follows from Definition

HERMANN AMANDUS SCHWARZ (1843–1921) *was a student of Weierstrass. Schwarz served as professor at several universities including Göttingen and in 1892 he succeeded Weierstrass at Berlin. Schwarz did great work in mathematical analysis and partial differential equations.*

5-2 that $|\alpha\mathbf{u} + \mathbf{v}| \geq 0$ and thus we also have $|\alpha\mathbf{u} + \mathbf{v}|^2 \geq 0$.

This last inequality may be expressed by

$$(\alpha\mathbf{u} + \mathbf{v}) \cdot (\alpha\mathbf{u} + \mathbf{v}) \geq 0 \qquad [7]$$

Making use of Example 1, part a, we see that $\alpha^2(\mathbf{u} \cdot \mathbf{u}) + 2\alpha(\mathbf{u} \cdot \mathbf{v}) + (\mathbf{v} \cdot \mathbf{v}) \geq 0$ for all real α.

The clue to the proof of Theorem 5-1 is that we can choose a particular value for α in the last inequality to yield exactly the result stated in Theorem 5-1. Using the fact that $\mathbf{u} \cdot \mathbf{u} = |\mathbf{u}|^2$ and that $\mathbf{v} \cdot \mathbf{v} = |\mathbf{v}|^2$, and the assumption $|\mathbf{u}| \neq 0$, we may write:

Completion of squares performed.

$$\begin{aligned}(\alpha\mathbf{u} + \mathbf{v}) \cdot (\alpha\mathbf{u} + \mathbf{v}) &= \alpha^2|\mathbf{u}|^2 + 2\alpha(\mathbf{u} \cdot \mathbf{v}) + |\mathbf{v}|^2 \\ &= |\mathbf{u}|^2\left[\alpha^2 + 2\alpha\frac{\mathbf{u} \cdot \mathbf{v}}{|\mathbf{u}|^2} + \frac{|\mathbf{v}|^2}{|\mathbf{u}|^2}\right] \\ &= |\mathbf{u}|^2\left[\left(\alpha + \frac{\mathbf{u} \cdot \mathbf{v}}{|\mathbf{u}|^2}\right)^2 - \frac{(\mathbf{u} \cdot \mathbf{v})^2}{|\mathbf{u}|^4} + \frac{|\mathbf{v}|^2}{|\mathbf{u}|^2}\right] \\ &= |\mathbf{u}|^2\left[\left(\alpha + \frac{\mathbf{u} \cdot \mathbf{v}}{|\mathbf{u}|^2}\right)^2 + \frac{|\mathbf{u}|^2|\mathbf{v}|^2 - (\mathbf{u} \cdot \mathbf{v})^2}{|\mathbf{u}|^4}\right]\end{aligned}$$

It follows from inequality 7 that

$$|\mathbf{u}|^2\left[\left(\alpha + \frac{\mathbf{u} \cdot \mathbf{v}}{|\mathbf{u}|^2}\right)^2 + \frac{|\mathbf{u}|^2|\mathbf{v}|^2 - (\mathbf{u} \cdot \mathbf{v})^2}{|\mathbf{u}|^4}\right] \geq 0 \qquad [8]$$

Since inequality 8 is valid for every real number α we may choose $\alpha = \dfrac{-\mathbf{u} \cdot \mathbf{v}}{|\mathbf{u}|^2}$. For this particular choice, inequality 8 reduces to

$$|\mathbf{u}|^2\left[\frac{|\mathbf{u}|^2|\mathbf{v}|^2 - (\mathbf{u} \cdot \mathbf{v})^2}{|\mathbf{u}|^4}\right] \geq 0$$

Note that $\sqrt{(\mathbf{u} \cdot \mathbf{v})^2} = |\mathbf{u} \cdot \mathbf{v}|$.

Since $|\mathbf{u}|^2$ and $|\mathbf{u}|^4$ are positive quantities it follows that $|\mathbf{u}|^2|\mathbf{v}|^2 - (\mathbf{u} \cdot \mathbf{v})^2 \geq 0$ and thus $(\mathbf{u} \cdot \mathbf{v})^2 \leq |\mathbf{u}|^2|\mathbf{v}|^2$. Taking square roots of both sides, we finally get

$$|\mathbf{u} \cdot \mathbf{v}| \leq |\mathbf{u}|\,|\mathbf{v}|$$

Corollary 5-1 Let $\mathbf{u}$ and $\mathbf{v}$ be nonzero vectors in $\mathbf{R}^n$. Then

$$-1 \leq \frac{\mathbf{u} \cdot \mathbf{v}}{|\mathbf{u}|\,|\mathbf{v}|} \leq 1 \qquad [9]$$

Proof Dividing both sides of the Cauchy-Schwarz inequality by $|\mathbf{u}|\,|\mathbf{v}|$ we obtain the inequality $\dfrac{|\mathbf{u} \cdot \mathbf{v}|}{|\mathbf{u}|\,|\mathbf{v}|} \leq 1$ which is equivalent to inequality 9.

Corollary 5-1 enables us to define the angle θ between any two nonzero vectors $\mathbf{u}$ and $\mathbf{v}$ in $\mathbf{R}^n$. Since the number $\frac{\mathbf{u}\cdot\mathbf{v}}{|\mathbf{u}|\,|\mathbf{v}|}$ lies between -1 and $+1$ we know that there exists a unique angle θ satisfying the condition $0° \leq \theta \leq 180°$ such that

$$\cos\theta = \frac{\mathbf{u}\cdot\mathbf{v}}{|\mathbf{u}|\,|\mathbf{v}|} \qquad \textbf{[10]}$$

Definition 5-3 The angle θ between any nonzero vectors $\mathbf{u}$ and $\mathbf{v}$ in $\mathbf{R}^n$ is defined by equation 10.

It is agreed that if θ is measured in degrees then we must have $0° \leq \theta \leq 180°$. If θ is measured in radians then $0 \leq \theta \leq \pi$.

Example 3 Consider the vectors $\mathbf{u} = (3,1,1,1)$ and $\mathbf{v} = (2,2,0,1)$ in $\mathbf{R}^4$. Use Definition 5-3 to find the angle between $\mathbf{u}$ and $\mathbf{v}$.

Solution We have $|\mathbf{u}| = [(3)^2 + (1)^2 + (1)^2 + (1)^2]^{1/2} = \sqrt{12}$, $|\mathbf{v}| = [(2)^2 + (2)^2 + (0)^2 + (1)^2]^{1/2} = \sqrt{9} = 3$, and $\mathbf{u}\cdot\mathbf{v} = (3)(2) + (1)(2) + (1)(0) + (1)(1) = 9$, so that

$$\cos\theta = \frac{9}{(\sqrt{12})(3)} = \frac{3}{\sqrt{12}} = \frac{3}{2\sqrt{3}} = \frac{\sqrt{3}}{2}$$

It follows that $\theta = 30°$ $\left(\text{or } \theta = \frac{\pi}{6} \text{ radians}\right)$.

Exercises

1 Find the length of each of the following vectors:
  a $(3,2)$ b $(4,1,3)$ c $(2,1,7,-3)$ d $(5,-4,3,-2,1)$

2 Find the distance between each of the following pairs of vectors:
  a $\mathbf{u} = (1,2)$, $\mathbf{v} = (-1,-3)$ b $\mathbf{u} = (2,0,3)$, $\mathbf{v} = (5,1,1)$
  c $\mathbf{u} = (4,-1,2,-3)$, $\mathbf{v} = (0,1,-1,2)$
  d $\mathbf{u} = (2,0,1,3,1)$, $\mathbf{v} = (5,1,0,4,0)$

3 Find $\mathbf{u}\cdot\mathbf{v}$ for each pair of vectors $\mathbf{u}$ and $\mathbf{v}$ in problem 2.

4 Find the cosine of the angle θ between $\mathbf{u}$ and $\mathbf{v}$ for each of the pairs of vectors in problem 2.

5 Show that each of the following pairs of vectors are perpendicular.
  a $(1,1)$ and $(1,-1)$ b $(2,3)$ and $(6,-4)$
  c $(2,-1,7)$ and $(9,4,-2)$ d $(6,3,1)$ and $(2,-5,3)$

6 A triangle has the vertices $(2,-1,1)$, $(1,-3,-5)$ and $(3,-4,-4)$. Find the cosine of each of the three angles of the triangle.

The reader should notice that $|\alpha|$ is the absolute value of α while $|\alpha\mathbf{v}|$ and $|\mathbf{v}|$ are the lengths of $\alpha\mathbf{v}$ and $\mathbf{v}$, respectively.

7 Let $\mathbf{v} \in \mathbf{R}^n$ and α be any real number. Show that $|\alpha\mathbf{v}| = |\alpha|\,|\mathbf{v}|$ by using Definition 5-2.

8 Let $\mathbf{u}$ and $\mathbf{v}$ be nonzero vectors in $\mathbf{R}^n$ such that $\mathbf{u} = \alpha\mathbf{v}$ for some real number α. Show that if $\alpha > 0$ then the angle θ between $\mathbf{u}$ and $\mathbf{v}$ satisfies the equation $\cos\theta = 1$ while if $\alpha < 0$ then θ satisfies the equation $\cos\theta = -1$. What is θ in each case?

9 Let $\mathbf{v}$ be a nonzero vector in $\mathbf{R}^n$. Show that the vectors $\dfrac{1}{|\mathbf{v}|}\mathbf{v}$ and $\dfrac{-1}{|\mathbf{v}|}\mathbf{v}$ have length 1.

10 Prove the vector identity $|\mathbf{u}+\mathbf{v}|^2 + |\mathbf{u}-\mathbf{v}|^2 = 2|\mathbf{u}|^2 + 2|\mathbf{v}|^2$.

11 Prove the vector identity $\frac{1}{4}|\mathbf{u}+\mathbf{v}|^2 - \frac{1}{4}|\mathbf{u}-\mathbf{v}|^2 = \mathbf{u}\cdot\mathbf{v}$.

5-2 Orthogonal and Orthonormal Sets

Objectives

1. Discuss the concept of orthogonal vectors in $\mathbf{R}^n$.
2. Introduce the concept of an orthonormal set of vectors in $\mathbf{R}^n$.
3. Introduce the concepts of scalar and vector projection of a vector $\mathbf{v}$ onto a vector $\mathbf{u}$ in $\mathbf{R}^n$.
4. Discuss properties of orthonormal bases for $\mathbf{R}^n$.

We have seen in the previous section that the formula

$$\cos\theta = \frac{\mathbf{u}\cdot\mathbf{v}}{|\mathbf{u}|\,|\mathbf{v}|} \qquad [1]$$

defines an angle θ for every two nonzero vectors $\mathbf{u}$ and $\mathbf{v}$ in $\mathbf{R}^n$. We know that if $\mathbf{u}$ and $\mathbf{v}$ are nonzero vectors in $\mathbf{R}^2$ or $\mathbf{R}^3$, then $\mathbf{u}\cdot\mathbf{v} = 0$ if and only if $\mathbf{u}$ and $\mathbf{v}$ are perpendicular to each other. It follows from equation 1 that for every nonzero vector $\mathbf{u}$ and $\mathbf{v}$ in $\mathbf{R}^n$ we have $\mathbf{u}\cdot\mathbf{v} = 0$ if and only if $\cos\theta = 0$. Thus we may now state that the nonzero vectors $\mathbf{u}$ and $\mathbf{v}$ in $\mathbf{R}^n$ are perpendicular if and only if $\mathbf{u}\cdot\mathbf{v} = 0$.

$\cos\theta = 0 \Longleftrightarrow \theta = 90°$.

The reader is probably aware of the fact that the definition of the angle between the vectors $\mathbf{u}$ and $\mathbf{v}$ requires that we deal with nonzero vectors $\mathbf{u}$ and $\mathbf{v}$. The case where one of the vectors is the zero vector of $\mathbf{R}^n$ needs special attention since equation 1 does not apply. The following agreement is accepted by mathematicians and is motivated by the fact that $\mathbf{u}\cdot\mathbf{v} = 0$ whenever $\mathbf{u}$ or $\mathbf{v}$ is the zero vector $(0, 0, \ldots, 0)$ in $\mathbf{R}^n$.

Agreement 5-1 The zero vector $(0, 0, \ldots, 0)$ of $\mathbf{R}^n$ is perpendicular to each vector of $\mathbf{R}^n$.

Based on Agreement 5-1 and equation 1 we now formalize the concept of **orthogonality** (which is synonymous with perpendicularity) in the following definition.

Definition 5-4 Let $\mathbf{u}$ and $\mathbf{v}$ be any vectors in $\mathbf{R}^n$. We say that $\mathbf{u}$ and $\mathbf{v}$ are **orthogonal** if and only if $\mathbf{u}\cdot\mathbf{v} = 0$.

Example 1 Determine if any of the following pairs of vectors are orthogonal:

a $(1,-1,2,3)$ and $(2,1,1,0)$
b $(1,1,2,-2)$ and $(2,-4,3,2)$

Solution **a** $(1,-1,2,3) \cdot (2,1,1,0) = (1)(2) + (-1)(1) + (2)(1) + (3)(0)$
$= 2 - 1 + 2 = 3 \neq 0$

Thus, the vectors $(1,-1,2,3)$ and $(2,1,1,0)$ are *not* orthogonal.

b $(1,1,2,-2) \cdot (2,-4,3,2) = (1)(2) + (1)(-4) + (2)(3) + (-2)(2)$
$= 2 - 4 + 6 - 4 = 0$

It follows from Definition 5-4 that the vectors $(1,1,2,-2)$ and $(2,-4,3,2)$ are orthogonal.

It is an easy matter to show that two nonzero vectors $\mathbf{u}$ and $\mathbf{v}$ which are orthogonal will necessarily be linearly independent. The following example introduces an important and interesting technique when we deal with such a problem.

Example 2 Let $\mathbf{u}$ and $\mathbf{v}$ be nonzero vectors in $\mathbf{R}^n$. Show that if $\mathbf{u}$ and $\mathbf{v}$ are orthogonal then they are linearly independent.

Solution We must show that the vector equation

$\mathbf{0}$ is the zero vector $(0, 0, \ldots, 0)$ in $\mathbf{R}^n$.

$$\alpha\mathbf{u} + \beta\mathbf{v} = \mathbf{0} \qquad [2]$$

can be satisfied only for $\alpha = \beta = 0$.

To show that α must be zero we take the dot product of the two sides of equation 2 by $\mathbf{u}$. We obtain

$$(\alpha\mathbf{u} + \beta\mathbf{v}) \cdot \mathbf{u} = \mathbf{0} \cdot \mathbf{u} \qquad [3]$$

But $(\alpha\mathbf{u} + \beta\mathbf{v}) \cdot \mathbf{u} = \alpha(\mathbf{u} \cdot \mathbf{u}) + \beta(\mathbf{v} \cdot \mathbf{u})$ and $\mathbf{0} \cdot \mathbf{u} = 0$. Thus, equation 3 reduces to

$$\alpha(\mathbf{u} \cdot \mathbf{u}) + \beta(\mathbf{v} \cdot \mathbf{u}) = 0 \qquad [4]$$

Since $\mathbf{u}$ and $\mathbf{v}$ are orthogonal, we have $\mathbf{u} \cdot \mathbf{v} = \mathbf{v} \cdot \mathbf{u} = 0$, and equation 4 reduces to

$$\alpha(\mathbf{u} \cdot \mathbf{u}) = 0 \qquad [5]$$

We conclude that $\alpha = 0$ because $\mathbf{u}$ is a nonzero vector (satisfying the condition $\mathbf{u} \cdot \mathbf{u} \neq 0$). The proof that β must also be zero is done in a similar manner and is left for the student to verify.

Thus, we have shown that if two nonzero vectors in $\mathbf{R}^n$ are orthogonal, then they must be linearly independent.

It is a natural tendency on the part of many mathematicians to check whether the converse of a true statement is also true and thus

Find two linearly independent vectors which are not orthogonal.

leads to further mathematical discovery. Unfortunately, the converse of the statement above is not a true statement because we can easily find two nonzero linearly independent vectors which are *not* orthogonal.

At this point we wish to introduce the idea of parallel vectors in $\mathbf{R}^n$.

Definition 5-5 Let $\mathbf{u}$ and $\mathbf{v}$ be any nonzero vectors in $\mathbf{R}^n$. We say that $\mathbf{u}$ is **parallel** to $\mathbf{v}$ if there exists a scalar α such that $\mathbf{u} = \alpha\mathbf{v}$.

We now pose a simple problem associated with the concept of parallel vectors.

Problem 2 Given a nonzero vector $\mathbf{v}$ in $\mathbf{R}^n$, find a vector $\mathbf{u}$ such that $\mathbf{u}$ is parallel to $\mathbf{v}$ and $|\mathbf{u}| = 1$.

Solution We are seeking a vector $\mathbf{u}$ such that

$$\mathbf{u} = \alpha\mathbf{v} \tag{6}$$

and

$$|\mathbf{u}| = 1 \tag{7}$$

It follows from equation 6 that

$$|\mathbf{u}| = |\alpha\mathbf{v}| = |\alpha|\,|\mathbf{v}|$$

Using equation 7, we are led to the equation

$$1 = |\alpha|\,|\mathbf{v}| \tag{8}$$

Solving for $|\alpha|$ in equation 8, we get

$$|\alpha| = \frac{1}{|\mathbf{v}|} \tag{9}$$

It is obvious that equation 9 has two different solutions, given by

$$\alpha = \frac{1}{|\mathbf{v}|} \quad \text{and} \quad \alpha = \frac{-1}{|\mathbf{v}|}$$

Thus, Problem 2 has two different solutions, given by

$$\mathbf{u} = \frac{1}{|\mathbf{v}|}\mathbf{v} \quad \text{and} \quad \mathbf{u} = \frac{-1}{|\mathbf{v}|}\mathbf{v}$$

In general, a vector $\mathbf{u}$ such that $|\mathbf{u}| = 1$ is called a **unit vector**. The process of converting a nonzero vector $\mathbf{v}$ into a unit vector $\frac{1}{|\mathbf{v}|}\mathbf{v}\left(\text{or } \frac{-1}{|\mathbf{v}|}\mathbf{v}\right)$ is referred to as **normalizing** the vector $\mathbf{v}$.

Example 3 Find a unit vector $\mathbf{u}$ parallel to the given vector $\mathbf{v} = (1,2,-2,3)$.

Solution $|\mathbf{v}| = [(1)^2 + (2)^2 + (-1)^2 + (3)^2]^{1/2} = \sqrt{15}$. Normalizing the nonzero vector $\mathbf{v}$, we obtain

$$\mathbf{u} = \frac{1}{\sqrt{15}}(1,2,-1,3) = \left(\frac{1}{\sqrt{15}}, \frac{2}{\sqrt{15}}, \frac{-1}{\sqrt{15}}, \frac{3}{\sqrt{15}}\right)$$

The reader can easily check that $\mathbf{u}$ is a unit vector.

We now focus our attention on some sets of vectors in $\mathbf{R}^n$ in connection with the concepts of orthogonality and normalization of vectors. We first need to define the concept of an *orthogonal set of vectors.*

Definition 5-6 Let $\mathbf{v}_1, \mathbf{v}_2, \ldots, \mathbf{v}_k$ be vectors in $\mathbf{R}^n$. We say that the set $\{\mathbf{v}_1, \mathbf{v}_2, \ldots, \mathbf{v}_k\}$ is **orthogonal** if and only if $\mathbf{v}_i \cdot \mathbf{v}_j = 0$ whenever $i \neq j$.

Example 4 Let $\mathbf{v}_1 = (1,1,1,1)$, $\mathbf{v}_2 = (2,-1,1,-2)$, and $\mathbf{v}_3 = (-1,1,1,-1)$. Show that the set $\{\mathbf{v}_1,\mathbf{v}_2,\mathbf{v}_3\}$ is orthogonal.

Solution
$$\begin{aligned}\mathbf{v}_1 \cdot \mathbf{v}_2 &= (1)(2) + (1)(-1) + (1)(1) + (1)(-2)\\ &= 2 - 1 + 1 - 2 = 0\\ \mathbf{v}_1 \cdot \mathbf{v}_3 &= (1)(-1) + (1)(1) + (1)(1) + (1)(-1)\\ &= -1 + 1 + 1 - 1 = 0\\ \mathbf{v}_2 \cdot \mathbf{v}_3 &= (2)(-1) + (-1)(1) + (1)(1) + (-2)(-1)\\ &= -2 - 1 + 1 + 2 = 0\end{aligned}$$

Since every pair of vectors here is orthogonal, it follows from Definition 5-6 that the set $\{\mathbf{v}_1,\mathbf{v}_2,\mathbf{v}_3\}$ is orthogonal.

It has been shown in Example 2 that any two nonzero vectors in $\mathbf{R}^n$ that are orthogonal are also linearly independent. We now generalize this result as follows.

Theorem 5-2 Let $\mathbf{v}_1, \mathbf{v}_2, \ldots, \mathbf{v}_k$ be nonzero vectors in $\mathbf{R}^n$. Let $S = \{\mathbf{v}_1, \mathbf{v}_2, \ldots, \mathbf{v}_k\}$ be an orthogonal set. Then S is linearly independent.

Proof We must show that the vector equation

$$\alpha_1\mathbf{v}_1 + \alpha_2\mathbf{v}_2 + \cdots + \alpha_i\mathbf{v}_i + \cdots + \alpha_k\mathbf{v}_k = \mathbf{0} \qquad [10]$$

has the unique solution $\alpha_1 = \alpha_2 = \cdots = \alpha_k = 0$.

Taking the dot product of the two sides of equation 10 with the vector $\mathbf{v}_i$, we obtain

$$(\alpha_1\mathbf{v}_1 + \alpha_2\mathbf{v}_2 + \cdots + \alpha_i\mathbf{v}_i + \cdots + \alpha_k\mathbf{v}_k) \cdot \mathbf{v}_i = \mathbf{0} \cdot \mathbf{v}_i$$

which can also be written as

$$\alpha_1(\mathbf{v}_1 \cdot \mathbf{v}_i) + \alpha_2(\mathbf{v}_2 \cdot \mathbf{v}_i) + \cdots + \alpha_i(\mathbf{v}_i \cdot \mathbf{v}_i) + \cdots + \alpha_k(\mathbf{v}_k \cdot \mathbf{v}_i) = 0 \qquad [11]$$

Since the set $S = \{\mathbf{v}_1, \mathbf{v}_2, \ldots, \mathbf{v}_k\}$ is orthogonal, it follows that all the dot products $\mathbf{v}_j \cdot \mathbf{v}_i$ where $i \neq j$ must be zero. Thus equation 11 reduces to

$$\alpha_i(\mathbf{v}_i \cdot \mathbf{v}_i) = 0 \qquad [12]$$

But $\mathbf{v}_i$ is a nonzero vector, so that $\mathbf{v}_i \cdot \mathbf{v}_i \neq 0$; therefore, we must have

$$\alpha_i = 0 \qquad [13]$$

An orthogonal set of vectors may contain the zero vector of $\mathbf{R}^n$, but then it is linearly dependent and Theorem 5-2 does not apply.

Since equation 13 holds for $i = 1, 2, \ldots, k$, it follows that equation 10 has the unique solution $\alpha_1 = \alpha_2 = \cdots = \alpha_k = 0$. We have therefore proved that the orthogonal set $S = \{\mathbf{v}_1, \mathbf{v}_2, \ldots, \mathbf{v}_k\}$ is linearly independent.

Example 5 Show that the set of vectors $(1,3)$, $(3,-1)$, $(0,0)$ is orthogonal and linearly dependent.

Solution $(1,3) \cdot (3,-1) = 0$, $(1,3) \cdot (0,0) = 0$, and $(3,-1) \cdot (0,0) = 0$, which proves that the set of vectors above is orthogonal. Next, we examine the vector equation

$$\alpha_1(1,3) + \alpha_2(3,-1) + \alpha_3(0,0) = (0,0) \qquad [14]$$

for a possible nonzero solution α_1, α_2, and α_3. It is easy to see by inspection that the choice $\alpha_1 = 0$, $\alpha_2 = 0$, and $\alpha_3 = 1$ is a nontrivial solution of equation 14. This proves that the set above is linearly dependent.

Now we shall deal with orthogonal sets in $\mathbf{R}^n$ which do not contain the zero vector of $\mathbf{R}^n$. Let $S = \{\mathbf{v}_1, \mathbf{v}_2, \ldots, \mathbf{v}_k\}$ be an orthogonal set such that $\mathbf{v}_i \neq \mathbf{0}$ for $i = 1, 2, \ldots, k$. We can normalize each vector $\mathbf{v}_i$ in S and obtain the set $T = \{\mathbf{u}_1, \mathbf{u}_2, \ldots, \mathbf{u}_k\}$ where $\mathbf{u}_i = \dfrac{1}{|\mathbf{v}_i|}\mathbf{v}_i$ for $i = 1, 2, \ldots, k$. Thus, in addition to orthogonality the set T also has the property that each of its vectors is a unit vector.

We refer to the set T as an *orthonormal* set according to the following definition.

Definition 5-7 Let $\mathbf{u}_1, \mathbf{u}_2, \ldots, \mathbf{u}_k$ be vectors in $\mathbf{R}^n$. We say that the set $\{\mathbf{u}_1, \mathbf{u}_2, \ldots, \mathbf{u}_k\}$ is **orthonormal** if and only if $\mathbf{u}_i \cdot \mathbf{u}_j = \delta_{ij}$.

Example 6 Construct two different orthonormal sets of vectors in $\mathbf{R}^2$ each of which contains two elements.

Solution The set $\{(1,0), (0,1)\}$ is obviously an orthonormal set in $\mathbf{R}^2$ consisting of two vectors.

In order to construct a second orthonormal set, we start with an orthogonal set $S = \{(1,1), (1,-1)\}$ and normalize each vector of S. We obtain the orthonormal set

$$T = \left\{\left(\frac{1}{\sqrt{2}}, \frac{1}{\sqrt{2}}\right), \left(\frac{1}{\sqrt{2}}, \frac{-1}{\sqrt{2}}\right)\right\}$$

Since every orthonormal set in $\mathbf{R}^n$ is also an orthogonal set of nonzero vectors, it follows from Theorem 5-2 that T is linearly independent. Thus, an orthonormal set in $\mathbf{R}^n$ consisting of n vectors will provide a basis for $\mathbf{R}^n$. Such a basis will be of special interest to us because the representation of each vector in $\mathbf{R}^n$ in terms of such a basis becomes a very simple matter. But before discussing the usefulness of an orthonormal basis, we must introduce the concepts of scalar projection and vector projection.

See Theorem 3-10.

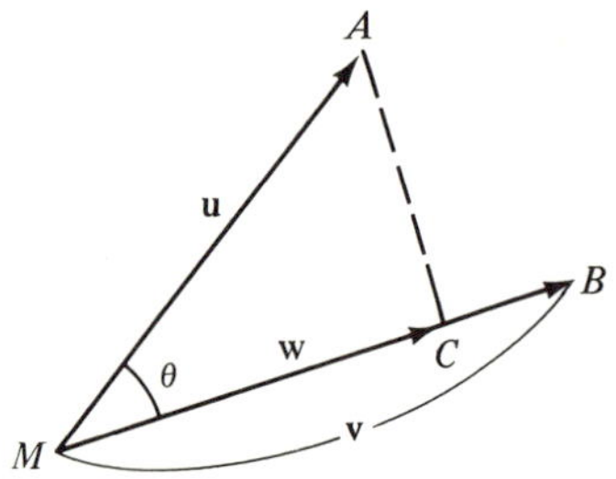

Figure 5-2

As a lead into the general definition, let us consider again the dot product $\mathbf{u} \cdot \mathbf{v}$ where $\mathbf{u}$ and $\mathbf{v}$ are nonzero vectors in $\mathbf{R}^2$. Let the vectors $\mathbf{u}$ and $\mathbf{v}$ be represented by directed line segments having the same initial point, M, and terminal points A and B, respectively (see Figure 5-2). Let θ be the angle between $\mathbf{u}$ and $\mathbf{v}$. Then we have

$$\mathbf{u} \cdot \mathbf{v} = |\mathbf{u}|\,|\mathbf{v}| \cos\theta \qquad [15]$$

Dividing both sides of equation 15 by $|\mathbf{v}|$ and using the fact that $\frac{1}{|\mathbf{v}|}(\mathbf{u} \cdot \mathbf{v}) = \mathbf{u} \cdot \left(\frac{1}{|\mathbf{v}|}\mathbf{v}\right)$, we get

$$\mathbf{u} \cdot \left(\frac{1}{|\mathbf{v}|}\mathbf{v}\right) = |\mathbf{u}| \cos\theta \qquad [16]$$

We know from elementary trigonometry that the number $|\mathbf{u}| \cos\theta$ measures the length of the segment MC where C is the foot of the perpendicular drawn from A to the segment MB. We refer to the length of the segment MC as the **scalar projection** of $\mathbf{u}$ onto $\mathbf{v}$.

By virtue of the left-hand side of equation 16 we may also describe the **scalar projection** of $\mathbf{u}$ onto $\mathbf{v}$ as the dot product of $\mathbf{u}$ with the unit vector $\frac{1}{|\mathbf{v}|}\mathbf{v}$ in the direction of $\mathbf{v}$.

The vector $\mathbf{w}$ having the initial point M and terminal point C is called the **vector projection** of $\mathbf{u}$ onto $\mathbf{v}$.

To find $\mathbf{w}$, we multiply the unit vector $\frac{1}{|\mathbf{v}|}\mathbf{v}$ (in the direction of $\mathbf{v}$) by the scalar projection $\mathbf{u} \cdot \left(\frac{1}{|\mathbf{v}|}\mathbf{v}\right)$. Thus, we obtain the expression

$$\mathbf{w} = \left[\mathbf{u} \cdot \left(\frac{1}{|\mathbf{v}|}\mathbf{v}\right)\right]\left(\frac{1}{|\mathbf{v}|}\mathbf{v}\right) \qquad [17]$$

Example 7 Consider the vectors $\mathbf{u} = (1,2)$ and $\mathbf{v} = (3,4)$ in $\mathbf{R}^2$. Find the scalar projection and the vector projection of $\mathbf{u}$ onto $\mathbf{v}$.

Solution $|\mathbf{v}| = [(3)^2 + (4)^2]^{1/2} = [9 + 16]^{1/2} = 5$. Thus, we obtain

$$\frac{1}{|\mathbf{v}|}\mathbf{v} = \frac{1}{5}(3,4) = \left(\frac{3}{5}, \frac{4}{5}\right)$$

The scalar projection of $\mathbf{u}$ onto $\mathbf{v}$ is therefore given by

$$\mathbf{u} \cdot \left(\frac{1}{|\mathbf{v}|}\mathbf{v}\right) = (1,2) \cdot \left(\frac{3}{5}, \frac{4}{5}\right) = \frac{3}{5} + \frac{8}{5} = \frac{11}{5}$$

In order to compute the vector projection of $\mathbf{u}$ onto $\mathbf{v}$, we use the right-hand side of equation 17 and obtain

$$\left[\mathbf{u} \cdot \left(\frac{1}{|\mathbf{v}|}\mathbf{v}\right)\right]\left(\frac{1}{|\mathbf{v}|}\mathbf{v}\right) = \frac{11}{5}\left(\frac{3}{5}, \frac{4}{5}\right) = \left(\frac{33}{25}, \frac{44}{25}\right)$$

We are now prepared to formulate the definition of scalar projection and vector projection in $\mathbf{R}^n$.

Definition 5-8 Let $\mathbf{u}$ and $\mathbf{v}$ be vectors in $\mathbf{R}^n$ and let $|\mathbf{v}| \neq 0$. The number $\mathbf{u} \cdot \left(\frac{1}{|\mathbf{v}|}\mathbf{v}\right)$ is called the **scalar projection** of $\mathbf{u}$ onto $\mathbf{v}$. The vector $\left[\mathbf{u} \cdot \left(\frac{1}{|\mathbf{v}|}\mathbf{v}\right)\right]\left(\frac{1}{|\mathbf{v}|}\mathbf{v}\right)$ is called the **vector projection** of $\mathbf{u}$ onto $\mathbf{v}$.

If $\mathbf{v}$ is already a unit vector then $|\mathbf{v}| = 1$, and the formulas for the scalar projection and vector projection of $\mathbf{u}$ onto $\mathbf{v}$ simplify to $\mathbf{u} \cdot \mathbf{v}$ and $(\mathbf{u} \cdot \mathbf{v})\mathbf{v}$, respectively.

Example 8 Consider the vectors $\mathbf{u} = (1,2,3,4)$ and $\mathbf{v} = \left(\frac{1}{3}, \frac{-2}{3}, 0, \frac{2}{3}\right)$ in $\mathbf{R}^4$. Find the scalar projection and vector projection of $\mathbf{u}$ onto $\mathbf{v}$.

Solution A simple computation shows that $\mathbf{v}$ is a unit vector. Thus, the scalar projection of $\mathbf{u}$ onto $\mathbf{v}$ is given by

$$\mathbf{u} \cdot \mathbf{v} = (1,2,3,4) \cdot \left(\frac{1}{3}, \frac{-2}{3}, 0, \frac{2}{3}\right) = \frac{1}{3} - \frac{4}{3} + 0 + \frac{8}{3} = \frac{5}{3}$$

and the vector projection of $\mathbf{u}$ onto $\mathbf{v}$ is given by

$$(\mathbf{u} \cdot \mathbf{v})\mathbf{v} = \frac{5}{3}\left(\frac{1}{3}, \frac{-2}{3}, 0, \frac{2}{3}\right) = \left(\frac{5}{9}, \frac{-10}{9}, 0, \frac{10}{9}\right)$$

The next theorem reveals the reason for our interest in orthonormal bases for $\mathbf{R}^n$.

Theorem 5-3 Let $S = \{\mathbf{v}_1, \mathbf{v}_2, \ldots, \mathbf{v}_n\}$ be an orthonormal basis for $\mathbf{R}^n$. If $\mathbf{u}$ is any vector in $\mathbf{R}^n$, then $\mathbf{u}$ is equal to the sum of its vector projections onto each of the vectors $\mathbf{v}_i$; that is,

$$\mathbf{u} = (\mathbf{u} \cdot \mathbf{v}_1)\mathbf{v}_1 + (\mathbf{u} \cdot \mathbf{v}_2)\mathbf{v}_2 + \cdots + (\mathbf{u} \cdot \mathbf{v}_n)\mathbf{v}_n \qquad [18]$$

Proof Since S is a basis for $\mathbf{R}^n$ it follows that $\mathbf{u}$ has a unique representation in terms of $\mathbf{v}_1, \mathbf{v}_2, \ldots, \mathbf{v}_n$, say

See Theorem 3-5.

$$\mathbf{u} = \alpha_1\mathbf{v}_1 + \alpha_2\mathbf{v}_2 + \cdots + \alpha_n\mathbf{v}_n \qquad [19]$$

To compute α_1 we take the dot product of each side of equation 19 with $\mathbf{v}_1$. After some simplification, we obtain

$$\mathbf{u} \cdot \mathbf{v}_1 = \alpha_1(\mathbf{v}_1 \cdot \mathbf{v}_1) + \alpha_2(\mathbf{v}_2 \cdot \mathbf{v}_1) + \cdots + \alpha_n(\mathbf{v}_n \cdot \mathbf{v}_1) \qquad [20]$$

Since S is an orthonormal set we have $\mathbf{v}_1 \cdot \mathbf{v}_1 = 1$ and $\mathbf{v}_i \cdot \mathbf{v}_1 = 0$ for $i \neq 1$.

Thus, equation 20 reduces to

$$\mathbf{u} \cdot \mathbf{v}_1 = \alpha_1 \qquad [21]$$

Repeating this process with $\mathbf{v}_2, \mathbf{v}_3, \ldots, \mathbf{v}_n$ we obtain

$$\mathbf{u} \cdot \mathbf{v}_2 = \alpha_2, \alpha_3, \ldots, \mathbf{u} \cdot \mathbf{v}_n = \alpha_n \qquad [22]$$

Substituting equations 21 and 22 into equation 19, we obtain equation 18. This proves the theorem.

The following example is a simple application of Theorem 5-3, showing the procedure of finding the coordinates of a vector relative to an orthonormal basis.

Example 9 Express the vector $\mathbf{u} = (1,1,1,1)$ as a linear combination of the vectors $\mathbf{v}_1 = (1,0,0,0)$, $\mathbf{v}_2 = \left(0,\frac{2}{3},\frac{1}{3},\frac{-2}{3}\right)$, $\mathbf{v}_3 = \left(0,\frac{2}{3},\frac{-2}{3},\frac{1}{3}\right)$, and $\mathbf{v}_4 = \left(0,\frac{1}{3},\frac{2}{3},\frac{2}{3}\right)$.

Solution

$$\mathbf{v}_1 \cdot \mathbf{v}_1 = \mathbf{v}_2 \cdot \mathbf{v}_2 = \mathbf{v}_3 \cdot \mathbf{v}_3 = \mathbf{v}_4 \cdot \mathbf{v}_4 = 1$$
$$\mathbf{v}_1 \cdot \mathbf{v}_2 = \mathbf{v}_1 \cdot \mathbf{v}_3 = \mathbf{v}_1 \cdot \mathbf{v}_4 = \mathbf{v}_2 \cdot \mathbf{v}_3 = \mathbf{v}_2 \cdot \mathbf{v}_4 = \mathbf{v}_3 \cdot \mathbf{v}_4 = 0$$

Thus, the set $S = \{\mathbf{v}_1,\mathbf{v}_2,\mathbf{v}_3,\mathbf{v}_4\}$ is orthonormal. It is also obvious that S is a basis for $\mathbf{R}^4$. Therefore, we may apply Theorem 5-3 (with $n = 4$):

If $\mathbf{u} = \alpha_1\mathbf{v}_1 + \alpha_2\mathbf{v}_2 + \alpha_3\mathbf{v}_3 + \alpha_4\mathbf{v}_4$ then we have

$$\alpha_1 = \mathbf{u} \cdot \mathbf{v}_1 = (1,1,1,1) \cdot (1,0,0,0) = 1$$
$$\alpha_2 = \mathbf{u} \cdot \mathbf{v}_2 = (1,1,1,1) \cdot \left(0,\frac{2}{3},\frac{1}{3},\frac{-2}{3}\right) = \frac{1}{3}$$
$$\alpha_3 = \mathbf{u} \cdot \mathbf{v}_3 = (1,1,1,1) \cdot \left(0,\frac{2}{3},\frac{-2}{3},\frac{1}{3}\right) = \frac{1}{3}$$
$$\alpha_4 = \mathbf{u} \cdot \mathbf{v}_4 = (1,1,1,1) \cdot \left(0,\frac{1}{3},\frac{2}{3},\frac{2}{3}\right) = \frac{5}{3}$$

Thus, the unique representation of $\mathbf{u}$ as a linear combination of $\mathbf{v}_1$, $\mathbf{v}_2$, $\mathbf{v}_3$, and $\mathbf{v}_4$ is given by

$$\mathbf{u} = 1(1,0,0,0) + \frac{1}{3}\left(0,\frac{2}{3},\frac{1}{3},\frac{-2}{3}\right) + \frac{1}{3}\left(0,\frac{2}{3},\frac{-2}{3},\frac{1}{3}\right) + \frac{5}{3}\left(0,\frac{1}{3},\frac{2}{3},\frac{2}{3}\right) \quad [23]$$

The reader should check that the right-hand side of equation 23 produces the vector $(1,1,1,1)$.

By using the same technique that we used in the proof of Theorem 5-3 we can also prove the following.

Theorem 5-4 Let $\{\mathbf{v}_1, \mathbf{v}_2, \ldots, \mathbf{v}_k\}$ be an orthonormal set of vectors in $\mathbf{R}^n$. Let $\mathbf{u}$ be an element of $\mathbf{R}^n$. Then $\mathbf{u}$ is a linear combination of $\mathbf{v}_1, \mathbf{v}_2, \ldots, \mathbf{v}_k$ if and only if

$$\mathbf{u} = (\mathbf{u}\cdot\mathbf{v}_1)\mathbf{v}_1 + (\mathbf{u}\cdot\mathbf{v}_2)\mathbf{v}_2 + \cdots + (\mathbf{v}\cdot\mathbf{v}_k)\mathbf{v}_k$$

Note: If $k < n$ then the set $\{\mathbf{v}_1, \mathbf{v}_2, \ldots, \mathbf{v}_k\}$ is not a basis for $\mathbf{R}^n$.

See problem 11.

The proof is left as an exercise for the reader.

We now pose a problem the solution of which may be obtained by using Theorem 5-4.

Problem 3 Determine whether the vector $\mathbf{u} = (2,3,4)$ belongs to the subspace of $\mathbf{R}^3$ spanned by the vectors $\mathbf{v}_1 = \frac{1}{\sqrt{14}}(1,2,3)$ and $\mathbf{v}_2 = \frac{1}{\sqrt{3}}(1,1,-1)$.

Solution A simple check shows that $\mathbf{v}_1\cdot\mathbf{v}_1 = \mathbf{v}_2\cdot\mathbf{v}_2 = 1$ and $\mathbf{v}_1\cdot\mathbf{v}_2 = 0$. Thus, the set of vectors $\{\mathbf{v}_1,\mathbf{v}_2\}$ is an orthonormal set in $\mathbf{R}^3$. In order to determine whether the vector $\mathbf{u} = (2,3,4)$ is a linear combination of $\mathbf{v}_1$ and $\mathbf{v}_2$, we apply Theorem 5-4. We must therefore check whether the linear combination $(\mathbf{v}\cdot\mathbf{v}_1)\mathbf{v}_1 + (\mathbf{u}\cdot\mathbf{v}_2)\mathbf{v}_2$ produces the given vector $\mathbf{u} = (2,3,4)$. We have

$$\mathbf{u}\cdot\mathbf{v}_1 = (2,3,4)\cdot\left[\frac{1}{\sqrt{14}}(1,2,3)\right] = \frac{20}{\sqrt{14}}$$

and

$$\mathbf{u}\cdot\mathbf{v}_2 = (2,3,4)\cdot\left[\frac{1}{\sqrt{3}}(1,1,-1)\right] = \frac{1}{\sqrt{3}}$$

Thus, we obtain

$$\begin{aligned}(\mathbf{u}\cdot\mathbf{v}_1)\mathbf{v}_1 + (\mathbf{u}\cdot\mathbf{v}_2)\mathbf{v}_2 &= \frac{20}{\sqrt{14}}\left(\frac{1}{\sqrt{14}}(1,2,3)\right) + \frac{1}{\sqrt{3}}\left(\frac{1}{\sqrt{3}}(1,1,-1)\right)\\ &= \frac{20}{14}(1,2,3) + \frac{1}{3}(1,1,-1)\\ &= \left(\frac{37}{21},\frac{67}{21},\frac{83}{21}\right)\end{aligned}$$

The resulting vector $\left(\frac{37}{21}, \frac{67}{21}, \frac{83}{21}\right)$ is obviously different from $\mathbf{u} = (2,3,4)$. We conclude that the vector $(2,3,4)$ is not a linear combination of $\mathbf{v}_1 = \frac{1}{\sqrt{14}}(1,2,3)$ and $\mathbf{v}_2 = \frac{1}{\sqrt{3}}(1,1,-1)$.

Exercises

1 Find all values of k for which the following pairs of vectors are orthogonal:
 a $(2,3)$ and $(-5,k)$ **b** $(1,2,3)$ and $(k^2,1,k)$
 c $(1,1,-1,-5)$ and (k^3,k^2,k,k)

2 Consider the vectors, $\mathbf{u} = (1,1,1)$ and $\mathbf{v} = (2,1,-1)$ in $\mathbf{R}^3$.
 a Show that the set of all vectors in $\mathbf{R}^3$ which are orthogonal to both $\mathbf{u}$ and $\mathbf{v}$ is a subspace of $\mathbf{R}^3$ and find a basis for it.
 b What is the dimension of the subspace mentioned in part a?
 c Find *all unit* vectors in $\mathbf{R}^3$ which are orthogonal to both $\mathbf{u}$ and $\mathbf{v}$.

3 Which of the following sets are orthogonal?
 a $\{(2,1), (1,-2), (0,0)\}$ **b** $\{(1,1), (-1,-1)\}$
 c $\{(1,1,-1), (-1,0,1), (1,0,1)\}$
 d $\{(1,1,1), (1,-2,1), (1,0,-1),(0,0,0)\}$
 e $\{(2,1,2,1), (1,1,0,-3), (-4,7,0,1), (-2,2,1,0)\}$

4 Is it possible to find an orthonormal set of vectors in $\mathbf{R}^4$ consisting of exactly 5 vectors? Explain your answer.

5 Which of the following sets are orthonormal?
 a $\left\{\left(\frac{1}{\sqrt{2}}, \frac{1}{\sqrt{2}}\right), \left(\frac{1}{\sqrt{2}}, \frac{-1}{\sqrt{2}}\right)\right\}$ **b** $\left\{\left(\frac{1}{3}, \frac{2}{3}\right), \left(\frac{2}{3}, \frac{-1}{3}\right)\right\}$
 c $\left\{\left(\frac{-2}{3}, \frac{1}{3}, \frac{2}{3}\right), \left(\frac{1}{3}, \frac{-2}{3}, \frac{2}{3}\right), \left(\frac{2}{3}, \frac{2}{3}, \frac{1}{3}\right)\right\}$
 d $\left\{\left(\frac{2}{\sqrt{10}}, \frac{1}{\sqrt{10}}, \frac{2}{\sqrt{10}}, \frac{1}{\sqrt{10}}\right), \left(\frac{1}{\sqrt{11}}, \frac{1}{\sqrt{11}}, 0, \frac{-3}{\sqrt{11}}\right),\right.$
 $\left.\left(\frac{-2}{3}, \frac{2}{3}, \frac{1}{3}, 0\right), \left(\frac{-4}{\sqrt{66}}, \frac{7}{\sqrt{66}}, 0, \frac{1}{\sqrt{66}}\right)\right\}$

6 Find the scalar projection of $\mathbf{u}$ onto $\mathbf{v}$ if:
 a $\mathbf{u} = (2,3)$ and $\mathbf{v} = (3,4)$
 b $\mathbf{u} = (2,1,-1)$ and $\mathbf{v} = (1,2,2)$
 c $\mathbf{u} = (1,-1,2,-2)$ and $\mathbf{v} = (2,0,1,-2)$

7 Find the vector projection of $\mathbf{u}$ onto $\mathbf{v}$ if:
 a $\mathbf{u} = (1,2)$ and $\mathbf{v} = (4,-3)$

b $\mathbf{u} = (1,-2,1)$ and $\mathbf{v} = (2,-2,1)$
c $\mathbf{u} = (4,2,3,1)$ and $\mathbf{v} = (1,0,2,-2)$

8 a Show that the vectors $\left(\frac{1}{\sqrt{5}}, \frac{2}{\sqrt{5}}\right)$ and $\left(\frac{2}{\sqrt{5}}, \frac{-1}{\sqrt{5}}\right)$ form an orthonormal basis for $\mathbf{R}^2$.

b Express the vector $(3,2)$ as a linear combination of the vectors $\left(\frac{1}{\sqrt{5}}, \frac{2}{\sqrt{5}}\right)$ and $\left(\frac{2}{\sqrt{5}}, \frac{-1}{\sqrt{5}}\right)$.

9 a Show that the vectors $\mathbf{u}_1 = \left(\frac{1}{\sqrt{3}}, \frac{1}{\sqrt{3}}, \frac{-1}{\sqrt{3}}\right)$, $\mathbf{u}_2 = \left(\frac{1}{\sqrt{2}}, 0, \frac{1}{\sqrt{2}}\right)$, and $\mathbf{u}_3 = \left(\frac{-1}{\sqrt{6}}, \frac{2}{\sqrt{6}}, \frac{1}{\sqrt{6}}\right)$ form an orthonormal basis for $\mathbf{R}^3$.

b Express the vector $(1,2,3)$ as a linear combination of $\mathbf{u}_1$, $\mathbf{u}_2$, and $\mathbf{u}_3$.

10 Show that the vector $(4,2,1)$ is *not* a linear combination of the vectors $\frac{1}{\sqrt{3}}(1,1,-1)$ and $\frac{1}{\sqrt{2}}(1,0,1)$.

11 Prove Theorem 5-4.

5-3 The Gram-Schmidt Process

Objective
Establish a technique of constructing orthonormal bases for $\mathbf{R}^n$.

ERHARD SCHMIDT (1876–1959) *was a professor of mathematics at several German universities. He did outstanding work on integral equations. Schmidt used methods originated by Schwarz in the latter's studies of potential theory. He simplified much of Hilbert's work on integral equations and generalized the Pythagorean theorem.*

It is easy to see that the vectors $(1,0)$ and $(0,1)$ form an orthonormal basis for the vector space $\mathbf{R}^2$. One of our objectives in this section is to show that $\mathbf{R}^2$ has infinitely many other orthonormal bases. We will exhibit a constructive method of producing an orthonormal basis by transforming a given basis of $\mathbf{R}^2$ in a special way.

Let $\mathbf{u}_1$ and $\mathbf{u}_2$ be a given basis for $\mathbf{R}^2$. We normalize $\mathbf{u}_1$, obtaining the vector $\mathbf{v}_1 = \frac{\mathbf{u}_1}{|\mathbf{u}_1|}$. We now subtract from $\mathbf{u}_2$ the vector projection of $\mathbf{u}_2$ onto $\mathbf{v}_1$ and obtain the vector

$$\mathbf{w} = \mathbf{u}_2 - (\mathbf{u}_2 \cdot \mathbf{v}_1)\mathbf{v}_1$$

A simple check shows that $\mathbf{w}$ is orthogonal to $\mathbf{v}_1$. We have

$$\mathbf{w} \cdot \mathbf{v}_1 = [\mathbf{u}_2 - (\mathbf{u}_2 \cdot \mathbf{v}_1)\mathbf{v}_1] \cdot \mathbf{v}_1 = \mathbf{u}_2 \cdot \mathbf{v}_1 - (\mathbf{u}_2 \cdot \mathbf{v}_1)(\mathbf{v}_1 \cdot \mathbf{v}_1)$$

Since $\mathbf{v}_1 \cdot \mathbf{v}_1 = 1$ we obtain

$$\mathbf{w} \cdot \mathbf{v}_1 = \mathbf{u}_2 \cdot \mathbf{v}_1 - \mathbf{u}_2 \cdot \mathbf{v}_1 = 0.$$

Thus, $\mathbf{w}$ is orthogonal to $\mathbf{v}_1$.

In order to complete the construction of an orthonormal basis for $\mathbf{R}^2$ which includes $\mathbf{v}_1$, we must normalize $\mathbf{w}$. This can be done provided we are sure that $\mathbf{w}$ is not the zero vector $(0,0)$ of $\mathbf{R}^2$. To show this, we write $\mathbf{w}$ in the form

$$\mathbf{w} = \mathbf{u}_2 - \left(\mathbf{u}_2 \cdot \frac{1}{|\mathbf{u}_1|}\mathbf{u}_1\right)\frac{1}{|\mathbf{u}_1|}\mathbf{u}_1 \qquad [1]$$

It is obvious from equation 1 that $\mathbf{w}$ is a linear combination of $\mathbf{u}_1$ and $\mathbf{u}_2$. Since $\mathbf{u}_1$ and $\mathbf{u}_2$ are linearly independent vectors, it is impossible to generate the zero vector $(0,0)$ by a linear combination of these vectors unless both coefficients of $\mathbf{u}_1$ and $\mathbf{u}_2$ are zero. Since the coefficient of $\mathbf{u}_2$ in equation 1 is equal to 1, it follows that $\mathbf{w}$ must be different from $(0,0)$.

We can now normalize $\mathbf{w}$ by writing $\mathbf{v}_2 = \dfrac{1}{|\mathbf{w}|}\mathbf{w}$ and obtain the orthonormal basis $\{\mathbf{v}_1,\mathbf{v}_2\}$ for $\mathbf{R}^2$. The reader should notice that $\mathbf{v}_1$ is a scalar multiple of $\mathbf{u}_1$ while $\mathbf{v}_2$ is a linear combination of $\mathbf{u}_1$ and $\mathbf{u}_2$.

The following example illustrates the procedure of transforming a given basis of $\mathbf{R}^2$ into an orthonormal basis for $\mathbf{R}^2$.

Example 1 Consider the basis for $\mathbf{R}^2$ given by the vectors $\mathbf{u}_1 = (1,2)$ and $\mathbf{u}_2 = (1,1)$. Construct an orthonormal basis $\{\mathbf{v}_1, \mathbf{v}_2\}$ for $\mathbf{R}^2$ such that $\mathbf{v}_1$ is parallel to $\mathbf{u}_1$.

Solution We normalize the vector $\mathbf{u}_1 = (1,2)$ and obtain

$$\mathbf{v}_1 = \frac{1}{|\mathbf{u}_1|}\mathbf{u}_1 = \frac{1}{\sqrt{5}}(1,2) = \left(\frac{1}{\sqrt{5}}, \frac{2}{\sqrt{5}}\right)$$

Following the procedure outlined above, we now construct the vector $\mathbf{w} = \mathbf{u}_2 - (\mathbf{u}_2 \cdot \mathbf{v}_1)\mathbf{v}_1$. Thus, we have

$$\begin{aligned}\mathbf{w} &= (1,1) - \left[(1,1)\cdot\left(\frac{1}{\sqrt{5}}, \frac{2}{\sqrt{5}}\right)\right]\left(\frac{1}{\sqrt{5}}, \frac{2}{\sqrt{5}}\right)\\ &= (1,1) - \left(\frac{1}{\sqrt{5}} + \frac{2}{\sqrt{5}}\right)\left(\frac{1}{\sqrt{5}}, \frac{2}{\sqrt{5}}\right)\\ &= (1,1) - \frac{3}{\sqrt{5}}\left(\frac{1}{\sqrt{5}}, \frac{2}{\sqrt{5}}\right) = (1,1) - \left(\frac{3}{5}, \frac{6}{5}\right) = \left(\frac{2}{5}, \frac{-1}{5}\right)\end{aligned}$$

In order to obtain $\mathbf{v}_2$ we normalize $\mathbf{w}$ and write

$$\mathbf{v}_2 = \frac{1}{|\mathbf{w}|}\mathbf{w} = \sqrt{5}\left(\frac{2}{5}, \frac{-1}{5}\right) = \left(\frac{2}{\sqrt{5}}, \frac{-1}{\sqrt{5}}\right)$$

An easy check shows that $|\mathbf{v}_1| = |\mathbf{v}_2| = 1$ and $\mathbf{v}_1 \cdot \mathbf{v}_2 = 0$. Thus, the vectors $\mathbf{v}_1 = \left(\dfrac{1}{\sqrt{5}}, \dfrac{2}{\sqrt{5}}\right)$ and $\mathbf{v}_2 = \left(\dfrac{2}{\sqrt{5}}, \dfrac{-1}{\sqrt{5}}\right)$ form an orthonormal basis for $\mathbf{R}^2$.

We now shift our attention to the vector space $\mathbf{R}^3$. It is a simple matter to see that the vectors (1,0,0), (0,1,0), and (0,0,1) form an orthonormal basis for $\mathbf{R}^3$. We shall now show how to construct additional orthonormal bases for $\mathbf{R}^3$.

Let $\mathbf{u}_1$, $\mathbf{u}_2$, and $\mathbf{u}_3$ be a given basis for $\mathbf{R}^3$. We normalize $\mathbf{u}_1$ and obtain the unit vector $\mathbf{v}_1 = \dfrac{1}{|\mathbf{u}_1|}\mathbf{u}_1$. Subtracting from $\mathbf{u}_2$ the vector projection of $\mathbf{u}_2$ onto $\mathbf{v}_1$ we obtain the vector

$$\mathbf{w} = \mathbf{u}_2 - (\mathbf{u}_2 \cdot \mathbf{v}_1)\mathbf{v}_1$$

The vector $\mathbf{w}$ has the following properties:

(a) $\mathbf{w}$ is orthogonal to $\mathbf{v}_1$

(b) $\mathbf{w}$ is different from the zero vector (0,0,0) of $\mathbf{R}^3$.

The proof of properties **a** and **b** is similar to the one discussed in the case of the vector space $\mathbf{R}^2$ and is therefore left for the reader as an exercise.

We proceed with our construction and normalize $\mathbf{w}$, obtaining the unit vector $\mathbf{v}_2 = \dfrac{1}{|\mathbf{w}|}\mathbf{w}$.

We now subtract from $\mathbf{u}_3$ the vector projection of $\mathbf{u}_3$ onto $\mathbf{v}_1$ and the vector projection of $\mathbf{u}_3$ onto $\mathbf{v}_2$. We obtain the vector

$$\mathbf{z} = \mathbf{u}_3 - (\mathbf{u}_3 \cdot \mathbf{v}_1)\mathbf{v}_1 - (\mathbf{u}_3 \cdot \mathbf{v}_2)\mathbf{v}_2$$

It turns out that the vector $\mathbf{z}$ has the following properties:

(a) $\mathbf{z}$ is orthogonal to $\mathbf{v}_1$

(b) $\mathbf{z}$ is orthogonal to $\mathbf{v}_2$

(c) $\mathbf{z} \neq (0,0,0)$

Property **a** follows from the equation

$$\begin{aligned}\mathbf{z} \cdot \mathbf{v}_1 &= [\mathbf{u}_3 - (\mathbf{u}_3 \cdot \mathbf{v}_1)\mathbf{v}_1 - (\mathbf{u}_3 \cdot \mathbf{v}_2)\mathbf{v}_2] \cdot \mathbf{v}_1\\ &= \mathbf{u}_3 \cdot \mathbf{v}_1 - (\mathbf{u}_3 \cdot \mathbf{v}_1)(\mathbf{v}_1 \cdot \mathbf{v}_1) - (\mathbf{u}_3 \cdot \mathbf{v}_2)(\mathbf{v}_2 \cdot \mathbf{v}_1)\\ &= \mathbf{u}_3 \cdot \mathbf{v}_1 - (\mathbf{u}_3 \cdot \mathbf{v}_1)(1) - (\mathbf{u}_3 \cdot \mathbf{v}_2)(0)\\ &= 0\end{aligned}$$

Property **b** follows from the equation

$$\begin{aligned}\mathbf{z} \cdot \mathbf{v}_2 &= [\mathbf{u}_3 - (\mathbf{u}_3 \cdot \mathbf{v}_1)\mathbf{v}_1 - (\mathbf{u}_3 \cdot \mathbf{v}_2)\mathbf{v}_2] \cdot \mathbf{v}_2\\ &= \mathbf{u}_3 \cdot \mathbf{v}_2 - (\mathbf{u}_3 \cdot \mathbf{v}_1)(\mathbf{v}_1 \cdot \mathbf{v}_2) - (\mathbf{u}_3 \cdot \mathbf{v}_2)(\mathbf{v}_2 \cdot \mathbf{v}_2)\\ &= \mathbf{u}_3 \cdot \mathbf{v}_2 - (\mathbf{u}_3 \cdot \mathbf{v}_1)(0) - (\mathbf{u}_3 \cdot \mathbf{v}_2)(1)\\ &= 0\end{aligned}$$

Property **c** follows from the fact that $\mathbf{z}$ is a linear combination of the basis vectors $\mathbf{u}_1$, $\mathbf{u}_2$, and $\mathbf{u}_3$ with the coefficient of $\mathbf{u}_3$ being 1; hence, $\mathbf{z}$ must be different from (0,0,0).

If we now normalize $\mathbf{z}$ we obtain $\mathbf{v}_3 = \frac{1}{|\mathbf{z}|}\mathbf{z}$. It follows that $\{\mathbf{v}_1,\mathbf{v}_2,\mathbf{v}_3\}$ is an orthonormal basis for $\mathbf{R}^3$. The following example illustrates the procedure of transforming a given basis of $\mathbf{R}^3$ into an orthonormal basis for $\mathbf{R}^3$.

Example 2 Consider the basis for $\mathbf{R}^3$ given by the vectors $\mathbf{u}_1 = (1,1,0)$, $\mathbf{u}_2 = (1,2,2)$, and $\mathbf{u}_3 = (2,1,1)$. Construct an orthonormal basis $\{\mathbf{v}_1,\mathbf{v}_2,\mathbf{v}_3\}$ for $\mathbf{R}^3$ such that $\mathbf{v}_1$ is parallel to $\mathbf{u}_1$.

Solution We normalize the vector $\mathbf{u}_1 = (1,1,0)$, obtaining the vector $\mathbf{v}_1 = \frac{1}{|\mathbf{u}_1|}\mathbf{u}_1 = \frac{1}{\sqrt{2}}(1,1,0) = \left(\frac{1}{\sqrt{2}},\frac{1}{\sqrt{2}},0\right)$. Now we construct the vector $\mathbf{w} = \mathbf{u}_2 - (\mathbf{u}_2 \cdot \mathbf{v}_1)\mathbf{v}_1$. We have

$$\begin{aligned}\mathbf{w} &= (1,2,2) - (1,2,2)\cdot\left(\frac{1}{\sqrt{2}},\frac{1}{\sqrt{2}},0\right)\left(\frac{1}{\sqrt{2}},\frac{1}{\sqrt{2}},0\right)\\ &= (1,2,2) - \frac{3}{\sqrt{2}}\left(\frac{1}{\sqrt{2}},\frac{1}{\sqrt{2}},0\right)\\ &= \left(\frac{-1}{2},\frac{1}{2},2\right)\end{aligned}$$

We normalize $\mathbf{w}$ and obtain the vector $\mathbf{v}_2$.

$$\mathbf{v}_2 = \frac{1}{|\mathbf{w}|}\mathbf{w} = \frac{\sqrt{2}}{3}\left(\frac{-1}{2},\frac{1}{2},2\right) = \left(\frac{-\sqrt{2}}{6},\frac{\sqrt{2}}{6},\frac{2}{3}\sqrt{2}\right)$$

Continuing with the procedure as described above, we now construct the vector $\mathbf{z} = \mathbf{u}_3 - (\mathbf{u}_3 \cdot \mathbf{v}_1)\mathbf{v}_1 - (\mathbf{u}_3 \cdot \mathbf{v}_2)\mathbf{v}_2$, obtaining

$$\begin{aligned}\mathbf{z} &= (2,1,1) - (2,1,1)\cdot\left(\frac{1}{\sqrt{2}},\frac{1}{\sqrt{2}},0\right)\left(\frac{1}{\sqrt{2}},\frac{1}{\sqrt{2}},0\right)\\ &\qquad - (2,1,1)\cdot\left(\frac{-\sqrt{2}}{6},\frac{\sqrt{2}}{6},\frac{2}{3}\sqrt{2}\right)\left(\frac{-\sqrt{2}}{6},\frac{\sqrt{2}}{6},\frac{2}{3}\sqrt{2}\right)\\ &= (2,1,1) - \frac{3}{\sqrt{2}}\left(\frac{1}{\sqrt{2}},\frac{1}{\sqrt{2}},0\right) - \frac{\sqrt{2}}{2}\left(\frac{-\sqrt{2}}{6},\frac{\sqrt{2}}{6},\frac{2}{3}\sqrt{2}\right)\\ &= (2,1,1) - \left(\frac{3}{2},\frac{3}{2},0\right) - \left(\frac{-1}{6},\frac{1}{6},\frac{2}{3}\right)\\ &= \left(\frac{2}{3},\frac{-2}{3},\frac{1}{3}\right)\end{aligned}$$

Since $\mathbf{z}$ is already a unit vector we define $\mathbf{v}_3$ by $\mathbf{v}_3 = \mathbf{z} = \left(\frac{2}{3},\frac{-2}{3},\frac{1}{3}\right)$.

An easy check confirms that the vectors $\mathbf{v}_1 = \left(\frac{1}{\sqrt{2}}, \frac{1}{\sqrt{2}}, 0\right)$, $\mathbf{v}_2 = \left(\frac{-\sqrt{2}}{6}, \frac{\sqrt{2}}{6}, \frac{2}{3}\sqrt{2}\right)$, $\mathbf{v}_3 = \left(\frac{2}{3}, \frac{-2}{3}, \frac{1}{3}\right)$ form an orthonormal basis for $\mathbf{R}^3$.

We are now in a position to generalize the process of converting an arbitrary basis into an orthonormal basis as described in Examples 1 and 2. We pose the following generalized problem:

Problem 4 Let U be a subspace of $\mathbf{R}^n$ such that $\dim U = m\ (1 \leq m \leq n)$. Let $\{\mathbf{u}_1, \mathbf{u}_2, \ldots, \mathbf{u}_m\}$ be a basis for U. Construct an orthonormal basis $\{\mathbf{v}_1, \mathbf{v}_2, \ldots, \mathbf{v}_m\}$ for U such that $\mathbf{v}_1$ is parallel to $\mathbf{u}_1$.

Solution Based on our previous discussion in this section we list here the sequence of steps that will produce an orthonormal basis for U as required.

Gram-Schmidt process.

STEP 1 Normalize $\mathbf{u}_1$ and obtain $\mathbf{v}_1 = \frac{1}{|\mathbf{u}_1|}\mathbf{u}_1$

STEP 2 Construct the vector $\mathbf{w}_1 = \mathbf{u}_2 - (\mathbf{u}_2 \cdot \mathbf{v}_1)\mathbf{v}_1$

STEP 3 Normalize $\mathbf{w}_1$ and obtain $\mathbf{v}_2 = \frac{1}{|\mathbf{w}_1|}\mathbf{w}_1$

STEP 4 Construct the vector $\mathbf{w}_2 = \mathbf{u}_3 - (\mathbf{u}_3 \cdot \mathbf{v}_1)\mathbf{v}_1 - (\mathbf{u}_3 \cdot \mathbf{v}_2)\mathbf{v}_2$

STEP 5 Normalize $\mathbf{w}_2$ and obtain $\mathbf{v}_3 = \frac{1}{|\mathbf{w}_2|}\mathbf{w}_2$

STEP 6 Construct the vector $\mathbf{w}_3 = \mathbf{u}_4 - (\mathbf{u}_4 \cdot \mathbf{v}_1)\mathbf{v}_1 - (\mathbf{u}_4 \cdot \mathbf{v}_2)\mathbf{v}_2 - (\mathbf{u}_4 \cdot \mathbf{v}_3)\mathbf{v}_3$

STEP 7 Normalize $\mathbf{w}_3$ and obtain $\mathbf{v}_4 = \frac{1}{|\mathbf{w}_3|}\mathbf{w}_3$

At this point we have an orthonormal set $\{\mathbf{v}_1,\mathbf{v}_2,\mathbf{v}_3,\mathbf{v}_4\}$. If $m > 4$, we continue in a similar manner to construct $\mathbf{v}_5$ by subtracting from $\mathbf{u}_5$ the vector projections of $\mathbf{u}_5$ onto $\mathbf{v}_1$, $\mathbf{v}_2$, $\mathbf{v}_3$, and $\mathbf{v}_4$ and normalizing the resulting vector. In general, if we already have an orthonormal set $\{\mathbf{v}_1, \mathbf{v}_2, \ldots, \mathbf{v}_k\}$ with $k < m$ then we may obtain $\mathbf{v}_{k+1}$ by normalizing the vector $\mathbf{u}_{k+1} - \left(\sum_{i=1}^{k} (\mathbf{u}_{k+1} \cdot \mathbf{v}_i)\mathbf{v}_i\right)$. After a finite number of steps we must have an orthonormal set $\{\mathbf{v}_1, \mathbf{v}_2, \ldots, \mathbf{v}_m\}$. Since such a set is linearly independent it follows that $\{\mathbf{v}_1, \mathbf{v}_2, \ldots, \mathbf{v}_m\}$ is an orthonormal basis for U.

Remember, if $\dim U = m$ then every basis of U must contain exactly m elements.

Exercises

1 Let $\mathbf{u}_1 = (3,1)$ and $\mathbf{u}_2 = (2,-1)$.

a Show that the set $\{\mathbf{u}_1,\mathbf{u}_2\}$ is a basis for $\mathbf{R}^2$.

b Find an orthonormal basis $\{\mathbf{v}_1,\mathbf{v}_2\}$ for $\mathbf{R}^2$ such that $\mathbf{v}_1$ is parallel to $\mathbf{u}_1$.

2 Let $\mathbf{u}_1 = (1,2,2)$ and $\mathbf{u}_2 = (2,1,2)$. Find an orthonormal basis $\{\mathbf{v}_1,\mathbf{v}_2\}$ for the subspace of $\mathbf{R}^3$ spanned by $\mathbf{u}_1$ and $\mathbf{u}_2$ such that $\mathbf{v}_1$ is parallel to $\mathbf{u}_1$.

3 Let $\mathbf{u}_1 = (1,1,1)$, $\mathbf{u}_2 = (-1,-1,1)$, and $\mathbf{u}_3 = (1,2,2)$.

a Show that the set $\{\mathbf{u}_1,\mathbf{u}_2,\mathbf{u}_3\}$ is a basis for $\mathbf{R}^3$.

b Find an orthonormal basis $\{\mathbf{v}_1,\mathbf{v}_2,\mathbf{v}_3\}$ for $\mathbf{R}^3$ such that $\mathbf{v}_1$ is parallel to $\mathbf{u}_1$.

4 Let $\mathbf{u}_1 = (1,0,1,0)$, $\mathbf{u}_2 = (2,0,0,1)$, and $\mathbf{u}_3 = (0,0,2,3)$. Find an orthonormal basis $\{\mathbf{v}_1,\mathbf{v}_2,\mathbf{v}_3\}$ for the subspace of $\mathbf{R}^4$ spanned by $\mathbf{u}_1$, $\mathbf{u}_2$, and $\mathbf{u}_3$ such that $\mathbf{v}_1$ is parallel to $\mathbf{u}_1$.

5 Let $\mathbf{u}_1 = (1,1,1,1)$, $\mathbf{u}_2 = (1,2,3,4)$, $\mathbf{u}_3 = (0,1,3,6)$, and $\mathbf{u}_4 = (0,0,1,4)$.

a Show that the set $\{\mathbf{u}_1,\mathbf{u}_2,\mathbf{u}_3,\mathbf{u}_4\}$ is a basis for $\mathbf{R}^4$.

b Find an orthonormal basis $\{\mathbf{v}_1,\mathbf{v}_2,\mathbf{v}_3,\mathbf{v}_4\}$ for $\mathbf{R}^4$ such that $\mathbf{v}_1$ is parallel to $\mathbf{u}_1$.

6 Let $\mathbf{u}_1$, $\mathbf{u}_2$, and $\mathbf{u}_3$ be a basis for $\mathbf{R}^3$.

Note: The symbol $\mathbf{w} \overset{\text{def}}{=}$ means "$\mathbf{w}$ is defined by."

a Show that $\mathbf{w} \overset{\text{def}}{=} \mathbf{u}_2 - (\mathbf{u}_2 \cdot \mathbf{v}_1)\mathbf{v}_1 \neq (0,0,0)$ where $\mathbf{v}_1 = \dfrac{1}{|\mathbf{u}_1|}\mathbf{u}_1$ and that $\mathbf{w}$ is orthogonal to $\mathbf{v}_1$.

b Show that $\mathbf{z} \overset{\text{def}}{=} \mathbf{u}_3 - (\mathbf{u}_3 \cdot \mathbf{v}_1)\mathbf{v}_1 - (\mathbf{u}_3 \cdot \mathbf{v}_2)\mathbf{v}_2 \neq (0,0,0)$ where $\mathbf{v}_2 = \dfrac{1}{|\mathbf{w}_1|}\mathbf{w}_1$.

c Show that the vector $\mathbf{z}$ defined in part b is orthogonal to $\mathbf{v}_1$ and $\mathbf{v}_2$.

Review of Chapter 5

1 Let $\mathbf{u} = [u_1, u_2 \ldots, u_n]$ and $\mathbf{v} = [v_1, v_2, \ldots, v_n]$ be vectors in $\mathbf{R}^n$. If the dot product $\mathbf{u} \cdot \mathbf{v} = 0$, does it follow that either $\mathbf{u} = \mathbf{0}$ or $\mathbf{v} = \mathbf{0}$?

2 Consider the vectors $\mathbf{u} = [1,2,3]$ and $\mathbf{v} = [4,1,-2]$ in $\mathbf{R}^3$. Are $\mathbf{u}$ and $\mathbf{v}$ orthogonal?

3 Let $\mathbf{u} = [u_1, u_2, \ldots, u_n]$ and $\mathbf{v} = [v_1, v_2, \ldots, v_n]$ be vectors in $\mathbf{R}^n$. Express the dot product $(\mathbf{u} + \mathbf{v}) \cdot (\mathbf{u} - \mathbf{v})$ in terms of $|\mathbf{u}|$ and $|\mathbf{v}|$.

4 Let $\mathbf{u}$ and $\mathbf{v}$ be vectors in $\mathbf{R}^n$. Show that if $|\mathbf{u}| = |\mathbf{v}|$ then $\mathbf{u} + \mathbf{v}$ and $\mathbf{u} - \mathbf{v}$ are orthogonal vectors in $\mathbf{R}^n$.

5 Let $\mathbf{u}$, $\mathbf{v}$, and $\mathbf{w}$ be vectors in $\mathbf{R}^3$ such that $\mathbf{u} \cdot \mathbf{v} = 0$ and $\mathbf{v} \cdot \mathbf{w} = 0$. Under what conditions can we state that the set $\{\mathbf{u},\mathbf{v},\mathbf{w}\}$ is linearly independent?

6 Let $S = \{\mathbf{v}_1, \mathbf{v}_2, \ldots, \mathbf{v}_k\}$ be an orthogonal set of vectors. Does it follow that S is an orthonormal set?

7 Let $T = \{\mathbf{u}_1, \mathbf{u}_2, \ldots, \mathbf{u}_m\}$ be an orthonormal set of vectors. Does it follow that T is an orthogonal set?

8 Can an orthogonal set of vectors be linearly dependent? Can an orthonormal set of vectors be linearly dependent?

9 Let $\mathbf{u}$ and $\mathbf{v}$ be vectors in $\mathbf{R}^n$ and let $\mathbf{v} \neq \mathbf{0}$. What is the difference between the scalar projection of $\mathbf{u}$ onto $\mathbf{v}$ and the vector projection of $\mathbf{u}$ onto $\mathbf{v}$?

10 Let $\{\mathbf{v}_1,\mathbf{v}_2,\mathbf{v}_3,\mathbf{v}_4\}$ be an orthonormal basis for $\mathbf{R}^4$. Let $\mathbf{v}$ be a vector in $\mathbf{R}^4$. Express $\mathbf{v}$ as a linear combination of $\mathbf{v}_1$, $\mathbf{v}_2$, $\mathbf{v}_3$, and $\mathbf{v}_4$.

11 Let $\{\mathbf{v}_1, \mathbf{v}_2, \ldots, \mathbf{v}_n\}$ be an orthonormal basis for $\mathbf{R}^n$. Show that if $\mathbf{u} \in \mathbf{R}^n$, then $|\mathbf{u}|^2 = (\mathbf{u} \cdot \mathbf{v}_1)^2 + (\mathbf{u} \cdot \mathbf{v}_2)^2 + \cdots + (\mathbf{u} \cdot \mathbf{v}_n)^2$.

12 Use your own words to describe the purpose of the Gram-Schmidt process.

Linear Transformations and Matrices

6

6-1 Definition and Properties of Linear Transformations

Objectives
1 Introduce linear transformations.
2 Show the existence of nonlinear transformations.
3 Gain familiarity with linear transformations through numerous examples.

The concept of a function was introduced in Section 0-2. Particular emphasis was given to functions whose domain and range are sets of real numbers. This type of function is the main concern of some topics in calculus. Functions also play a major role in the study of linear algebra. Here we deal with **vector-valued** functions, whose domain (and range) are vector spaces. Thus, if we write $\mathbf{w} = T(\mathbf{v})$, we mean that T is a function which assigns to each vector $\mathbf{v}$ in a vector space V a unique vector $\mathbf{w}$ in a vector space W. The vector space V is the domain of the function T, and the vector space W contains the range of T.

The following example illustrates the idea of a vector-valued function.

Example 1 Let $\mathbf{v} = (x,y,z)$ be a vector in $\mathbf{R}^3$ and let T be the vector-valued function defined by

$$T(\mathbf{v}) = (x + y, x + z) \qquad [1]$$

It follows from equation 1 that T is a function that associates with every vector (x,y,z) in $\mathbf{R}^3$ the vector $(x + y, x + z)$ in $\mathbf{R}^2$. Thus, the domain of T is $\mathbf{R}^3$ and the range of T is $\mathbf{R}^2$. In particular, we have from equation 1

$$T(1,2,3) = (1 + 2, 1 + 3) = (3,4) \qquad [2]$$

It is customary to refer to the function T as a **transformation** and to say that the transformation T maps $\mathbf{R}^3$ into $\mathbf{R}^2$. We describe equation 2 by saying that the vector $(3,4)$ is the image of the vector $(1,2,3)$ under the transformation T.

Our main objective is to study a special set of functions, called *linear transformations,* and to investigate their relationship to matrices.

Definition 6-1 Let V and W be vector spaces and let T be a transformation from V into W. Then T is called a **linear transformation** if the following conditions hold:

a $T(\mathbf{v}_1 + \mathbf{v}_2) = T(\mathbf{v}_1) + T(\mathbf{v}_2)$ for all vectors $\mathbf{v}_1$ and $\mathbf{v}_2$ in V
b $T(\alpha\mathbf{v}) = \alpha T(\mathbf{v})$ for all vectors $\mathbf{v}$ in V and all real numbers α

Let us now show that the transformation T in Example 1, defined by

$$T(x,y,z) = (x + y, x + z) \qquad [3]$$

is linear.

Let $\mathbf{v}_1 = (x_1,y_1,z_1)$ and $\mathbf{v}_2 = (x_2,y_2,z_2)$. Then we have

$$\begin{aligned}T(\mathbf{v}_1 + \mathbf{v}_2) &= T(x_1 + x_2, y_1 + y_2, z_1 + z_2)\\ &= ((x_1 + x_2) + (y_1 + y_2), (x_1 + x_2) + (z_1 + z_2))\end{aligned}$$

and

$$\begin{aligned}T(\mathbf{v}_1) + T(\mathbf{v}_2) &= (x_1 + y_1, x_1 + z_1) + (x_2 + y_2, x_2 + z_2)\\ &= ((x_1 + y_1) + (x_2 + y_2), (x_1 + z_1) + (x_2 + z_2))\end{aligned}$$

Since

$$(x_1 + x_2) + (y_1 + y_2) = (x_1 + y_1) + (x_2 + y_2)$$

and

$$(x_1 + x_2) + (z_1 + z_2) = (x_1 + z_1) + (x_2 + z_2)$$

it follows that $T(\mathbf{v}_1 + \mathbf{v}_2) = T(\mathbf{v}_1) + T(\mathbf{v}_2)$.

Let $\mathbf{v} = (x, y, z)$ and let α be any real number. Then we have

$$T(\alpha\mathbf{v}) = T(\alpha x, \alpha y, \alpha z) = (\alpha x + \alpha y, \alpha x + \alpha z)$$

and

$$\alpha T(\mathbf{v}) = \alpha(x + y, x + z) = (\alpha(x + y), \alpha(x + z))$$

Since $\alpha x + \alpha y = \alpha(x + y)$ and $\alpha x + \alpha z = \alpha(x + z)$, it follows that $T(\alpha\mathbf{v}) = \alpha T(\mathbf{v})$. We conclude that the transformation T defined by equation 3 is linear.

Before proceeding with a variety of examples of linear transformations we wish to give here an example of a nonlinear transformation.

Example 2 Let a transformation T from $\mathbf{R}^2$ into $\mathbf{R}^3$ be defined by the formula

$$T(x, y) = (x, y, 1) \qquad [4]$$

Show that T is not linear.

Solution Let (x_1, y_1) and (x_2, y_2) be vectors in $\mathbf{R}^2$. Then we have

$$\begin{aligned}T((x_1, y_1) + (x_2, y_2)) &= T(x_1 + x_2, y_1 + y_2)\\ &= (x_1 + x_2, y_1 + y_2, 1)\end{aligned}$$

and

$$\begin{aligned}T(x_1, y_1) + T(x_2, y_2) &= (x_1, y_1, 1) + (x_2, y_2, 1)\\ &= (x_1 + x_2, y_1 + y_2, 2)\end{aligned}$$

Note: The transformation T does not satisfy the second condition of linearity either—that is, $T[\alpha(x, y)] \neq \alpha T(x, y)$ for every (x, y) in $\mathbf{R}^2$. The student is encouraged to check this.

Since

$$(x_1 + x_2, y_1 + y_2, 1) \neq (x_1 + x_2, y_1 + y_2, 2)$$

it follows that

$$T[(x_1,y_1) + (x_2,y_2)] \neq T(x_1,y_1) + T(x_2,y_2)$$

and T is not linear.

We wish to emphasize here that one could show that T is not linear by simply providing a counterexample. If we let $(x,y) = (1,1)$ and $\alpha = 2$, then we have

$$T[2(1,1)] = T(2,2) = (2,2,1) \text{ and } 2T(1,1) = 2(1,1,1) = (2,2,2)$$

Since $(2,2,1) \neq (2,2,2)$ it follows that $T[2(1,1)] \neq 2T(1,1)$ and T is not linear.

Example 3 Let T be the transformation from $\mathbf{R}^n$ into $\mathbf{R}^1$ defined by the formula

$$T(\mathbf{v}) = \mathbf{v} \cdot \mathbf{u} \qquad [5]$$

where $\mathbf{u}$ is a fixed vector in $\mathbf{R}^n$ and $\mathbf{v}$ is an arbitrary vector in $\mathbf{R}^n$. Show that T is linear.

Solution The linearity of T follows from properties of the dot product. We have

$$(\mathbf{v}_1 + \mathbf{v}_2) \cdot \mathbf{u} = \mathbf{v}_1 \cdot \mathbf{u} + \mathbf{v}_2 \cdot \mathbf{u}$$

and

$$(\alpha\mathbf{v}) \cdot \mathbf{u} = \alpha(\mathbf{v} \cdot \mathbf{u})$$

for every $\mathbf{v}$, $\mathbf{v}_1$, $\mathbf{v}_2$, and $\mathbf{u}$ in $\mathbf{R}^n$. Therefore we may write

$$T(\mathbf{v}_1 + \mathbf{v}_2) = (\mathbf{v}_1 + \mathbf{v}_2) \cdot \mathbf{u} = \mathbf{v}_1 \cdot \mathbf{u} + \mathbf{v}_2 \cdot \mathbf{u} = T(\mathbf{v}_1) + T(\mathbf{v}_2)$$

and

$$T(\alpha\mathbf{v}) = (\alpha\mathbf{v}) \cdot \mathbf{u} = \alpha(\mathbf{v} \cdot \mathbf{u}) = \alpha T(\mathbf{v}).$$

It follows that the transformation T defined by equation 5 is linear.

Example 4 Let P be the transformation from $\mathbf{R}^n$ into $\mathbf{R}^n$ defined by the formula

$$P(\mathbf{v}) = (\mathbf{v} \cdot \mathbf{u})\mathbf{u} \qquad [6]$$

where $\mathbf{u}$ is a fixed unit vector in $\mathbf{R}^n$ and $\mathbf{v}$ is an arbitrary vector in $\mathbf{R}^n$. Show that P is linear.

Solution The reader should recognize the expression $(\mathbf{v} \cdot \mathbf{u})\mathbf{u}$ in equation 6 as the vector projection of a vector $\mathbf{v}$ onto the unit vector $\mathbf{u}$. Thus, P is the "projection" transformation which associates with every vector $\mathbf{v}$ in $\mathbf{R}^n$ its vector projection onto the unit vector $\mathbf{u}$. To prove linearity we must show that

$$P(\mathbf{v}_1 + \mathbf{v}_2) = P(\mathbf{v}_1) + P(\mathbf{v}_2) \qquad [7]$$

and

$$P(\alpha \mathbf{v}) = \alpha P(\mathbf{v}) \qquad [8]$$

for all $\mathbf{v}_1$, $\mathbf{v}_2$, and $\mathbf{v}$ in $\mathbf{R}^n$.

Using properties of the dot product and equation 6 we have

$$\begin{aligned} P(\mathbf{v}_1 + \mathbf{v}_2) &= [(\mathbf{v}_1 + \mathbf{v}_2) \cdot \mathbf{u}]\mathbf{u} = [(\mathbf{v}_1 \cdot \mathbf{u}) + (\mathbf{v}_2 \cdot \mathbf{u})]\mathbf{u} \\ &= (\mathbf{v}_1 \cdot \mathbf{u})\mathbf{u} + (\mathbf{v}_2 \cdot \mathbf{u})\mathbf{u} = P(\mathbf{v}_1) + P(\mathbf{v}_2) \\ P(\alpha \mathbf{v}) &= [(\alpha \mathbf{v}) \cdot \mathbf{u}]\mathbf{u} = [\alpha(\mathbf{v} \cdot \mathbf{u})]\mathbf{u} = \alpha[(\mathbf{v} \cdot \mathbf{u})\mathbf{u}] = \alpha P(\mathbf{v}) \end{aligned}$$

Thus, the projection transformation defined by equation 6 is linear.

Example 5 Let A be a given $m \times n$ matrix. Viewing vectors in $\mathbf{R}^n$ and $\mathbf{R}^m$ as matrices of order $n \times 1$ and $m \times 1$, respectively, we define a transformation T from $\mathbf{R}^n$ into $\mathbf{R}^m$ by the formula

$$\mathbf{x} = \begin{bmatrix} x_1 \\ x_2 \\ \vdots \\ x_n \end{bmatrix}$$

$$T(\mathbf{x}) = A\mathbf{x} \qquad [9]$$

for every $\mathbf{x} \in \mathbf{R}^n$. Show that T is a linear transformation.

Solution Recall that multiplying an $m \times n$ matrix by an $n \times 1$ matrix results in an $m \times 1$ matrix. Thus, the transformation T defined by equation 9 associates with every vector $\mathbf{x} \in \mathbf{R}^n$ a vector $A\mathbf{x} \in \mathbf{R}^m$. The linearity of T follows from properties of matrix multiplication. In particular, it follows from Theorem 1-8 that

$$A(\mathbf{x} + \mathbf{y}) = A\mathbf{x} + A\mathbf{y} \qquad (\mathbf{x}, \mathbf{y} \in \mathbf{R}^n) \qquad [10]$$

and from Theorem 1-7 that

$$A(\alpha \mathbf{x}) = \alpha(A\mathbf{x}) \qquad (\alpha \text{ real}) \qquad [11]$$

Thus, it follows from equations 9, 10, and 11 that

$$T(\mathbf{x} + \mathbf{y}) = A(\mathbf{x} + \mathbf{y}) = A\mathbf{x} + A\mathbf{y} = T(\mathbf{x}) + T(\mathbf{y}) \qquad [12]$$

and

$$T(\alpha \mathbf{x}) = A(\alpha \mathbf{x}) = \alpha(A\mathbf{x}) = \alpha T(\mathbf{x}) \qquad [13]$$

Equations 12 and 13 establish the linearity of T defined by equation 9. This transformation is sometimes referred to as a **matrix transformation.** We will have more to say about such transformations in the next section.

Example 6 Let V and U be any two vector spaces. Let T be the transformation from V into U defined by the formula

$\mathbf{0}$ in equation 14 is the zero vector in U.

$$T(\mathbf{v}) = \mathbf{0} \qquad [14]$$

for all $\mathbf{v} \in V$. Show that T is linear.

Solution Since T matches every vector $\mathbf{v}$ in the vector space V with the zero vector $\mathbf{0}$ of the vector space U, we must have

$$T(\mathbf{v}_1 + \mathbf{v}_2) = \mathbf{0}, \qquad T(\mathbf{v}_1) = \mathbf{0}, \qquad T(\mathbf{v}_2) = \mathbf{0} \tag{15}$$

It follows immediately from equations 15 that

$\mathbf{0} = \mathbf{0} + \mathbf{0}$ holds in every vector space.

$$T(\mathbf{v}_1 + \mathbf{v}_2) = \mathbf{0} = \mathbf{0} + \mathbf{0} = T(\mathbf{v}_1) + T(\mathbf{v}_1) \tag{16}$$

for all $\mathbf{v}_1, \mathbf{v}_2 \in V$. We also have

$$T(\mathbf{v}) = \mathbf{0}, \qquad T(\alpha\mathbf{v}) = \mathbf{0} \tag{17}$$

for every $\mathbf{v} \in V$ and an arbitrary scalar α. Thus,

$\mathbf{0} = \alpha\mathbf{0}$ holds in every vector space for an arbitrary scalar α.

$$T(\alpha\mathbf{v}) = \mathbf{0} = \alpha\mathbf{0} = \alpha T(\mathbf{v}) \tag{18}$$

Therefore, we conclude from equations 16 and 18 that T is a linear transformation. The transformation defined by equation 14 is called the **zero transformation.**

Example 7 Let V be a vector space. Let T be the transformation defined by the formula

$$T(\mathbf{v}) = \mathbf{v} \tag{19}$$

for all $\mathbf{v} \in V$. Show that T is linear.

Solution It follows from equation 19 that the transformation T matches every vector $\mathbf{v} \in V$ with itself. Thus, we have

$$T(\mathbf{v}_1 + \mathbf{v}_2) = \mathbf{v}_1 + \mathbf{v}_2, \qquad T(\mathbf{v}_1) = \mathbf{v}_1, \qquad T(\mathbf{v}_2) = \mathbf{v}_2 \tag{20}$$

Making use of equations 20 we obtain

$$T(\mathbf{v}_1 + \mathbf{v}_2) = \mathbf{v}_1 + \mathbf{v}_2 = T(\mathbf{v}_1) + T(\mathbf{v}_2) \tag{21}$$

for all $\mathbf{v}_1, \mathbf{v}_2 \in V$.

It also follows from equation 19 that

$$T(\alpha\mathbf{v}) = \alpha\mathbf{v}, \qquad T(\mathbf{v}) = \mathbf{v} \tag{22}$$

Hence, from equations 22 we have

$$T(\alpha\mathbf{v}) = \alpha\mathbf{v} = \alpha T(\mathbf{v}) \tag{23}$$

for all $\mathbf{v} \in V$ and arbitrary real α. Therefore we conclude from equations 21 and 23 that T is linear. The transformation T defined by equation 19 is called the **identity transformation** and is usually denoted by I.

Example 8 Let $U = C[0,1]$ be the vector space of all real-valued functions which are continuous on the interval $[0,1]$. Let V be the subspace

(For students with a calculus background.)

One can easily show that $C[0,1]$ is a vector space by using properties of continuous functions.

of $C[0,1]$ consisting of all functions having continuous first derivatives on the interval $[0,1]$. Let D be the transformation from V into U that matches every function f with its derivative f', namely,

$$D(f) = f' \tag{24}$$

Show that D is a linear transformation.

Solution

The derivative of the sum is the sum of the derivatives.

Using properties of differentiation and equation 24, we have

$$D(f + g) = (f + g)' = f' + g' = D(f) + D(g) \tag{25}$$

and

$$D(\alpha f) = (\alpha f)' = \alpha f' = \alpha D(f) \tag{26}$$

for any f and g in V and any real number α. Equations 25 and 26 prove the linearity of D.

Example 9

(For students with a calculus background.)

R is the vector space of all real numbers.

Let $U = C[0,1]$ be as in Example 8, and let T be the transformation from U into $\mathbf{R}$ defined by

$$T(f) = \int_0^1 f(x)\,dx \tag{27}$$

Show that T is a linear transformation.

Solution

Using properties of integration and equation 27, we have

$$\begin{aligned} T(f + g) &= \int_0^1 [f(x) + g(x)]\,dx = \int_0^1 f(x)\,dx + \int_0^1 g(x)\,dx \\ &= T(f) + T(g) \end{aligned} \tag{28}$$

and

$$T(\alpha f) = \int_0^1 \alpha f(x)\,dx = \alpha \int_0^1 f(x)\,dx = \alpha T(f) \tag{29}$$

for every f and g in U and any real number α. The linearity of T follows immediately from equations 28 and 29.

We close this section with the following theorem, which will bring out some of the basic properties of linear transformations.

Theorem 6-1 Let $T: V \longrightarrow W$ be a linear transformation. Then

a $T(\mathbf{0}) = \mathbf{0}$

b $T(-\mathbf{v}) = -T(\mathbf{v})$

c $T(\mathbf{v} - \mathbf{u}) = T(\mathbf{v}) - T(\mathbf{u})$ for all $\mathbf{v}$ and $\mathbf{u}$ in V

Proof The equation $T(\mathbf{0}) = \mathbf{0}$ means that the linear transformation T matches the zero vector of V with the zero vector of W. To prove part a, we use the fact that $\mathbf{0} = 0\mathbf{v}$ for any vector $\mathbf{v}$ in V. Using the linearity of T, we obtain

See Theorem 2-1.

$$T(\mathbf{0}) = T(0\mathbf{v}) = 0(T(\mathbf{v})) = \mathbf{0}$$

which proves part a.

To prove part b, we use the fact that $-\mathbf{v} = (-1)\mathbf{v}$ for any vector $\mathbf{v}$ in V. The linerarity of T implies that we have

See Theorem 2-1.

$$T(-\mathbf{v}) = T((-1)\mathbf{v}) = (-1)T(\mathbf{v}) = -T(\mathbf{v})$$

which proves part b.

Finally, to prove part c we write $\mathbf{v} - \mathbf{u} = \mathbf{v} + (-1)\mathbf{u}$. Using the linearity of T, we obtain

$$T(\mathbf{v} - \mathbf{u}) = T(\mathbf{v} + (-1)\mathbf{u}) = T(\mathbf{v}) + T((-1)\mathbf{u}) = T(\mathbf{v}) + (-1)T(\mathbf{u})$$
$$= T(\mathbf{v}) - T(\mathbf{u})$$

This concludes the proof of Theorem 6-1.

Exercises

1 Let $T: V \longrightarrow W$ be a mapping of a vector space V into a vector space W. Prove that T is a linear transformation if and only if

$$T(\alpha\mathbf{v} + \beta\mathbf{u}) = \alpha T(\mathbf{v}) + \beta T(\mathbf{u}) \qquad [30]$$

for any real numbers α, β and any vectors $\mathbf{v}, \mathbf{u} \in V$.

The reader should note that equation 30 is an alternate way of defining a linear transformation.

2 Let $T: \mathbf{R}^2 \longrightarrow \mathbf{R}^2$ be defined by

$$T(x,y) = (x,-y)$$

Show that T is a linear transformation.

3 Let $T: \mathbf{R}^3 \longrightarrow \mathbf{R}^2$ be defined by

$$T(x,y,z) = (x,y)$$

Show that T is a linear transformation.

4 Let $T: \mathbf{R}^2 \longrightarrow \mathbf{R}^2$ be defined by

$$T(x,y) = (1,2)$$

Show that T is *not* linear.

5 We describe the transformation T in problem 2 geometrically by saying that $T(x,y) = (x,-y)$ is a **reflection** in the x-axis (see Figure 6-1). Construct a transformation $L: \mathbf{R}^2 \longrightarrow \mathbf{R}^2$ such that L maps each point in the plane into its reflection about the y-axis

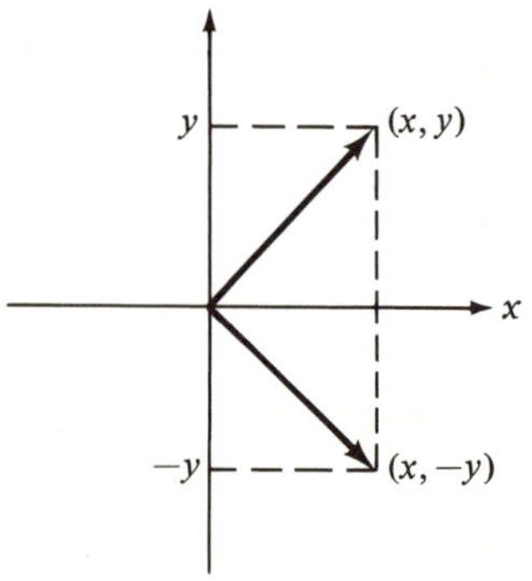

Figure 6-1

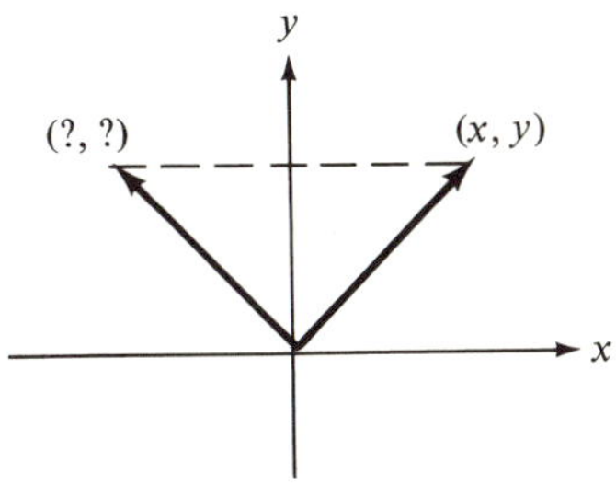

Figure 6-2

(see Figure 6-2). Using the formula you constructed for L, show that L is a linear transformation.

6 Let $T: \mathbf{R}^2 \longrightarrow \mathbf{R}^2$ be defined by

$$T(x,y) = (y,x)$$

Show that T is linear. Describe T geometrically.

7 Let V be any vector space and γ a fixed real number. Let $T: V \longrightarrow V$ be defined by

$$T(\mathbf{v}) = \gamma\mathbf{v}$$

Show that T is a linear transformation.

8 Let $T: \mathbf{R}^2 \longrightarrow \mathbf{R}$ be defined by the following formulas. Determine for each formula if T is linear.

a $T(x,y) = x$ **b** $T(x,y) = y^2$
c $T(x,y) = 0$ **d** $T(x,y) = \sqrt[3]{x}$
e $T(x,y) = 2x - y$ **f** $T(x,y) = x + y - 1$

9 Let $T: P_3 \longrightarrow \mathbf{R}$ be defined by the following formulas. Determine for each formula if T is linear.

a $T(a_3x^3 + a_2x^2 + a_1x + a_0) = a_3 + a_2 + a_1 + a_0$
b $T(a_3x^3 + a_2x^2 + a_1x + a_0) = a_3$
c $T(a_3x^3 + a_2x^2 + a_1x + a_0) = a_2 + 1$
d $T(a_3x^3 + a_2x^2 + a_1x + a_0) = 0$

10 Let $\{\mathbf{v}_1, \mathbf{v}_2, \ldots, \mathbf{v}_k\}$ be a basis for a vector space V. Let $T: V \longrightarrow V$ be a linear transformation and let $\mathbf{v}$ be any vector in V. Show that $T(\mathbf{v})$ is a linear combination of $T(\mathbf{v}_1)$, $T(\mathbf{v}_2)$, $\ldots$, $T(\mathbf{v}_k)$.

11 Let $\{\mathbf{v}_1, \mathbf{v}_2, \ldots, \mathbf{v}_k\}$ be a basis for a vector space V. Let $T: V \longrightarrow V$ be a linear transformation such that $T(\mathbf{v}_1) = \mathbf{v}_1$, $T(\mathbf{v}_2) = \mathbf{v}_2$, $\ldots$, $T(\mathbf{v}_k) = \mathbf{v}_k$. Show that T is the identity transformation on V, that is, that $T(\mathbf{v}) = \mathbf{v}$ for all $\mathbf{v} \in V$.

6-2 Matrix Representation of a Linear Transformation

Objective
Show that every linear transformation from $\mathbf{R}^n$ into $\mathbf{R}^m$ has a matrix representation with respect to the standard basis of $\mathbf{R}^n$.

We have seen in the previous section (see Example 5) that if A is an $m \times n$ matrix then the transformation T given by the formula

$$T(\mathbf{x}) = A\mathbf{x} \qquad [1]$$

is a linear transformation from $\mathbf{R}^n$ into $\mathbf{R}^m$. We are interested in investigating here whether the converse of this last statement is true. Therefore we raise the following question:

Suppose S is a linear transformation from $\mathbf{R}^n$ into $\mathbf{R}^m$. Can S be represented by an equation like equation 1 with a properly chosen matrix A?

The answer to this question is in the affirmative and is stated in the following theorem.

Theorem 6-2 Let S be a linear transformation from $\mathbf{R}^n$ into $\mathbf{R}^m$. Then there exists a unique $m \times n$ matrix A such that

$$S(\mathbf{x}) = A\mathbf{x} \qquad [2]$$

for all $\mathbf{x} \in \mathbf{R}^n$.

Proof We shall prove the validity of this theorem for the case $m = 2$, $n = 3$ and leave the general proof as an exercise for the reader.

See problem 7.

The proof will be a constructive one. Let $\begin{bmatrix} x \\ y \\ z \end{bmatrix}$ be any element of $\mathbf{R}^3$. Then we may write

$$\begin{bmatrix} x \\ y \\ z \end{bmatrix} = x\begin{bmatrix} 1 \\ 0 \\ 0 \end{bmatrix} + y\begin{bmatrix} 0 \\ 1 \\ 0 \end{bmatrix} + z\begin{bmatrix} 0 \\ 0 \\ 1 \end{bmatrix}$$

Since S is a linear transformation, it follows that

$$S\begin{bmatrix} x \\ y \\ z \end{bmatrix} = xS\begin{bmatrix} 1 \\ 0 \\ 0 \end{bmatrix} + yS\begin{bmatrix} 0 \\ 1 \\ 0 \end{bmatrix} + zS\begin{bmatrix} 0 \\ 0 \\ 1 \end{bmatrix} \qquad [3]$$

The meaning of equation 3 is that the image of $\begin{bmatrix} x \\ y \\ z \end{bmatrix}$ under S is completely determined by the effect which the linear transformation S has on the standard basis $\begin{bmatrix} 1 \\ 0 \\ 0 \end{bmatrix}$, $\begin{bmatrix} 0 \\ 1 \\ 0 \end{bmatrix}$, and $\begin{bmatrix} 0 \\ 0 \\ 1 \end{bmatrix}$ of $\mathbf{R}^3$. Since $S\begin{bmatrix} 1 \\ 0 \\ 0 \end{bmatrix}$, $S\begin{bmatrix} 0 \\ 1 \\ 0 \end{bmatrix}$, and $S\begin{bmatrix} 0 \\ 0 \\ 1 \end{bmatrix}$ are vectors in $\mathbf{R}^2$, we may write

$$S\begin{bmatrix} 1 \\ 0 \\ 0 \end{bmatrix} = \begin{bmatrix} a_{11} \\ a_{21} \end{bmatrix}, \qquad S\begin{bmatrix} 0 \\ 1 \\ 0 \end{bmatrix} = \begin{bmatrix} a_{12} \\ a_{22} \end{bmatrix}, \qquad S\begin{bmatrix} 0 \\ 0 \\ 1 \end{bmatrix} = \begin{bmatrix} a_{13} \\ a_{23} \end{bmatrix} \qquad [4]$$

Combining equations 3 and 4, we get

$$S\begin{bmatrix} x \\ y \\ z \end{bmatrix} = x\begin{bmatrix} a_{11} \\ a_{21} \end{bmatrix} + y\begin{bmatrix} a_{12} \\ a_{22} \end{bmatrix} + z\begin{bmatrix} a_{13} \\ a_{23} \end{bmatrix} \qquad [5]$$

But equation 5 can also be written as

See equation 14 in Section 3-6.

$$S\begin{bmatrix} x \\ y \\ z \end{bmatrix} = \begin{bmatrix} a_{11} & a_{12} & a_{13} \\ a_{21} & a_{22} & a_{23} \end{bmatrix}\begin{bmatrix} x \\ y \\ z \end{bmatrix} \qquad [6]$$

Thus, if $A = \begin{bmatrix} a_{11} & a_{12} & a_{13} \\ a_{21} & a_{22} & a_{23} \end{bmatrix}$ and $\mathbf{v} = \begin{bmatrix} x \\ y \\ z \end{bmatrix}$, equation 6 may be written as

$$S(\mathbf{v}) = A\mathbf{v}$$

This completes the proof for the case $m = 2$, $n = 3$. Note that we obtained the columns of the matrix A by finding the images of the standard basis $\begin{bmatrix} 1 \\ 0 \\ 0 \end{bmatrix}$, $\begin{bmatrix} 0 \\ 1 \\ 0 \end{bmatrix}$, and $\begin{bmatrix} 0 \\ 0 \\ 1 \end{bmatrix}$ of $\mathbf{R}^3$ under the linear transformation S. This suggests that to find the matrix representation of any linear transformation S from $\mathbf{R}^n$ into $\mathbf{R}^m$ we must compute the images (under S) of the standard basis for $\mathbf{R}^n$ and arrange the resulting vectors of $\mathbf{R}^m$ as n columns forming a matrix A of order $m \times n$. The following examples will clarify this procedure.

Example 1 Let T be the transformation given by

$$T\begin{bmatrix} x \\ y \\ z \end{bmatrix} = \begin{bmatrix} x + y \\ x + z \end{bmatrix} \qquad [7]$$

See Example 1 in Section 6-1.

It was shown in Section 6-1 that T defined by equation 7 is a linear transformation from $\mathbf{R}^3$ into $\mathbf{R}^2$. Find a matrix representation for T.

Solution Let A be the matrix such that

$$T\begin{bmatrix} x \\ y \\ z \end{bmatrix} = A\begin{bmatrix} x \\ y \\ z \end{bmatrix}$$

for all $\begin{bmatrix} x \\ y \\ z \end{bmatrix}$ in $\mathbf{R}^3$. The columns of A are given by the vectors $T\begin{bmatrix} 1 \\ 0 \\ 0 \end{bmatrix}$, $T\begin{bmatrix} 0 \\ 1 \\ 0 \end{bmatrix}$, and $T\begin{bmatrix} 0 \\ 0 \\ 1 \end{bmatrix}$. A simple computation using equation 7 yields

$$T\begin{bmatrix}1\\0\\0\end{bmatrix}=\begin{bmatrix}1\\1\end{bmatrix},\qquad T\begin{bmatrix}0\\1\\0\end{bmatrix}=\begin{bmatrix}1\\0\end{bmatrix},\qquad T\begin{bmatrix}0\\0\\1\end{bmatrix}=\begin{bmatrix}0\\1\end{bmatrix}$$

Thus, the matrix A has the form

$$A=\begin{bmatrix}1&1&0\\1&0&1\end{bmatrix}$$

The reader can easily check that we have indeed

$$T\begin{bmatrix}x\\y\\z\end{bmatrix}=\begin{bmatrix}1&1&0\\1&0&1\end{bmatrix}\begin{bmatrix}x\\y\\z\end{bmatrix}$$

Example 2 Let P be the projection transformation from $\mathbf{R}^2$ into $\mathbf{R}^2$ defined by the formula

$$P(\mathbf{v})=(\mathbf{v}\cdot\mathbf{u})\mathbf{u} \qquad \textbf{[8]}$$

See Example 4 in Section 6-1.

where $\mathbf{u}=\begin{bmatrix}\frac{1}{\sqrt{2}}\\\frac{1}{\sqrt{2}}\end{bmatrix}$ and $\mathbf{v}$ is an arbitrary vector in $\mathbf{R}^2$. It was shown in Section 6-1 that P defined by equation 8 is a linear transformation. Find a matrix representation for P.

Solution Let A be the matrix such that

$$P\begin{bmatrix}x\\y\end{bmatrix}=A\begin{bmatrix}x\\y\end{bmatrix} \qquad \textbf{[9]}$$

for all $\mathbf{v}=\begin{bmatrix}x\\y\end{bmatrix}$ in $\mathbf{R}^2$.

$\begin{bmatrix}1\\0\end{bmatrix}$ and $\begin{bmatrix}0\\1\end{bmatrix}$ form the standard basis for $\mathbf{R}^2$.

In order to find the columns of A we must compute $P\begin{bmatrix}1\\0\end{bmatrix}$ and $P\begin{bmatrix}0\\1\end{bmatrix}$. Using the particular unit vector $\mathbf{u}=\begin{bmatrix}\frac{1}{\sqrt{2}}\\\frac{1}{\sqrt{2}}\end{bmatrix}$ we obtain from equation 8

$$P\begin{bmatrix}1\\0\end{bmatrix}=\frac{1}{\sqrt{2}}\begin{bmatrix}\frac{1}{\sqrt{2}}\\\frac{1}{\sqrt{2}}\end{bmatrix}=\begin{bmatrix}\frac{1}{2}\\\frac{1}{2}\end{bmatrix},\qquad P\begin{bmatrix}0\\1\end{bmatrix}=\frac{1}{\sqrt{2}}\begin{bmatrix}\frac{1}{\sqrt{2}}\\\frac{1}{\sqrt{2}}\end{bmatrix}=\begin{bmatrix}\frac{1}{2}\\\frac{1}{2}\end{bmatrix}$$

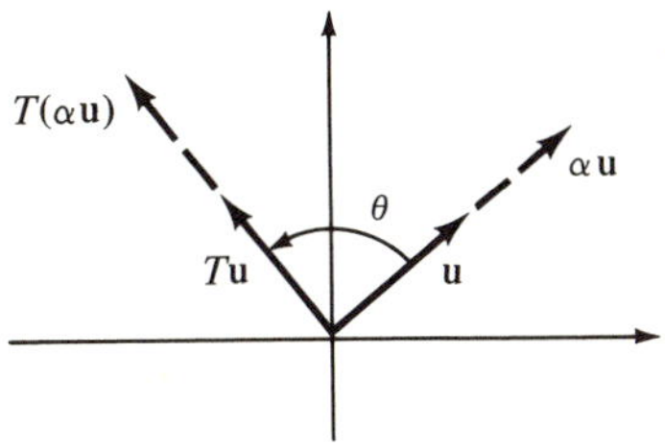

Figure 6-3

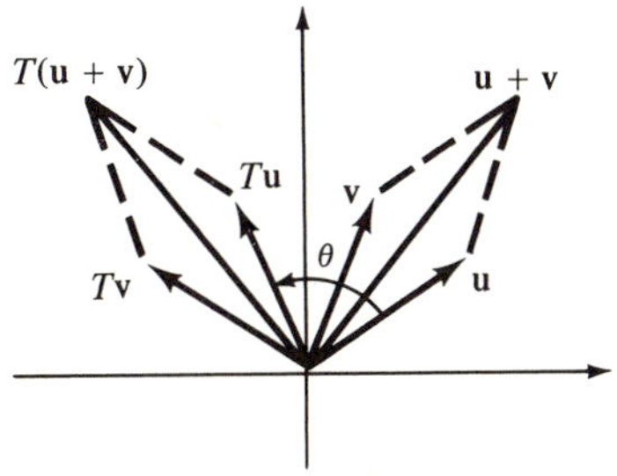

Figure 6-4

It follows that the matrix A in equation 9 must have the form

$$A = \begin{bmatrix} \frac{1}{2} & \frac{1}{2} \\ \frac{1}{2} & \frac{1}{2} \end{bmatrix}$$

Thus, we have

$$P\begin{bmatrix} x \\ y \end{bmatrix} = \begin{bmatrix} \frac{1}{2} & \frac{1}{2} \\ \frac{1}{2} & \frac{1}{2} \end{bmatrix}\begin{bmatrix} x \\ y \end{bmatrix} \quad \textbf{[10]}$$

and we may compute the projection of any vector in $\mathbf{R}^2$ onto the unit vector $\mathbf{u} = \begin{bmatrix} \frac{1}{\sqrt{2}} \\ \frac{1}{\sqrt{2}} \end{bmatrix}$ by using equation 10 rather than equation 8.

Example 3 Let T be a counterclockwise rotation of vectors in $\mathbf{R}^2$ through the angle θ. Since rotation preserves the length of each vector and also the angle between any two vectors, it is easy to see that T is a linear transformation from $\mathbf{R}^2$ into $\mathbf{R}^2$ (see Figures 6-3 and 6-4). Find a matrix representation for the rotation T.

Solution Let A be the 2×2 matrix such that

$$T\begin{bmatrix} x \\ y \end{bmatrix} = A\begin{bmatrix} x \\ y \end{bmatrix} \quad \textbf{[11]}$$

for all $\mathbf{v} = \begin{bmatrix} x \\ y \end{bmatrix}$ in $\mathbf{R}^2$.

In order to find the columns of A, we must determine $T\begin{bmatrix} 1 \\ 0 \end{bmatrix}$ and $T\begin{bmatrix} 0 \\ 1 \end{bmatrix}$. Using elementary trigonometry (see Figure 6-5), we obtain

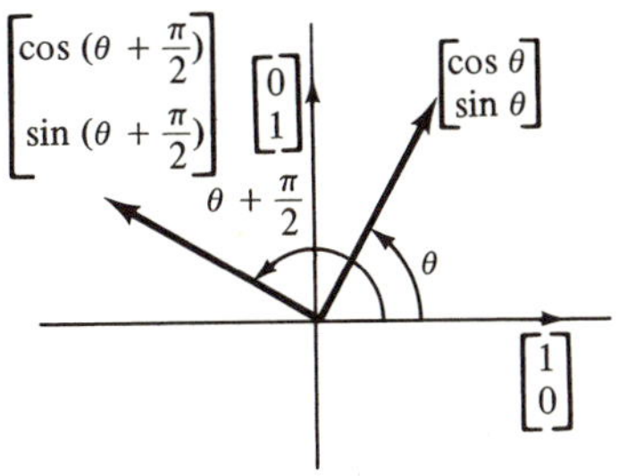

Figure 6-5

$$T\begin{bmatrix} 1 \\ 0 \end{bmatrix} = \begin{bmatrix} \cos\theta \\ \sin\theta \end{bmatrix}, \; T\begin{bmatrix} 0 \\ 1 \end{bmatrix} = \begin{bmatrix} \cos\left(\theta + \frac{\pi}{2}\right) \\ \sin\left(\theta + \frac{\pi}{2}\right) \end{bmatrix} \quad \textbf{[12]}$$

Since $\cos\left(\theta + \frac{\pi}{2}\right) = -\sin\theta$ and $\sin\left(\theta + \frac{\pi}{2}\right) = \cos\theta$, we can write equations 12 in the form

$$T\begin{bmatrix}1\\0\end{bmatrix} = \begin{bmatrix}\cos\theta\\ \sin\theta\end{bmatrix}, \quad T\begin{bmatrix}0\\1\end{bmatrix} = \begin{bmatrix}-\sin\theta\\ \cos\theta\end{bmatrix} \qquad \textbf{[13]}$$

It follows that the matrix A in equation 11 has the form

$$A = \begin{bmatrix}\cos\theta & -\sin\theta\\ \sin\theta & \cos\theta\end{bmatrix} \qquad \textbf{[14]}$$

Combining equations 11 and 14, we obtain the formula of rotation of vectors in $\mathbf{R}^2$ by the angle θ:

$$T\begin{bmatrix}x\\y\end{bmatrix} = \begin{bmatrix}\cos\theta & -\sin\theta\\ \sin\theta & \cos\theta\end{bmatrix}\begin{bmatrix}x\\y\end{bmatrix} \qquad \textbf{[15]}$$

Equation 15 enables us to compute the image of any vector $\begin{bmatrix}x\\y\end{bmatrix}$ for a given angle of rotation θ.

Example 4 Let $T: \mathbf{R}^4 \longrightarrow \mathbf{R}^3$ be the transformation defined by

The reader should have no difficulty in showing that equation 16 defines a linear transformation.

$$T\begin{bmatrix}x\\y\\z\\w\end{bmatrix} = \begin{bmatrix}y\\y\\y\end{bmatrix} \qquad \textbf{[16]}$$

Find a matrix representation for T.

Solution We must find a 3×4 matrix A such that

$$T\begin{bmatrix}x\\y\\z\\w\end{bmatrix} = A\begin{bmatrix}x\\y\\z\\w\end{bmatrix} \qquad \textbf{[17]}$$

The columns of A are given by $T\begin{bmatrix}1\\0\\0\\0\end{bmatrix}$, $T\begin{bmatrix}0\\1\\0\\0\end{bmatrix}$, $T\begin{bmatrix}0\\0\\1\\0\end{bmatrix}$, and $T\begin{bmatrix}0\\0\\0\\1\end{bmatrix}$

Making use of equation 16, we obtain

$$T\begin{bmatrix}1\\0\\0\\0\end{bmatrix} = \begin{bmatrix}0\\0\\0\end{bmatrix}, \quad T\begin{bmatrix}0\\1\\0\\0\end{bmatrix} = \begin{bmatrix}1\\1\\1\end{bmatrix}, \quad T\begin{bmatrix}0\\0\\1\\0\end{bmatrix} = \begin{bmatrix}0\\0\\0\end{bmatrix}, \quad T\begin{bmatrix}0\\0\\0\\1\end{bmatrix} = \begin{bmatrix}0\\0\\0\end{bmatrix}$$

It follows that the matrix A in equation 17 has the form

$$A = \begin{bmatrix} 0 & 1 & 0 & 0 \\ 0 & 1 & 0 & 0 \\ 0 & 1 & 0 & 0 \end{bmatrix}$$

Thus, we may rewrite equation 17 as

$$T\begin{bmatrix} x \\ y \\ z \\ w \end{bmatrix} = \begin{bmatrix} 0 & 1 & 0 & 0 \\ 0 & 1 & 0 & 0 \\ 0 & 1 & 0 & 0 \end{bmatrix}\begin{bmatrix} x \\ y \\ z \\ w \end{bmatrix}$$

Remark The procedure of finding a matrix representation for a given linear transformation T as exhibited in Examples 1 through 4 may be summarized as follows:

If $T: \mathbf{R}^n \longrightarrow \mathbf{R}^m$ is a linear transformation, then there exists a matrix A of order $m \times n$ such that

$$T(\mathbf{v}) = A\mathbf{v}$$

for all $\mathbf{v} \in \mathbf{R}^n$. The columns of the matrix A are given by $T(\mathbf{e}_1)$, $T(\mathbf{e}_2)$, $\ldots$, $T(\mathbf{e}_n)$ where $\mathbf{e}_1, \mathbf{e}_2, \ldots, \mathbf{e}_n$ is the standard basis for $\mathbf{R}^n$.

$$\mathbf{e}_1 = \begin{bmatrix} 1 \\ 0 \\ 0 \\ \vdots \\ 0 \end{bmatrix}, \mathbf{e}_2 = \begin{bmatrix} 0 \\ 1 \\ 0 \\ \vdots \\ 0 \end{bmatrix}, \ldots, \mathbf{e}_n = \begin{bmatrix} 0 \\ \vdots \\ 0 \\ 0 \\ 1 \end{bmatrix}$$

Exercises

1 Find the matrix of each of the following linear transformations:

a $T\begin{bmatrix} x \\ y \end{bmatrix} = \begin{bmatrix} x + y \\ x - y \end{bmatrix}$ **b** $T\begin{bmatrix} x \\ y \end{bmatrix} = \begin{bmatrix} y \\ x \end{bmatrix}$

c $T\begin{bmatrix} x \\ y \end{bmatrix} = \begin{bmatrix} x \\ y \end{bmatrix}$ **d** $T\begin{bmatrix} x \\ y \end{bmatrix} = \begin{bmatrix} 0 \\ 0 \end{bmatrix}$

2 Use formula 15 in Example 3 to find the matrix of a counterclockwise rotation in $\mathbf{R}^2$ through the angle θ if θ is given by:

a $\frac{\pi}{6}$ **b** $\frac{\pi}{4}$ **c** $\frac{\pi}{3}$ **d** $\frac{\pi}{2}$ **e** $\frac{2}{3}\pi$ **f** π

3 Find the matrix of each of the following linear transformations:

a $T\begin{bmatrix} x \\ y \\ z \end{bmatrix} = \begin{bmatrix} x + y + z \\ x - 2y - z \\ 3x + y - 2z \\ 2x - 3y + 4z \end{bmatrix}$ **b** $T\begin{bmatrix} x \\ y \\ z \\ w \end{bmatrix} = \begin{bmatrix} x + 2y \\ 2z - w \end{bmatrix}$

c $T\begin{bmatrix} x \\ y \\ z \end{bmatrix} = [2x - y + 3z]$

4 Let $P:\mathbf{R}^2 \longrightarrow \mathbf{R}^2$ be the projection transformation onto the unit vector $\mathbf{u} = \frac{1}{5}\begin{bmatrix}3\\4\end{bmatrix}$. Find the matrix of P.

5 Let $P:\mathbf{R}^2 \longrightarrow \mathbf{R}^2$ be the projection transformation onto a unit vector $\mathbf{u} = \begin{bmatrix}u_1\\u_2\end{bmatrix}$. Show that the matrix of P is singular.

See problem 7 in Section 6-1.

6 Let $T:\mathbf{R}^4 \longrightarrow \mathbf{R}^4$ be the linear transformation defined by

$$T(\mathbf{v}) = \gamma\mathbf{v}$$

where γ is a fixed real number and $\mathbf{v} \in \mathbf{R}^4$. Find the matrix representation of T.

7 Prove Theorem 6-2 in its full generality.

Hint: Express $\begin{bmatrix}1\\0\end{bmatrix}$ and $\begin{bmatrix}0\\1\end{bmatrix}$ as a linear combination of the basis $\begin{bmatrix}1\\2\end{bmatrix}$ and $\begin{bmatrix}2\\1\end{bmatrix}$ of $\mathbf{R}^2$. Use the linearity of T to compute $T\begin{bmatrix}1\\0\end{bmatrix}$ and $T\begin{bmatrix}0\\1\end{bmatrix}$.

8 Let $T:\mathbf{R}^2 \longrightarrow \mathbf{R}^1$ be a linear transformation such that $T\begin{bmatrix}1\\2\end{bmatrix} = -1$ and $T\begin{bmatrix}2\\1\end{bmatrix} = 1$. Find a matrix representation for T.

9 Let $T:\mathbf{R}^2 \longrightarrow \mathbf{R}^2$ be a linear transformation such that $T\begin{bmatrix}1\\1\end{bmatrix} = \begin{bmatrix}3\\2\end{bmatrix}$ and $T\begin{bmatrix}1\\2\end{bmatrix} = \begin{bmatrix}4\\5\end{bmatrix}$. Find a matrix representation for T.

10 Let $T:\mathbf{R}^2 \longrightarrow \mathbf{R}^2$ be a linear transformation. Show that T is completely determined if we prescribe the effect of T on any basis for $\mathbf{R}^2$.

6-3 Operations on Linear Transformations

Objectives

1 Discuss the operations of sums and scalar products of linear transformations.

2 Discuss the operation of composition of linear transformations.

In the previous two sections we have discussed properties of linear transformations. In this section we proceed to define some operations on linear transformations and show that the new functions generated by these operations are themselves linear transformations. We start with the definition of the sum and scalar multiplication of linear transformations.

Definition 6-2 Let V and W be vector spaces. Let $T: V \longrightarrow W$ and $S: V \longrightarrow W$ be linear transformations, and let α be an arbitrary real number.

The **sum** $T + S$ is defined by the formula

$$(T + S)(\mathbf{v}) = T(\mathbf{v}) + S(\mathbf{v}) \qquad [1]$$

and the **scalar product** αT is defined by the formula

$$(\alpha T)(\mathbf{v}) = \alpha T(\mathbf{v}) \qquad [2]$$

for all $\mathbf{v}$ in V.

For students who find the definitions of $T + S$ and αT too formal, we add the following explanation.

Equation 1 should be interpreted as saying that $T + S$ is a transformation from V into W such that the image of $\mathbf{v}$ under $T + S$ is equal to the sum of the vectors $T(\mathbf{v})$ and $S(\mathbf{v})$ in W.

Similarly, equation 2 says that αT is a transformation from V into W such that the image of $\mathbf{v}$ under αT is equal to the scalar product of α and $T(\mathbf{v})$.

Example 1 Let $T\begin{bmatrix} x \\ y \end{bmatrix} = \begin{bmatrix} x + y \\ x - y \\ x \end{bmatrix}$ and $S\begin{bmatrix} x \\ y \end{bmatrix} = \begin{bmatrix} 2x \\ 3y \\ y \end{bmatrix}$ and let $\alpha = 3$. Find $(T + S)\begin{bmatrix} 1 \\ 2 \end{bmatrix}$ and $(3T)\begin{bmatrix} 1 \\ 2 \end{bmatrix}$.

Solution Using equation 1, we obtain

The reader should have no difficulty in showing that T and S are linear transformations from $\mathbf{R}^2$ into $\mathbf{R}^3$.

$$(T + S)\begin{bmatrix} 1 \\ 2 \end{bmatrix} = T\begin{bmatrix} 1 \\ 2 \end{bmatrix} + S\begin{bmatrix} 1 \\ 2 \end{bmatrix} = \begin{bmatrix} 1 + 2 \\ 1 - 2 \\ 1 \end{bmatrix} + \begin{bmatrix} 2 \cdot 1 \\ 3 \cdot 2 \\ 2 \end{bmatrix} = \begin{bmatrix} 5 \\ 5 \\ 3 \end{bmatrix}$$

Using equation 2, we obtain

$$(3T)\begin{bmatrix} 1 \\ 2 \end{bmatrix} = 3T\begin{bmatrix} 1 \\ 2 \end{bmatrix} = 3\begin{bmatrix} 1 + 2 \\ 1 - 2 \\ 1 \end{bmatrix} = \begin{bmatrix} 9 \\ -3 \\ 3 \end{bmatrix}$$

We can also compute $(T + S)\begin{bmatrix} 1 \\ 2 \end{bmatrix}$ and $(3T)\begin{bmatrix} 1 \\ 2 \end{bmatrix}$ by first obtaining formulas for $(T + S)\begin{bmatrix} x \\ y \end{bmatrix}$ and $(3T)\begin{bmatrix} x \\ y \end{bmatrix}$ and then substituting the values $x = 1$ and $y = 2$ in these formulas. Using Definition 6-2, we obtain

$$(T + S)\begin{bmatrix} x \\ y \end{bmatrix} = T\begin{bmatrix} x \\ y \end{bmatrix} + S\begin{bmatrix} x \\ y \end{bmatrix} = \begin{bmatrix} x + y \\ x - y \\ x \end{bmatrix} + \begin{bmatrix} 2x \\ 3y \\ y \end{bmatrix} \tag{3}$$

$$= \begin{bmatrix} 3x + y \\ x + 2y \\ x + y \end{bmatrix}$$

and

$$(3T)\begin{bmatrix} x \\ y \end{bmatrix} = 3T\begin{bmatrix} x \\ y \end{bmatrix} = 3\begin{bmatrix} x + y \\ x - y \\ x \end{bmatrix} = \begin{bmatrix} 3x + 3y \\ 3x - 3y \\ 3x \end{bmatrix} \tag{4}$$

Substituting the vector $\begin{bmatrix} x \\ y \end{bmatrix} = \begin{bmatrix} 1 \\ 2 \end{bmatrix}$ in equations 3 and 4 yields the same results we have already obtained.

A close examination of formulas 3 and 4 reveals that $T + S$ as well as $3T$ are linear transformations. This is not accidental; in fact, we have the following general result:

Theorem 6-3 Let T and S be linear transformations from V into W and let α be any real number. Then $T + S$ and αT are also linear transformations from V into W.

Proof In order to show that $T + S$ is linear, we must prove the following two properties:

$$(T + S)(\mathbf{v} + \mathbf{u}) = (T + S)(\mathbf{v}) + (T + S)(\mathbf{u}) \quad \textbf{[5]}$$

$$(T + S)(\beta\mathbf{v}) = \beta(T + S)(\mathbf{v}) \quad \textbf{[6]}$$

See problem 1.

We shall prove equation 5 and leave the proof of equation 6 as an exercise for the reader.

$$\begin{aligned}(T + S)(\mathbf{v} + \mathbf{u}) &= T(\mathbf{v} + \mathbf{u}) + S(\mathbf{v} + \mathbf{u}) && \text{(by Definition 6-2)}\\ &= T(\mathbf{v}) + T(\mathbf{u}) + S(\mathbf{v}) + S(\mathbf{u}) && \text{(linearity of } S \text{ and } T)\\ &= T(\mathbf{v}) + S(\mathbf{v}) + T(\mathbf{u}) + S(\mathbf{u}) && \text{(vector space axioms)}\\ &= (T + S)(\mathbf{v}) + (T + S)\mathbf{u} && \text{(Definition 6-2)}\end{aligned}$$

The reader should note the structure of the proof of equation 5. The first step uses the definition of the sum $T + S$. The third step uses a combination of associativity and commutativity of the vector space W in order to rearrange the vectors $T(\mathbf{v})$, $S(\mathbf{v})$, $T(\mathbf{u})$, and $S(\mathbf{u})$. The final step consists of using Definition 6-2 again (in reverse) to obtain the vector $(T + S)(\mathbf{v})$ and $(T + S)\mathbf{u}$. A similar sequence of steps will yield the proof of equation 6.

To prove equation 6 the student must use $T(\beta\mathbf{v}) = \beta T(\mathbf{v})$ and $S(\beta\mathbf{v}) = \beta S(\mathbf{v})$.

In order to prove that αT is linear, we must show the following two properties:

$$(\alpha T)(\mathbf{v} + \mathbf{u}) = (\alpha T)(\mathbf{v}) + (\alpha T)(\mathbf{u}) \quad \textbf{[7]}$$

$$(\alpha T)(\beta\mathbf{v}) = \beta[(\alpha T)\mathbf{v}] \quad \textbf{[8]}$$

See problem 2.

We shall prove equation 8 and leave the proof of equation 7 as an exercise for the reader.

$$\begin{aligned}(\alpha T)(\beta\mathbf{v}) &= \alpha T(\beta\mathbf{v}) && \text{(by Definition 6-2)}\\ &= \alpha[\beta T(\mathbf{v})] && \text{(linearity of } T)\\ &= (\alpha\beta)T(\mathbf{v}) && \text{(vector space axiom)}\\ &= (\beta\alpha)T(\mathbf{v}) && \text{(commutativity of real numbers)}\\ &= \beta[\alpha T(\mathbf{v})] && \text{(vector space axiom)}\\ &= \beta(\alpha T)(\mathbf{v}) && \text{(Definition 6-2)}\end{aligned}$$

The results of Theorem 6-3 may also be stated in the following way:

Assume that V and W are fixed vector spaces. Then the set of all linear transformations from V into W is closed under addition and scalar multiplication as defined in Definition 6-2.

See Definition 2-1.

Since the closure properties constitute 2 (out of 10) of the axioms for a vector space, one is tempted to find out whether the remaining axioms are also satisfied. The answer to this question is given in the following theorem.

Theorem 6-4 Let V and W be fixed vector spaces. Let addition and scalar multiplication be defined on the set L of all linear transformations from V into W according to Definition 6-2. Then the set L is a vector space. (Each vector in L is a linear transformation.)

The reader should recall other examples of vector spaces where the vectors were functions. See Example 4 in Section 2-2.

The closure properties were proved in Theorem 6-3. The proof of the remaining 8 axioms of a vector space, which does not depend on the linearity of the transformations, is left as an exercise for the diligent reader.

See problem 3.

Theorem 6-4 guarantees that every linear combination of linear transformations is again a linear transformation. Thus, $\alpha T + \beta S$ represents a linear transformation whenever T and S are linear and α, β are any real numbers.

This fact is used in the following example in our discussion of the geometric operation of reflection of vectors with respect to a given line in the plane.

Example 2 Let l be a line through the origin of a coordinate system x-y in the plane. Let $\mathbf{u}$ be a unit vector having its initial point at the origin and lying along l. Show that the reflection of each vector $\mathbf{v} \in \mathbf{R}^2$ with respect to the line l is a linear transformation $T: \mathbf{R}^2 \longrightarrow \mathbf{R}^2$, given by

$$T(\mathbf{v}) = 2P(\mathbf{v}) - \mathbf{v} \tag{9}$$

where $P(\mathbf{v})$ is the vector projection of $\mathbf{v}$ onto $\mathbf{u}$.

Solution The reflection $T(\mathbf{v})$ of any vector $\mathbf{v}$ is obtained as follows (see also Figure 6-6):

Let $\mathbf{v}$ have initial point 0 and terminal point A. Construct a perpendicular to the line l passing through A. Let B denote the point of intersection of the two lines. Now construct a vector equal to the vector $\overrightarrow{AB}$ having B as its initial point. Let C denote the terminal point of this new vector. Then the vector $T(\mathbf{v}) = \overrightarrow{OC}$ is called the reflection of $\mathbf{v} = \overrightarrow{OA}$ with respect to the line l.

In order to establish the validity of equation 9 we use the fact that the vector $\overrightarrow{OB}$ is actually the vector projection $P(\mathbf{v})$ of $\mathbf{v}$ onto $\mathbf{u}$. The vector $\overrightarrow{AB}$ is given by $P(\mathbf{v}) - \mathbf{v}$ and since $\overrightarrow{AB} = \overrightarrow{BC}$ we also have

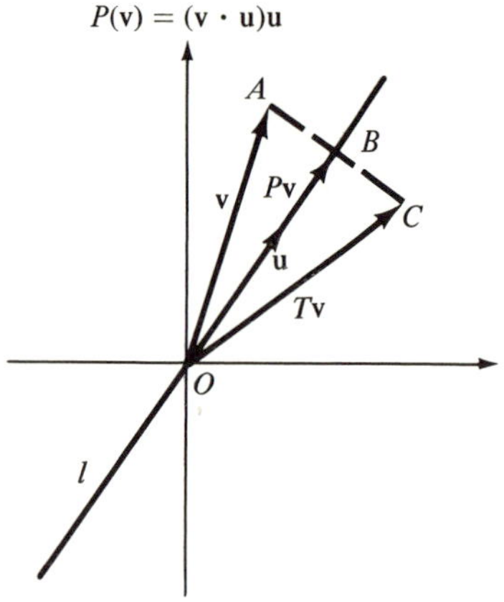

Figure 6-6

$\overrightarrow{BC} = P(\mathbf{v}) - \mathbf{v}$. Substituting these results into the equation $\overrightarrow{OC} = \overrightarrow{OB} + \overrightarrow{BC}$, we obtain

$$T(\mathbf{v}) = P(\mathbf{v}) + P(\mathbf{v}) - \mathbf{v} = 2P(\mathbf{v}) - \mathbf{v}$$

which establishes equation 9.

Since we have $-\mathbf{v} = (-1)\mathbf{v}$ and $\mathbf{v} = I(\mathbf{v})$ where I is the identity transformation (see Example 7 in Section 6-1), we may write equation 9 in the form

$$T(\mathbf{v}) = 2P(\mathbf{v}) + (-1)I(\mathbf{v}) \qquad \textbf{[10]}$$

It follows from equation 10 that T is a linear combination of two linear transformations, the projection P and the identity transformation I. Hence, by Theorem 6-4, we conclude that the reflection T is also a linear transformation.

The following example makes use of equation 9 in deriving matrix representations for two different transformations of reflection.

Example 3 Let $T:\mathbf{R}^2 \longrightarrow \mathbf{R}^2$ be reflection with respect to the x-axis and let $S:\mathbf{R}^2 \longrightarrow \mathbf{R}^2$ be reflection with respect to the line whose equation is $y = x$. Find matrix representations for T and S.

Solution Using the fact that $P(\mathbf{v}) = (\mathbf{v} \cdot \mathbf{u})\mathbf{u}$ (see Example 4 in Section 6-1), we may rewrite equation 9 in the form

$$T(\mathbf{v}) = 2(\mathbf{v} \cdot \mathbf{u})\mathbf{u} - \mathbf{v} \qquad \textbf{[11]}$$

To find a matrix representation for T, we take $\mathbf{u} = \begin{bmatrix}1\\0\end{bmatrix}$. (Note that $\mathbf{u} = \begin{bmatrix}1\\0\end{bmatrix}$ is a unit vector along the x-axis.) Computing $T\begin{bmatrix}1\\0\end{bmatrix}$ and $T\begin{bmatrix}0\\1\end{bmatrix}$, we obtain

$$T\begin{bmatrix}1\\0\end{bmatrix} = (2)(1)\begin{bmatrix}1\\0\end{bmatrix} - \begin{bmatrix}1\\0\end{bmatrix} = \begin{bmatrix}1\\0\end{bmatrix},$$

$$T\begin{bmatrix}0\\1\end{bmatrix} = (2)(0)\begin{bmatrix}1\\0\end{bmatrix} - \begin{bmatrix}0\\1\end{bmatrix} = \begin{bmatrix}0\\-1\end{bmatrix}$$

Thus, the matrix representation of T is given by

$$T\begin{bmatrix}x\\y\end{bmatrix} = \begin{bmatrix}1 & 0\\0 & -1\end{bmatrix}\begin{bmatrix}x\\y\end{bmatrix} \qquad \textbf{[12]}$$

Multiplying the matrices on the right side of equation 12, we obtain

$$T\begin{bmatrix}x\\y\end{bmatrix} = \begin{bmatrix}x\\-y\end{bmatrix}$$

To find a matrix representation for S, we take $\mathbf{u} = \begin{bmatrix}\dfrac{1}{\sqrt{2}}\\[2mm]\dfrac{1}{\sqrt{2}}\end{bmatrix}$, which is a unit vector along the line $y = x$. Computing $S\begin{bmatrix}1\\0\end{bmatrix}$ and $S\begin{bmatrix}0\\1\end{bmatrix}$, we obtain

$$S\begin{bmatrix}1\\0\end{bmatrix} = (2)\left(\frac{1}{\sqrt{2}}\right)\begin{bmatrix}\frac{1}{\sqrt{2}}\\ \frac{1}{\sqrt{2}}\end{bmatrix} - \begin{bmatrix}1\\0\end{bmatrix} = \begin{bmatrix}0\\1\end{bmatrix}$$

$$S\begin{bmatrix}0\\1\end{bmatrix} = (2)\frac{1}{(\sqrt{2})}\begin{bmatrix}\frac{1}{\sqrt{2}}\\ \frac{1}{\sqrt{2}}\end{bmatrix} - \begin{bmatrix}0\\1\end{bmatrix} = \begin{bmatrix}1\\0\end{bmatrix}$$

Thus, the matrix representation of S is given by

$$S\begin{bmatrix}x\\y\end{bmatrix} = \begin{bmatrix}0 & 1\\1 & 0\end{bmatrix}\begin{bmatrix}x\\y\end{bmatrix} \qquad [13]$$

Multiplying the matrices on the right side of equation 13 we obtain

$$S\begin{bmatrix}x\\y\end{bmatrix} = \begin{bmatrix}y\\x\end{bmatrix}$$

The operation of composition of functions (see Chapter 0) is widely used in many branches of mathematics and is of special interest in linear algebra. Here we shall deal with composition of transformations and discuss some important properties of this operation.

Definition 6-3 Let V, U, and W be vector spaces. Let T be a transformation from V into U and let S be a transformation from U into W.

The **composite** ST is a transformation from V into W defined by the formula

$$(ST)(\mathbf{v}) = S(T(\mathbf{v})) \qquad [14]$$

for all $\mathbf{v}$ in V.

The composite transformation of S by T is sometimes denoted by $S \circ T$. We may safely omit the symbol $\circ$ and simply write ST because we will not introduce any other kind of product for transformations in linear algebra.

A few words of explanation about Definition 6-3 will help to clarify it further.

It is important for the reader to realize that the transformations T and S give rise to the new transformation ST. The domain of ST is the vector space V (which is also the domain of T) and its range is contained in the vector space W. In order to find the image of any $\mathbf{v}$ in V under the transformation ST we proceed in the following way (see also Figure 6-7): First, we find the image of $\mathbf{v}$ under T, i.e., $T(\mathbf{v})$. Second, we apply the transformation S to the vector $T(\mathbf{v})$, obtaining $S(T(\mathbf{v}))$. This is possible because $T(\mathbf{v})$ lies in the domain of S, namely, U. Thus, the transformation ST associates with every vector $\mathbf{v}$ in V the vector $S(T(\mathbf{v}))$ in W.

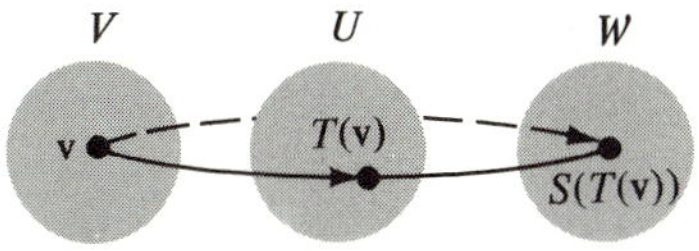

Figure 6-7

The following example shows how one obtains the composite transformation ST from the given transformations T and S.

Example 4 Let $T\begin{bmatrix} x \\ y \end{bmatrix} = \begin{bmatrix} x \\ -y \end{bmatrix}$ and $S\begin{bmatrix} x \\ y \end{bmatrix} = \begin{bmatrix} y \\ x \end{bmatrix}$. Find a formula for the composite transformation ST.

Solution The reader should note that the domain and range of each of the transformations T and S are the same, namely, $\mathbf{R}^2$. Thus, in the context of Definition 6-3 we have $V = U = W = \mathbf{R}^2$. Using formula 14, we get

$$(ST)\begin{bmatrix} x \\ y \end{bmatrix} = S\left(T\begin{bmatrix} x \\ y \end{bmatrix}\right) \qquad \textbf{[15]}$$

Since

$$T\begin{bmatrix} x \\ y \end{bmatrix} = \begin{bmatrix} x \\ -y \end{bmatrix}$$

we may write

$$S\left(T\begin{bmatrix} x \\ y \end{bmatrix}\right) = S\begin{bmatrix} x \\ -y \end{bmatrix} \qquad \textbf{[16]}$$

We now interpret the formula $S\begin{bmatrix} x \\ y \end{bmatrix} = \begin{bmatrix} y \\ x \end{bmatrix}$ by saying that the transformation S switches the components of every vector in $\mathbf{R}^2$. Thus, we have

$$S\begin{bmatrix} x \\ -y \end{bmatrix} = \begin{bmatrix} -y \\ x \end{bmatrix} \qquad \textbf{[17]}$$

Combining equations 15, 16, and 17, we obtain the formula

$$(ST)\begin{bmatrix} x \\ y \end{bmatrix} = \begin{bmatrix} -y \\ x \end{bmatrix} \qquad \textbf{[18]}$$

Using results from Example 3, we can describe the transformation ST geometrically by saying that in order to obtain $(ST)\begin{bmatrix} x \\ y \end{bmatrix}$ we reflect the vector $\begin{bmatrix} x \\ y \end{bmatrix}$ in the x-axis, obtaining the vector $\begin{bmatrix} x \\ -y \end{bmatrix}$, and then reflect the vector $\begin{bmatrix} x \\ -y \end{bmatrix}$ in the line $y = x$, obtaining the vector $\begin{bmatrix} -y \\ x \end{bmatrix}$.

The transformations T and S in Example 4 were shown to be linear. (See Examples 2 and 3.) The reader can easily prove that the composite transformation ST given by equation 18 is also linear. We have the following general result.

Theorem 6-5 Let $T: V \longrightarrow U$ be linear and let $S: U \longrightarrow W$ be linear. Then $(ST): V \longrightarrow W$ is also linear.

Proof We must show that for all $\mathbf{v}$, $\mathbf{v}_1$, and $\mathbf{v}_2$ in V and any real number α we have

$$(ST)(\mathbf{v}_1 + \mathbf{v}_2) = (ST)(\mathbf{v}_1) + (ST)(\mathbf{v}_2) \tag{19}$$

and

$$(ST)(\alpha\mathbf{v}) = \alpha(ST)(\mathbf{v}) \tag{20}$$

See problem 8.

We shall prove equation 19 and leave the proof of equation 20 as an exercise for the reader.

$$\begin{aligned} (ST)(\mathbf{v}_1 + \mathbf{v}_2) &= S(T(\mathbf{v}_1 + \mathbf{v}_2)) && \text{(definition of } ST\text{)} \\ &= S(T(\mathbf{v}_1) + T(\mathbf{v}_2)) && \text{(linearity of } T\text{)} \\ &= S(T(\mathbf{v}_1)) + S(T(\mathbf{v}_2)) && \text{(linearity of } S\text{)} \\ &= (ST)(\mathbf{v}_1) + (ST)(\mathbf{v}_2) && \text{(definition of } ST\text{)} \end{aligned}$$

Thus, equation 19 is established. Using a similar sequence of steps the reader can easily prove equation 20.

In the definition of the composite transformation of T and S we should pay special attention to the order in which the transformations are applied. Thus, according to Definition 6-3 the expression $(ST)(\mathbf{v})$ means that we first apply T to obtain $T(\mathbf{v})$ and then apply S to obtain the vector $S(T(\mathbf{v}))$. It is important to note that the composition TS might not be defined even though ST is. The following example illustrates such a possibility.

Note: The composition TS is the composite transformation of T and S.

Example 5 Find transformations S and T such that ST is defined but TS is not.

Solution Let $T\begin{bmatrix} x \\ y \\ z \end{bmatrix} = \begin{bmatrix} x \\ y \end{bmatrix}$ and $S\begin{bmatrix} x \\ y \end{bmatrix} = \begin{bmatrix} x \\ -y \end{bmatrix}$. Since T is a transformation from $\mathbf{R}^3$ into $\mathbf{R}^2$ and S is a transformation from $\mathbf{R}^2$ into $\mathbf{R}^2$, ST is a transformation from $\mathbf{R}^3$ into $\mathbf{R}^2$. The composite ST is given by the formula

$$(ST)\begin{bmatrix} x \\ y \\ z \end{bmatrix} = \begin{bmatrix} x \\ -y \end{bmatrix}$$

The composition TS is not defined because $S\begin{bmatrix} x \\ y \end{bmatrix}$ is a vector which is not in the domain of T (which is $\mathbf{R}^3$).

Even in examples where both composites ST and TS are defined, they need not have the same effect.

If we define a transformation T by $T\begin{bmatrix} x \\ y \end{bmatrix} = \begin{bmatrix} x \\ -y \end{bmatrix}$ and a transformation S by $S\begin{bmatrix} x \\ y \end{bmatrix} = \begin{bmatrix} y \\ x \end{bmatrix}$, then both ST and TS are also defined and we have

$$(ST)\begin{bmatrix} x \\ y \end{bmatrix} = S\left(T\begin{bmatrix} x \\ y \end{bmatrix}\right) = S\begin{bmatrix} x \\ -y \end{bmatrix} = \begin{bmatrix} -y \\ x \end{bmatrix}$$ (see equation 18)

and

$$(TS)\begin{bmatrix} x \\ y \end{bmatrix} = T\left(S\begin{bmatrix} x \\ y \end{bmatrix}\right) = T\begin{bmatrix} y \\ x \end{bmatrix} = \begin{bmatrix} y \\ -x \end{bmatrix}$$

Note: We sometimes describe the statement $ST \neq TS$ by saying that T and S do not commute.

Since $\begin{bmatrix} -y \\ x \end{bmatrix} \neq \begin{bmatrix} y \\ -x \end{bmatrix}$ (unless $x = y = 0$) it follows that $ST \neq TS$.

An important property of composition of transformations is the associativity of this operation. Let us consider three transformations $T: V \longrightarrow U$, $S: U \longrightarrow W$, and $P: W \longrightarrow Y$. It is understood of course that V, U, W, and Y are vector spaces. We have the following result.

Theorem 6-6 If T, S, and P are transformations such that $P(ST)$ is defined, then $(PS)T$ is also defined and we have

$$P(ST) = (PS)T \qquad [21]$$

Proof The domain of ST is V and its range is contained in W. Since P maps W into Y it follows that $P(ST)$ has domain V and its range is contained in Y. The domain of (PS) is U and its range is contained in Y. Since T maps V into U it follows that $(PS)T$ is defined and has domain V while its range is contained in Y. Note that both transformations $P(ST)$ and $(PS)T$ have the same domain V and for each the range is contained in Y.

To prove equality of $P(ST)$ and $(PS)T$ we check the effect of each transformation on an arbitrary vector $\mathbf{v}$ in V. We have by Definition 6-3

$$[P(ST)](\mathbf{v}) = P[(ST)\mathbf{v}] = P[S(T(\mathbf{v}))] \qquad [22]$$

and

$$[(PS)T](\mathbf{v}) = (PS)[T(\mathbf{v})] = P(S[T(\mathbf{v})]) \qquad [23]$$

It follows from equations 22 and 23 that

$$[P(ST)](\mathbf{v}) = [(PS)T](\mathbf{v}) = P(S[T(\mathbf{v})])$$

Thus, equation 21 is proved.

Note that linearity of transformations does not play any role in the property of associativity just proved.

We conclude this section by posing the following problem and providing its solution.

Problem 1 Let $T: \mathbf{R}^n \longrightarrow \mathbf{R}^m$ and $S: \mathbf{R}^m \longrightarrow \mathbf{R}^k$ be linear transformations. It follows from Theorem 6-5 that $ST: \mathbf{R}^n \longrightarrow \mathbf{R}^k$ is linear too. Theorem 6-2 guarantees that each of the transformations T, S, and ST has a matrix representation. Thus, we have for all $\mathbf{v} \in \mathbf{R}^n$, $\mathbf{u} \in \mathbf{R}^m$

$$T(\mathbf{v}) = A\mathbf{v}, \qquad S(\mathbf{u}) = B\mathbf{u}, \qquad \text{and} \qquad (ST)(\mathbf{v}) = C\mathbf{v}$$

where A is an $m \times n$ matrix, B is a $k \times m$ matrix, and C is a $k \times n$ matrix. The problem is to find the relationship among the matrices A, B, and C.

Solution Let us examine the columns of the matrix C. In order to obtain them we have to apply the transformation ST to each of the vectors in the standard basis for $\mathbf{R}^n$. Thus, the first column of C is given by

$$(ST)\begin{bmatrix} 1 \\ 0 \\ \vdots \\ 0 \end{bmatrix} \quad \text{which by Definition 6-3 is equal to} \quad S\left(T\begin{bmatrix} 1 \\ 0 \\ \vdots \\ 0 \end{bmatrix}\right)$$

Since $T\begin{bmatrix} 1 \\ 0 \\ \vdots \\ 0 \end{bmatrix}$ is actually the first column of the matrix A, we may write

$$S\left(T\begin{bmatrix} 1 \\ 0 \\ \vdots \\ 0 \end{bmatrix}\right) = S\begin{bmatrix} a_{11} \\ a_{21} \\ \vdots \\ a_{m1} \end{bmatrix}$$

Using the matrix representation for the transformation S, we have

$$S\left(T\begin{bmatrix} 1 \\ 0 \\ \vdots \\ 0 \end{bmatrix}\right) = S\begin{bmatrix} a_{11} \\ a_{21} \\ \vdots \\ a_{m1} \end{bmatrix} = B\begin{bmatrix} a_{11} \\ a_{21} \\ \vdots \\ a_{m1} \end{bmatrix}$$

Since the product of B and $\begin{bmatrix} a_{11} \\ a_{21} \\ \vdots \\ a_{m1} \end{bmatrix}$ produces the first column of the matrix product BA, it follows that $(ST)\begin{bmatrix} 1 \\ 0 \\ \vdots \\ 0 \end{bmatrix}$ is obtained by multiplying the matrix B by the first column of A. In a similar fashion, it follows that $(ST)\begin{bmatrix} 0 \\ 1 \\ 0 \\ \vdots \\ 0 \end{bmatrix}$ is obtained by multiplying the matrix B by the second column of A. Continuing in this fashion, we obtain all the columns of the matrix C representing the transformation ST. Thus, we conclude that the matrix C is equal to the product BA.

Exercises

1 Let T and S be linear transformations from V into W and let β be any real number. Show that $(T + S)(\beta\mathbf{v}) = \beta(T + S)(\mathbf{v})$ for every $\mathbf{v}$ in V. (This will complete the proof of the linearity of $T + S$.)

2 Let T be a linear transformation from V into W and let α be any real number. Show that $(\alpha T)(\mathbf{v} + \mathbf{u}) = (\alpha T)(\mathbf{v}) + (\alpha T)(\mathbf{u})$. (This will complete the proof of the linearity of αT.)

3 Complete the proof of Theorem 6-4.

4 Let T be the linear transformation of reflecting each vector $\mathbf{v}$ in $\mathbf{R}^2$ with respect to the line $y = -x$. Find a matrix representation for T.

$\mathbf{u} = \begin{bmatrix} u_1 \\ u_2 \end{bmatrix}$

5 Let T be the linear transformation of reflecting each vector $\mathbf{v}$ in $\mathbf{R}^2$ with respect to a line l through the origin. Let $\mathbf{u}$ be a unit vector along l. Derive a matrix representation for T in terms of the components u_1 and u_2 of $\mathbf{u}$ and show that this matrix is nonsingular.

Hint: Use problem 5.

6 Let a linear transformation T be given by $T\begin{bmatrix} x \\ y \end{bmatrix} = \begin{bmatrix} -\frac{3}{5} & \frac{4}{5} \\ \frac{4}{5} & \frac{3}{5} \end{bmatrix}\begin{bmatrix} x \\ y \end{bmatrix}$. Show that T represents a transformation of reflection in a line l through the origin and find a unit vector $\mathbf{u}$ along the line l.

7 Let $T\begin{bmatrix} x \\ y \end{bmatrix} = \begin{bmatrix} 2x - y \\ 3y \end{bmatrix}$ and $S\begin{bmatrix} x \\ y \end{bmatrix} = \begin{bmatrix} 0 \\ x + 2y \end{bmatrix}$. Find formulas for

$(ST)\begin{bmatrix} x \\ y \end{bmatrix}$ and $(TS)\begin{bmatrix} x \\ y \end{bmatrix}$ by using Definition 6-3. Also find matrix representations for T, S, ST, and TS.

8 Complete the proof of Theorem 6-5 by showing that $(ST)(\alpha\mathbf{v}) = \alpha(ST)(\mathbf{v})$.

9 Let S and T be transformations from V into U and let P be a linear transformation from U into W. Show that $P(S + T) = PS + PT$.

Note that we require only P to be a linear transformation in problem 9.

10 Let P be a transformation from V into U and let S and T be transformations from U into W. Show that $(S + T)P = SP + TP$.

Note that no linearity requirement is necessary in problem 10.

6-4 Null Space and Range of a Linear Transformation

Objectives

1 Study the concepts of null space and range of a linear transformation.
2 Continue the study of the structure of a linear transformation.

We introduced the concept of the null space of a matrix when we discussed the structure of the set of all solutions of the homogeneous system of equations $A\mathbf{x} = \mathbf{0}$ (see Section 3-6). If we regard the $m \times n$ matrix A as representing the transformation T from $\mathbf{R}^n$ into $\mathbf{R}^m$ given by $T(\mathbf{x}) = A\mathbf{x}$, then we know that the set of all vectors $\mathbf{x}$ in $\mathbf{R}^n$ which T maps onto the zero vector $\mathbf{0}$ of $\mathbf{R}^m$ is a subspace of $\mathbf{R}^n$ called the null space of the matrix A. (See Theorem 3-19.)

We are now interested in generalizing the concept of null space to an arbitrary linear transformation. Let us therefore assume that T is a linear transformation from V into W. The following theorem is a direct outcome of the linearity of T.

Theorem 6-7 Let $T: V \longrightarrow W$ be a linear transformation. Then the subset N of V containing all vectors $\mathbf{v}$ such that $T(\mathbf{v}) = \mathbf{0}$ is a subspace of V.

Proof We must show that N is closed under the addition and scalar multiplication defined on the vector space V. Let $\mathbf{v}_1$ and $\mathbf{v}_2$ be any elements of N. This implies that $T(\mathbf{v}_1) = \mathbf{0}$ and $T(\mathbf{v}_2) = \mathbf{0}$. Thus, we have

$$\begin{aligned} T(\mathbf{v}_1 + \mathbf{v}_2) &= T(\mathbf{v}_1) + T(\mathbf{v}_2) && \text{(by linearity of } T\text{)} \\ &= \mathbf{0} + \mathbf{0} \\ &= \mathbf{0} \end{aligned}$$

It follows that $\mathbf{v}_1 + \mathbf{v}_2$ is also an element of N. Thus, N is closed under addition. In order to prove that N is closed under scalar multiplication, let $\mathbf{v}$ be any element of N and let α be any real number. Then we have

$$\begin{aligned} T(\alpha\mathbf{v}) &= \alpha T(\mathbf{v}) && \text{(by linearity of } T\text{)} \\ &= \alpha\mathbf{0} && (T(\mathbf{v}) = \mathbf{0}) \\ &= \mathbf{0} \end{aligned}$$

It follows that $\alpha\mathbf{v}$ is also an element of N and therefore N is closed under scalar multiplication. We conclude that N is a subspace of V.

Definition 6-4 Let $T: V \longrightarrow W$ be a linear transformation. The subspace N of V containing all vectors which T maps into the zero vector $\mathbf{0}$ of W is called the null space of T.

The null space is also called the kernel of T.

The size of the null space of a linear transformation may vary from one vector (which must be the zero vector of V) to the whole space V. The following examples deal with these two extremes.

Example 1 Find a linear transformation $T: V \longrightarrow W$ such that the null space of T is the whole space V.

Solution Let V and W be any vector spaces. Let T be the zero transformation, defined by $T(\mathbf{v}) = \mathbf{0}$ for every $\mathbf{v}$ in V. It is obvious that the null space of T is the whole space V.

See Example 6 in Section 6-1.

Example 2 Find a linear transformation $T: V \longrightarrow W$ such that the null space of T consists of one vector only.

Solution Let $V = W = \mathbf{R}^2$. Let T be the identity transformation, defined by $T\begin{bmatrix} x \\ y \end{bmatrix} = \begin{bmatrix} x \\ y \end{bmatrix}$. It is clear that the unique vector which T maps into $\begin{bmatrix} 0 \\ 0 \end{bmatrix}$ is the vector $\begin{bmatrix} 0 \\ 0 \end{bmatrix}$ itself. Thus, the null space of the identity transformation consists of the unique vector $\begin{bmatrix} 0 \\ 0 \end{bmatrix}$.

See Example 7 in Section 6-1.

Let us now look at the set of vectors obtained by applying a linear transformation T to all the vectors in V. The following theorem deals with such a set.

Theorem 6-8 Let $T: V \longrightarrow W$ be a linear transformation. Then the set of all vectors in W that are images of at least one vector in V is a subspace of W.

Proof Let $\mathbf{w}_1$ and $\mathbf{w}_2$ be vectors in W which are images of $\mathbf{v}_1$ and $\mathbf{v}_2$, respectively, under T. Thus, we have $T(\mathbf{v}_1) = \mathbf{w}_1$ and $T(\mathbf{v}_2) = \mathbf{w}_2$. We must show that $\mathbf{w}_1 + \mathbf{w}_2$ is also the image of some vector in V. The linearity of T suggests that $\mathbf{v}_1 + \mathbf{v}_2$ could be such a vector. Indeed, we have

$$T(\mathbf{v}_1 + \mathbf{v}_2) = T(\mathbf{v}_1) + T(\mathbf{v}_2) = \mathbf{w}_1 + \mathbf{w}_2$$

To complete the proof of the theorem we must show that if $\mathbf{w}$ is a vector in W which is an image of $\mathbf{v}$ in V, then $\alpha\mathbf{w}$ is also an image of some vector in V for any real number α. The linearity of T suggests that $\alpha\mathbf{v}$ is such a vector. We have

$$T(\alpha\mathbf{v}) = \alpha T(\mathbf{v}) = \alpha\mathbf{w}$$

Thus, the set of all vectors in W that are images of at least one vector in V (under T) is closed under addition and scalar multiplication, and hence is a subspace of W. This completes the proof of the theorem.

The subspace of W mentioned in Theorem 6-8 is given a special name and notation by the following definition.

Definition 6-5 Let $T: V \longrightarrow W$ be a linear transformation. The subspace of W consisting of all the vectors that are images of at least one vector in V is called the **range** of T and is denoted by $R(T)$.

Note that the definition of the range of a linear transformation is in agreement with the concept of the range of a function mentioned in Definition 0-7.

The following example shows how one determines the range of a given linear transformation.

Example 3 Let $T: \mathbf{R}^3 \longrightarrow \mathbf{R}^2$ be defined by $T\begin{bmatrix} x \\ y \\ z \end{bmatrix} = A\begin{bmatrix} x \\ y \\ z \end{bmatrix}$ where $A = \begin{bmatrix} 2 & 1 & 3 \\ 1 & 0 & 4 \end{bmatrix}$. Find the range of T.

Solution To find $R(T)$ we let the matrix A operate on all vectors $\begin{bmatrix} x \\ y \\ z \end{bmatrix}$ in $\mathbf{R}^3$. Thus, $R(T)$ consists of all vectors in $\mathbf{R}^2$ which are obtained from the product $\begin{bmatrix} 2 & 1 & 3 \\ 1 & 0 & 4 \end{bmatrix}\begin{bmatrix} x \\ y \\ z \end{bmatrix}$. This last expression can be written as

$$x\begin{bmatrix} 2 \\ 1 \end{bmatrix} + y\begin{bmatrix} 1 \\ 0 \end{bmatrix} + z\begin{bmatrix} 3 \\ 4 \end{bmatrix} \qquad [1]$$

It follows from expression 1 that $R(T)$ consists of all linear combinations of the columns of the matrix A, namely, $\begin{bmatrix} 2 \\ 1 \end{bmatrix}$, $\begin{bmatrix} 1 \\ 0 \end{bmatrix}$, and $\begin{bmatrix} 3 \\ 4 \end{bmatrix}$. Since the vectors $\begin{bmatrix} 2 \\ 1 \end{bmatrix}$, $\begin{bmatrix} 1 \\ 0 \end{bmatrix}$ form a basis for $\mathbf{R}^2$, it follows that $R(T)$ is the whole vector space $\mathbf{R}^2$.

Since the null space and the range of a linear transformation are subspaces of given vector spaces, we may speak of their dimensions. The following definition gives special names to these dimensions.

Definition 6-6 Let $T: V \longrightarrow W$ be a linear transformation. The dimension of the null space of T is called the **nullity** of T and the dimension of the range of T is called the **rank** of T.

In order to determine the nullity and rank of a linear transformation we make use of the methods introduced in Section 3-4. The following examples illustrate the use of these methods.

Example 4 Let T be the linear transformation defined by $T\begin{bmatrix} x \\ y \end{bmatrix} = \begin{bmatrix} x \\ x+y \\ y \end{bmatrix}$.

Find the nullity of T and the rank of T.

Solution In order to determine the nullity of T, we must find the null space of T and determine its dimension. Since T is a linear transformation from $\mathbf{R}^2$ into $\mathbf{R}^3$ we set $T\begin{bmatrix} x \\ y \end{bmatrix} = \begin{bmatrix} 0 \\ 0 \\ 0 \end{bmatrix}$. Thus, we obtain the system of equations

$$\begin{aligned} x &= 0 \\ x + y &= 0 \\ y &= 0 \end{aligned} \qquad [2]$$

System 2 has the unique solution $x = 0$, $y = 0$. Thus, the null space of T consists of the unique vector $\begin{bmatrix} 0 \\ 0 \end{bmatrix}$. It follows that $\dim$ (null space of T) $= 0$ and thus the nullity of T is equal to 0.

In order to determine the rank of T, we must find the range of T and determine its dimension.

The matrix representation of T is given by

$$T\begin{bmatrix} x \\ y \end{bmatrix} = \begin{bmatrix} 1 & 0 \\ 1 & 1 \\ 0 & 1 \end{bmatrix}\begin{bmatrix} x \\ y \end{bmatrix} \qquad [3]$$

Expressing the right-hand side of equation 3 as a linear combination of the columns $\begin{bmatrix} 1 \\ 1 \\ 0 \end{bmatrix}$ and $\begin{bmatrix} 0 \\ 1 \\ 1 \end{bmatrix}$, we obtain

$$T\begin{bmatrix} x \\ y \end{bmatrix} = x\begin{bmatrix} 1 \\ 1 \\ 0 \end{bmatrix} + y\begin{bmatrix} 0 \\ 1 \\ 1 \end{bmatrix} \qquad [4]$$

Thus, it follows from equation 4 that the range of T consists of all linear combinations of the vectors $\begin{bmatrix} 1 \\ 1 \\ 0 \end{bmatrix}$ and $\begin{bmatrix} 0 \\ 1 \\ 1 \end{bmatrix}$. Since $\begin{bmatrix} 1 \\ 1 \\ 0 \end{bmatrix}$ and $\begin{bmatrix} 0 \\ 1 \\ 1 \end{bmatrix}$ are linearly independent vectors in $\mathbf{R}^3$, they form a basis for the range of T. Therefore, it follows that the rank of T is equal to 2.

Example 5 Let $T: \mathbf{R}^3 \longrightarrow \mathbf{R}^2$ be the linear transformation defined by $T\begin{bmatrix} x \\ y \\ z \end{bmatrix} = \begin{bmatrix} x \\ y \end{bmatrix}$. Find the nullity of T and the rank of T.

Solution The null space of T consists of all vectors $\begin{bmatrix} x \\ y \\ z \end{bmatrix}$ such that $T\begin{bmatrix} x \\ y \\ z \end{bmatrix} = \begin{bmatrix} 0 \\ 0 \end{bmatrix}$. This means that we must have $x = 0$ and $y = 0$. Since z is arbitrary, it follows that the null space of T consists of all vectors of the form $\begin{bmatrix} 0 \\ 0 \\ z \end{bmatrix}$. A basis for the null space is therefore given by the vector $\begin{bmatrix} 0 \\ 0 \\ 1 \end{bmatrix}$. It follows that the nullity of T is 1.

Remark: A close look at Examples 4 and 5 shows that we have the following relationship between the nullity of T, the rank of T, and the dimension of the domain of T: (nullity of T) + (rank of T) = dim (domain of T).

The range of T consists of all vectors of the form $\begin{bmatrix} x \\ y \end{bmatrix}$ where x and y are arbitrary. This is obviously the vector space $\mathbf{R}^2$. Thus the rank of T is 2.

We have the following general result:

Theorem 6-9 Let $T: V \longrightarrow W$ be a linear transformation. Let V be an n-dimensional vector space. Then we have

$$(\text{nullity of } T) + (\text{rank of } T) = n \quad (= \dim V) \qquad [5]$$

Proof If T is the zero transformation mapping every vector $\mathbf{v}$ in V into the zero vector of W, then the null space of T is the whole vector space V and the range of T is the unique zero vector of W. Since the nullity of T is equal to the dimension of V and the rank of T is equal to 0, it follows that equation 5 is satisfied.

$\dim V = n.$

If T is different from the zero transformation, then the null space of T is a proper subspace of V. Let $\dim$ (null space of T) $= k$. We must show that $\dim$ (range of T) $= n - k$. Let $\{\mathbf{v}_1, \mathbf{v}_2, \ldots, \mathbf{v}_k\}$ be a basis for the null space of T. Since $\{\mathbf{v}_1, \mathbf{v}_2, \ldots, \mathbf{v}_k\}$ is a linearly independent set in V we can expand it into a basis for V by adding $n - k$ vectors from V (see Theorem 3-12). Let $\mathbf{v}_{k+1}, \mathbf{v}_{k+2}, \ldots, \mathbf{v}_n$ be chosen such that the expanded set $\{\mathbf{v}_1, \mathbf{v}_2, \ldots, \mathbf{v}_k, \mathbf{v}_{k+1}, \ldots, \mathbf{v}_n\}$ is a basis for V. We will show that the set $L = \{T(\mathbf{v}_{k+1}), \ldots, T(\mathbf{v}_n)\}$ is a basis for the range of T.

In order to show that L spans the range of T we must prove that if $\mathbf{w} \in R(T)$ then $\mathbf{w}$ is a linear combination of $T(\mathbf{v}_{k+1}), \ldots, T(\mathbf{v}_n)$. The assumption $\mathbf{w} \in R(T)$ implies that there exists a vector $\mathbf{v} \in V$ such that $T(\mathbf{v}) = \mathbf{w}$. The vector $\mathbf{v}$ has a unique representation in terms of the basis $\{\mathbf{v}_1, \mathbf{v}_2, \ldots, \mathbf{v}_k, \mathbf{v}_{k+1}, \ldots, \mathbf{v}_n\}$. Thus, there exist scalars $\alpha_1, \alpha_2, \ldots, \alpha_k, \alpha_{k+1}, \ldots, \alpha_n$ such that

$$\mathbf{v} = \alpha_1\mathbf{v}_1 + \alpha_2\mathbf{v}_2 + \cdots + \alpha_k\mathbf{v}_k + \alpha_{k+1}\mathbf{v}_{k+1} + \cdots + \alpha_n\mathbf{v}_n \qquad [6]$$

Applying T to both sides of equation 6, we obtain

$$\begin{aligned} T(\mathbf{v}) &= T(\alpha_1\mathbf{v}_1 + \alpha_2\mathbf{v}_2 + \cdots + \alpha_k\mathbf{v}_k + \alpha_{k+1}\mathbf{v}_{k+1} + \cdots + \alpha_n\mathbf{v}_n) \\ &= \alpha_1 T(\mathbf{v}_1) + \alpha_2 T(\mathbf{v}_2) + \cdots + \alpha_k T(\mathbf{v}_k) + \alpha_{k+1}T(\mathbf{v}_{k+1}) \\ &\qquad + \cdots + \alpha_n T(\mathbf{v}_n) \\ &= \alpha_1\mathbf{0} + \alpha_2\mathbf{0} + \cdots + \alpha_k\mathbf{0} + \alpha_{k+1}T(\mathbf{v}_{k+1}) + \cdots + \alpha_n T(\mathbf{v}_n) \end{aligned}$$

Since $T(\mathbf{v}) = \mathbf{w}$ we obtain the equation

$$\mathbf{w} = \alpha_{k+1}T(\mathbf{v}_{k+1}) + \cdots + \alpha_n T(\mathbf{v}_n) \qquad [7]$$

It follows from equation 7 that the set L spans the range of T.

In order to complete the proof we must show that L is linearly independent. Let

$$\beta_{k+1}T(\mathbf{v}_{k+1}) + \cdots + \beta_n T(\mathbf{v}_n) = \mathbf{0} \qquad [8]$$

Since T is linear we may write equation 8 as

$$T(\beta_{k+1}\mathbf{v}_{k+1} + \cdots + \beta_n\mathbf{v}_n) = \mathbf{0} \qquad [9]$$

It follows from equation 9 that the vector $\beta_{k+1}\mathbf{v}_{k+1} + \cdots + \beta_n\mathbf{v}_n$ is in the null space of T. Therefore it may be expressed as a linear combination of $\mathbf{v}_1, \mathbf{v}_2, \ldots, \mathbf{v}_k$, which form a basis for the null space of T. We have

$$\beta_{k+1}\mathbf{v}_{k+1} + \cdots + \beta_n\mathbf{v}_n = \gamma_1\mathbf{v}_1 + \gamma_2\mathbf{v}_2 + \cdots + \gamma_k\mathbf{v}_k \qquad [10]$$

But equation 10 may also be written as

$$-\gamma_1\mathbf{v}_1 - \gamma_2\mathbf{v}_2 \cdots - \gamma_k\mathbf{v}_k + \beta_{k+1}\mathbf{v}_{k+1} + \cdots + \beta_n\mathbf{v}_n = \mathbf{0} \qquad [11]$$

Since the set $\{\mathbf{v}_1, \mathbf{v}_2, \ldots, \mathbf{v}_n\}$ is linearly independent, it follows that $\gamma_1 = \gamma_2 = \cdots = \gamma_k = \beta_{k+1} = \cdots = \beta_n = 0$. Since all the β's are zero it follows from equation 8 that the set $L = \{T(\mathbf{v}_{k+1}), \ldots, T(\mathbf{v}_n)\}$ is linearly independent.

We conclude that the set $L = \{T(\mathbf{v}_{k+1}), \ldots, T(\mathbf{v}_n)\}$ is a basis for the range of T. Since L has $n - k$ vectors it follows that the rank of T is $n - k$. Thus we have

$$\text{nullity } (T) + \text{rank } (T) = n = \dim V$$

We have already seen that every linear transformation T from $\mathbf{R}^n$ into $\mathbf{R}^m$ has a matrix representation which is easily obtained by finding the effect of T on the standard basis of $\mathbf{R}^n$. We can actually say that the linear transformation T is completely determined when its effect on the standard basis of $\mathbf{R}^n$ is known.

The same is true with respect to any finite-dimensional vector space, as the following theorem indicates.

Theorem 6-10 Let V be an n-dimensional vector space and let $\{\mathbf{v}_1, \mathbf{v}_2, \ldots, \mathbf{v}_n\}$ be a basis for V. Let W be a vector space and let $\mathbf{w}_1, \mathbf{w}_2, \ldots, \mathbf{w}_n$ be any vectors in W. Then there is a unique linear transformation T from V into W such that

The $\mathbf{w}_i$'s do not have to be distinct.

$$T(\mathbf{v}_i) = \mathbf{w}_i \qquad (i = 1, 2, \ldots, n)$$

Proof Rather than give a complete proof, we shall only indicate how to construct the transformation T, leaving to the reader the task of proving the linearity and uniqueness of T as an exercise.

Let $\mathbf{v}$ be any vector in V. Thus $\mathbf{v}$ has a unique representation as a linear combination of the vectors in the basis $\{\mathbf{v}_1, \mathbf{v}_2, \ldots, \mathbf{v}_n\}$. Thus, there exist n real numbers $\alpha_1, \alpha_2, \ldots, \alpha_n$ such that

$$\mathbf{v} = \alpha_1\mathbf{v}_1 + \alpha_2\mathbf{v}_2 + \cdots + \alpha_n\mathbf{v}_n$$

For this vector $\mathbf{v}$ we define T by the formula

$$T(\mathbf{v}) = \alpha_1\mathbf{w}_1 + \alpha_2\mathbf{w}_2 + \cdots + \alpha_n\mathbf{w}_n \qquad [12]$$

It can be shown that equation 12 defines a linear transformation T such that

$$T(\mathbf{v}_i) = \mathbf{w}_i \qquad (i = 1, 2, \ldots, n) \qquad [13]$$

See problem 8.

and that T is the unique linear transformation which satisfies equation 13. These details are left as an exercise for the reader.

The following example illustrates the construction of a linear transformation given by its effect on a basis of a vector space.

Example 6 Find a linear transformation $T: \mathbf{R}^2 \longrightarrow \mathbf{R}^2$ such that

$$T\begin{bmatrix}1\\2\end{bmatrix} = \begin{bmatrix}1\\1\end{bmatrix}, \qquad T\begin{bmatrix}2\\1\end{bmatrix} = \begin{bmatrix}2\\3\end{bmatrix} \tag{14}$$

Solution The vectors $\begin{bmatrix}1\\2\end{bmatrix}$ and $\begin{bmatrix}2\\1\end{bmatrix}$ form a basis for $\mathbf{R}^2$. Thus, any vector $\begin{bmatrix}x\\y\end{bmatrix}$ of $\mathbf{R}^2$ has a unique representation having the form

$$\begin{bmatrix}x\\y\end{bmatrix} = \alpha\begin{bmatrix}1\\2\end{bmatrix} + \beta\begin{bmatrix}2\\1\end{bmatrix}$$

Solving for α and β, we find $\alpha = \dfrac{2y - x}{3}$ and $\beta = \dfrac{2x - y}{3}$. Thus, we have

$$\begin{bmatrix}x\\y\end{bmatrix} = \left(\frac{2y - x}{3}\right)\begin{bmatrix}1\\2\end{bmatrix} + \left(\frac{2x - y}{3}\right)\begin{bmatrix}2\\1\end{bmatrix}$$

We now define T by the formula

$$T\begin{bmatrix}x\\y\end{bmatrix} = \left(\frac{2y - x}{3}\right)T\begin{bmatrix}1\\2\end{bmatrix} + \left(\frac{2x - y}{3}\right)T\begin{bmatrix}2\\1\end{bmatrix} \tag{15}$$

Using equation 14, we may rewrite equation 15 as

$$T\begin{bmatrix}x\\y\end{bmatrix} = \left(\frac{2y - x}{3}\right)\begin{bmatrix}1\\1\end{bmatrix} + \left(\frac{2x - y}{3}\right)\begin{bmatrix}2\\3\end{bmatrix} \tag{16}$$

Simplifying equation 16, we obtain

$$T\begin{bmatrix}x\\y\end{bmatrix} = \begin{bmatrix}x\\ \dfrac{5x - y}{3}\end{bmatrix} \tag{17}$$

It follows from Theorem 6-10 that equation 17 defines the unique linear transformation T such that

$$T\begin{bmatrix}1\\2\end{bmatrix} = \begin{bmatrix}1\\1\end{bmatrix} \quad \text{and} \quad T\begin{bmatrix}2\\1\end{bmatrix} = \begin{bmatrix}2\\3\end{bmatrix}$$

Exercises

1 Let $T: \mathbf{R}^2 \longrightarrow \mathbf{R}^2$ be the linear transformation defined by $T\begin{bmatrix} x \\ y \end{bmatrix} = \begin{bmatrix} x - y \\ x - y \end{bmatrix}$.

a Describe the null space of T.

b Describe the range of T.

c Find the nullity of T and the rank of T.

2 Let $T: \mathbf{R}^3 \longrightarrow \mathbf{R}^3$ be the linear transformation defined by

$$T\begin{bmatrix} x \\ y \\ z \end{bmatrix} = \begin{bmatrix} 2x + 4y + 6z \\ 3x + y + 2z \\ y - z \end{bmatrix}.$$

a Describe the null space of T.

b Describe the range of T.

c Find the nullilty of T and the rank of T.

3 Let $T: \mathbf{R}^4 \longrightarrow \mathbf{R}^2$ be the linear transformation defined by

$$T\begin{bmatrix} x \\ y \\ z \\ w \end{bmatrix} = \begin{bmatrix} x + y + z + w \\ x - y - z + w \end{bmatrix}.$$

a Describe the null space of T.

b Describe the range of T.

c Find the nullity of T and the rank of T.

4 Let $T: P_1 \longrightarrow P_2$ be the transformation defined by $T(a_1 x + a_0) = x(a_1 x + a_0)$.

a Show that T is a linear transformation.

b Describe the null space of T.

c Describe the range of T.

d Find the nullity of T and the rank of T.

Hint: Use Theorem 6-9.

5 Let V be an n-dimensional vector space. Let $T: V \longrightarrow V$ be a linear transformation whose null space and range are identical. Show that n must be even.

6 Find a linear transformation $T: \mathbf{R}^2 \longrightarrow \mathbf{R}^2$ such that

$$T\begin{bmatrix} 1 \\ 1 \end{bmatrix} = \begin{bmatrix} 2 \\ 2 \end{bmatrix} \text{ and } T\begin{bmatrix} 2 \\ 1 \end{bmatrix} = \begin{bmatrix} 4 \\ 2 \end{bmatrix}.$$

7 Find a linear transformation $T: \mathbf{R}^3 \longrightarrow \mathbf{R}^2$ such that

$$T\begin{bmatrix} 1 \\ -3 \\ -1 \end{bmatrix} = \begin{bmatrix} 1 \\ 1 \end{bmatrix},\ T\begin{bmatrix} 0 \\ 4 \\ -2 \end{bmatrix} = \begin{bmatrix} 1 \\ 1 \end{bmatrix},\ \text{and } T\begin{bmatrix} 2 \\ 6 \\ 3 \end{bmatrix} = \begin{bmatrix} 1 \\ 1 \end{bmatrix}$$

8 Complete the proof of Theorem 6-10 by showing the linearity and uniqueness of T defined by equation 12.

9 (For students with a calculus background:) Let $D: P_2 \longrightarrow P_1$ be the differentiation transformation defined by $D(\mathbf{p}) = \mathbf{p}'$ for all $\mathbf{p} \in P_2$.
a Show that D is a linear transformation.
b Describe the null space of D.
c Describe the range of D.
d Find the nullity of D and the rank of D.

10 (For students with a calculus background:) Let $D^2: P_3 \longrightarrow P_1$ be the double differentiation transformation defined by $D^2(\mathbf{p}) = \mathbf{p}''$ for all $\mathbf{p} \in P_3$.
a Show that D^2 is a linear transformation.
b Describe the null space of D^2.
c Describe the range of D^2.
d Find the nullity of D^2 and the rank of D^2.

6-5 Change of Basis

Objectives
1 Introduce the concept of a coordinate vector with respect to an ordered basis.
2 Find how coordinate vectors transform under a change of basis.

Let V be an n-dimensional vector space. This means that V has a basis containing n vectors (see Definition 3-7). Until now we have disregarded the order in which we listed the elements of any such basis. It becomes necessary in this section to talk about ordered bases. Thus, if we view a basis $B = \{\mathbf{v}_1, \mathbf{v}_2, \ldots, \mathbf{v}_n\}$ as an ordered set having a first element $\mathbf{v}_1$, a second element $\mathbf{v}_2$, and so on, then the basis $B_1 = \{\mathbf{v}_2, \mathbf{v}_1, \ldots, \mathbf{v}_n\}$ obtained from B by interchanging the position of $\mathbf{v}_1$ and $\mathbf{v}_2$ is different from the basis B, despite the fact that B and B_1 contain the same n vectors. Therefore, whenever we talk about an ordered basis for a vector space, we must take into account not only the vectors contained in the basis but also the order in which they are listed. This new approach will make it possible for us to introduce the concept of a coordinate vector with respect to an ordered basis.

Let $B = \{\mathbf{v}_1, \mathbf{v}_2, \ldots, \mathbf{v}_n\}$ be an ordered basis for an n-dimensional vector space V. We know that every vector $\mathbf{v}$ in V has a unique representation in terms of the vectors in the basis B (see Theorem 3-5). This means that there exist n real numbers $\alpha_1, \alpha_2, \ldots, \alpha_n$ such that

$$\mathbf{v} = \alpha_1\mathbf{v}_1 + \alpha_2\mathbf{v}_2 + \cdots + \alpha_n\mathbf{v}_n \qquad [1]$$

and the representation of $\mathbf{v}$ by equation 1 is unique. Thus, the real numbers $\alpha_1, \alpha_2, \ldots, \alpha_n$ determine the vector $\mathbf{v}$ completely. We refer to

$$[\mathbf{v}]_B = \begin{bmatrix} \alpha_1 \\ \alpha_2 \\ \vdots \\ \alpha_n \end{bmatrix}$$

as the **coordinate vector** of $\mathbf{v}$ with respect to the ordered basis B. The entries of $[\mathbf{v}]_B$ are called the *coordinates* of $\mathbf{v}$ with respect to B. The following examples illustrate the concept of a coordinate vector.

Example 1 Let P_2 be the vector space of all polynomials of degree ≤ 2 and the zero polynomial. Let $B = \{x^2, x, 1\}$ and $B' = \{x^2 - 2x,\ x + 1,\ x - 1\}$ be two ordered bases for P_2. Find the coordinate vector of the polynomial $\mathbf{p} = 2x^2 + 4x + 3$ with respect to each of the bases B and B'.

See Example 4 in Section 3-3.

Solution The coordinate vector of the polynomial $\mathbf{p}$ with respect to the ordered basis $B = \{x^2, x, 1\}$ is obtained easily from the representation of $\mathbf{p}$ given by

$$\mathbf{p} = (2)x^2 + (4)x + (3)1 \qquad [2]$$

Thus, we have from equation 2

$$[\mathbf{p}]_B = \begin{bmatrix} 2 \\ 4 \\ 3 \end{bmatrix}$$

In order to find the coordinate vector of the polynomial $\mathbf{p}$ with respect to the ordered basis $B' = \{x^2 - 2x, x + 1, x - 1\}$, we must find the representation of $\mathbf{p}$ as a linear combination of the elements in B'. If we set

$$2x^2 + 4x + 3 = \alpha(x^2 - 2x) + \beta(x + 1) + \gamma(x - 1)$$

then we find that $\alpha = 2$, $\beta = \frac{11}{2}$, and $\gamma = \frac{5}{2}$. Thus, we have

$$\mathbf{p} = (2)(x^2 - 2x) + (\tfrac{11}{2})(x + 1) + (\tfrac{5}{2})(x - 1) \qquad [3]$$

It follows from equation 3 that the coordinate vector of $\mathbf{p}$ with respect to the ordered basis $B' = \{x^2 - 2x, x + 1, x - 1\}$ is given by

$$[\mathbf{p}]_{B'} = \begin{bmatrix} 2 \\ \frac{11}{2} \\ \frac{5}{2} \end{bmatrix}$$

Example 2 Let $B = \left\{\begin{bmatrix} 1 \\ 0 \end{bmatrix}, \begin{bmatrix} 0 \\ 1 \end{bmatrix}\right\}$ and $B' = \left\{\begin{bmatrix} 1 \\ 1 \end{bmatrix}, \begin{bmatrix} 1 \\ 2 \end{bmatrix}\right\}$ be two ordered bases for $\mathbf{R}^2$. Find the coordinate vector of $\mathbf{v} = \begin{bmatrix} 5 \\ 7 \end{bmatrix}$ with respect to each of the bases B and B'.

Solution The representation of $\mathbf{v} = \begin{bmatrix} 5 \\ 7 \end{bmatrix}$ in terms of the ordered basis B is obviously given by

$$\mathbf{v} = 5\begin{bmatrix} 1 \\ 0 \end{bmatrix} + 7\begin{bmatrix} 0 \\ 1 \end{bmatrix} \quad \textbf{[4]}$$

Thus, it follows from equation 4 that we have

$$[\mathbf{v}]_B = \begin{bmatrix} 5 \\ 7 \end{bmatrix}$$

To find the representation of $\mathbf{v} = \begin{bmatrix} 5 \\ 7 \end{bmatrix}$ in terms of the ordered basis B', we write $\mathbf{v}$ as a linear combination of the vectors $\begin{bmatrix} 1 \\ 1 \end{bmatrix}$ and $\begin{bmatrix} 1 \\ 2 \end{bmatrix}$. Thus, we have

$$\begin{bmatrix} 5 \\ 7 \end{bmatrix} = \alpha\begin{bmatrix} 1 \\ 1 \end{bmatrix} + \beta\begin{bmatrix} 1 \\ 2 \end{bmatrix} \quad \textbf{[5]}$$

Solving equation 5 for α and β, we find $\alpha = 3$ and $\beta = 2$. Therefore, the representation of $\mathbf{v}$ in terms of the vectors $\begin{bmatrix} 1 \\ 1 \end{bmatrix}$ and $\begin{bmatrix} 1 \\ 2 \end{bmatrix}$ is given by

$$\mathbf{v} = 3\begin{bmatrix} 1 \\ 1 \end{bmatrix} + 2\begin{bmatrix} 1 \\ 2 \end{bmatrix} \quad \textbf{[6]}$$

Thus, it follows from equation 6 that we have

$$[\mathbf{v}]_{B'} = \begin{bmatrix} 3 \\ 2 \end{bmatrix}$$

We have seen in Examples 1 and 2 that when using different bases B and B' for the vector space under discussion, we have obtained different coordinate vectors for a given vector. We now wish to establish a relationship between any two such coordinate vectors. For the sake of simplicity let us consider a 2-dimensional vector space V. Let $B = \{\mathbf{v}_1,\mathbf{v}_2\}$ and $B' = \{\mathbf{u}_1,\mathbf{u}_2\}$ be the two different ordered bases for V, and let $\mathbf{v}$ be any vector in V. Using the basis B we may write

$$\mathbf{v} = \alpha_1\mathbf{v}_1 + \alpha_2\mathbf{v}_2 \tag{7}$$

Thus, we have $[\mathbf{v}]_B = \begin{bmatrix}\alpha_1\\ \alpha_2\end{bmatrix}$.

Since B' is also an ordered basis for V we may express $\mathbf{v}_1$ and $\mathbf{v}_2$ as linear combinations of $\mathbf{u}_1$ and $\mathbf{u}_2$. Thus, there exist scalars a_{11}, a_{21}, a_{12}, and a_{22} such that

$$\mathbf{v}_1 = a_{11}\mathbf{u}_1 + a_{21}\mathbf{u}_2, \qquad \mathbf{v}_2 = a_{12}\mathbf{u}_1 + a_{22}\mathbf{u}_2 \tag{8}$$

Substituting equation 8 into equation 7, we obtain

$$\mathbf{v} = \alpha_1(a_{11}\mathbf{u}_1 + a_{21}\mathbf{u}_2) + \alpha_2(a_{12}\mathbf{u}_1 + a_{22}\mathbf{u}_2) \tag{9}$$

Computing the coefficients of $\mathbf{u}_1$ and $\mathbf{u}_2$ in equation 9, we get

$$\mathbf{v} = (\alpha_1 a_{11} + \alpha_2 a_{12})\mathbf{u}_1 + (\alpha_1 a_{21} + \alpha_2 a_{22})\mathbf{u}_2 \tag{10}$$

It follows from equation 10 that the coordinate vector of $\mathbf{v}$ with respect to the ordered basis $B' = \{\mathbf{u}_1,\mathbf{u}_2\}$ is given by

$$[\mathbf{v}]_{B'} = \begin{bmatrix}\alpha_1 a_{11} + \alpha_2 a_{12}\\ \alpha_1 a_{21} + \alpha_2 a_{22}\end{bmatrix} \tag{11}$$

A simple observation shows that we can also write equation 11 as

$$[\mathbf{v}]_{B'} = \begin{bmatrix}a_{11} & a_{12}\\ a_{21} & a_{22}\end{bmatrix}\begin{bmatrix}\alpha_1\\ \alpha_2\end{bmatrix} \tag{12}$$

Thus, the coordinate vector of $\mathbf{v}$ with respect to the ordered basis B' is obtained from $[\mathbf{v}]_B = \begin{bmatrix}\alpha_1\\ \alpha_2\end{bmatrix}$ by multiplying the coordinate vector $[\mathbf{v}]_B$ from the left by the matrix $\begin{bmatrix}a_{11} & a_{12}\\ a_{21} & a_{22}\end{bmatrix}$. This matrix is called the *transition matrix* from the ordered basis $B = \{\mathbf{v}_1,\mathbf{v}_2\}$ to the ordered basis $B' = \{\mathbf{u}_1,\mathbf{u}_2\}$. The matrix has a very simple structure. The first column $\begin{bmatrix}a_{11}\\ a_{21}\end{bmatrix}$ of the matrix is actually $[\mathbf{v}_1]_{B'}$ while the second column $\begin{bmatrix}a_{12}\\ a_{22}\end{bmatrix}$ of the matrix is $[\mathbf{v}_2]_{B'}$.

This leads to the following general definition of the transition matrix from an ordered basis B to an ordered basis B':

Definition 6-7 Let $B = \{\mathbf{v}_1, \mathbf{v}_2, \ldots, \mathbf{v}_n\}$ and $B' = \{\mathbf{u}_1, \mathbf{u}_2, \ldots, \mathbf{u}_n\}$ be ordered bases for an n-dimensional vector space V. The **transition matrix** from B

to B' is the $n \times n$ matrix

$$A = \begin{bmatrix} a_{11} & a_{12} & \cdots & a_{1n} \\ a_{21} & a_{22} & \cdots & a_{2n} \\ \vdots & \vdots & & \vdots \\ a_{n1} & a_{n2} & \cdots & a_{nn} \end{bmatrix}$$

where

$$\begin{bmatrix} a_{11} \\ a_{21} \\ \vdots \\ a_{n1} \end{bmatrix} = [\mathbf{v}_1]_{B'}, \qquad \begin{bmatrix} a_{12} \\ a_{22} \\ \vdots \\ a_{n2} \end{bmatrix} = [\mathbf{v}_2]_{B'}, \ldots, \qquad \begin{bmatrix} a_{1n} \\ a_{2n} \\ \vdots \\ a_{nn} \end{bmatrix} = [\mathbf{v}_n]_{B'}$$

It follows from Definition 6-7 that in order to construct the transition matrix from an ordered basis B to an ordered basis B' we must express each $\mathbf{v}_i$ in B as a linear combination of the $\mathbf{u}$'s in B' and use the coefficients for building up the columns of the matrix. Thus, if $\mathbf{v}_i = a_{1i}\mathbf{u}_1 + a_{2i}\mathbf{u}_2 + \cdots + a_{ni}\mathbf{u}_n$ then the ith column of the transition matrix A is given by

$$[\mathbf{v}_i]_{B'} = \begin{bmatrix} a_{1i} \\ a_{2i} \\ \vdots \\ a_{ni} \end{bmatrix}$$

Example 3 Let $B = \left\{\begin{bmatrix} 1 \\ 0 \end{bmatrix}, \begin{bmatrix} 0 \\ 1 \end{bmatrix}\right\}$ and $B' = \left\{\begin{bmatrix} 1 \\ 1 \end{bmatrix}, \begin{bmatrix} 1 \\ 2 \end{bmatrix}\right\}$ be two ordered bases for $\mathbf{R}^2$. Find the transition matrix A from B to B' and show that if $\mathbf{v} = \begin{bmatrix} 5 \\ 7 \end{bmatrix}$ then $[\mathbf{v}]_{B'} = A[\mathbf{v}]_B$.

Use results from Example 2.

Solution In order to obtain the first column of A we must find the coordinate vector of $\begin{bmatrix} 1 \\ 0 \end{bmatrix}$ with respect to the ordered basis B'. If we write

$$\begin{bmatrix} 1 \\ 0 \end{bmatrix} = \alpha_1 \begin{bmatrix} 1 \\ 1 \end{bmatrix} + \alpha_2 \begin{bmatrix} 1 \\ 2 \end{bmatrix}$$

then we find $\alpha_1 = 2$ and $\alpha_2 = -1$. Thus, the first column of A is given by

$$\begin{bmatrix} a_{11} \\ a_{21} \end{bmatrix} = \begin{bmatrix} 2 \\ -1 \end{bmatrix}$$

In order to obtain the second column of A we must find the coordinate vector of $\begin{bmatrix} 0 \\ 1 \end{bmatrix}$ with respect to the ordered basis B'. If we write

$$\begin{bmatrix}0\\1\end{bmatrix} = \beta_1\begin{bmatrix}1\\1\end{bmatrix} + \beta_2\begin{bmatrix}1\\2\end{bmatrix}$$

then we find $\beta_1 = -1$ and $\beta_2 = 1$. Thus, the second column of A is given by

$$\begin{bmatrix}a_{12}\\a_{22}\end{bmatrix} = \begin{bmatrix}-1\\1\end{bmatrix}$$

The transition matrix A is therefore given by

$$A = \begin{bmatrix}2 & -1\\-1 & 1\end{bmatrix}$$

We have seen in Example 2 that the coordinate vectors of $\mathbf{v} = \begin{bmatrix}5\\7\end{bmatrix}$ with respect to the ordered bases B and B' were given by $[\mathbf{v}]_B = \begin{bmatrix}5\\7\end{bmatrix}$ and $[\mathbf{v}]_{B'} = \begin{bmatrix}3\\2\end{bmatrix}$. If we now multiply $[\mathbf{v}]_B$ from the left by the matrix $A = \begin{bmatrix}2 & -1\\-1 & 1\end{bmatrix}$ we obtain

$$A[\mathbf{v}]_B = \begin{bmatrix}2 & -1\\-1 & 1\end{bmatrix}\begin{bmatrix}5\\7\end{bmatrix} = \begin{bmatrix}3\\2\end{bmatrix} = [\mathbf{v}]_{B'}$$

This establishes the equation

$$[\mathbf{v}]_{B'} = A[\mathbf{v}]_B$$

for the vector $\mathbf{v} = \begin{bmatrix}5\\7\end{bmatrix}$.

See Example 1.

Example 4 Let $B = \{x^2,x,1\}$ and $B' = \{x^2 - 2x, x + 1, x - 1\}$ be two ordered bases for P_2. Find the transition matrix A from B to B' and show that if $\mathbf{p} = 2x^2 + 4x + 3$ then $[\mathbf{p}]_{B'} = A[\mathbf{p}]_B$.

Solution We must find the coordinate vectors of x^2, x, and 1 with respect to the ordered basis B'. If we write

$$x^2 = \alpha_1(x^2 - 2x) + \alpha_2(x + 1) + \alpha_3(x - 1) \qquad [13]$$

then we find $\alpha_1 = 1$, $\alpha_2 = 1$, and $\alpha_3 = 1$. If we write

$$x = \beta_1(x^2 - 2x) + \beta_2(x + 1) + \beta_3(x - 1) \qquad [14]$$

then we find $\beta_1 = 0$, $\beta_2 = \frac{1}{2}$, and $\beta_3 = \frac{1}{2}$. If we write

$$1 = \gamma_1(x^2 - 2x) + \gamma_2(x + 1) + \gamma_3(x - 1) \qquad [15]$$

then we find $\gamma_1 = 0$, $\gamma_2 = \frac{1}{2}$, and $\gamma_3 = -\frac{1}{2}$.

Thus, it follows from equations 13, 14, and 15 that we have

$$[\mathbf{x}^2]_{B'} = \begin{bmatrix} 1 \\ 1 \\ 1 \end{bmatrix}, \qquad [\mathbf{x}]_{B'} = \begin{bmatrix} 0 \\ \frac{1}{2} \\ \frac{1}{2} \end{bmatrix}, \qquad [\mathbf{1}]_{B'} = \begin{bmatrix} 0 \\ \frac{1}{2} \\ -\frac{1}{2} \end{bmatrix}$$

and therefore the transition matrix A from the ordered basis B to the ordered basis B' is given by

$$A = \begin{bmatrix} 1 & 0 & 0 \\ 1 & \frac{1}{2} & \frac{1}{2} \\ 1 & \frac{1}{2} & -\frac{1}{2} \end{bmatrix}$$

We have seen in Example 1 that the coordinate vectors of $\mathbf{p} = 2x^2 + 4x + 3$ with respect to the ordered bases B and B' were given by $[\mathbf{p}]_B = \begin{bmatrix} 2 \\ 4 \\ 3 \end{bmatrix}$ and $[\mathbf{p}]_{B'} = \begin{bmatrix} 2 \\ \frac{11}{2} \\ \frac{5}{2} \end{bmatrix}$. If we now multiply $[\mathbf{p}]_B$ from the left by the matrix $A = \begin{bmatrix} 1 & 0 & 0 \\ 1 & \frac{1}{2} & \frac{1}{2} \\ 1 & \frac{1}{2} & -\frac{1}{2} \end{bmatrix}$ we obtain

$$A[\mathbf{p}]_B = \begin{bmatrix} 1 & 0 & 0 \\ 1 & \frac{1}{2} & \frac{1}{2} \\ 1 & \frac{1}{2} & -\frac{1}{2} \end{bmatrix} \begin{bmatrix} 2 \\ 4 \\ 3 \end{bmatrix} = \begin{bmatrix} 2 \\ \frac{11}{2} \\ \frac{5}{2} \end{bmatrix} = [\mathbf{p}]_{B'}$$

This establishes the equation

$$[\mathbf{p}]_{B'} = A[\mathbf{p}]_B$$

for the polynomial $\mathbf{p} = 2x^2 + 4x + 3$.

We have the following general result:

Theorem 6-11 Let A be the transition matrix from an ordered basis B to an ordered basis B' for an n-dimensional vector space V. Let $\mathbf{v}$ be any vector in V; then the coordinate vectors $[\mathbf{v}]_B$ and $[\mathbf{v}]_{B'}$ satisfy the equation

$$[\mathbf{v}]_{B'} = A[\mathbf{v}]_B$$

See problem 13.

Earlier in this section we showed the validity of Theorem 6-11 for the case of a 2-dimensional vector space V. The general proof is similar in nature and is left as an exercise for the diligent student.

In Theorem 6-11 we dealt with the transition matrix A from an ordered basis B to an ordered basis B'. Suppose we reverse the roles of B and B' and deal with the transition matrix from B' to B.

Let us denote this transition matrix by A'. A natural question arises. Are the matrices A and A' related, and if so, what is the nature of this relationship? The answer to this question is given in the following theorem.

Theorem 6-12 Let A be the transition matrix from an ordered basis B to an ordered basis B' of an n-dimensional vector space V. Let A' be the transition matrix from B' to B. Then the matrices A and A' are invertible and satisfy the relation

$$A' = A^{-1} \qquad [16]$$

Proof The transition matrices A and A' are square matrices of order n. Therefore, to establish equation 16 we must show that $A'A = I$.

It follows from Theorem 6-11 that we have

$$[\mathbf{v}]_{B'} = A[\mathbf{v}]_B \qquad [17]$$

for all $\mathbf{v}$ in V.
Exchanging the roles of B and B' in Theorem 6-11, we get

$$[\mathbf{v}]_B = A'[\mathbf{v}]_{B'} \qquad [18]$$

for all $\mathbf{v}$ in V. Substituting for $[\mathbf{v}]_{B'}$ in equation 18 the expression $A[\mathbf{v}]_B$ from equation 17, we obtain

$$[\mathbf{v}]_B = A'A[\mathbf{v}]_B \qquad [19]$$

for all $\mathbf{v}$ in V.

I is the identity matrix of order n.

Since we always have $[\mathbf{v}]_B = I[\mathbf{v}]_B$, we may rewrite equation 19 in the form

$$(A'A - I)[\mathbf{v}]_B = \mathbf{0} \qquad [20]$$

The reader should note that equation 20 holds for all $\mathbf{v}$ in V. Now, let the ordered basis B be given by $B = \{\mathbf{v}_1, \mathbf{v}_2, \ldots, \mathbf{v}_n\}$.

$\mathbf{v}_1 = 1\mathbf{v}_1 + 0\mathbf{v}_2 + \cdots + 0\mathbf{v}_n$

Choosing $\mathbf{v} = \mathbf{v}_1$, we get $[\mathbf{v}]_B = [\mathbf{v}_1]_B = \begin{bmatrix} 1 \\ 0 \\ \vdots \\ 0 \end{bmatrix}$. Using this result in equation 20, we obtain

$$(A'A - I)[\mathbf{v}_1]_B = (A'A - I)\begin{bmatrix} 1 \\ 0 \\ \vdots \\ 0 \end{bmatrix} = \begin{bmatrix} 0 \\ 0 \\ \vdots \\ 0 \end{bmatrix} \qquad [21]$$

The expression $(A'A - I)\begin{bmatrix}1\\0\\\vdots\\0\end{bmatrix}$ may be interpreted as the linear combination of the columns of the matrix $A'A - I$ with the coefficients $1, 0, \ldots, 0$. Thus, it follows that

$(A'A - I)\begin{bmatrix}1\\0\\\vdots\\0\end{bmatrix}$ is actually the first column of the matrix $A'A - I$.

We conclude from equation 21 that this column consists of zeros. In similar fashion, using the vectors $\mathbf{v}_2, \mathbf{v}_3, \ldots, \mathbf{v}_n$, we find that the remaining columns of $A'A - I$ are all zero columns. The fact that $A'A - I$ is the zero matrix leads to the equation

$$A'A = I \qquad [22]$$

It follows from equation 22 that $A' = A^{-1}$ and thus equation 16 is proved.

Exercises

1 Let P_2 be the vector space of all polynomials of degree ≤ 2 and the zero polynomial. Find the coordinate vector of the polynomial $\mathbf{p} = 3x^2 - 3x + 1$ with respect to each of the ordered bases $B = \{x^2, x, 1\}$ and $B' = \{x^2 - 2x, x + 1, x - 1\}$ of P_2.

2 Let A be the transition matrix from the ordered basis B to the ordered basis B' where B and B' are the bases from problem1. Let A' be the transition matrix from B' to B. Find the matrices A and A' and show that $A'A = I$.

This shows that $A' = A^{-1}$.

3 Use your results from problems 1 and 2 to verify the following equations for $\mathbf{p} = 3x^2 - 3x + 1$.

a $[\mathbf{p}]_{B'} = A[\mathbf{p}]_B$ **b** $[\mathbf{p}]_B = A'[\mathbf{p}]_{B'}$

4 Consider the ordered bases $B = \left\{\begin{bmatrix}1\\0\end{bmatrix}, \begin{bmatrix}1\\1\end{bmatrix}\right\}$ and $B' = \left\{\begin{bmatrix}1\\2\end{bmatrix}, \begin{bmatrix}2\\1\end{bmatrix}\right\}$ for $\mathbf{R}^2$. Find the coordinate vector of $\mathbf{v} = \begin{bmatrix}-2\\1\end{bmatrix}$ with respect to each of the bases B and B' of $\mathbf{R}^2$.

5 **a** Find the transition matrix A from the ordered basis B to the ordered basis B' where B and B' are the bases of $\mathbf{R}^2$ from problem 4.

b Find the transition matrix A' from the ordered basis B' to the ordered basis B.

c Show that $A'A = I$.

6 Use your results from problems 4 and 5 to show that the coor-

dinate vectors $[\mathbf{v}]_B$ and $[\mathbf{v}]_{B'}$ of $\mathbf{u} = \begin{bmatrix} -2 \\ 1 \end{bmatrix}$ satisfy the equations $[\mathbf{v}]_{B'} = A[\mathbf{v}]_B$ and $[\mathbf{v}]_B = A'[\mathbf{v}]_{B'}$.

7 Let P_1 be the vector space of all polynomials of degree ≤ 1 and the zero polynomial. Find the coordinate vector of the polynomial $\mathbf{p} = 5x + 2$ with respect to each of the ordered bases $B = \{x + 1, x - 1\}$ and $B' = \{2x + 1, x - 1\}$ of P_1.

8 **a** Find the transition matrix A from the ordered basis B to the ordered basis B' where B and B' are the bases of P_1 from problem 7.

b Find the transition matrix A' from the ordered basis B' to the ordered basis B.

c Show that $A'A = I$.

9 Use your results from problems 7 and 8 to show that the coordinate vectors $[\mathbf{p}]_B$ and $[\mathbf{p}]_{B'}$ of $\mathbf{p} = 5x + 2$ satisfy the equations $[\mathbf{p}]_{B'} = A[\mathbf{p}]_B$ and $[\mathbf{p}]_B = A'[\mathbf{p}]_{B'}$.

10 Consider the ordered bases $B = \{\mathbf{v}_1, \mathbf{v}_2, \mathbf{v}_3\}$ and $B' = \{\mathbf{u}_1, \mathbf{u}_2, \mathbf{u}_3\}$ for $\mathbf{R}^3$ where $\mathbf{v}_1 = \begin{bmatrix} 1 \\ 0 \\ 0 \end{bmatrix}$, $\mathbf{v}_2 = \begin{bmatrix} 1 \\ 1 \\ 0 \end{bmatrix}$, $\mathbf{v}_3 = \begin{bmatrix} 1 \\ 1 \\ 1 \end{bmatrix}$, $\mathbf{u}_1 = \begin{bmatrix} 0 \\ 0 \\ 1 \end{bmatrix}$, $\mathbf{u}_2 = \begin{bmatrix} 0 \\ 1 \\ 1 \end{bmatrix}$, and $\mathbf{u}_3 = \begin{bmatrix} 1 \\ 0 \\ 1 \end{bmatrix}$. Find the coordinate vector of $\mathbf{v} = \begin{bmatrix} 1 \\ 2 \\ 3 \end{bmatrix}$ with respect to each of the bases B and B'.

11 **a** Find the transition matrix A from B to B' where B and B' are the bases of $\mathbf{R}^3$ from problem 10.

b Find the transition matrix A' from B' to B.

c Show that $A'A = I$.

12 Use your results from problems 10 and 11 to show that the coordinate vectors $[\mathbf{v}]_B$ and $[\mathbf{v}]_{B'}$ of $\mathbf{v} = \begin{bmatrix} 1 \\ 2 \\ 3 \end{bmatrix}$ satisfy the equations $[\mathbf{v}]_{B'} = A[\mathbf{v}]_B$ and $[\mathbf{v}]_B = A'[\mathbf{v}]_{B'}$.

13 Prove Theorem 6-11.

6-6 More on Matrix Representation of a Linear Transformation

We have seen in Section 6-2 that any linear transformation T from $\mathbf{R}^n$ into $\mathbf{R}^m$ has a matrix representation of the form

$$T(\mathbf{v}) = A\mathbf{v} \tag{1}$$

for all $\mathbf{v}$ in $\mathbf{R}^n$.

Objectives

1. Introduce matrix representations with respect to given ordered bases of finite-dimensional vector spaces.
2. Construct matrices of linear operators defined on finite-dimensional vector spaces.

In this section we generalize the concept of the matrix representation of a linear transformation and show that any linear transformation T from an n-dimensional vector space V into an m-dimensional vector space W can be regarded as a matrix transformation. The generalization is achieved in the following way.

Let $B = \{\mathbf{v}_1, \mathbf{v}_2, \ldots, \mathbf{v}_n\}$ and $B' = \{\mathbf{u}_1, \mathbf{u}_2, \ldots, \mathbf{u}_m\}$ be ordered bases for the vector spaces V and W, respectively. If T is a linear transformation from V into W, then $T(\mathbf{v})$ is a vector in W for every vector $\mathbf{v}$ in V. Guided by our results from Section 6-2 (see Theorem 6-2), we now proceed to relate the coordinate vectors $[\mathbf{v}]_B$ and $[T(\mathbf{v})]_{B'}$. This means that we now try to find a matrix A of order $m \times n$ such that

$$[T(\mathbf{v})]_{B'} = A[\mathbf{v}]_B \tag{2}$$

for all $\mathbf{v}$ in V.

The reader should note that if $V = \mathbf{R}^n$, $W = \mathbf{R}^m$ and the ordered bases B and B' are taken to be the standard bases for $\mathbf{R}^n$ and $\mathbf{R}^m$, respectively, then equation 2 reduces to equation 1. The reader should also note that a matrix A satisfying equation 2 will necessarily depend on the two bases B and B'. Thus, if we change these bases, we should expect the matrix A satisfying equation 2 to change also. In order to determine the entries of the matrix A, we substitute successively for $\mathbf{v}$ in equation 2 the vectors $\mathbf{v}_1, \mathbf{v}_2, \ldots, \mathbf{v}_n$ from the ordered basis B of V. Thus, for $\mathbf{v} = \mathbf{v}_1$ we obtain

$$[\mathbf{v}_1]_B = \begin{bmatrix} 1 \\ 0 \\ \vdots \\ 0 \end{bmatrix}$$

and equation 2 reduces to

$$[T(\mathbf{v}_1)]_{B'} = A \begin{bmatrix} 1 \\ 0 \\ \vdots \\ 0 \end{bmatrix} \tag{3}$$

Since $A \begin{bmatrix} 1 \\ 0 \\ \vdots \\ 0 \end{bmatrix}$ represents the first column of A, we may write equation 3 in the form

$$[T(\mathbf{v}_1)]_{B'} = \begin{bmatrix} a_{11} \\ a_{21} \\ \vdots \\ a_{m1} \end{bmatrix}$$

Now choose $\mathbf{v} = \mathbf{v}_2$. Using the fact that $[\mathbf{v}_2]_B = \begin{bmatrix} 0 \\ 1 \\ 0 \\ \vdots \\ 0 \end{bmatrix}$ we obtain from equation 2

$$[T(\mathbf{v}_2)]_{B'} = A \begin{bmatrix} 0 \\ 1 \\ 0 \\ \vdots \\ 0 \end{bmatrix} \tag{4}$$

Since $A \begin{bmatrix} 0 \\ 1 \\ 0 \\ \vdots \\ 0 \end{bmatrix}$ represents the second column of A, we may write equation 4 in the form

$$[T(\mathbf{v}_2)]_{B'} = \begin{bmatrix} a_{12} \\ a_{22} \\ \vdots \\ a_{m2} \end{bmatrix}$$

Proceeding in a similar fashion, we obtain the following general result: The ith column of the matrix A in equation 2 is given by

$$\begin{bmatrix} a_{1i} \\ a_{2i} \\ \vdots \\ a_{mi} \end{bmatrix} = [T(\mathbf{v}_i)]_{B'} \tag{5}$$

Thus, in order to construct the ith column of the matrix in equation 3 we proceed as follows:

a Find the image of $\mathbf{v}_i$ from the basis B under T.

b Find the coordinate vector of $T(\mathbf{v}_i)$ with respect to the basis B' of W.

It follows that equation 2 is satisfied for all the n vectors in the basis $B = \{\mathbf{v}_1, \mathbf{v}_2, \ldots, \mathbf{v}_n\}$ if the columns of the matrix A are determined by equation 5. Using the linearity of T and the linearity of the matrix A one can easily show that equation 2 holds for all $\mathbf{v}$ in V with the matrix A determined by 5.

We now formalize the role of the matrix A in the following:

Definition 6-8 Let $B = \{\mathbf{v}_1, \mathbf{v}_2, \ldots, \mathbf{v}_n\}$ be an ordered basis for an n-dimensional vector space V and let $B' = \{\mathbf{u}_1, \mathbf{u}_2, \ldots, \mathbf{u}_m\}$ be an ordered basis for an m-dimensional vector space W. Let T be a linear transformation from V into W. The matrix of T with respect to the bases B and B' is the $m \times n$ matrix A whose ith column is defined by

$$\begin{bmatrix} a_{1i} \\ a_{2i} \\ \vdots \\ a_{mi} \end{bmatrix} = [T(\mathbf{v}_i)]_{B'}$$

Example 1 Let $T: \mathbf{R}^2 \longrightarrow \mathbf{R}^3$ be the linear transformation defined by

Example 1 will clarify how one constructs the matrix of a linear transformation with respect to two given bases B and B' of vector spaces V and W, respectively.

$$T\begin{bmatrix} x \\ y \end{bmatrix} = \begin{bmatrix} x - y \\ x + y \\ y \end{bmatrix} \quad [6]$$

Let $B = \{\mathbf{v}_1, \mathbf{v}_2\}$ and $B' = \{\mathbf{u}_1, \mathbf{u}_2, \mathbf{u}_3\}$ be ordered bases for $\mathbf{R}^2$ and $\mathbf{R}^3$, respectively, given by

$$\mathbf{v}_1 = \begin{bmatrix} 1 \\ 1 \end{bmatrix} \quad \mathbf{v}_2 = \begin{bmatrix} 1 \\ 2 \end{bmatrix} \quad \mathbf{u}_1 = \begin{bmatrix} 1 \\ 0 \\ 0 \end{bmatrix} \quad \mathbf{u}_2 = \begin{bmatrix} 1 \\ 1 \\ 0 \end{bmatrix} \quad \mathbf{u}_3 = \begin{bmatrix} 1 \\ 1 \\ 1 \end{bmatrix}$$

Find the matrix of T with respect to the bases B and B'.

Solution The columns of the desired matrix A are determined by using equation 5.

It follows from equation 6 that we have

$$T(\mathbf{v}_1) = \begin{bmatrix} 0 \\ 2 \\ 1 \end{bmatrix} \qquad T(\mathbf{v}_2) = \begin{bmatrix} -1 \\ 3 \\ 2 \end{bmatrix}$$

We must now find the coordinate vectors of $T(\mathbf{v}_1)$ and $T(\mathbf{v}_2)$ with respect to the basis $B' = \{\mathbf{u}_1, \mathbf{u}_2, \mathbf{u}_3\}$. Expressing $T(\mathbf{v}_1)$ as a linear combination of $\mathbf{u}_1$, $\mathbf{u}_2$, and $\mathbf{u}_3$, we obtain

$$T(\mathbf{v}_1) = \begin{bmatrix} 0 \\ 2 \\ 1 \end{bmatrix} = \alpha_1 \begin{bmatrix} 1 \\ 0 \\ 0 \end{bmatrix} + \alpha_2 \begin{bmatrix} 1 \\ 1 \\ 0 \end{bmatrix} + \alpha_3 \begin{bmatrix} 1 \\ 1 \\ 1 \end{bmatrix} \quad [7]$$

The vector equation 7 leads to the system of equations

$$\begin{aligned} \alpha_1 + \alpha_2 + \alpha_3 &= 0 \\ \alpha_2 + \alpha_3 &= 2 \\ \alpha_3 &= 1 \end{aligned}$$

Solving this system, we obtain

$$\alpha_1 = -2 \qquad \alpha_2 = 1 \qquad \alpha_3 = 1$$

Thus, the first column of A is given by

$$\begin{bmatrix} a_{11} \\ a_{21} \\ a_{31} \end{bmatrix} = [T(\mathbf{v}_1)]_{B'} = \begin{bmatrix} -2 \\ 1 \\ 1 \end{bmatrix} \tag{8}$$

In order to determine the second column of A, we write

$$T(\mathbf{v}_2) = \begin{bmatrix} -1 \\ 3 \\ 2 \end{bmatrix} = \beta_1 \begin{bmatrix} 1 \\ 0 \\ 0 \end{bmatrix} + \beta_2 \begin{bmatrix} 1 \\ 1 \\ 0 \end{bmatrix} + \beta_3 \begin{bmatrix} 1 \\ 1 \\ 1 \end{bmatrix} \tag{9}$$

The vector equation 9 leads to the system

$$\begin{aligned} \beta_1 + \beta_2 + \beta_3 &= -1 \\ \beta_2 + \beta_3 &= 3 \\ \beta_3 &= 2 \end{aligned}$$

Solving this system, we obtain

$$\beta_1 = -4 \qquad \beta_2 = 1 \qquad \beta_3 = 2$$

Thus, the second column of A is given by

$$\begin{bmatrix} a_{12} \\ a_{22} \\ a_{32} \end{bmatrix} = [T(\mathbf{v}_2)]_{B'} = \begin{bmatrix} -4 \\ 1 \\ 2 \end{bmatrix} \tag{10}$$

Combining equations 8 and 10, we obtain

$$A = \begin{bmatrix} -2 & -4 \\ 1 & 1 \\ 1 & 2 \end{bmatrix}$$

which is the matrix of the linear transformation T with respect to the ordered bases B and B'.

It is important for the reader to see the difference between equations 1 and 2, namely,

$$T(\mathbf{v}) = A\mathbf{v}, \qquad [T(\mathbf{v})]_{B'} = A[\mathbf{v}]_B$$

a The equation $T(\mathbf{v}) = A\mathbf{v}$ deals with vector spaces of the form $\mathbf{R}^n$ and $\mathbf{R}^m$ only, while the equation $[T(\mathbf{v})]_{B'} = A[\mathbf{v}]_B$ deals with arbitrary n-dimensional and m-dimensional vector spaces (which include $\mathbf{R}^n$ and $\mathbf{R}^m$ as special cases).

b The equation $T(\mathbf{v}) = A\mathbf{v}$ provides a direct way to compute $T(\mathbf{v})$ for any given $\mathbf{v}$, while the equation $[T(\mathbf{v})]_{B'} = A[\mathbf{v}]_B$ provides an *indirect* way of computing $T(\mathbf{v})$ from a given $\mathbf{v}$. One deals with the coordinate vectors $[\mathbf{v}]_B$ and $[T(\mathbf{v})]_{B'}$ rather than with the vectors $\mathbf{v}$ and $T(\mathbf{v})$ themselves. In order to find $T(\mathbf{v})$ from a given $\mathbf{v}$ we must reconstruct the vector $T(\mathbf{v})$ from its coordinate vector $[T(\mathbf{v})]_{B'}$. The following example exhibits this procedure.

Example 2 Let T, B, and B' be the linear transformation and the ordered bases from Example 1. Use the matrix representation of T found in Example 1 to compute $T(\mathbf{v})$ by means of the equation $[T(\mathbf{v})]_{B'} = A[\mathbf{v}]_B$ for $\mathbf{v} = \begin{bmatrix} 4 \\ 5 \end{bmatrix}$.

Solution We must first determine the coordinate vector of $\mathbf{v} = \begin{bmatrix} 4 \\ 5 \end{bmatrix}$ with respect to the basis B. Expressing $\mathbf{v}$ as a linear combination of $\mathbf{v}_1 = \begin{bmatrix} 1 \\ 1 \end{bmatrix}$ and $\mathbf{v}_2 = \begin{bmatrix} 1 \\ 2 \end{bmatrix}$, we obtain

$$\mathbf{v} = \begin{bmatrix} 4 \\ 5 \end{bmatrix} = \alpha_1 \begin{bmatrix} 1 \\ 1 \end{bmatrix} + \alpha_2 \begin{bmatrix} 1 \\ 2 \end{bmatrix} \qquad \textbf{[11]}$$

The vector equation 11 leads to the system

$$\begin{aligned} \alpha_1 + \alpha_2 &= 4 \\ \alpha_1 + 2\alpha_2 &= 5 \end{aligned}$$

Solving this system, we obtain

$$\alpha_1 = 3 \qquad \alpha_2 = 1$$

Thus, we have

$$[\mathbf{v}]_B = \begin{bmatrix} 3 \\ 1 \end{bmatrix}$$

Using the matrix A that we found in Example 1, we get

$$[T(\mathbf{v})]_{B'} = A[\mathbf{v}]_B = \begin{bmatrix} -2 & -4 \\ 1 & 1 \\ 1 & 2 \end{bmatrix} \begin{bmatrix} 3 \\ 1 \end{bmatrix} = \begin{bmatrix} -10 \\ 4 \\ 5 \end{bmatrix}$$

We now reconstruct the vector $T(\mathbf{v})$ from its coordinate vector $\begin{bmatrix} -10 \\ 4 \\ 5 \end{bmatrix}$ with respect to the ordered basis B'. We have

$$T(\mathbf{v}) = -10\mathbf{u}_1 + 4\mathbf{u}_2 + 5\mathbf{u}_3$$

Substituting for $\mathbf{u}_1$, $\mathbf{u}_2$, $\mathbf{u}_3$ from Example 1, we obtain

$$T(\mathbf{v}) = -10\begin{bmatrix}1\\0\\0\end{bmatrix} + 4\begin{bmatrix}1\\1\\0\end{bmatrix} + 5\begin{bmatrix}1\\1\\1\end{bmatrix} = \begin{bmatrix}-1\\9\\5\end{bmatrix}$$

In deriving the equation

$$[T(\mathbf{v})]_{B'} = A[\mathbf{v}]_B \tag{12}$$

it has been assumed that $T: V \longrightarrow W$ is a linear transformation and that B and B' are bases for V and W, respectively. We will now deal with the special case in which $V = W$. In this case the linear transformation $T: V \longrightarrow V$ is called a **linear operator** on V. If we also choose $B' = B$ (this is possible because $V = W$), then equation 12 reduces to

$$[T(\mathbf{v})]_B = A[\mathbf{v}]_B \tag{13}$$

The matrix A in equation (13) is called the **matrix of *T* with respect to the basis *B***.

In the following example we deal with such a matrix.

Example 3 Let $T: \mathbf{R}^2 \longrightarrow \mathbf{R}^2$ be the linear operator defined by

$$T\begin{bmatrix}x\\y\end{bmatrix} = \begin{bmatrix}x + 2y\\2x - y\end{bmatrix} \tag{14}$$

Find the matrix of T with respect to the ordered basis $B = \{\mathbf{v}_1,\mathbf{v}_2\}$ where $\mathbf{v}_1 = \begin{bmatrix}2\\1\end{bmatrix}$ and $\mathbf{v}_2 = \begin{bmatrix}1\\1\end{bmatrix}$. Check your result by finding $T(\mathbf{v})$ for $\mathbf{v} = \begin{bmatrix}7\\5\end{bmatrix}$ using equation 14 first and equation 13 second.

Solution Let A be the matrix of T with respect to the ordered basis $B = \{\mathbf{v}_1,\mathbf{v}_2\}$. The first column of A is given by $[T(\mathbf{v}_1)]_B$. In order to determine it, we write

$$T(\mathbf{v}_1) = T\begin{bmatrix}2\\1\end{bmatrix} = \alpha_1\begin{bmatrix}2\\1\end{bmatrix} + \alpha_2\begin{bmatrix}1\\1\end{bmatrix} \tag{15}$$

Using equation 14, we get

$$T\begin{bmatrix}2\\1\end{bmatrix} = \begin{bmatrix}4\\3\end{bmatrix}$$

Thus, we may write equation 15 in the form

$$T(\mathbf{v}_1) = \begin{bmatrix} 4 \\ 3 \end{bmatrix} = \alpha_1 \begin{bmatrix} 2 \\ 1 \end{bmatrix} + \alpha_2 \begin{bmatrix} 1 \\ 1 \end{bmatrix} \quad \textbf{[16]}$$

The vector equation 16 leads to the system

$$\begin{aligned} 2\alpha_1 + \alpha_2 &= 4 \\ \alpha_1 + \alpha_2 &= 3 \end{aligned}$$

the solution of which is given by

$$\alpha_1 = 1 \qquad \alpha_2 = 2$$

Thus, we have

$$[T(\mathbf{v}_1)]_B = \begin{bmatrix} 1 \\ 2 \end{bmatrix}$$

The second column of A is given by $[T(\mathbf{v}_2)]_B$. Expressing $T(\mathbf{v}_2)$ as a linear combination of $\mathbf{v}_1$ and $\mathbf{v}_2$, we obtain

$$T(\mathbf{v}_2) = T\begin{bmatrix} 1 \\ 1 \end{bmatrix} = \beta_1 \begin{bmatrix} 2 \\ 1 \end{bmatrix} + \beta_2 \begin{bmatrix} 1 \\ 1 \end{bmatrix} \quad \textbf{[17]}$$

Using equation 14, we get

$$T\begin{bmatrix} 1 \\ 1 \end{bmatrix} = \begin{bmatrix} 3 \\ 1 \end{bmatrix}$$

It follows that we may write equation 17 in the form

$$T(\mathbf{v}_2) = \begin{bmatrix} 3 \\ 1 \end{bmatrix} = \beta_1 \begin{bmatrix} 2 \\ 1 \end{bmatrix} + \beta_2 \begin{bmatrix} 1 \\ 1 \end{bmatrix} \quad \textbf{[18]}$$

The vector equation 18 leads to the system

$$\begin{aligned} 2\beta_1 + \beta_2 &= 3 \\ \beta_1 + \beta_2 &= 1 \end{aligned}$$

the solution of which is given by

$$\beta_1 = 2 \qquad \beta_2 = -1$$

Thus we have

$$[T(\mathbf{v}_2)]_B = \begin{bmatrix} 2 \\ -1 \end{bmatrix}$$

It follows that the matrix A is given by

$$A = \begin{bmatrix} 1 & 2 \\ 2 & -1 \end{bmatrix}$$

Checking our result for $\mathbf{v} = \begin{bmatrix} 7 \\ 5 \end{bmatrix}$, we find from equation 14 that $T\begin{bmatrix} 7 \\ 5 \end{bmatrix} = \begin{bmatrix} 17 \\ 9 \end{bmatrix}$.

Since $\mathbf{v} = \begin{bmatrix} 7 \\ 5 \end{bmatrix} = 2\begin{bmatrix} 2 \\ 1 \end{bmatrix} + 3\begin{bmatrix} 1 \\ 1 \end{bmatrix}$, we have $[\mathbf{v}]_B = \begin{bmatrix} 2 \\ 3 \end{bmatrix}$. Using equation 13, we obtain

$$[T(\mathbf{v})]_B = \begin{bmatrix} 1 & 2 \\ 2 & -1 \end{bmatrix}\begin{bmatrix} 2 \\ 3 \end{bmatrix} = \begin{bmatrix} 8 \\ 1 \end{bmatrix}$$

Thus, $T(\mathbf{v}) = 8\begin{bmatrix} 2 \\ 1 \end{bmatrix} + 1\begin{bmatrix} 1 \\ 1 \end{bmatrix} = \begin{bmatrix} 17 \\ 9 \end{bmatrix}$, the same result obtained above.

Exercises

1 Let $T:\mathbf{R}^2 \longrightarrow \mathbf{R}^3$ be the linear transformation defined by $T\begin{bmatrix} x \\ y \end{bmatrix} = \begin{bmatrix} x \\ x - y \\ x + y \end{bmatrix}$. Let $B = \{\mathbf{v}_1,\mathbf{v}_2\}$ and $B' = \{\mathbf{u}_1,\mathbf{u}_2,\mathbf{u}_3\}$ be ordered bases for $\mathbf{R}^2$ and $\mathbf{R}^3$, respectively, given by $\mathbf{v}_1 = \begin{bmatrix} 1 \\ 2 \end{bmatrix}$, $\mathbf{v}_2 = \begin{bmatrix} 2 \\ 1 \end{bmatrix}$, $\mathbf{u}_1 = \begin{bmatrix} 0 \\ 0 \\ 1 \end{bmatrix}$, $\mathbf{u}_2 = \begin{bmatrix} 0 \\ 1 \\ 1 \end{bmatrix}$, and $\mathbf{u}_3 = \begin{bmatrix} 1 \\ 0 \\ 1 \end{bmatrix}$. Find the matrix of T with respect to the bases B and B'.

2 Let T, B, and B' be the linear transformation and the ordered bases from problem 1. Find $[\mathbf{v}]_B$ and $[T(\mathbf{v})]_{B'}$ for $\mathbf{v} = \begin{bmatrix} 1 \\ 8 \end{bmatrix}$ and show that the equation $[T(\mathbf{v})]_{B'} = A[\mathbf{v}]_B$ is satisfied where A is the matrix found in problem 1.

3 Let $T:P_1 \longrightarrow P_2$ be the linear transformation defined by $T(p(x)) = xp(x)$. Let $B = \{x - 1, x + 1\}$ and $B' = \{1,x,x^2\}$ be ordered bases for P_1 and P_2, respectively. Find the matrix of T with respect to the bases B and B'.

4 Let T, B, and B' be the linear transformation and the ordered bases from problem 3. Find $[\mathbf{p}]_B$ and $[T(\mathbf{p})]_{B'}$ for $\mathbf{p} = 2x - 2$ and show that they satisfy the equation $[T(\mathbf{p})]_{B'} = A[\mathbf{p}]_B$ where A is the matrix found in problem 3.

5 Let $T:\mathbf{R}^2 \longrightarrow \mathbf{R}^2$ be the linear operator defined by $T\begin{bmatrix} x \\ y \end{bmatrix} =$

$\begin{bmatrix} 2x - y \\ x + 2y \end{bmatrix}$. Find the matrix of T with respect to the ordered basis $B = \{\mathbf{v}_1,\mathbf{v}_2\}$ where $\mathbf{v}_1 = \begin{bmatrix} 1 \\ 3 \end{bmatrix}$ and $\mathbf{v}_2 = \begin{bmatrix} 2 \\ 1 \end{bmatrix}$.

6 Let T and B be the linear operator and the ordered basis from problem 5. Find $[\mathbf{v}]_B$ and $[T(\mathbf{v})]_B$ for $\mathbf{v} = \begin{bmatrix} -1 \\ 7 \end{bmatrix}$ and show that they satisfy the equation $[T(\mathbf{v})]_B = A[\mathbf{v}]_B$ where A is the matrix found in problem 5.

7 Let $T: \mathbf{R}^3 \longrightarrow \mathbf{R}^3$ be the linear operator defined by $T\begin{bmatrix} x \\ y \\ z \end{bmatrix} = \begin{bmatrix} y + z \\ x + z \\ y + x \end{bmatrix}$. Find the matrix of T with respect to the ordered basis $B = \{\mathbf{v}_1,\mathbf{v}_2,\mathbf{v}_3\}$ where $\mathbf{v}_1 = \begin{bmatrix} 1 \\ 1 \\ 0 \end{bmatrix}$, $\mathbf{v}_2 = \begin{bmatrix} 1 \\ 0 \\ 1 \end{bmatrix}$, and $\mathbf{v}_3 = \begin{bmatrix} 1 \\ 1 \\ 1 \end{bmatrix}$.

8 Let T and B be the linear operator and the ordered basis from problem 7. Find $[\mathbf{v}]_B$ and $[T(\mathbf{v})]_B$ for $\mathbf{v} = \begin{bmatrix} 5 \\ 7 \\ 2 \end{bmatrix}$ and show that they satisfy the equation $[T(\mathbf{v})]_B = A[\mathbf{v}]_B$ where A is the matrix found in problem 7.

9 Let $T: V \longrightarrow W$ be a linear transformation and let $B = \{\mathbf{v}_1, \mathbf{v}_2, \ldots, \mathbf{v}_n\}$ and $B' = \{\mathbf{u}_1, \mathbf{u}_2, \ldots, \mathbf{u}_m\}$ be ordered bases for V and W, respectively. Show that if the matrix of T with respect to the bases B and B' is the zero matrix, then T is the zero transformation.

10 Let $T: V \longrightarrow V$ be a linear operator and let $B = \{\mathbf{v}_1, \mathbf{v}_2, \ldots, \mathbf{v}_n\}$ be an ordered basis for V. Show that if the matrix of T with respect to the basis B is the identity matrix I_n, then T is the identity transformation; that is, $T(\mathbf{v}) = \mathbf{v}$ for all $\mathbf{v}$ in V.

P_3 is the space of all polynomials of degree ≤ 3 and the zero polynomial.

11 (For students with a calculus background:) Let $D: P_3 \longrightarrow P_3$ be the differentiation operator.

a Find the matrix of D with respect to the ordered basis $B = \{1,x,x^2,x^3\}$ of P_3.

b Find the matrix of D with respect to the ordered basis $B' = \{1 - x, 1 + x, x^2 - x^3, x^2 + x^3\}$.

6-7 Similar Matrices

Objectives

1 Introduce the concept of similarity of matrices.
2 Show how the similarity relation can be used to solve matrix equations of the form $X^k = A$ (k is a positive integer) provided a solution exists.

Let V be an n-dimensional vector space and let $B = \{\mathbf{v}_1, \mathbf{v}_2, \ldots, \mathbf{v}_n\}$ be an ordered basis for V.

We have seen in Section 6-6 that if T is a linear operator on V, then for each $\mathbf{v}$ in V the coordinate vectors $[\mathbf{v}]_B$ and $[T(\mathbf{v})]_B$ are related by the equation

$$[T(\mathbf{v})]_B = A[\mathbf{v}]_B \qquad [1]$$

where A is the matrix of T with respect to the basis B.

The matrix A in equation 1 depends on the ordered basis B. Thus, if we change the basis B and deal instead with an ordered basis $B' = \{\mathbf{u}_1, \mathbf{u}_2, \ldots, \mathbf{u}_n\}$, then the new coordinate vectors $[\mathbf{v}]_{B'}$ and $[T(\mathbf{v})]_{B'}$ satisfy the equation

$$[T(\mathbf{v})]_{B'} = A'[\mathbf{v}]_{B'} \qquad [2]$$

where A' is the (new) matrix of T with respect to the basis B'.

The natural question arising at this point is the following: What is the relationship between the matrix A in equation 1 and the matrix A' in equation 2?

In order to answer this question, let us recall that if $\mathbf{v} \in V$, then the coordinate vectors $[\mathbf{v}]_B$ and $[\mathbf{v}]_{B'}$ satisfy the equation

See Theorem 6-11.

$$[\mathbf{v}]_{B'} = P[\mathbf{v}]_B \qquad [3]$$

where P is the transition matrix from the ordered basis B to the ordered basis B'. Since T is a linear operator, it follows that $T(\mathbf{v})$ is a vector in V and must also satisfy equation 3. Thus, replacing $\mathbf{v}$ with $T(\mathbf{v})$ in equation 3, we obtain

$$[T(\mathbf{v})]_{B'} = P[T(\mathbf{v})]_B \qquad [4]$$

If in equation 2 we substitute $P[T(\mathbf{v})]_B$ and $P[\mathbf{v}]_B$ from equations 4 and 3 for $[T(\mathbf{v})]_{B'}$ and $[\mathbf{v}]_{B'}$, then we obtain

$$P[T(\mathbf{v})]_B = A'P[\mathbf{v}]_B \qquad [5]$$

Now in equation 5 we substitute the expression $A[\mathbf{v}]_B$ from equation 1 for $[T(\mathbf{v})]_B$, obtaining

$$PA[\mathbf{v}]_B = A'P[\mathbf{v}]_B \qquad [6]$$

Since equation 6 holds for all $\mathbf{v}$ in V it follows that

$$PA = A'P \qquad [7]$$

See Theorem 6-12.

The reader should recall that the transition matrix P is invertible. Therefore, multiplying equation 7 by P^{-1} from the left, we obtain

$$A = P^{-1}A'P \qquad [8]$$

We have thus proved the following theorem.

Theorem 6-13 Let V be a finite-dimensional vector space and let $B = \{\mathbf{v}_1, \mathbf{v}_2, \ldots, \mathbf{v}_n\}$ and $B' = \{\mathbf{u}_1, \mathbf{u}_2, \ldots, \mathbf{u}_n\}$ be ordered bases for V. Let T be a linear operator on V. If A is the matrix of T with respect to the basis B and A' is the matrix of T with respect to the basis B', then we have

$$A = P^{-1}A'P$$

where P is the transition matrix from B to B'.

Example 1 Let $T: \mathbf{R}^2 \longrightarrow \mathbf{R}^2$ be the linear operator defined by

$$T\begin{bmatrix} x \\ y \end{bmatrix} = \begin{bmatrix} x + 2y \\ 2x - y \end{bmatrix} \qquad [9]$$

Let $B = \{\mathbf{v}_1, \mathbf{v}_2\}$ and $B' = \{\mathbf{u}_1, \mathbf{u}_2\}$ be ordered bases for $\mathbf{R}^2$ where

$$\mathbf{v}_1 = \begin{bmatrix} 2 \\ 1 \end{bmatrix} \qquad \mathbf{v}_2 = \begin{bmatrix} 1 \\ 1 \end{bmatrix} \qquad \mathbf{u}_1 = \begin{bmatrix} 1 \\ 2 \end{bmatrix} \qquad \mathbf{u}_2 = \begin{bmatrix} 0 \\ 1 \end{bmatrix}$$

Let A and A' be the matrices of T with respect to the bases B and B', respectively, and let P be the transition matrix from B to B'. Find the matrices A', P, and P^{-1} and use Theorem 6.13 to find the matrix A.

Solution The columns of the matrix A' are given by $[T(\mathbf{u}_1)]_{B'}$ and $[T(\mathbf{u}_2)]_{B'}$. In order to determine these coordinate vectors we express $T(\mathbf{u}_1)$ and $T(\mathbf{u}_2)$ as linear combinations of $\mathbf{u}_1$ and $\mathbf{u}_2$. We obtain the equations

The equations $T\begin{bmatrix} 1 \\ 2 \end{bmatrix} = \begin{bmatrix} 5 \\ 0 \end{bmatrix}$ and $T\begin{bmatrix} 0 \\ 1 \end{bmatrix} = \begin{bmatrix} 2 \\ -1 \end{bmatrix}$ follow from equation 9.

$$T(\mathbf{u}_1) = T\begin{bmatrix} 1 \\ 2 \end{bmatrix} = \begin{bmatrix} 5 \\ 0 \end{bmatrix} = \alpha_1\begin{bmatrix} 1 \\ 2 \end{bmatrix} + \alpha_2\begin{bmatrix} 0 \\ 1 \end{bmatrix} \qquad [10]$$

and

$$T(\mathbf{u}_2) = T\begin{bmatrix} 0 \\ 1 \end{bmatrix} = \begin{bmatrix} 2 \\ -1 \end{bmatrix} = \beta_1\begin{bmatrix} 1 \\ 2 \end{bmatrix} + \beta_2\begin{bmatrix} 0 \\ 1 \end{bmatrix} \qquad [11]$$

The solution of equation 10 yields

$$\alpha_1 = 5 \qquad \alpha_2 = -10$$

and the solution of equation 11 yields

$$\beta_1 = 2 \qquad \beta_2 = -5$$

Thus, we have

$$[T(\mathbf{u}_1)]_{B'} = \begin{bmatrix} 5 \\ -10 \end{bmatrix} \qquad [T(\mathbf{u}_2)]_{B'} = \begin{bmatrix} 2 \\ -5 \end{bmatrix}$$

It follows that the matrix A' is given by

$$A' = \begin{bmatrix} 5 & 2 \\ -10 & -5 \end{bmatrix} \tag{12}$$

Now recall that the columns of the matrix P are given by $[\mathbf{v}_1]_{B'}$ and $[\mathbf{v}_2]_{B'}$. In order to determine these coordinate vectors, we express $\mathbf{v}_1$ and $\mathbf{v}_2$ as linear combinations of $\mathbf{u}_1$ and $\mathbf{u}_2$. We obtain the equations

$$\mathbf{v}_1 = \begin{bmatrix} 2 \\ 1 \end{bmatrix} = \gamma_1 \begin{bmatrix} 1 \\ 2 \end{bmatrix} + \gamma_2 \begin{bmatrix} 0 \\ 1 \end{bmatrix} \tag{13}$$

and

$$\mathbf{v}_2 = \begin{bmatrix} 1 \\ 1 \end{bmatrix} = \delta_1 \begin{bmatrix} 1 \\ 2 \end{bmatrix} + \delta_2 \begin{bmatrix} 0 \\ 1 \end{bmatrix} \tag{14}$$

The solution of equation 13 is given by

$$\gamma_1 = 2 \qquad \gamma_2 = -3$$

and the solution of equation 14 is given by

$$\delta_1 = 1 \qquad \delta_2 = -1$$

Thus, we have

$$[\mathbf{v}_1]_{B'} = \begin{bmatrix} 2 \\ -3 \end{bmatrix} \qquad [\mathbf{v}_2]_{B'} = \begin{bmatrix} 1 \\ -1 \end{bmatrix}$$

It follows that the matrix P is given by

$$P = \begin{bmatrix} 2 & 1 \\ -3 & -1 \end{bmatrix}$$

See Section 1-5.

A simple computation using elementary row operations yields the inverse of P. We have

$$P^{-1} = \begin{bmatrix} -1 & -1 \\ 3 & 2 \end{bmatrix}$$

We now apply Theorem 6-12 and obtain

$$\begin{aligned} A = P^{-1}A'P &= \begin{bmatrix} -1 & -1 \\ 3 & 2 \end{bmatrix}\begin{bmatrix} 5 & 2 \\ -10 & -5 \end{bmatrix}\begin{bmatrix} 2 & 1 \\ -3 & -1 \end{bmatrix} \\ &= \begin{bmatrix} 5 & 3 \\ -5 & -4 \end{bmatrix}\begin{bmatrix} 2 & 1 \\ -3 & -1 \end{bmatrix} = \begin{bmatrix} 1 & 2 \\ 2 & -1 \end{bmatrix} \end{aligned}$$

The relationship between the matrices A and A' in Example 1 and Theorem 6-13 motivates the following.

Definition 6-9 Let A and B be square $n \times n$ matrices. We say that **B is similar to A** if there is an invertible $n \times n$ matrix P such that $B = P^{-1}AP$.

Similarity of matrices.

It follows easily from Definition 6-9 that if B is similar to A then also A is similar to B. In order to establish this, we must show that there is an invertible matrix Q such that $A = Q^{-1}BQ$. Since B is similar to A we have $B = P^{-1}AP$ where P is some invertible matrix. Multiplying the last equation by P from the left and by P^{-1} from the right, we obtain

$$PBP^{-1} = A \tag{15}$$

Since we know that $(P^{-1})^{-1} = P$ we can rewrite equation 15 to obtain

$$A = (P^{-1})^{-1}BP^{-1} \tag{16}$$

If we now set $P^{-1} = Q$ then $(P^{-1})^{-1} = Q^{-1}$ and equation 16 may be written as

$$A = Q^{-1}BQ \tag{17}$$

This shows that A is similar to B. It follows that similarity of matrices is a symmetric relationship and we simply say that A and B are similar matrices.

It is important that the reader realize that there exist pairs of matrices that are *not* similar. The following example provides such a pair.

Example 2 Find two 2×2 matrices that are *not similar* to each other.

Solution Let one of the matrices be the identity matrix I_2. If P is any invertible 2×2 matrix, then we have $P^{-1}I_2P = P^{-1}(I_2P) = P^{-1}P = I_2$. Thus, it follows that the identity matrix I_2 is similar only to itself. Therefore, choose a 2×2 matrix A such that $A \neq I_2$, say $A = \begin{bmatrix} 1 & 2 \\ 3 & 4 \end{bmatrix}$. It follows that A and I_2 are not similar matrices.

$I_2 = \begin{bmatrix} 1 & 0 \\ 0 & 1 \end{bmatrix}$.

See Chapter 8.

Similar matrices play a significant role in some applications of linear algebra. It is therefore important that the reader be familiar with some basic properties shared by similar matrices. One such property is the subject of the following theorem.

Theorem 6-14 Let A and B be similar matrices. Then, either both A and B are singular matrices or both are invertible matrices.

Proof First we show that similarity of A and B implies that $\det A = \det B$.

Since A and B are similar matrices, it follows that there exists an invertible matrix P such that

$$A = P^{-1}BP \qquad [18]$$

Taking the determinant of both sides of equation 18, we get

$$\det A = \det (P^{-1}BP) \qquad [19]$$

See Theorem 4-6.

Since $\det (P^{-1}BP) = (\det P^{-1})(\det B)(\det P)$, we may rewrite equation 19, obtaining

$$\det A = (\det P^{-1})(\det B)(\det P) \qquad [20]$$

But $\det P^{-1}$, $\det B$, and $\det P$ are real numbers, and therefore they commute in the product $(\det P^{-1})(\det B)(\det P)$. We may thus write

$$\det A = (\det P^{-1})(\det P)(\det B) \qquad [21]$$

Since $(\det P^{-1})(\det P) = \det (P^{-1}P)$, we can write equation 21 as

$$\det A = [\det (P^{-1}P)][\det B] \qquad [22]$$

But $\det (P^{-1}P) = \det (I_n) = 1$, which leads to the equation

$$\det A = \det B \qquad [23]$$

Thus, similar matrices have equal determinants.

See Theorem 4-7.

Since A is singular if and only if $\det A = 0$, it follows from equation 23 that A is singular if and only if B is singular.

As a consequence of Theorem 6-14 we have the following:

Theorem 6-15 Let A be a nonsingular matrix which is similar to a matrix B. Then B is nonsingular and A^{-1} is similar to B^{-1}.

Proof Since A is nonsingular and similar to B, it follows from Theorem 6-14 that B is also nonsingular. Thus, the matrices A^{-1} and B^{-1} exist. In order to show that A^{-1} is similar to B^{-1} we use the similarity of A and B. We have

$$A = PBP^{-1} \qquad [24]$$

where P is some nonsingular matrix.

Taking the inverse of both sides of equation 24, we obtain

$$A^{-1} = (PBP^{-1})^{-1} \qquad [25]$$

Since $(PBP^{-1})^{-1} = (P^{-1})^{-1}B^{-1}P^{-1}$ we can rewrite equation 25 and obtain

$$A^{-1} = (P^{-1})^{-1}B^{-1}P^{-1} \qquad [26]$$

If we let $P^{-1} = Q$, then equation 26 yields

$$A^{-1} = Q^{-1}B^{-1}Q \qquad [27]$$

It follows from equation 27 that A^{-1} and B^{-1} are similar matrices.

The similarity relation has a variety of applications in many branches of mathematics. We now present an example showing how the similarity relation of matrices can be used to solve some matrix equations of the form $X^k = A$ where A is a given $n \times n$ matrix and k is a positive integer greater than 1.

First, we wish to emphasize that there exist matrix equations of the form above which possess *no* solution at all. If we take $A = \begin{bmatrix} 0 & 1 \\ 0 & 0 \end{bmatrix}$, then the equation

$$X^2 = \begin{bmatrix} 0 & 1 \\ 0 & 0 \end{bmatrix} \tag{28}$$

has no solution. In other words, there is *no* 2×2 matrix X such that equation 28 is satisfied. To see this, we set $X = \begin{bmatrix} a & b \\ c & d \end{bmatrix}$ and substitute it into equation 28. We obtain

$$\begin{bmatrix} a & b \\ c & d \end{bmatrix}\begin{bmatrix} a & b \\ c & d \end{bmatrix} = \begin{bmatrix} 0 & 1 \\ 0 & 0 \end{bmatrix} \tag{29}$$

Multiplying the matrices on the left-hand side of equation 29, we obtain

$$\begin{bmatrix} a^2 + bc & (a + d)b \\ (a + d)c & cb + d^2 \end{bmatrix} = \begin{bmatrix} 0 & 1 \\ 0 & 0 \end{bmatrix} \tag{30}$$

It follows from equation 30 that we have the following equations for a, b, c, and d:

Note that the equations in System 31 constitute a *nonlinear* system for the unknowns a, b, c, and d. Such systems have not previously been encountered in this text.

$$a^2 + bc = 0, \quad (a + d)b = 1, \quad (a + d)c = 0, \quad cb + d^2 = 0 \tag{31}$$

But the system of equations 31 does not have a solution for a, b, c, and d. To show the inconsistency of system 31, we start with the equation $(a + d)b = 1$. It follows from this equation that $a + d \neq 0$. From the third equation of system 31, namely, $(a + d)c = 0$ we conclude that $c = 0$. Thus, the equations $a^2 + bc = 0$ and $cb + d^2 = 0$ from system 31 reduce to $a^2 = 0$ and $d^2 = 0$. The last two equations imply that $a = 0$ and $d = 0$. But this is impossible because we have found that $a + d \neq 0$. Thus, system 31 is inconsistent and equation 28 does *not* have a solution.

In order to discuss methods of solving the matrix equation $X^k = A$ by use of the similarity relation, we need to introduce the concept of a diagonal matrix.

Definition 6-10 Let A be a square $n \times n$ matrix. We say that A is a **diagonal** matrix if and only if $a_{ij} = 0$ whenever $i \neq j$. The entries a_{ii} for $i = 1, 2, \ldots, n$ are said to form the main diagonal of the matrix A.

It follows from Definition 6-10 that if A is a diagonal matrix of order 2×2, then it has the form

$$A = \begin{bmatrix} a_{11} & 0 \\ 0 & a_{22} \end{bmatrix}$$

and if B is a diagonal 3×3 matrix, then we have

$$B = \begin{bmatrix} b_{11} & 0 & 0 \\ 0 & b_{22} & 0 \\ 0 & 0 & b_{33} \end{bmatrix}$$

Diagonal matrices have properties which enable us to operate with them very easily. For example, the product of the two matrices

$$C = \begin{bmatrix} c_{11} & 0 \\ 0 & c_{22} \end{bmatrix} \quad \text{and} \quad D = \begin{bmatrix} d_{11} & 0 \\ 0 & d_{22} \end{bmatrix}$$

is given by

$$CD = \begin{bmatrix} c_{11}d_{11} & 0 \\ 0 & c_{22}d_{22} \end{bmatrix}$$

and the kth power of the matrix C is given by

$$C^k = \begin{bmatrix} (c_{11})^k & 0 \\ 0 & (c_{22})^k \end{bmatrix} \qquad [32]$$

It follows from equation 32 that the matrix equation

$$X^k = \begin{bmatrix} a_{11} & 0 \\ 0 & a_{22} \end{bmatrix} \qquad [33]$$

has at least one solution, given by

$$X = \begin{bmatrix} \sqrt[k]{a_{11}} & 0 \\ 0 & \sqrt[k]{a_{22}} \end{bmatrix}$$

The solution matrix X will contain complex entries if a_{11} or a_{22} are negative and k is even.

We now use the simplicity of finding solutions for equation 33 in order to solve a more difficult matrix equation.

Example 3 Let $A = \begin{bmatrix} 4 & 0 \\ 0 & 1 \end{bmatrix}$, $B = \begin{bmatrix} -5 & -3 \\ 18 & 10 \end{bmatrix}$, and $P = \begin{bmatrix} -1 & -1 \\ 3 & 2 \end{bmatrix}$.

a Prove that A is similar to B by showing that $A = P^{-1}BP$.

b Use the similarity of A and B to solve the matrix equation $X^2 = B$.

Solution A simple computation using elementary row operations yields the inverse of P. We have

$$P^{-1} = \begin{bmatrix} 2 & 1 \\ -3 & -1 \end{bmatrix}$$

Computing the matrix product $P^{-1}BP$, we obtain

$$(P^{-1}B)P = \left(\begin{bmatrix} 2 & 1 \\ -3 & -1 \end{bmatrix}\begin{bmatrix} -5 & -3 \\ 18 & 10 \end{bmatrix}\right)\begin{bmatrix} -1 & -1 \\ 3 & 2 \end{bmatrix}$$
$$= \begin{bmatrix} 8 & 4 \\ -3 & -1 \end{bmatrix}\begin{bmatrix} -1 & -1 \\ 3 & 2 \end{bmatrix} = \begin{bmatrix} 4 & 0 \\ 0 & 1 \end{bmatrix} = A$$

It follows that we have

$$A = P^{-1}BP \tag{34}$$

which proves part a.

We are now ready to solve the matrix equation $X^2 = B$. Multiplying equation 34 by P (from the left) and by P^{-1} (from the right), we obtain the equation

$$PAP^{-1} = B \tag{35}$$

The matrix $A = \begin{bmatrix} 4 & 0 \\ 0 & 1 \end{bmatrix}$ can also be expressed as follows: $A = \begin{bmatrix} 2 & 0 \\ 0 & -1 \end{bmatrix}^2$, $A = \begin{bmatrix} -2 & 0 \\ 0 & 1 \end{bmatrix}^2$, and $A = \begin{bmatrix} -2 & 0 \\ 0 & -1 \end{bmatrix}^2$.

We now use the fact that the diagonal matrix $A = \begin{bmatrix} 4 & 0 \\ 0 & 1 \end{bmatrix}$ can be expressed as a product

$$A = \begin{bmatrix} 2 & 0 \\ 0 & 1 \end{bmatrix}\begin{bmatrix} 2 & 0 \\ 0 & 1 \end{bmatrix} = \begin{bmatrix} 2 & 0 \\ 0 & 1 \end{bmatrix}$$

which enables us to rewrite equation 35 in the form

$$P\begin{bmatrix} 2 & 0 \\ 0 & 1 \end{bmatrix}\begin{bmatrix} 2 & 0 \\ 0 & 1 \end{bmatrix}P^{-1} = B \tag{36}$$

We now use a simple "trick" that will enable us to express the left-hand side of equation 36 as a product of two equal matrices. We separate the two factors $\begin{bmatrix} 2 & 0 \\ 0 & 1 \end{bmatrix}$ and $\begin{bmatrix} 2 & 0 \\ 0 & 1 \end{bmatrix}$ in equation 36 by introducing the identity matrix in the form of $P^{-1}P$. After placing parentheses appropriately, we obtain

$$\left(P\begin{bmatrix} 2 & 0 \\ 0 & 1 \end{bmatrix}P^{-1}\right)\left(P\begin{bmatrix} 2 & 0 \\ 0 & 1 \end{bmatrix}P^{-1}\right) = B \tag{37}$$

It follows from equation 37 that the matrix

$$X = P\begin{bmatrix} 2 & 0 \\ 0 & 1 \end{bmatrix}P^{-1} \tag{38}$$

is a solution for the equation $X^2 = B$.

If we substitute the proper matrices for P and P^{-1}, we obtain from equation 38

$$X = \left(\begin{bmatrix} -1 & -1 \\ 3 & 2 \end{bmatrix}\begin{bmatrix} 2 & 0 \\ 0 & 1 \end{bmatrix}\right)\begin{bmatrix} 2 & 1 \\ -3 & -1 \end{bmatrix}$$

$$= \begin{bmatrix} -2 & -1 \\ 6 & 2 \end{bmatrix}\begin{bmatrix} 2 & 1 \\ -3 & -1 \end{bmatrix} = \begin{bmatrix} -1 & -1 \\ 6 & 4 \end{bmatrix}$$

We have already indicated above that the matrix $A = \begin{bmatrix} 4 & 0 \\ 0 & 1 \end{bmatrix}$ can also be expressed in the forms

$$A = \begin{bmatrix} 2 & 0 \\ 0 & -1 \end{bmatrix}^2, \qquad A = \begin{bmatrix} -2 & 0 \\ 0 & 1 \end{bmatrix}^2, \qquad A = \begin{bmatrix} -2 & 0 \\ 0 & -1 \end{bmatrix}^2$$

As a result of this fact it follows that the matrices

$$P\begin{bmatrix} 2 & 0 \\ 0 & -1 \end{bmatrix}P^{-1}, \qquad P\begin{bmatrix} -2 & 0 \\ 0 & 1 \end{bmatrix}P^{-1}, \qquad P\begin{bmatrix} -2 & 0 \\ 0 & -1 \end{bmatrix}P^{-1}$$

are also solutions of the matrix equation $X^2 = B$.

Exercises

1 Let $T: \mathbf{R}^2 \longrightarrow \mathbf{R}^2$ be defined by $T\begin{bmatrix} x \\ y \end{bmatrix} = \begin{bmatrix} x - y \\ x + 2y \end{bmatrix}$. Let $B = \{\mathbf{v}_1, \mathbf{v}_2\}$ and $B' = \{\mathbf{u}_1, \mathbf{u}_2\}$ be ordered bases for $\mathbf{R}^2$ where $\mathbf{v}_1 = \begin{bmatrix} 1 \\ 1 \end{bmatrix}$, $\mathbf{v}_2 = \begin{bmatrix} 1 \\ 2 \end{bmatrix}$, $\mathbf{u}_1 = \begin{bmatrix} 2 \\ 1 \end{bmatrix}$, and $\mathbf{u}_2 = \begin{bmatrix} 1 \\ 0 \end{bmatrix}$. Let A and A' be the matrices of T with respect to the bases B and B', respectively, and let P be the transition matrix from B to B'. Find the matrices A', P, and P^{-1}. Use Theorem 6-13 to find the matrix A.

2 Repeat problem 1 for $T: \mathbf{R}^2 \longrightarrow \mathbf{R}^2$ defined by

$$T\begin{bmatrix} x \\ y \end{bmatrix} = \begin{bmatrix} 3x - y \\ x + 3y \end{bmatrix}.$$

3 Let $T: \mathbf{R}^3 \longrightarrow \mathbf{R}^3$ be defined by $T\begin{bmatrix} x \\ y \\ z \end{bmatrix} = \begin{bmatrix} x + y + z \\ x - y \\ y - z \end{bmatrix}$. Let $B = \{\mathbf{v}_1, \mathbf{v}_2, \mathbf{v}_3\}$ and $B' = \{\mathbf{u}_1, \mathbf{u}_2, \mathbf{u}_3\}$ be ordered bases for $\mathbf{R}^3$ where $\mathbf{v}_1 = \begin{bmatrix} 1 \\ 1 \\ 1 \end{bmatrix}$, $\mathbf{v}_2 = \begin{bmatrix} 1 \\ 1 \\ 0 \end{bmatrix}$, $\mathbf{v}_3 = \begin{bmatrix} 1 \\ 0 \\ 0 \end{bmatrix}$, $\mathbf{u}_1 = \begin{bmatrix} 0 \\ 0 \\ 1 \end{bmatrix}$, $\mathbf{u}_2 = \begin{bmatrix} 0 \\ 1 \\ 1 \end{bmatrix}$, and $\mathbf{u}_3 = \begin{bmatrix} 1 \\ 0 \\ 1 \end{bmatrix}$. Let A and A' be the matrices of T with respect to the bases B and B', respectively, and let P be the transition matrix from B to B'. Find the matrices A', P, and P^{-1}. Use Theorem 6-13 to find the matrix A.

4 Repeat problem 3 for $T: \mathbf{R}^3 \longrightarrow \mathbf{R}^3$ defined by

$$T\begin{bmatrix} x \\ y \\ z \end{bmatrix} = \begin{bmatrix} 2x + y - z \\ x + 2y + z \\ x - y - z \end{bmatrix}.$$

5 Show that every $n \times n$ matrix A is similar to itself.

6 Let A, B, and C be $n \times n$ matrices. If A is similar to B and B is similar to C, show that A is similar to C.

A^t is the transpose of A.

7 If A is similar to B, show that A^t is similar to B^t.

8 Let A be similar to B.

a Show that A^2 is similar to B^2.

b Show that A^k is similar to B^k for any positive integer k.

9 Let A and B be $n \times n$ matrices. Define the **trace** of A by $\text{tr}(A) = a_{11} + a_{22} + \cdots + a_{nn}$.

a Show that $\text{tr}(AB) = \text{tr}(BA)$.

b If A is similar to B, show that $\text{tr}(A) = \text{tr}(B)$.

10 Find *all* diagonal solutions of the matrix equation $X^2 = A$ where $A = \begin{bmatrix} 9 & 0 \\ 0 & 4 \end{bmatrix}$.

11 Find all diagonal solutions of the matrix equation $X^3 = A$ where $A = \begin{bmatrix} -1 & 0 \\ 0 & 8 \end{bmatrix}$.

12 Find all diagonal solutions of the matrix equation $X^4 = A$ where $A = \begin{bmatrix} 16 & 0 & 0 \\ 0 & 1 & 0 \\ 0 & 0 & 81 \end{bmatrix}$.

13 Let $A = \begin{bmatrix} 4 & 0 \\ 0 & 9 \end{bmatrix}$, $B = \begin{bmatrix} -1 & 10 \\ -5 & 14 \end{bmatrix}$, and $P = \begin{bmatrix} 2 & 1 \\ 1 & 1 \end{bmatrix}$.

a Prove that A is similar to B by showing that $A = P^{-1} BP$.

See Example 3 for solutions of $X^2 = B$.

b Use the similarity of A and B to solve the matrix equation $X^2 = B$.

14 Let $A = \begin{bmatrix} 4 & 0 & 0 \\ 0 & 1 & 0 \\ 0 & 0 & 9 \end{bmatrix}$, $B = \frac{1}{3}\begin{bmatrix} 29 & -2 & -13 \\ 20 & 7 & -10 \\ -6 & 6 & 6 \end{bmatrix}$, and

$$P = \begin{bmatrix} 1 & 1 & 1 \\ 2 & 0 & 1 \\ 1 & 2 & 0 \end{bmatrix}.$$

a Prove that A is similar to B by showing that $A = P^{-1}BP$.

b Use the similarity of A and B to solve the matrix equation $X^2 = B$.

15 Give an example of two matrices A and B such that $\det A = \det B$ but A is not similar to B.

Review of Chapter 6

1 Let T be a transformation from $\mathbf{R}^2$ into $\mathbf{R}$ such that $T\begin{bmatrix} 0 \\ 0 \end{bmatrix} = 1$. Is T linear?

2 Let T be a transformation from $\mathbf{R}^3$ into $\mathbf{R}^2$ such that

$$T\begin{bmatrix} -1 \\ -2 \\ -3 \end{bmatrix} = -4T\begin{bmatrix} 1 \\ 2 \\ 3 \end{bmatrix}.$$ Is T linear?

3 Let $T: \mathbf{R}^2 \longrightarrow \mathbf{R}^2$ be defined by

$$T\begin{bmatrix} x \\ y \end{bmatrix} = \begin{bmatrix} -x \\ -y \end{bmatrix}$$

Is T linear? Describe T geometrically.

4 Let $T: \mathbf{R}^3 \longrightarrow \mathbf{R}$ be a linear transformation such that

$$T\begin{bmatrix} 1 \\ 2 \\ 3 \end{bmatrix} = 6 \text{ and } T\begin{bmatrix} 4 \\ 1 \\ 2 \end{bmatrix} = 7. \text{ Find } T\left(2\begin{bmatrix} 1 \\ 2 \\ 3 \end{bmatrix} + 4\begin{bmatrix} 4 \\ 1 \\ 2 \end{bmatrix}\right).$$

5 Let $T: \mathbf{R}^2 \longrightarrow \mathbf{R}^3$ be defined by

$$T\begin{bmatrix} x \\ y \end{bmatrix} = \begin{bmatrix} 0 \\ 0 \\ 0 \end{bmatrix}$$

Find a matrix representation of T.

6 Let $T: \mathbf{R}^3 \longrightarrow \mathbf{R}^3$ be defined by

$$T\begin{bmatrix} x \\ y \\ z \end{bmatrix} = \begin{bmatrix} x \\ y \\ z \end{bmatrix}$$

Find a matrix representation of T.

7 Let $T\begin{bmatrix} x \\ y \end{bmatrix} = \begin{bmatrix} x + 2y \\ 2x - y \\ 3x + y \end{bmatrix}$ and $S\begin{bmatrix} x \\ y \end{bmatrix} = \begin{bmatrix} 3x \\ 2y \\ x - y \end{bmatrix}$. Find $(T + S)\begin{bmatrix} 4 \\ 5 \end{bmatrix}$ and $(6T)\begin{bmatrix} 4 \\ 5 \end{bmatrix}$.

8 Let T and S be linear transformations from V into W. Does it follow that $T + S$ is linear?

9 Let T and S be transformations from V into W such that $T + S$ is linear. Does it follow that T and S must be linear?

10 Let $T\begin{bmatrix} x \\ y \end{bmatrix} = \begin{bmatrix} x - 2y \\ x + 2y \end{bmatrix}$ and $S\begin{bmatrix} x \\ y \end{bmatrix} = \begin{bmatrix} y \\ x \end{bmatrix}$. Find formulas for $(ST)\begin{bmatrix} x \\ y \end{bmatrix}$ and $(TS)\begin{bmatrix} x \\ y \end{bmatrix}$.

11 Is it possible for the null space of a linear transformation T to consist of exactly one vector? Explain.

12 Is it possible for the range of a linear transformation T to consist of exactly one vector? Explain.

13 Is it possible for the null space as well as the range of a linear transformation to consist of exactly one vector? Explain.

14 Let $T: \mathbf{R}^4 \longrightarrow \mathbf{R}^2$ be the linear transformation defined by

$$T\begin{bmatrix} x \\ y \\ z \\ w \end{bmatrix} = \begin{bmatrix} x - y + z - w \\ x + y - z - w \end{bmatrix}$$

Describe the null space of T and the range of T, and find the nullity and the rank of T.

15 Let $T: V \longrightarrow W$ be a linear transformation and let V be an n-dimensional vector space. Is there any relation between the nullity of T, the rank of T, and the dimension of V?

16 Find a linear transformation $T: \mathbf{R}^2 \longrightarrow \mathbf{R}^2$ such that $T\begin{bmatrix} 1 \\ 2 \end{bmatrix} = \begin{bmatrix} 0 \\ -5 \end{bmatrix}$ and $T\begin{bmatrix} 2 \\ 1 \end{bmatrix} = \begin{bmatrix} 3 \\ -1 \end{bmatrix}$. How many such linear transformations are there?

17 Let $B = \{\mathbf{v}_1, \mathbf{v}_2, \ldots, \mathbf{v}_n\}$ be an ordered basis for a vector space V and let $\mathbf{v} \in V$. What is meant by the phrase "the coordinate vector of $\mathbf{v}$ with respect to the basis B"?

18 Let $B = \{\mathbf{v}_1, \mathbf{v}_2, \ldots, \mathbf{v}_n\}$ and $B' = \{\mathbf{u}_1, \mathbf{u}_2, \ldots, \mathbf{u}_n\}$ be ordered bases for a vector space V. Explain how one obtains the transition matrix A from the ordered basis B to the ordered basis B'.

19 If A' is the transition matrix from the ordered basis B' to the ordered basis B in review problem 18, what is the relation between A' and the transition matrix A mentioned in review problem 18?

20 Let $\mathbf{v}$ be any vector in an n-dimensional vector space V. Let B and B' be ordered bases for V. What is the relationship between the coordinate vectors $[\mathbf{v}]_B$ and $[\mathbf{v}]_{B'}$?

21 Let $B = \{\mathbf{v}_1,\mathbf{v}_2,\mathbf{v}_3\}$ be an ordered basis for a 3-dimensional vector space V and let $B' = \{\mathbf{u}_1,\mathbf{u}_2\}$ be an ordered basis for a 2-dimensional vector space W. Let T be a linear transformation from V into W. Explain how to construct the matrix A of the linear transformation T with respect to the bases B and B'.

22 Let V be a finite-dimensional vector space and let $B = \{\mathbf{v}_1, \mathbf{v}_2, \ldots, \mathbf{v}_n\}$ and $B' = \{\mathbf{u}_1, \mathbf{u}_2, \ldots, \mathbf{u}_n\}$ be ordered bases for V. Let T be a linear operator on V. If A is the matrix of T with respect to the basis B and A' is the matrix of T with respect to the basis B', what is the relationship between A and A'?

23 Let A be a 3×3 matrix that is similar to I_3. Determine all the entries of the matrix A.

24 Does there exist a matrix that is not similar to any matrix?

25 Does there exist a matrix that is similar *only* to itself?

26 Let A be an $n \times n$ nonsingular matrix and let B be an $n \times n$ singular matrix. Is it possible for A and B to be similar?

27 Let A and B be similar matrices. If $\det A = 2$, what is $\det B$?

28 Show that the matrices $\begin{bmatrix} 1 & 2 \\ 3 & 4 \end{bmatrix}$ and $\begin{bmatrix} 2 & 1 \\ 3 & 4 \end{bmatrix}$ are *not* similar.

Eigenvalues and Eigenvectors

7

7-1 Definition of Eigenelements

Objectives

1 Introduce the concepts of eigenvector and eigenvalue of a linear transformation.
2 Develop a technique of finding eigenelements of a linear transformation.

Let V be a finite-dimensional vector space and T a linear operator on V. We have seen in Section 6-6 that for each ordered basis of V, the linear operator T has a matrix representation which depends on that basis. We have also seen, in Section 6-7, that if A and A' are matrices of T with respect to the ordered bases B and B', respectively, then A and A' are similar.

The examples at the end of Section 6-7 have no doubt shown that computations with matrices become very simple whenever these matrices are diagonal. We are therefore led to the following natural questions.

1 Among all possible matrices representing a given linear operator T, does there exist one which is of a diagonal type?
2 If such a diagonal matrix exists, how does one find it?

It is obvious that in order to find such a matrix (if it exists) we must be able to determine an ordered basis of V such that the matrix of T with respect to that basis is diagonal.

In this chapter we discuss the existence of such a basis, and a method of finding it (whenever it exists).

The problem of finding a diagonal matrix representing a given linear operator T on V is directly connected with the problem of finding all nonzero vectors $\mathbf{v}$ satisfying the equation

$$T\mathbf{v} = \lambda\mathbf{v} \qquad (\lambda \text{ real}) \tag{1}$$

Definition 7-1 Let T be a linear operator on an n-dimensional vector space V. We say that a **nonzero** vector $\mathbf{v}$ is an **eigenvector** of T and that the real number λ is an **eigenvalue** of T associated with $\mathbf{v}$ if the vector $\mathbf{v}$ and the scalar λ satisfy equation 1.

Remark: Some mathematicians prefer to use the terms **characteristic vector** and **characteristic value** as substitutes for the terms eigenvector and eigenvalue, respectively.

We sometimes refer to eigenvectors and eigenvalues by the common word **eigenelements.**

The equation $T\mathbf{0} = \lambda\mathbf{0}$ is satisfied for every scalar λ.

The reader should note that although the zero vector of V satisfies the equation $T\mathbf{v} = \lambda\mathbf{v}$ we do not refer to it as an eigenvector of T. The reason for this is that we are interested only in nonzero vectors satisfying equation 1. The reader should note also that according to Definition 7-1 an eigenvalue λ must be a real number. Thus, we exclude the possibility of complex eigenvalues.

It is important for the reader to realize that there exist linear transformations having no eigenvectors at all. The following example provides such a linear transformation.

Example 1 Find a linear operator $T: \mathbf{R}^2 \longrightarrow \mathbf{R}^2$ such that T has *no* eigenvectors.

Solution Let T be the linear operator performing a counterclockwise 45° rotation of $\mathbf{R}^2$. It is obvious from the geometrical nature of T that the vector $T(\mathbf{v})$ is not a scalar multiple of $\mathbf{v}$ unless $\mathbf{v}$ is the zero vector of $\mathbf{R}^2$. Thus, T possesses no eigenvectors.

Remark: The reader should note that any rotation of $\mathbf{R}^2$ that is not an integral multiple of 180° constitutes an example of a linear operator T having no eigenvectors.

While some linear operators have no eigenvectors at all, there are others which have a multitude of them. We exhibit such an operator in the following example.

Example 2 Find a linear operator T on a finite-dimensional vector space V such that each nonzero vector of V is an eigenvector of T.

Solution Let $T: V \longrightarrow V$ be the linear operator defined by

$$T(\mathbf{v}) = 3\mathbf{v}$$

It is obvious from Definition 7-1 that every nonzero vector of V is an eigenvector of T belonging to the eigenvalue $\lambda = 3$.

Examples 1 and 2 are trivial in the sense that no elaborate method is necessary to determine eigenvectors and corresponding eigenvalues.

In the following example we show how one can determine the eigenvalues and their associated eigenvectors for a 2×2 matrix. This example will guide us in developing the general technique of finding eigenvalues and associated eigenvectors of any given $n \times n$ matrix.

Example 3 Let $A = \begin{bmatrix} 3 & 4 \\ 2 & 1 \end{bmatrix}$. Find the eigenvalues of A and their associated eigenvectors.

Solution Viewing the matrix A as a linear operator on $\mathbf{R}^2$, we wish to find its eigenvalues and associated eigenvectors from the equation

$$A\mathbf{v} = \lambda\mathbf{v} \qquad [2]$$

If $\mathbf{v} = \begin{bmatrix} v_1 \\ v_2 \end{bmatrix}$, then equation 2 can be written as

$$\begin{bmatrix} 3 & 4 \\ 2 & 1 \end{bmatrix}\begin{bmatrix} v_1 \\ v_2 \end{bmatrix} = \lambda\begin{bmatrix} v_1 \\ v_2 \end{bmatrix} \qquad [3]$$

We now use the fact that $\lambda\begin{bmatrix} v_1 \\ v_2 \end{bmatrix} = \lambda\begin{bmatrix} 1 & 0 \\ 0 & 1 \end{bmatrix}\begin{bmatrix} v_1 \\ v_2 \end{bmatrix}$ to rewrite equation 3 in the form

$$\begin{bmatrix} 3 & 4 \\ 2 & 1 \end{bmatrix}\begin{bmatrix} v_1 \\ v_2 \end{bmatrix} - \lambda\begin{bmatrix} 1 & 0 \\ 0 & 1 \end{bmatrix}\begin{bmatrix} v_1 \\ v_2 \end{bmatrix} = \begin{bmatrix} 0 \\ 0 \end{bmatrix} \qquad [4]$$

Using the distributive property of matrix multiplication, we obtain

$$\left(\begin{bmatrix} 3 & 4 \\ 2 & 1 \end{bmatrix} - \lambda \begin{bmatrix} 1 & 0 \\ 0 & 1 \end{bmatrix}\right)\begin{bmatrix} v_1 \\ v_2 \end{bmatrix} = \begin{bmatrix} 0 \\ 0 \end{bmatrix} \qquad [5]$$

Equation 5 represents a linear system with a matrix of coefficients given by

$$A - \lambda I = \begin{bmatrix} 3 & 4 \\ 2 & 1 \end{bmatrix} - \lambda \begin{bmatrix} 1 & 0 \\ 0 & 1 \end{bmatrix} \qquad [6]$$

We are searching for nontrivial solutions $\begin{bmatrix} v_1 \\ v_2 \end{bmatrix}$ of equation 5. Since the homogeneous system 5 has nontrivial solutions if and only if the matrix of coefficients $A - \lambda I$ is singular, we obtain the condition

$|A - \lambda I| = \det(A - \lambda I)$.

$$|A - \lambda I| = 0 \qquad [7]$$

Combining equations 6 and 7, we have

$$\det\left(\begin{bmatrix} 3 & 4 \\ 2 & 1 \end{bmatrix} - \lambda \begin{bmatrix} 1 & 0 \\ 0 & 1 \end{bmatrix}\right) = 0 \qquad [8]$$

Equation 8 may also be written as follows:

$$\begin{vmatrix} 3 - \lambda & 4 \\ 2 & 1 - \lambda \end{vmatrix} = 0 \qquad [9]$$

Expanding the determinant in equation 9, we obtain $(3 - \lambda)(1 - \lambda) - (4)(2) = 0$, which leads to the polynomial equation

$$\lambda^2 - 4\lambda - 5 = 0 \qquad [10]$$

The solutions of equation 10 are given by $\lambda = -1$ and $\lambda = 5$.

We now have two different eigenvalues of the matrix $\begin{bmatrix} 3 & 4 \\ 2 & 1 \end{bmatrix}$, given by $\lambda = -1$ and $\lambda = 5$. In order to determine the associated eigenvectors, we use equation 5 first with $\lambda = -1$ and then with $\lambda = 5$.

For $\lambda = -1$, we obtain the system

$$\left(\begin{bmatrix} 3 & 4 \\ 2 & 1 \end{bmatrix} - (-1) \begin{bmatrix} 1 & 0 \\ 0 & 1 \end{bmatrix}\right)\begin{bmatrix} v_1 \\ v_2 \end{bmatrix} = \begin{bmatrix} 0 \\ 0 \end{bmatrix} \qquad [11]$$

Simplifying equation 11, we obtain

$$\begin{bmatrix} 4 & 4 \\ 2 & 2 \end{bmatrix}\begin{bmatrix} v_1 \\ v_2 \end{bmatrix} = \begin{bmatrix} 0 \\ 0 \end{bmatrix} \qquad [12]$$

A nonzero solution of equation 12 is given by $\begin{bmatrix} v_1 \\ v_2 \end{bmatrix} = \begin{bmatrix} 1 \\ -1 \end{bmatrix}$. Thus, the vector $\begin{bmatrix} 1 \\ -1 \end{bmatrix}$ is an eigenvector of the matrix $\begin{bmatrix} 3 & 4 \\ 2 & 1 \end{bmatrix}$ corresponding to the eigenvalue $\lambda = -1$.

Substituting $\lambda = 5$ into equation 5, we obtain

$$\left(\begin{bmatrix} 3 & 4 \\ 2 & 1 \end{bmatrix} - 5\begin{bmatrix} 1 & 0 \\ 0 & 1 \end{bmatrix}\right)\begin{bmatrix} v_1 \\ v_2 \end{bmatrix} = \begin{bmatrix} 0 \\ 0 \end{bmatrix} \qquad [13]$$

Simplifying equation 13, we obtain

$$\begin{bmatrix} -2 & 4 \\ 2 & -4 \end{bmatrix}\begin{bmatrix} v_1 \\ v_2 \end{bmatrix} = \begin{bmatrix} 0 \\ 0 \end{bmatrix} \qquad [14]$$

A nonzero solution of equation 14 is given by $\begin{bmatrix} v_1 \\ v_2 \end{bmatrix} = \begin{bmatrix} 2 \\ 1 \end{bmatrix}$. Thus, the vector $\begin{bmatrix} 2 \\ 1 \end{bmatrix}$ is an eigenvector of the matrix $\begin{bmatrix} 3 & 4 \\ 2 & 1 \end{bmatrix}$ corresponding to the eigenvalue $\lambda = 5$.

Remark The student should note that the vector $\begin{bmatrix} v_1 \\ v_2 \end{bmatrix} = \begin{bmatrix} 1 \\ -1 \end{bmatrix}$ is not a unique solution of system 12. It is easy to see that all the vectors having the form $\alpha\begin{bmatrix} 1 \\ -1 \end{bmatrix}$ are also solutions of system 12. Thus, all the vectors $\alpha\begin{bmatrix} 1 \\ -1 \end{bmatrix}$, where $\alpha \neq 0$, can serve as eigenvectors of the matrix $\begin{bmatrix} 3 & 4 \\ 2 & 1 \end{bmatrix}$ associated with the eigenvalue $\lambda = -1$.

A similar remark holds also for the vector $\begin{bmatrix} 2 \\ 1 \end{bmatrix}$ in the example above.

Guided by Example 3, we shall now derive a general procedure for finding the eigenvalues and their associated eigenvectors for $n \times n$ matrices.

$\mathbf{v} = \begin{bmatrix} v_1 \\ v_2 \\ \vdots \\ v_n \end{bmatrix}$

Let A be an $n \times n$ matrix. A nonzero vector $\mathbf{v}$ in $\mathbf{R}^n$ is an eigenvector of A belonging to the eigenvalue λ of A if it satisfies the equation

$$A\mathbf{v} = \lambda\mathbf{v} \qquad [15]$$

Equation 15 may be written as

$$A\mathbf{v} - \lambda\mathbf{v} = \mathbf{0} \qquad [16]$$

I_n is the identity matrix of order n.

We can use the distributive property of matrices to factor $\mathbf{v}$. In order to do so, we write $\mathbf{v} = I_n\mathbf{v}$ in equation 16, obtaining

$$A\mathbf{v} - \lambda I_n\mathbf{v} = \mathbf{0} \tag{17}$$

We proceed now with the factorization of $\mathbf{v}$ in equation 17 and get

$$(A - \lambda I_n)\mathbf{v} = \mathbf{0} \tag{18}$$

Equation 18 represents a linear system of n equations in the n unknowns $\mathbf{v}_1, \mathbf{v}_2, \ldots, \mathbf{v}_n$.

The homogeneous system 18 has nonzero solutions if and only if the matrix of coefficients $A - \lambda I_n$ is singular. This condition may be expressed by the determinant equation

$$|A - \lambda I_n| = 0 \tag{19}$$

Writing equation 19 in detail, we get

$$\begin{vmatrix} a_{11} - \lambda & a_{12} & \cdots & a_{1n} \\ a_{21} & a_{22} - \lambda & \cdots & a_{2n} \\ \vdots & \vdots & \cdots & \vdots \\ a_{n1} & a_{n2} & \cdots & a_{nn} - \lambda \end{vmatrix} = 0 \tag{20}$$

The expansion of the determinant in equation 20 results in a polynomial $f(\lambda)$ of degree n in λ. The zeros of the polynomial $f(\lambda)$ are the eigenvalues of the matrix A. The discussion above leads to the following definition:

Definition 7-2 Let A be an $n \times n$ matrix. The determinant $|A - \lambda I_n|$ is called the **characteristic polynomial** of the matrix A, and the equation $|A - \lambda I_n| = 0$ is called the **characteristic equation** of the matrix A.

Example 4 Find the characteristic polynomial of a 2×2 matrix A.

Solution The characteristic polynomial $f(\lambda)$ of A is given by

$$f(\lambda) = \begin{vmatrix} a_{11} - \lambda & a_{12} \\ a_{21} & a_{22} - \lambda \end{vmatrix} \tag{21}$$

The expansion of the determinant yields

$$f(\lambda) = \lambda^2 - \lambda(a_{11} + a_{22}) + (a_{11}a_{22} - a_{12}a_{21}) \tag{22}$$

Note that the constant term in equation 22, given by $a_{11}a_{22} - a_{12}a_{21}$, is exactly $|A|$.

Example 5 Find the characteristic polynomial of a 3×3 matrix A.

Solution The characteristic polynomial $f(\lambda)$ of the 3×3 matrix A is given by

$$f(\lambda) = \begin{vmatrix} a_{11} - \lambda & a_{12} & a_{13} \\ a_{21} & a_{22} - \lambda & a_{23} \\ a_{31} & a_{32} & a_{33} - \lambda \end{vmatrix} \quad [23]$$

The expansion of the determinant in equation 23 yields

$$f(\lambda) = -\lambda^3 + \lambda^2(a_{11} + a_{22} + a_{33}) - \lambda\left(\begin{vmatrix} a_{11} & a_{12} \\ a_{21} & a_{22} \end{vmatrix} + \begin{vmatrix} a_{11} & a_{13} \\ a_{31} & a_{33} \end{vmatrix} + \begin{vmatrix} a_{22} & a_{23} \\ a_{32} & a_{33} \end{vmatrix}\right) + |A| \quad [24]$$

Note again that the constant term in equation 24 is given by $|A|$.

We are now in a position to describe the general method of finding eigenvectors and their associated eigenvalues for a given $n \times n$ matrix A.

STEP 1 Construct the characteristic equation of A, namely, $|A - \lambda I_n| = 0$.

STEP 2 Find all the zeros of the equation $|A - \lambda I_n| = 0$. The zeros obtained are the eigenvalues of the matrix A.

STEP 3 For each eigenvalue λ found in Step 2, solve the system $(A - \lambda I_n)\mathbf{v} = \mathbf{0}$ by obtaining a basis for the null space of $A - \lambda I_n$.

The following examples illustrate the procedure outlined in Steps 1–3.

Example 6 Let $A = \begin{bmatrix} 1 & 1 & -1 \\ 2 & 3 & -4 \\ 4 & 1 & -4 \end{bmatrix}$. Find the eigenvalues of A and their associated eigenvectors.

Solution The characteristic equation of A is given by

$$\begin{vmatrix} 1 - \lambda & 1 & -1 \\ 2 & 3 - \lambda & -4 \\ 4 & 1 & -4 - \lambda \end{vmatrix} = 0 \quad [25]$$

Expanding the determinant in equation 25 and simplifying, we obtain

$$-\lambda^3 + 7\lambda - 6 = 0 \quad [26]$$

Multiplying equation 26 by -1, we obtain

$$\lambda^3 - 7\lambda + 6 = 0 \quad [27]$$

Inspection shows that $\lambda = 1$ is a solution of equation 27. By a well-known theorem from the algebra of polynomials, we conclude that $\lambda - 1$ is a factor of the polynomial $\lambda^3 - 7\lambda + 6$. Using division of polynomials, we find $\frac{\lambda^3 - 7\lambda + 6}{\lambda - 1} = \lambda^2 + \lambda - 6.$ Thus we have $\lambda^3 - 7\lambda + 6 = (\lambda - 1)(\lambda^2 + \lambda - 6) = (\lambda - 1)(\lambda - 2)(\lambda + 3)$.

Since

$$\lambda^3 - 7\lambda + 6 = (\lambda - 1)(\lambda - 2)(\lambda + 3) \quad \textbf{[28]}$$

we combine equations 27 and 28 to obtain

$$(\lambda - 1)(\lambda - 2)(\lambda + 3) = 0 \quad \textbf{[29]}$$

Thus, the characteristic equation 25 reduces to equation 29, from which we obtain the eigenvalues of the matrix A. We have $\lambda = 1$, $\lambda = 2$, and $\lambda = -3$.

In order to find an eigenvector of A associated with the eigenvalue $\lambda = 1$ we must determine a basis for the null space of $A - (1)I_3$. Solving the system $(A - (1)I_3)\mathbf{v} = \mathbf{0}$, we obtain $v_1 = v_2 = v_3$. Thus, a basis for the null space of $A - (1)I_3$ is $\mathbf{v} = \begin{bmatrix} 1 \\ 1 \\ 1 \end{bmatrix}$.

An eigenvector of A associated with the eigenvalue $\lambda = 2$ is determined by finding a basis for the null space of $A - 2I_3$. The solution of the system $(A - 2I_3)\mathbf{v} = \mathbf{0}$ is given by $v_1 = v_3$ and $v_2 = 2v_1$. Thus, a basis for the null space of $A - 2I_3$ is $\mathbf{v} = \begin{bmatrix} 1 \\ 2 \\ 1 \end{bmatrix}$.

Finally, an eigenvector of A associated with the eigenvalue $\lambda = -3$ is determined by finding a basis for the null space of $A - (-3)I_3$. The solution of the system $(A + 3I_3)\mathbf{v} = \mathbf{0}$ is given by $v_1 = \frac{1}{11}v_3$ and $v_2 = \frac{7}{11}v_3$. Thus, a basis for the null space of $A + 3I_3$ is $\mathbf{v} = \begin{bmatrix} 1 \\ 7 \\ 11 \end{bmatrix}$.

Example 7 Let $A = \begin{bmatrix} 0 & 1 & 2 & 1 \\ 0 & 1 & 0 & 2 \\ 0 & 0 & 1 & 0 \\ 0 & 0 & 0 & 2 \end{bmatrix}$. Find the eigenvalues of A and their associated eigenvectors.

Solution The reader should note that the matrix A is a triangular matrix in which all the entries below the main diagonal are zero. The eigenvalues of such a matrix are simply the entries of the main diagonal. This follows easily from the fact that the determinant of a triangular matrix is equal to the product of the entries in the main diagonal. In this instance we have

$$|A - \lambda I_4| = \begin{vmatrix} -\lambda & 1 & 2 & 1 \\ 0 & 1-\lambda & 0 & 2 \\ 0 & 0 & 1-\lambda & 0 \\ 0 & 0 & 0 & 2-\lambda \end{vmatrix} = -\lambda(1 - \lambda)^2(2 - \lambda)$$

Thus, the characteristic polynomial of the matrix A is given by

$$f(\lambda) = |A - \lambda I_4| = -\lambda(1-\lambda)^2(2-\lambda) \qquad [30]$$

The zeros of $f(\lambda)$ are obviously given by $\lambda = 0$, $\lambda = 1$, and $\lambda = 2$.

Thus, the eigenvalues of A are $\lambda = 0$, $\lambda = 1$, and $\lambda = 2$. The reader should note that these are exactly the entries of the main diagonal of the triangular matrix A. An eigenvector of A associated with the eigenvalue $\lambda = 0$ is obtained by finding a basis for the null space of $A - 0I_4$ (which is equal to A). The solution of the system $(A - 0I_4)\mathbf{v} = \mathbf{0}$ is given by $v_2 = v_3 = v_4 = 0$ and v_1 is an arbitrary real number. Thus, a basis for the null space of $A - 0I_4$ is $\mathbf{v} = \begin{bmatrix} 1 \\ 0 \\ 0 \\ 0 \end{bmatrix}$.

The reader should note that the eigenvalue of A given by $\lambda = 1$ is a zero of multiplicity 2 for the characteristic polynomial of A. Whenever an eigenvalue of a matrix has multiplicity greater than 1, it *may* happen that two or more linearly independent eigenvectors will be associated with the same eigenvalue. As a matter of fact, the eigenvalue $\lambda = 1$ has two linearly independent eigenvectors associated with it. In order to see this, we solve the system

$$(A - 1I_4)\mathbf{v} = \mathbf{0} \qquad [31]$$

Writing equation 31 in detail, we have

$$\begin{aligned} -v_1 + v_2 + 2v_3 + v_4 &= 0 \\ 2v_4 &= 0 \\ 0 &= 0 \\ v_4 &= 0 \end{aligned} \qquad [32]$$

System 32 is equivalent to the system

$$\begin{aligned} -v_1 + v_2 + 2v_3 &= 0 \\ v_4 &= 0 \end{aligned} \qquad [33]$$

A basis for the solution of system 33 is given by the vectors $\begin{bmatrix} 1 \\ 1 \\ 0 \\ 0 \end{bmatrix}$ and $\begin{bmatrix} 2 \\ 0 \\ 1 \\ 0 \end{bmatrix}$. These are linearly independent eigenvectors associated with the eigenvalue $\lambda = 1$.

Finally, an eigenvector of A associated with the eigenvalue $\lambda = 2$ is obtained by finding a basis for the null space of $A - 2I_4$. The solution of the system $(A - 2I_4)\mathbf{v} = \mathbf{0}$ is given by $v_1 = \frac{3}{2}v_4$, $v_2 = 2v_4$, and $v_3 = 0$. Thus, a basis for the null space of $A - 2I_4$ is $\mathbf{v} = \begin{bmatrix} 3 \\ 4 \\ 0 \\ 2 \end{bmatrix}$.

We wish to conclude this section with an important remark concerning the eigenvalues of a linear operator T on a finite-dimensional vector space V. If A is the matrix representation of T with respect to any ordered basis B, then *the eigenvalues of T are the eigenvalues of the matrix A*. To justify this statement, we must show that if A' is the matrix of T with respect to an ordered basis B', then the eigenvalues of A' are identical to the eigenvalues of A.

Since we know that the matrices A and A' are similar, it is sufficient to show that any two similar matrices have the same characteristic polynomial. This is the subject of the following theorem.

Theorem 7-1 Let A and A' be similar matrices; then they have the same characteristic polynomials.

Proof Since A and A' are similar, there exists a nonsingular matrix P such that

$$A' = P^{-1}AP \qquad [34]$$

The characteristic polynomials of A and A' are $|A - \lambda I|$ and $|A' - \lambda I|$, respectively. We must show that

$$|A' - \lambda I| = |A - \lambda I| \qquad [35]$$

Using equation 34, we obtain

$$|A' - \lambda I| = |P^{-1}AP - \lambda I| \qquad [36]$$

Since we may write $\lambda I = \lambda P^{-1}IP$, we can rewrite equation 36 in the form

$$|A' - \lambda I| = |P^{-1}AP - \lambda P^{-1}IP| \qquad [37]$$

Using the distributive properties of matrix multiplication, we obtain from equation 37 the following:

$$|A' - \lambda I| = |P^{-1}(A - \lambda I)P| \qquad [38]$$

Taking the determinant of the product $P^{-1}(A - \lambda I)P$ in equation 38, we obtain

$$|A' - \lambda I| = |P^{-1}|\,|A - \lambda I|\,|P| \qquad [39]$$

Since $|P^{-1}|$ and $|P|$ are real numbers such that $|P^{-1}|\,|P| = 1$, we finally obtain from equation 39 that $|A' - \lambda I| = |A - \lambda I|$. This proves equation 35.

It follows from Theorem 7-1 that in order to find the eigenvalues of a linear transformation $T: V \longrightarrow V$, where V is an n-dimensional vector space, we may select any matrix representation of T and find its eigenvalues. In determining the eigenvectors of T it can be shown that if A is the matrix representing T with respect to an ordered basis B, then a vector $\mathbf{v}$ in V is an eigenvector of T corresponding to the eigenvalue λ if and only if its coordinate vector $[\mathbf{v}]_B$ is an eigenvector of A associated with λ. Thus, after finding an eigenvector $[\mathbf{v}]_B$ of the matrix A representing T with respect to an ordered basis B, we are able to find $\mathbf{v}$ by reconstructing it from its coordinate vector $[\mathbf{v}]_B$.

Exercises

1 Show that the vector $\mathbf{v} = \begin{bmatrix} -2 \\ 1 \end{bmatrix}$ is an eigenvector for the matrix $A = \begin{bmatrix} 5 & 4 \\ -1 & 1 \end{bmatrix}$. What is the eigenvalue associated with the vector $\mathbf{v}$?

2 Show that the vectors $\begin{bmatrix} 1 \\ 1 \end{bmatrix}$ and $\begin{bmatrix} 3 \\ 2 \end{bmatrix}$ are eigenvectors for the matrix $A = \begin{bmatrix} 6 & -3 \\ 2 & 1 \end{bmatrix}$. What are the eigenvalues associated with each of these eigenvectors?

3 Find the characteristic equation, eigenvalues, and corresponding eigenvectors for each of the following matrices:

a $\begin{bmatrix} 2 & 1 \\ 0 & 3 \end{bmatrix}$ **b** $\begin{bmatrix} 4 & 0 \\ 0 & 4 \end{bmatrix}$ **c** $\begin{bmatrix} 0 & 0 \\ 0 & 0 \end{bmatrix}$

d $\begin{bmatrix} 3 & 2 \\ 6 & -1 \end{bmatrix}$ **e** $\begin{bmatrix} 1 & 2 \\ 3 & 2 \end{bmatrix}$

4 Find the characteristic equation of the matrix $A = \begin{bmatrix} 2 & -1 \\ 9 & 2 \end{bmatrix}$ and show that A does not have (real) eigenvalues.

5 Find the characteristic equation, eigenvalues, and corresponding eigenvectors for each of the following matrices:

a $\begin{bmatrix} 2 & 1 & 0 \\ 0 & 2 & 0 \\ 2 & 3 & 1 \end{bmatrix}$ **b** $\begin{bmatrix} 4 & 2 & -2 \\ -5 & 3 & 2 \\ -2 & 4 & 1 \end{bmatrix}$

c $\begin{bmatrix} 1 & 2 & 3 \\ 0 & 4 & 5 \\ 0 & 0 & 6 \end{bmatrix}$ **d** $\begin{bmatrix} 1 & -1 & -1 \\ 1 & 3 & 1 \\ -3 & 1 & -1 \end{bmatrix}$

6 Let $T: P_2 \longrightarrow P_2$ be the linear transformation defined by $T(a_2x^2 + a_1x + a_0) = (a_2 - a_1 - a_0)x^2 + (a_2 + 3a_1 + a_0)x - 3a_2 + a_1 - a_0$.

a Find the matrix representation of T with respect to the ordered basis $\{x^2, x, 1\}$ of P_2.

b Find the eigenvalues of T by using part a.

c Find the eigenvectors corresponding to each of the eigenvalues of T.

7 Show that if λ is an eigenvalue of a matrix A and its corresponding eigenvector is $\mathbf{v}$, then λ^2 is an eigenvalue of A^2 and its corresponding eigenvector is also $\mathbf{v}$.

8 Let k be a positive integer. Show that if λ is an eigenvalue of A, then λ^k is an eigenvalue of A^k.

9 What are the eigenvalues of A^5 if $A = \begin{bmatrix} 1 & 0 & 0 & 0 \\ 2 & -1 & 0 & 0 \\ 5 & 3 & 2 & 0 \\ 7 & 4 & 6 & -2 \end{bmatrix}$.

10 Show that the square matrices A and A^t have the same eigenvalues.

11 Give an example of a 2×2 matrix A such that an eigenvector of A is not an eigenvector of A^t.

12 Show that $\lambda = 0$ is an eigenvalue of a matrix A if and only if A is singular.

13 Let A be a nonsingular matrix having an eigenvalue λ and a corresponding eigenvector $\mathbf{v}$. Show that λ^{-1} is an eigenvalue of A^{-1} and that its corresponding eigenvector is $\mathbf{v}$.

7-2 Diagonalization

We are now in a position to provide answers to questions raised in the previous section. For the convenience of the reader we restate the questions in the form of a problem.

Problem 1 Let $T: V \longrightarrow V$ be a linear operator on an n-dimensional vector space.

Objectives
1 Establish conditions for diagonalizability of a matrix.
2 Establish a technique for diagonalizing a matrix.

a Does there exist an ordered basis for B with respect to which the matrix representation of T is diagonal?

b If such a diagonal matrix representation exists, how does one find it?

Let A be the matrix representation of T with respect to an ordered basis B of V. Any matrix representation of T with respect to another ordered basis B' is given by $P^{-1}AP$ where P is the transition matrix from B' to B.

See Theorem 6-13.

To solve Problem 1 we must determine first whether there exists an invertible matrix P such that $P^{-1}AP$ is diagonal, and secondly provide the technique of finding P whenever it exists.

We may thus restate Problem 1 in its equivalent matrix form as follows:

Problem 1′ Let A be a square matrix.

a Does there exist an invertible matrix P such that $P^{-1}AP$ is diagonal?

b How does one find P if it exists?

We will show in the sequel that the answer to both parts of Problem 1′ depends on the nature of the eigenvectors of the matrix A. But first we will introduce the concept of a *diagonalizable* matrix.

Definition 7-3 Let A be a square matrix. We say that A is **diagonalizable** if there exists an invertible matrix P such that $P^{-1}AP$ is a diagonal matrix.

In order to solve Problem 1′ we must find a sufficient condition under which a given matrix A is diagonalizable. One possible way of doing this is to search first for a necessary condition and then check whether this condition is also a sufficient one. Let us therefore assume that A is an $n \times n$ matrix which is diagonalizable. It follows from Definition 7-3 that there exists an invertible matrix P such that $P^{-1}AP$ is diagonal. Denoting the matrix $P^{-1}AP$ by D, we have

$$P^{-1}AP = D \tag{1}$$

where

$$D = \begin{bmatrix} \lambda_1 & 0 & \cdots & 0 \\ 0 & \lambda_2 & \cdots & 0 \\ \vdots & \vdots & & \vdots \\ 0 & 0 & \cdots & \lambda_n \end{bmatrix} \quad \text{and} \quad P = \begin{bmatrix} p_{11} & p_{12} & \cdots & p_{1n} \\ p_{21} & p_{22} & \cdots & p_{2n} \\ \vdots & \vdots & & \vdots \\ p_{n1} & p_{n2} & \cdots & p_{nn} \end{bmatrix}$$

Multiplying equation 1 by P from the left yields

$$AP = PD \tag{2}$$

But a simple matrix multiplication yields

$$PD = \begin{bmatrix} p_{11} & p_{12} & \cdots & p_{1n} \\ p_{21} & p_{22} & \cdots & p_{2n} \\ \vdots & \vdots & & \vdots \\ p_{n1} & p_{n2} & \cdots & p_{nn} \end{bmatrix} \begin{bmatrix} \lambda_1 & 0 & \cdots & 0 \\ 0 & \lambda_2 & \cdots & 0 \\ \vdots & \vdots & & \vdots \\ 0 & 0 & \cdots & \lambda_n \end{bmatrix}$$

$$= \begin{bmatrix} \lambda_1 p_{11} & \lambda_2 p_{12} & \cdots & \lambda_n p_{1n} \\ \lambda_1 p_{21} & \lambda_2 p_{22} & \cdots & \lambda_n p_{2n} \\ \vdots & \vdots & & \vdots \\ \lambda_1 p_{n1} & \lambda_2 p_{n2} & \cdots & \lambda_n p_{nn} \end{bmatrix}$$

Thus we may rewrite equation 2 in the form

$$AP = \begin{bmatrix} \lambda_1 p_{11} & \lambda_2 p_{12} & \cdots & \lambda_n p_{1n} \\ \lambda_1 p_{21} & \lambda_2 p_{22} & \cdots & \lambda_n p_{2n} \\ \vdots & \vdots & & \vdots \\ \lambda_1 p_{n1} & \lambda_2 p_{n2} & \cdots & \lambda_n p_{nn} \end{bmatrix} \qquad [3]$$

If we view the columns of the matrix P as vectors in $\mathbf{R}^n$ and denote them by $\mathbf{p}_1, \mathbf{p}_2, \ldots, \mathbf{p}_n$, then it follows from equation 3 that we have

$$A\mathbf{p}_1 = \lambda_1\mathbf{p}_1, \qquad A\mathbf{p}_2 = \lambda_2\mathbf{p}_2, \ldots, \qquad A\mathbf{p}_n = \lambda_n\mathbf{p}_n \qquad [4]$$

Equations 4 suggest the possibility that $\mathbf{p}_1, \mathbf{p}_2, \ldots, \mathbf{p}_n$ are eigenvectors of the matrix A having corresponding eigenvalues $\lambda_1, \lambda_2, \ldots, \lambda_n$, respectively. For the last statement to be proved valid we must demonstrate that all the vectors $\mathbf{p}_1, \mathbf{p}_2, \ldots, \mathbf{p}_n$ are nonzero vectors. But this follows from the fact that the vectors $\mathbf{p}_1, \mathbf{p}_2, \ldots, \mathbf{p}_n$ are column vectors of a nonsingular matrix P. Such a set of vectors is linearly independent and therefore cannot contain the zero vector. We have therefore proved the following theorem.

Theorem 7-2 Let A be an $n \times n$ matrix. If A is diagonalizable, then A has n linearly independent eigenvectors.

We may express Theorem 7-2 in an equivalent form by saying that the condition stipulating that A has n linearly independent eigenvectors is a necessary condition for the diagonalizability of A. The natural question here is whether this condition is also a sufficient one. In other words, if A has n linearly independent eigenvectors, is it guaranteed that A is diagonalizable?

The answer to this question is found in the following theorem.

Theorem 7-3 Let A be an $n \times n$ matrix. If A has n linearly independent eigenvectors, then A is diagonalizable.

Proof Let $\mathbf{p}_1, \mathbf{p}_2, \ldots, \mathbf{p}_n$ denote n linearly independent eigenvectors of the matrix A. Construct a matrix P such that its first column is $\mathbf{p}_1$, its second column is $\mathbf{p}_2$, and so on up to its last column, which is $\mathbf{p}_n$. Thus, the columns of the matrix P are given by

$$\mathbf{p}_1 = \begin{bmatrix} p_{11} \\ p_{21} \\ \vdots \\ p_{n1} \end{bmatrix}, \qquad \mathbf{p}_2 = \begin{bmatrix} p_{12} \\ p_{22} \\ \vdots \\ p_{n2} \end{bmatrix}, \ldots, \qquad \mathbf{p}_n = \begin{bmatrix} p_{1n} \\ p_{2n} \\ \vdots \\ p_{nn} \end{bmatrix}$$

Since $\mathbf{p}_1, \mathbf{p}_2, \ldots, \mathbf{p}_n$ are eigenvectors of the matrix A, we have

$$A\begin{bmatrix} p_{11} \\ p_{21} \\ \vdots \\ p_{n1} \end{bmatrix} = \begin{bmatrix} \lambda_1 p_{11} \\ \lambda_1 p_{21} \\ \vdots \\ \lambda_1 p_{n1} \end{bmatrix}, \quad A\begin{bmatrix} p_{12} \\ p_{22} \\ \vdots \\ p_{n2} \end{bmatrix} = \begin{bmatrix} \lambda_2 p_{12} \\ \lambda_2 p_{22} \\ \vdots \\ \lambda_2 p_{n2} \end{bmatrix}, \ldots, \quad A\begin{bmatrix} p_{1n} \\ p_{2n} \\ \vdots \\ p_{nn} \end{bmatrix} = \begin{bmatrix} \lambda_n p_{1n} \\ \lambda_n p_{2n} \\ \vdots \\ \lambda_n p_{nn} \end{bmatrix} \qquad [5]$$

where $\lambda_1, \lambda_2, \ldots, \lambda_n$ are the corresponding eigenvalues of $\mathbf{p}_1, \mathbf{p}_2, \ldots, \mathbf{p}_n$, respectively.

Equations 5 may be condensed into one matrix equation:

$$AP = \begin{bmatrix} \lambda_1 p_{11} & \lambda_2 p_{12} & \cdots & \lambda_n p_{1n} \\ \lambda_1 p_{21} & \lambda_2 p_{22} & \cdots & \lambda_n p_{2n} \\ \vdots & \vdots & & \vdots \\ \lambda_1 p_{n1} & \lambda_2 p_{n2} & \cdots & \lambda_n p_{nn} \end{bmatrix} \qquad [6]$$

Since the right-hand side of equation 6 may be written as a product of the two matrices

$$P = \begin{bmatrix} p_{11} & p_{12} & \cdots & p_{1n} \\ p_{21} & p_{22} & \cdots & p_{2n} \\ \vdots & \vdots & & \vdots \\ p_{n1} & p_{n2} & \cdots & p_{nn} \end{bmatrix} \quad \text{and} \quad D = \begin{bmatrix} \lambda_1 & 0 & \cdots & 0 \\ 0 & \lambda_2 & \cdots & 0 \\ \vdots & \vdots & & \vdots \\ 0 & 0 & \cdots & \lambda_n \end{bmatrix}$$

we obtain from equation 6 the following:

$$AP = PD \qquad [7]$$

The assumption that $\mathbf{p}_1, \mathbf{p}_2, \ldots, \mathbf{p}_n$ are linearly independent implies that the matrix P is invertible. Thus, multiplying equation 7 by P^{-1} from the left, we obtain

$$P^{-1}AP = D \qquad [8]$$

which proves that A is diagonalizable.

The proof of Theorem 7-3 essentially includes the method of diagonalizing a given $n \times n$ matrix A and the condition under which it can be done. The method of diagonalization is described in the following steps.

STEP 1 Find n linearly independent eigenvectors of the $n \times n$ matrix A.

STEP 2 If $\mathbf{p}_1, \mathbf{p}_2, \ldots, \mathbf{p}_n$ denote the eigenvectors found in Step 1, form a matrix P such that $\mathbf{p}_1$ is its first column, $\mathbf{p}_2$ its second column, and so on up to its last column, which is the eigenvector $\mathbf{p}_n$.

STEP 3 Find the matrix P^{-1} using the matrix P that was constructed in Step 2.

STEP 4 Form the product $P^{-1}AP$ and carry out the matrix multiplication. You will get a diagonal matrix whose main diagonal entries are the corresponding eigenvalues of the eigenvectors found in Step 1.

The reader should note that Step 1 is *the crucial step* in an attempt to diagonalize a given matrix A.

It follows from Theorem 7-2 that if the $n \times n$ matrix A does not have n linearly independent eigenvectors, then it is *not* diagonalizable. The following example exhibits such a matrix.

Example 1 Let $A = \begin{bmatrix} 4 & -1 \\ 1 & 2 \end{bmatrix}$. Show that A is not diagonalizable.

Solution The characteristic equation of A is given by

$$|A - \lambda I| = \begin{vmatrix} 4-\lambda & -1 \\ 1 & 2-\lambda \end{vmatrix} = (4-\lambda)(2-\lambda) - (-1)(1) \qquad [9]$$
$$= (\lambda - 3)^2 = 0$$

It follows from equation 9 that $\lambda = 3$ is an eigenvalue of A (of multiplicity 2).

To find all eigenvectors of A corresponding to $\lambda = 3$ we must solve the system of equations

$$(A - 3I)\mathbf{v} = \begin{bmatrix} 4-3 & -1 \\ 1 & 2-3 \end{bmatrix}\begin{bmatrix} v_1 \\ v_2 \end{bmatrix} = \begin{bmatrix} 1 & -1 \\ 1 & -1 \end{bmatrix}\begin{bmatrix} v_1 \\ v_2 \end{bmatrix} = \begin{bmatrix} 0 \\ 0 \end{bmatrix} \qquad [10]$$

It follows from system 10 that we must have $v_1 = v_2$. Thus, a basis for the null space of the matrix $\begin{bmatrix} 1 & -1 \\ 1 & -1 \end{bmatrix}$ is given by $\mathbf{v} = \begin{bmatrix} 1 \\ 1 \end{bmatrix}$. We conclude that the matrix $A = \begin{bmatrix} 4 & -1 \\ 1 & 2 \end{bmatrix}$ does *not* have *any two* linearly independent eigenvectors and therefore is not diagonalizable.

The following example exhibits the diagonalization procedure as outlined in Steps 1 through 4 above.

Example 2 Let $A = \begin{bmatrix} -1 & 1 & 2 \\ 0 & -2 & 1 \\ 0 & 0 & -3 \end{bmatrix}$. Find an invertible matrix P such that $P^{-1}AP$ is a diagonal matrix.

Solution

See Example 7 in Section 7-1.

Since A is a triangular matrix, it follows that the entries of its main diagonal are the eigenvalues of A.

Thus, the eigenvalues of A are given by $\lambda_1 = -1$, $\lambda_2 = -2$, and $\lambda_3 = -3$. The eigenvectors of A corresponding to the eigenvalues λ_1, λ_2, and λ_3 are obtained as nonzero solutions to the systems $(A - \lambda_1 I)\mathbf{p}_1 = \mathbf{0}$, $(A - \lambda_2 I)\mathbf{p}_2 = \mathbf{0}$, and $(A - \lambda_3 I)\mathbf{p}_3 = \mathbf{0}$, respectively. We obtain easily

$$\mathbf{p}_1 = \begin{bmatrix} 1 \\ 0 \\ 0 \end{bmatrix}, \qquad \mathbf{p}_2 = \begin{bmatrix} 1 \\ -1 \\ 0 \end{bmatrix}, \qquad \mathbf{p}_3 = \begin{bmatrix} 1 \\ 2 \\ -2 \end{bmatrix} \tag{11}$$

We now construct the matrix P using the eigenvectors $\mathbf{p}_1$, $\mathbf{p}_2$, and $\mathbf{p}_3$ from equation 11 as the first, second, and third columns of P. Thus, we have

$$P = \begin{bmatrix} 1 & 1 & 1 \\ 0 & -1 & 2 \\ 0 & 0 & -2 \end{bmatrix} \tag{12}$$

The linear independence of $\mathbf{p}_1$, $\mathbf{p}_2$, and $\mathbf{p}_3$ from equations 11 will follow immediately if we show that P in equation 12 is an invertible matrix. Thus, inverting P we obtain

$$P^{-1} = \begin{bmatrix} 1 & 1 & \frac{3}{2} \\ 0 & -1 & -1 \\ 0 & 0 & -\frac{1}{2} \end{bmatrix} \tag{13}$$

To diagonalize the matrix A we now form the matrix product $P^{-1}AP$. We obtain

$$P^{-1}AP = \begin{bmatrix} 1 & 1 & \frac{3}{2} \\ 0 & -1 & -1 \\ 0 & 0 & -\frac{1}{2} \end{bmatrix} \begin{bmatrix} -1 & 1 & 2 \\ 0 & -2 & 1 \\ 0 & 0 & -3 \end{bmatrix} \begin{bmatrix} 1 & 1 & 1 \\ 0 & -1 & 2 \\ 0 & 0 & -2 \end{bmatrix}$$

$$= \begin{bmatrix} -1 & 0 & 0 \\ 0 & -2 & 0 \\ 0 & 0 & -3 \end{bmatrix} \tag{14}$$

It follows from equation 14 that $P^{-1}AP$ is a diagonal matrix having the eigenvalues of the matrix A as the entries of its main diagonal.

After a careful study of Examples 1 and 2 the reader might get the impression that we have two contrasting situations here. On the one hand, we have the matrix $\begin{bmatrix} 4 & -1 \\ 1 & 2 \end{bmatrix}$ in Example 1 whose characteristic polynomial $p(\lambda) = (\lambda - 3)^2$ has a zero $\lambda = 3$ of multiplicity 2. This matrix is not diagonalizable because the eigenvalue $\lambda = 3$ does not lead to the required two linearly independent eigenvectors for the matrix $\begin{bmatrix} 4 & -1 \\ 1 & 2 \end{bmatrix}$. On the other hand, in Example 2 we have a matrix whose characteristic polynomial $p(\lambda) = (\lambda + 1)(\lambda + 2)(\lambda + 3)$ has three different zeros which lead to three linearly independent eigenvectors for this matrix and to the diagonalizability of it. We wish to emphasize that it is *not* correct to conclude that matrices whose characteristic polynomials have zeros of multiplicity greater than 1 are necessarily not diagonalizable. The following example exhibits a matrix whose characteristic polynomial has a zero of multiplicity 2 but is nevertheless a diagonalizable matrix.

Example 3 Let $A = \begin{bmatrix} 0 & 1 & 2 & 1 \\ 0 & 1 & 0 & 2 \\ 0 & 0 & 1 & 0 \\ 0 & 0 & 0 & 2 \end{bmatrix}$. Show that A is diagonalizable.

Solution The characteristic polynomial of A is given by $p(\lambda) = \lambda(\lambda - 1)^2(\lambda - 2)$. This follows immediately from the fact that A is a triangular matrix. Note that $\lambda = 1$ is a zero of the characteristic polynomial of A having multiplicity 2. In Example 7 in Section 7-1 we found the eigenvectors of A associated with each of the eigenvalues $\lambda = 0$, $\lambda = 1$, and $\lambda = 2$. The eigenvector associated with $\lambda = 0$ was given by $\mathbf{p}_1 = \begin{bmatrix} 1 \\ 0 \\ 0 \\ 0 \end{bmatrix}$. The two eigenvectors associated with $\lambda = 1$ were given by $\mathbf{p}_2 = \begin{bmatrix} 1 \\ 1 \\ 0 \\ 0 \end{bmatrix}$ and $\mathbf{p}_3 = \begin{bmatrix} 2 \\ 0 \\ 1 \\ 0 \end{bmatrix}$. The eigenvector associated with $\lambda = 2$ was given by $\mathbf{p}_4 = \begin{bmatrix} 3 \\ 4 \\ 0 \\ 2 \end{bmatrix}$. Constructing the matrix

P whose columns are $\mathbf{p}_1$, $\mathbf{p}_2$, $\mathbf{p}_3$, and $\mathbf{p}_4$, we obtain an invertible matrix which shows that $\mathbf{p}_1$, $\mathbf{p}_2$, $\mathbf{p}_3$, and $\mathbf{p}_4$ are linearly independent. We have

$$P = \begin{bmatrix} 1 & 1 & 2 & 3 \\ 0 & 1 & 0 & 4 \\ 0 & 0 & 1 & 0 \\ 0 & 0 & 0 & 2 \end{bmatrix}, \quad P^{-1} = \begin{bmatrix} 1 & -1 & -2 & \frac{1}{2} \\ 0 & 1 & 0 & -2 \\ 0 & 0 & 1 & 0 \\ 0 & 0 & 0 & \frac{1}{2} \end{bmatrix}$$

A simple calculation shows that we have

$$P^{-1}AP = \begin{bmatrix} 1 & -1 & -2 & \frac{1}{2} \\ 0 & 1 & 0 & -2 \\ 0 & 0 & 1 & 0 \\ 0 & 0 & 0 & \frac{1}{2} \end{bmatrix} \begin{bmatrix} 0 & 1 & 2 & 1 \\ 0 & 1 & 0 & 2 \\ 0 & 0 & 1 & 0 \\ 0 & 0 & 0 & 2 \end{bmatrix} \begin{bmatrix} 1 & 1 & 2 & 3 \\ 0 & 1 & 0 & 4 \\ 0 & 0 & 1 & 0 \\ 0 & 0 & 0 & 2 \end{bmatrix}$$

$$= \begin{bmatrix} 0 & 0 & 0 & 0 \\ 0 & 1 & 0 & 0 \\ 0 & 0 & 1 & 0 \\ 0 & 0 & 0 & 2 \end{bmatrix}$$

Based on Examples 1 and 3, we conclude that the existence of a zero of multiplicity greater than 1 for the characteristic polynomial of a given matrix A is not an indication of the diagonalizability or nondiagonalizability of this matrix.

The situation is quite different if an $n \times n$ matrix A has n distinct eigenvalues. In this case the matrix A will be shown to have n linearly independent eigenvectors corresponding to the n distinct eigenvalues and thus be diagonalizable. The general result leading to such a conclusion is given in the following theorem.

See Theorem 7-3.

Theorem 7-4 Let $\mathbf{v}_1, \mathbf{v}_2, \ldots, \mathbf{v}_k$ be eigenvectors of a square matrix A with associated eigenvalues $\lambda_1, \lambda_2, \ldots, \lambda_k$, respectively. If the eigenvalues $\lambda_1, \lambda_2, \ldots, \lambda_k$ are distinct, then the set $\{\mathbf{v}_1, \mathbf{v}_2, \ldots, \mathbf{v}_k\}$ is linearly independent.

Proof We will show that the assumption that the vectors $\mathbf{v}_1, \mathbf{v}_2, \ldots, \mathbf{v}_k$ are linearly dependent will lead to a contradiction, and thus that $\mathbf{v}_1, \mathbf{v}_2, \ldots, \mathbf{v}_k$ must be linearly independent.

We start our proof by assuming that the vectors $\mathbf{v}_1, \mathbf{v}_2, \ldots, \mathbf{v}_k$ are linearly dependent. To obtain a contradiction we take the following approach: Starting with $\mathbf{v}_1$, which as a nonzero vector constitutes a linearly independent set $\{\mathbf{v}_1\}$, we construct a maximal linearly independent set $\{\mathbf{v}_1, \mathbf{v}_2, \ldots, \mathbf{v}_s\}$. This means that s is a positive integer satisfying the condition $1 \leq s < k$ such that the set $\{\mathbf{v}_1, \mathbf{v}_2,$

$\ldots, \mathbf{v}_s\}$ is linearly independent, while the set $\{\mathbf{v}_1, \mathbf{v}_2, \ldots, \mathbf{v}_s, \mathbf{v}_{s+1}\}$, containing the additional vector $\mathbf{v}_{s+1}$, is linearly dependent. The existence of such a number s follows from the assumption that $\mathbf{v}_1, \mathbf{v}_2, \ldots, \mathbf{v}_k$ are linearly dependent. The linear dependence of the set $\{\mathbf{v}_1, \mathbf{v}_2, \ldots, \mathbf{v}_s, \mathbf{v}_{s+1}\}$ implies that there exist scalars $\alpha_1, \alpha_2, \ldots, \alpha_s, \alpha_{s+1}$ *not all zero* such that

$$\alpha_1\mathbf{v}_1 + \alpha_2\mathbf{v}_2 + \cdots + \alpha_s\mathbf{v}_s + \alpha_{s+1}\mathbf{v}_{s+1} = \mathbf{0} \qquad [15]$$

Multiplying equation 15 by A, we obtain

$$A(\alpha_1\mathbf{v}_1 + \alpha_2\mathbf{v}_2 + \cdots + \alpha_s\mathbf{v}_s + \alpha_{s+1}\mathbf{v}_{s+1}) = A\mathbf{0}$$

which after simplification and use of the equations

$$A\mathbf{v}_i = \lambda_i\mathbf{v}_i \qquad \text{for } i = 1, 2, \ldots, s, s+1$$

becomes

$$\alpha_1\lambda_1\mathbf{v}_1 + \alpha_2\lambda_2\mathbf{v}_2 + \cdots + \alpha_s\lambda_s\mathbf{v}_s + \alpha_{s+1}\lambda_{s+1}\mathbf{v}_{s+1} = \mathbf{0} \qquad [16]$$

Now we multiply equation 15 by λ_{s+1} and obtain

$$\alpha_1\lambda_{s+1}\mathbf{v}_1 + \alpha_2\lambda_{s+1}\mathbf{v}_2 + \cdots + \alpha_s\lambda_{s+1}\mathbf{v}_s + \alpha_{s+1}\lambda_{s+1}\mathbf{v}_{s+1} = \mathbf{0} \qquad [17]$$

Subtracting equation 17 from equation 16 yields the equation

$$\alpha_1(\lambda_1 - \lambda_{s+1})\mathbf{v}_1 + \alpha_2(\lambda_2 - \lambda_{s+1})\mathbf{v}_2 + \cdots + \alpha_s(\lambda_s - \lambda_{s+1})\mathbf{v}_s = \mathbf{0} \qquad [18]$$

Since the set $\mathbf{v}_1, \mathbf{v}_2, \ldots, \mathbf{v}_s\}$ is linearly independent, it follows from equation 18 that we have

$$\alpha_1(\lambda_1 - \lambda_{s+1}) = \alpha_2(\lambda_2 - \lambda_{s+1}) = \cdots = \alpha_s(\lambda_s - \lambda_{s+1}) = 0 \qquad [19]$$

But $\lambda_1 - \lambda_{s+1} \neq 0$, $\lambda_2 - \lambda_{s+1} \neq 0$, $\ldots$, $\lambda_s - \lambda_{s+1} \neq 0$ because all the λ_i's are distinct. It follows from equation 19 that we must have

$$\alpha_1 = \alpha_2 = \cdots = \alpha_s = 0 \qquad [20]$$

Using equations 20 in equation 15 reduces the latter to

$$\alpha_{s+1}\mathbf{v}_{s+1} = \mathbf{0} \qquad [21]$$

Note: $\mathbf{v}_{s+1}$ is an eigenvector.

Since $\mathbf{v}_{s+1}$ is a nonzero vector we must conclude from equation 21 that

$$\alpha_{s+1} = 0 \qquad [22]$$

See equation 15.

Equations 20 and 22 provide us with a contradiction because we assumed earlier that the scalars $\alpha_1, \alpha_2, \ldots, \alpha_s, \alpha_{s+1}$ were *not all zero*. We therefore conclude that the set $\{\mathbf{v}_1, \mathbf{v}_2, \ldots, \mathbf{v}_k\}$ is linearly independent.

As an immediate consequence of Theorem 7-4 we obtain the previously announced result, which we may state as follows:

Theorem 7-5 Let A be an $n \times n$ matrix. If A has n distinct eigenvalues, then A is diagonalizable.

Proof Let $\mathbf{v}_1, \mathbf{v}_2, \ldots, \mathbf{v}_n$ be the eigenvectors of A associated with the distinct eigenvalues $\lambda_1, \lambda_2, \ldots, \lambda_n$, respectively. It follows from Theorem 7-4 that the set $\{\mathbf{v}_1, \mathbf{v}_2, \ldots, \mathbf{v}_n\}$ is linearly independent. Since the matrix A has n linearly independent eigenvectors, it follows from Theorem 7-3 that it is diagonalizable.

We conclude this section with an example which shows how to find an ordered basis for a vector space V such that a linear operator defined on V will have a diagonal matrix representation with respect to that basis.

Example 4 Let $T: P_1 \longrightarrow P_1$ be the linear operator defined by

P_1 is the vector space of all polynomials of degree ≤ 1 and the zero polynomial.

$$T(a_0 + a_1x) = a_0 + (2a_0 - a_1)x$$

Find an ordered basis for P_1 with respect to which the matrix of T is diagonal.

Solution Let A be the matrix of T with respect to the ordered basis $B = \{1,x\}$ of the vector space P_1. We know that the first and second columns of A are given by

$$[T(1)]_B = [1 + 2x]_B = \begin{bmatrix} 1 \\ 2 \end{bmatrix} \quad \text{and} \quad [T(x)]_B = [-x]_B = \begin{bmatrix} 0 \\ -1 \end{bmatrix}$$

respectively. Thus, we have

$$A = \begin{bmatrix} 1 & 0 \\ 2 & -1 \end{bmatrix}$$

The eigenvalues of A are obviously $\lambda_1 = 1$ and $\lambda_2 = -1$ and the corresponding eigenvectors are $\mathbf{p}_1 = \begin{bmatrix} 1 \\ 1 \end{bmatrix}$ and $\mathbf{p}_2 = \begin{bmatrix} 0 \\ 1 \end{bmatrix}$, respectively. Using $\mathbf{p}_1$ and $\mathbf{p}_2$ as the columns of a matrix P (which will diagonalize A), we have

$$P = \begin{bmatrix} 1 & 0 \\ 1 & 1 \end{bmatrix}$$

A simple computation yields

$$P^{-1} = \begin{bmatrix} 1 & 0 \\ -1 & 1 \end{bmatrix}$$

Computing the product $P^{-1}AP$, we obtain

$$P^{-1}AP = \begin{bmatrix} 1 & 0 \\ -1 & 1 \end{bmatrix}\begin{bmatrix} 1 & 0 \\ 2 & -1 \end{bmatrix}\begin{bmatrix} 1 & 0 \\ 1 & 1 \end{bmatrix} = \begin{bmatrix} 1 & 0 \\ 0 & -1 \end{bmatrix} \qquad \textbf{[23]}$$

It follows from equation 23 that the matrix $A' = P^{-1}AP$ is diagonal and may be regarded as the matrix of T with respect to some ordered basis B'. The polynomials in the basis B' must be the eigenvectors of T.

We obtain these polynomials from $\mathbf{p}_1 = \begin{bmatrix} 1 \\ 1 \end{bmatrix}$ and $\mathbf{p}_2 = \begin{bmatrix} 0 \\ 1 \end{bmatrix}$, which represent the coordinate vectors of the polynomials with respect to the ordered basis $B = \{1,x\}$.

The vector $\mathbf{p}_1 = \begin{bmatrix} 1 \\ 1 \end{bmatrix}$ leads to the polynomial $1 + x$ and the vector $\mathbf{p}_2 = \begin{bmatrix} 0 \\ 1 \end{bmatrix}$ leads to the polynomial x. Thus, the ordered basis B' is given by $B' = \{1 + x, x\}$.

An easy check shows that we have

$$T(1 + x) = 1 + x, \qquad T(x) = (-1)x \qquad \textbf{[24]}$$

It follows from equations 24 that the matrix of T with respect to the ordered basis B' is diagonal.

Exercises

1 Let $A = \begin{bmatrix} 1 & -1 \\ -1 & 1 \end{bmatrix}$. Find a matrix P such that $P^{-1}AP$ is a diagonal matrix. Compute the product $P^{-1}AP$ to check that you obtain a diagonal matrix.

2 Repeat problem 1 for $A = \begin{bmatrix} 13 & 3 \\ 9 & 7 \end{bmatrix}$.

3 Let $A = \begin{bmatrix} 3 & -1 \\ 4 & -1 \end{bmatrix}$. Show that A is not diagonalizable.

4 Repeat problem 3 for $A = \begin{bmatrix} 5 & 4 \\ -1 & 1 \end{bmatrix}$.

5 Diagonalize the matrix A given by $A = \begin{bmatrix} 1 & -1 & -1 \\ -1 & 1 & -1 \\ -1 & -1 & 1 \end{bmatrix}$.

6 Let $A = \begin{bmatrix} 0 & 1 & 1 \\ 0 & 0 & 1 \\ 0 & 0 & 0 \end{bmatrix}$. Show that A is not diagonalizable.

7 Diagonalize the matrix A given by $A = \begin{bmatrix} 1 & -1 & -1 \\ 1 & 3 & 1 \\ -3 & 1 & -1 \end{bmatrix}$.

8 Let $A = \begin{bmatrix} a & b \\ c & d \end{bmatrix}$. Use Theorem 7-5 to show that if $(a - d)^2 + 4bc > 0$, then A is diagonalizable.

9 Show that

a $(P^{-1}AP)^2 = P^{-1}A^2P$,

b $(P^{-1}AP)^3 = P^{-1}A^3P$, and in general

c $(P^{-1}AP)^k = P^{-1}A^kP$ for every positive integer k.

10 Let $A = \begin{bmatrix} 14 & 12 \\ -16 & -14 \end{bmatrix}$. Compute A^8 using the following method:

a Find a matrix Q such that $Q^{-1}AQ = D$ where D is a diagonal matrix.

b Use problem 9 to find A^8.

11 Let $A = \begin{bmatrix} a & b & c \\ 0 & d & e \\ 0 & 0 & f \end{bmatrix}$. Show that if $(a - d)(a - f)(d - f) \neq 0$, then A is diagonalizable.

12 We say that a square matrix A is **nilpotent** if $A^k = 0$ for some positive integer k. Show that if A is a nilpotent matrix and λ is its eigenvalue then $\lambda = 0$.

Hint: Use problem 12.

13 Show that a nonzero nilpotent matrix A is not diagonalizable.

14 Use problem 13 to show that the following matrices are not diagonalizable.

a $\begin{bmatrix} 0 & 1 \\ 0 & 0 \end{bmatrix}$ b $\begin{bmatrix} 0 & 1 & 1 \\ 0 & 0 & 1 \\ 0 & 0 & 0 \end{bmatrix}$ c $\begin{bmatrix} 0 & 1 & 1 & 1 \\ 0 & 0 & 1 & 1 \\ 0 & 0 & 0 & 1 \\ 0 & 0 & 0 & 0 \end{bmatrix}$

15 Let A and B be similar matrices. Show that if A is diagonalizable then so is B.

Hint: Show that AB and BA are similar and use problem 15.

16 Let A and B be $n \times n$ matrices. If B is nonsingular and AB is diagonalizable, show that BA is also diagonalizable.

17 Show that if A is diagonalizable then A^t is also diagonalizable.

18 Use problem 9 to show that if A is diagonalizable then A^k is diagonalizable for every positive integer k.

19 Let $T: P_1 \longrightarrow P_1$ be the linear operator defined by $T(a_0 + a_1x) = a_0 + (a_0 - 3a_1)x$. Find an ordered basis for P_1 with respect to which the matrix of T is diagonal.

7-3 Symmetric Matrices and Diagonalization

Objectives
1. Introduce orthogonal matrices
2. Diagonalize symmetric matrices by means of orthogonal matrices.

We have seen in the previous section that the problem of determining the diagonalizability of a given $n \times n$ matrix A is quite involved. Theorem 7-5 asserts that if A has n distinct (real) eigenvalues, then A is diagonalizable. In order to be able to make use of this result we have to compute all the eigenvalues of A, and they must all turn out to be distinct. If any of the eigenvalues is a repeated zero of the characteristic polynomial of A, as in Examples 1 and 3 of Section 7-2, further investigation is necessary. To guarantee the diagonalizability of A we must show that an eigenvalue of multiplicity k will have k corresponding linearly independent eigenvectors so that in the final count we will have n linearly independent eigenvectors of A. Thus, in general, we must go through a substantial part of the diagonalizing procedure before we can determine if a given matrix is diagonalizable at all.

See Theorem 7-3.

A particular class of matrices which does not share this disadvantage is the class consisting of all $n \times n$ matrices A satisfying the condition $A = A^t$ where A^t is the transpose of A. We refer to this class of matrices as the class of *symmetric matrices.*

See Definition 3-12.

Linear transformations with symmetric matrices occur quite commonly in many problems in dynamics, geometry, and the theory of relativity (to mention just a few areas), and therefore the study of symmetric matrices and their diagonalizability is of major practical importance. In this section we discuss in detail the diagonalization of symmetric matrices.

We formalize the concept of symmetry in matrices by the following definition.

Definition 7-4

Symmetric matrices.

Let A be a square matrix. We say that A is **symmetric** if it satisfies the condition

$$A = A^t$$

where A^t is the transpose of A.

Example 1 Determine which of the following matrices are symmetric:

$$A = \begin{bmatrix} 1 & 2 \\ 2 & 3 \end{bmatrix}, \qquad B = \begin{bmatrix} 0 & 1 \\ -1 & 2 \end{bmatrix},$$

$$C = \begin{bmatrix} 1 & 2 & 3 \\ 2 & 4 & 5 \\ 3 & -5 & 6 \end{bmatrix}, \qquad D = \begin{bmatrix} 1 & 4 & 5 \\ 4 & 2 & 6 \\ 5 & 6 & 3 \end{bmatrix}$$

Solution The matrices A and D are symmetric, since $A = A^t$ and $D = D^t$. The matrix B is not symmetric, since $b_{12} = -1$ and $b_{21} = 1$; therefore $B \neq B^t$. Similarly, the matrix C is not symmetric.

The simplest way to check whether a given $n \times n$ matrix A is symmetric or not is to examine each entry a_{ij} above the main diagonal against the entry a_{ji} below the main diagonal. These entries are said to be symmetrically located with respect to the main diagonal. Thus, if $a_{ij} = a_{ji}$ for $i, j = 1, 2, \ldots, n$, then A is a symmetric matrix. There is no condition imposed on the entries a_{ii} in the main diagonal.

Symmetric matrices have some special properties which bear directly on their diagonalizability.

If we temporarily abandon our agreement to deal only with real numbers in this text and (just for this discussion) allow eigenvalues to be complex numbers by making the proper change in Definition 7-1, then we can show that there exist matrices with real entries whose eigenvalues however are complex numbers. One such example is furnished by the matrix $A = \begin{bmatrix} 0 & 1 \\ -1 & 0 \end{bmatrix}$. The characteristic polynomial of A is $f(\lambda) = \lambda^2 + 1$, and its zeros are given by $\lambda = \sqrt{-1} = i$ and $\lambda = -\sqrt{-1} = -i$.

Since $i^2 = (-i)^2 = -1$ we have $i^2 + 1 = 0$ and $(-i)^2 + 1 = 0$.

This situation can *never* happen in dealing with a symmetric matrix because we have the following result:

Theorem 7-6 If A is a symmetric matrix then *all* its eigenvalues are real.

The proof, which involves manipulation with complex numbers, is omitted.

For a proof of Theorem 7-6, see *Introduction to Linear Algebra* by F. Hohn, p. 268.

Example 2 Let $A = \begin{bmatrix} 1 & 2 \\ 2 & 3 \end{bmatrix}$. Show that all the eigenvalues of A are real.

Solution

Since the matrix A is symmetric, it follows from Theorem 7-6 that all its eigenvalues are real. This fact can be checked easily by direct computation.

The reader should have no difficulty in showing that the characteristic polynomial of A is given by $f(\lambda) = \lambda^2 - 3\lambda - 2$ and that its zeros are the real numbers $\lambda = \dfrac{3 + \sqrt{17}}{2}$ and $\lambda = \dfrac{3 - \sqrt{17}}{2}$.

Another property of symmetric matrices which essentially guarantees their diagonalizability is stated in the following theorem.

Theorem 7-7 Let λ_1 be an eigenvalue of a symmetric matrix A. If λ_1 is a zero of the characteristic polynomial of A of multiplicity k, then the matrix A has k linearly independent eigenvectors associated with the eigenvalue λ_1.

λ_1 is a zero of the polynomial $f(\lambda)$ of multiplicity k if and only if the factorization of $f(\lambda)$ contains the factor $\lambda - \lambda_1$ exactly k times.

The proof, which is beyond the scope of this text, is omitted. The important fact that follows from Theorems 7-6 and 7-7 is that every $n \times n$ symmetric matrix A has n real linearly independent eigenvectors. Therefore, it follows from Theorem 7-3 that A is diagonalizable. The reader should recall that in order to carry out the diagonalization process we must have n linearly independent eigenvectors of A and

For a proof of Theorem 7-7, see *Introduction to Linear Algebra* by F. Hohn, p. 272.

then proceed with Steps 1 through 4 as outlined in Section 7-2. Thus, if P is an $n \times n$ matrix whose columns are the n linearly independent eigenvectors of A mentioned above, then we know that the matrix $P^{-1}AP$ is a diagonal matrix. We wish to show now that the matrix P which diagonalizes the symmetric matrix A can be modified so that it has a very special character. For this purpose we need to introduce the concept of an *orthogonal* matrix.

Definition 7-5 Let A be an $n \times n$ matrix. We say that A is an **orthogonal** matrix if the column vectors in A form an orthonormal set in $\mathbf{R}^n$.

Example 3 Let $A = \begin{bmatrix} \frac{1}{\sqrt{2}} & \frac{-1}{\sqrt{2}} \\ \frac{1}{\sqrt{2}} & \frac{1}{\sqrt{2}} \end{bmatrix}$. Show that A is orthogonal.

Solution The column vectors of A are given by

$$\mathbf{a}_1 = \begin{bmatrix} \frac{1}{\sqrt{2}} \\ \frac{1}{\sqrt{2}} \end{bmatrix} \quad \text{and} \quad \mathbf{a}_2 = \begin{bmatrix} \frac{-1}{\sqrt{2}} \\ \frac{1}{\sqrt{2}} \end{bmatrix}$$

Since $\mathbf{a}_1 \cdot \mathbf{a}_1 = 1$, $\mathbf{a}_2 \cdot \mathbf{a}_2 = 1$, and $\mathbf{a}_1 \cdot \mathbf{a}_2 = 0$, we conclude that the set $\{\mathbf{a}_1,\mathbf{a}_2\}$ forms an orthonormal set in $\mathbf{R}^2$. By Definition 7-5, A is an orthogonal matrix.

Example 4 Let $A = \begin{bmatrix} \frac{1}{3} & 0 & \frac{-4}{\sqrt{18}} \\ \frac{2}{3} & \frac{1}{\sqrt{2}} & \frac{1}{\sqrt{18}} \\ \frac{2}{3} & \frac{-1}{\sqrt{2}} & \frac{1}{\sqrt{18}} \end{bmatrix}$. Show that A is orthogonal.

Solution The column vectors of A are given by

$$\mathbf{a}_1 = \begin{bmatrix} \frac{1}{3} \\ \frac{2}{3} \\ \frac{2}{3} \end{bmatrix}, \quad \mathbf{a}_2 = \begin{bmatrix} 0 \\ \frac{1}{\sqrt{2}} \\ \frac{-1}{\sqrt{2}} \end{bmatrix}, \quad \text{and} \quad \mathbf{a}_3 = \begin{bmatrix} \frac{-4}{\sqrt{18}} \\ \frac{1}{\sqrt{18}} \\ \frac{1}{\sqrt{18}} \end{bmatrix}$$

A simple computation yields the equations

$$\mathbf{a}_1 \cdot \mathbf{a}_1 = \mathbf{a}_2 \cdot \mathbf{a}_2 = \mathbf{a}_3 \cdot \mathbf{a}_3 = 1$$

and

$$\mathbf{a}_1 \cdot \mathbf{a}_2 = \mathbf{a}_1 \cdot \mathbf{a}_3 = \mathbf{a}_2 \cdot \mathbf{a}_3 = 0$$

Thus, it follows that $\{\mathbf{a}_1,\mathbf{a}_2,\mathbf{a}_3\}$ is an orthonormal set in $\mathbf{R}^3$. Therefore, by Definition 7-5, the matrix A is orthogonal.

Orthogonal matrices have some interesting properties. If we construct A^t from the matrix A in Example 3 and form the product A^tA, we obtain

$$A^tA = \begin{bmatrix} \frac{1}{\sqrt{2}} & \frac{1}{\sqrt{2}} \\ \frac{-1}{\sqrt{2}} & \frac{1}{\sqrt{2}} \end{bmatrix} \begin{bmatrix} \frac{1}{\sqrt{2}} & \frac{-1}{\sqrt{2}} \\ \frac{1}{\sqrt{2}} & \frac{1}{\sqrt{2}} \end{bmatrix} = \begin{bmatrix} 1 & 0 \\ 0 & 1 \end{bmatrix} = I_2 \qquad [1]$$

Thus, it follows from equation 1 that the matrix $A = \begin{bmatrix} \frac{1}{\sqrt{2}} & \frac{-1}{\sqrt{2}} \\ \frac{1}{\sqrt{2}} & \frac{1}{\sqrt{2}} \end{bmatrix}$ is invertible and that its inverse A^{-1} is given by the transpose of A, namely $A^t = \begin{bmatrix} \frac{1}{\sqrt{2}} & \frac{1}{\sqrt{2}} \\ \frac{-1}{\sqrt{2}} & \frac{1}{\sqrt{2}} \end{bmatrix}$. The property expressed by the equation $A^{-1} = A^t$ holds for every orthogonal matrix A. As a matter of fact we have the following result:

Theorem 7-8 Let A be an $n \times n$ matrix. Then A is orthogonal if and only if $A^{-1} = A^t$.

Proof Let $\mathbf{a}_1, \mathbf{a}_2, \ldots, \mathbf{a}_n$ denote the columns of the matrix A. If we assume that A is orthogonal, then $\{\mathbf{a}_1, \mathbf{a}_2, \ldots, \mathbf{a}_n\}$ is an orthonormal set and we have

$$\mathbf{a}_i \cdot \mathbf{a}_j = \delta_{ij} \qquad (i,j = 1, 2, \ldots, n) \qquad [2]$$

In order to show that $A^{-1} = A^t$ we examine the matrix product A^tA. The (i,j) entry of the matrix A^tA is the dot product of the ith row of A^t with the jth column of A. Since the ith row of A^t is actually the ith column of A, it follows that the (i,j) entry of A^tA is given by $\mathbf{a}_i \cdot \mathbf{a}_j$. We conclude from equation 2 that

$$A^tA = I_n \qquad [3]$$

It follows from equation 3 that $A^t = A^{-1}$.

To complete the proof of this theorem one must show that the condition $A^t = A^{-1}$ implies the orthogonality of A. This part of the proof is left as an exercise for the reader.

See problem 11.

In order to discover another property of orthogonal matrices, let us focus our attention on the rows of an orthogonal matrix. It is easily verified that the row vectors of the orthogonal matrix

See Example 3.

$$\begin{bmatrix} \frac{1}{\sqrt{2}} & \frac{-1}{\sqrt{2}} \\ \frac{1}{\sqrt{2}} & \frac{1}{\sqrt{2}} \end{bmatrix}$$

form an orthonormal set of vectors in $\mathbf{R}^2$.

Similarly, the row vectors of the matrix

See Example 4.

$$\begin{bmatrix} \frac{1}{3} & 0 & \frac{-4}{\sqrt{18}} \\ \frac{2}{3} & \frac{1}{\sqrt{2}} & \frac{1}{\sqrt{18}} \\ \frac{2}{3} & \frac{-1}{\sqrt{2}} & \frac{1}{\sqrt{18}} \end{bmatrix}$$

form an orthonormal set of vectors in $\mathbf{R}^3$. As a matter of fact we have the following general result:

Theorem 7-9 Let A be an $n \times n$ matrix. Then A is orthogonal if and only if the row vectors of A form an orthonormal set of vectors in $\mathbf{R}^n$.

Proof Assume that A is orthogonal. Then it follows from Theorem 7-8 that $A^{-1} = A^t$. Taking the transpose of both sides of this equation, we obtain

$$(A^{-1})^t = (A^t)^t \qquad [4]$$

See problem 13 in Section 3-5.

Since $(A^{-1})^t = (A^t)^{-1}$ we may rewrite equation 4 as

$$(A^t)^{-1} = (A^t)^t \qquad [5]$$

See Definition 7-5.

Remark: The equation $A^{-1} = A^t$ which characterizes orthogonal matrices can be used in the

Equation 5 states that the matrix A^t has the property that its inverse is equal to its transpose. Thus, by Theorem 7-8, A^t is an orthogonal matrix. This means that the column vectors of A^t form an orthonormal set of vectors in $\mathbf{R}^n$. But the column vectors of A^t are the row vectors of A, and hence the row vectors of A form an ortho-

following two ways:
a To compute the inverse of a given orthogonal matrix A, we simply find the transpose of A.
b To check whether a given matrix is orthogonal, we form the product A^tA. We know that A is orthogonal if and only if $A^tA = I$.

normal set of vectors in $\mathbf{R}^n$. Thus, we have proved that if A is orthogonal then the row vectors of A form an orthonormal set of vectors in $\mathbf{R}^n$.

The proof of the converse is similar in nature and is left as an exercise for the reader. (See problem 12.)

Now that we are familiar with the concept of an orthogonal matrix we pose the following problem.

Problem 2 Let A be an $n \times n$ matrix and T an $n \times n$ orthogonal matrix such that $T^{-1}AT$ is diagonal. What can we conclude about the symmetry of A?

Let us denote by D the diagonal matrix $T^{-1}AT$. Thus we have the equation

$$D = T^{-1}AT \quad [6]$$

Multiplying equation 6 from the left by T and from the right by T^{-1}, we obtain the equation

$$TDT^{-1} = A \quad [7]$$

Since T is orthogonal we have $T^{-1} = T^t$ and equation 7 may be rewritten as

$$TDT^t = A \quad [8]$$

Note: $(ABC)^t = C^tB^tA^t$ whenever the matrix product ABC is defined.

Taking the transpose of both sides of equation 8 and using the fact that $(TDT^t)^t = (T^t)^tD^tT^t$, we obtain

$$(T^t)^tD^tT^t = A^t \quad [9]$$

$D^t = D$ because D is a diagonal matrix.

Since $(T^t)^t = T$ and $D^t = D$, we obtain from equation 9 the result that

$$TDT^t = A^t \quad [10]$$

Combining equations 8 and 10, we conclude that

$$A = A^t \quad [11]$$

It follows from equation 11 that A is a symmetric matrix. We have therefore proved the following theorem.

Theorem 7-10 Let A be an $n \times n$ matrix. If A is diagonalizable by an orthogonal matrix, then A is symmetric.

We have already discussed the fact that each symmetric matrix is diagonalizable. We have also hinted that the diagonalization may be performed by means of a special matrix. The introduction of the concept of an orthogonal matrix was done with this fact in mind.

Therefore we are naturally led to investigate the validity of the converse of Theorem 7-10, which we now state as:

Problem 3 Given that A is a symmetric matrix, does there exist an orthogonal matrix T such that the matrix $T^{-1}AT$ is diagonal?

Before we answer the question raised in Problem 3 we shall first establish the following important result.

Theorem 7-11 Let A be a symmetric matrix. Let λ_1 and λ_2 be eigenvalues of A with associated eigenvectors $\mathbf{v}_1$ and $\mathbf{v}_2$, respectively. If $\lambda_1 \neq \lambda_2$, then $\mathbf{v}_1$ and $\mathbf{v}_2$ are orthogonal.

Proof We must show that $\mathbf{v}_1 \cdot \mathbf{v}_2 = 0$. In order to do this, we write the dot product $\mathbf{v}_1 \cdot \mathbf{v}_2$ as a matrix product. This can easily be done by regarding each of the vectors $\mathbf{v}_1$ and $\mathbf{v}_2$ as a column matrix. Thus, if we write

$$\mathbf{v}_1 = \begin{bmatrix} \alpha_1 \\ \alpha_2 \\ \vdots \\ \alpha_n \end{bmatrix} \quad \text{and} \quad \mathbf{v}_2 = \begin{bmatrix} \beta_1 \\ \beta_2 \\ \vdots \\ \beta_n \end{bmatrix}$$

then we have

$\mathbf{v}_1{}^t = [\alpha_1, \alpha_2, \ldots, \alpha_n]$

$$\mathbf{v}_1{}^t\mathbf{v}_2 = [\alpha_1, \alpha_2, \ldots, \alpha_n]\begin{bmatrix} \beta_1 \\ \beta_2 \\ \vdots \\ \beta_n \end{bmatrix} = \mathbf{v}_1 \cdot \mathbf{v}_2 \qquad [12]$$

We now consider the matrix product $\mathbf{v}_1{}^tA\mathbf{v}_2$. Since the product $\mathbf{v}_1{}^tA\mathbf{v}_2$ is actually a 1×1 matrix, we have

$$(\mathbf{v}_1{}^tA\mathbf{v}_2)^t = \mathbf{v}_1{}^tA\mathbf{v}_2 \qquad [13]$$

Using the fact that $(\mathbf{v}_1{}^tA\mathbf{v}_2)^t = \mathbf{v}_2{}^tA^t(\mathbf{v}_1{}^t)^t$ we may rewrite equation 13, obtaining

$$\mathbf{v}_2{}^tA^t(\mathbf{v}_1{}^t)^t = \mathbf{v}_1{}^tA\mathbf{v}_2 \qquad [14]$$

Since A is symmetric and $(\mathbf{v}_1{}^t)^t = \mathbf{v}_1$, we obtain from equation 14 the equation

$$\mathbf{v}_2{}^tA\mathbf{v}_1 = \mathbf{v}_1{}^tA\mathbf{v}_2 \qquad [15]$$

Using the fact that $A\mathbf{v}_1 = \lambda_1\mathbf{v}_1$ and $A\mathbf{v}_2 = \lambda_2\mathbf{v}_2$, we have

$$\mathbf{v}_2{}^t(A\mathbf{v}_1) = \mathbf{v}_2{}^t(\lambda_1\mathbf{v}_1) = \lambda_1\mathbf{v}_2{}^t\mathbf{v}_1 = \lambda_1(\mathbf{v}_2 \cdot \mathbf{v}_1) \qquad [16]$$

and

$$\mathbf{v}_1^t(A\mathbf{v}_2) = \mathbf{v}_1^t(\lambda_2\mathbf{v}_2) = \lambda_2\mathbf{v}_1^t\mathbf{v}_2 = \lambda_2(\mathbf{v}_1 \cdot \mathbf{v}_2) \quad [17]$$

Substituting from equation 16 and equation 17 into equation 15, we obtain

$$\lambda_1(\mathbf{v}_2 \cdot \mathbf{v}_1) = \lambda_2(\mathbf{v}_1 \cdot \mathbf{v}_2) \quad [18]$$

Since $\mathbf{v}_1 \cdot \mathbf{v}_2 = \mathbf{v}_2 \cdot \mathbf{v}_1$ we may rewrite equation 18 in the form

$$(\lambda_1 - \lambda_2)(\mathbf{v}_1 \cdot \mathbf{v}_2) = 0 \quad [19]$$

If $\lambda_1 \neq \lambda_2$ then $\lambda_1 - \lambda_2 \neq 0$ and we must have $\mathbf{v}_1 \cdot \mathbf{v}_2 = 0$. This means that $\mathbf{v}_1$ and $\mathbf{v}_2$ are orthogonal vectors.

Example 4 Let $A = \begin{bmatrix} 2 & 1 \\ 1 & 2 \end{bmatrix}$. Show that the matrix A has eigenvectors $\mathbf{v}_1$ and $\mathbf{v}_2$ which are orthogonal.

Solution The equation $|A - \lambda I| = 0$ leads to the eigenvalues $\lambda_1 = 1$ and $\lambda_2 = 3$. An eigenvector $\mathbf{v}_1$ of A associated with the eigenvalue $\lambda_1 = 1$ is given by

The student is encouraged to show that in fact $\lambda_1 = 1$ and $\lambda_2 = 3$.

$$\mathbf{v}_1 = \begin{bmatrix} 1 \\ -1 \end{bmatrix}$$

An eigenvector $\mathbf{v}_2$ of A associated with the eigenvalue $\lambda_2 = 3$ is given by $\mathbf{v}_2 = \begin{bmatrix} 1 \\ 1 \end{bmatrix}$. We obviously have $\mathbf{v}_1 \cdot \mathbf{v}_2 = 0$. This result has been guaranteed by Theorem 7-11, because A is a symmetric matrix whose eigenvalues λ_1 and λ_2 are different.

The main result of this section, which is a solution to Problem 3, is now stated as the converse of Theorem 7-10.

Theorem 7-12 Let A be a symmetric matrix. Then there exists an orthogonal matrix T such that $T^{-1}AT$ is diagonal.

For a proof of Theorem 7-12, see *Introduction to Linear Algebra* by F. Hohn, pp. 272–73.

The proof of the theorem is beyond the scope of this book and is therefore omitted. In order to show the applicability of Theorem 7-12, we outline the steps that one should use in constructing an orthogonal matrix T which will diagonalize a given $n \times n$ symmetric matrix A.

STEP 1 Find all eigenvalues of the given symmetric matrix A and determine their multiplicity as zeros of the characteristic polynomial of A.

STEP 2 If λ is an eigenvalue of A having multiplicity k, find a linearly independent set of k eigenvectors of A associated with λ.

The existence of such a set of k vectors is guaranteed by Theorem 7-7.

STEP 3 Replace each linearly independent set of eigenvectors found in Step 2 with an orthonormal set of eigenvectors by applying the Gram-Schmidt process. The totality of all such eigenvectors will constitute an orthonormal set of n eigenvectors.

See Section 5-3.

STEP 4 Construct the matrix T whose columns are the vectors of the orthonormal set of n eigenvectors obtained in Step 3, and check that the matrix $T^{-1}AT$ is in fact diagonal.

Example 5 Let $A = \begin{bmatrix} 1 & -1 \\ -1 & 1 \end{bmatrix}$. Find an orthogonal matrix T such that $T^{-1}AT$ is a diagonal matrix.

Note that A is symmetric.

Solution The characteristic polynomial of A is given by

$$f(\lambda) = |A - \lambda I| = \begin{vmatrix} 1-\lambda & -1 \\ -1 & 1-\lambda \end{vmatrix} = (1+\lambda)^2 - (-1)^2 \qquad [20]$$

$$= \lambda^2 - 2\lambda$$

It follows from equation 20 that the eigenvalues of A are $\lambda_1 = 0$ and $\lambda_2 = 2$. Since the factorization of the characteristic polynomial of A is given by $f(\lambda) = \lambda(\lambda - 2)$, it follows that the multiplicity of each eigenvalue is 1.

Having completed Step 1, we proceed now to Step 2 and find one eigenvector associated with each of the eigenvalues $\lambda_1 = 0$ and $\lambda_2 = 2$.

Solving the systems $(A - \lambda_1 I)\mathbf{v}_1 = \mathbf{0}$ and $(A - \lambda_2 I)\mathbf{v}_2 = \mathbf{0}$, we obtain the eigenvectors $\mathbf{v}_1 = \begin{bmatrix} 1 \\ 1 \end{bmatrix}$ and $\mathbf{v}_2 = \begin{bmatrix} 1 \\ -1 \end{bmatrix}$ associated with $\lambda_1 = 0$ and $\lambda_2 = 2$, respectively.

We proceed now to Step 3 and normalize each of the vectors $\mathbf{v}_1$ and $\mathbf{v}_2$. We obtain

$$\mathbf{u}_1 = \frac{1}{|\mathbf{v}_1|}\mathbf{v}_1 = \frac{1}{\sqrt{2}}\begin{bmatrix} 1 \\ 1 \end{bmatrix} = \begin{bmatrix} \frac{1}{\sqrt{2}} \\ \frac{1}{\sqrt{2}} \end{bmatrix},$$

$$\mathbf{u}_2 = \frac{1}{|\mathbf{v}_2|}\mathbf{v}_2 = \frac{1}{\sqrt{2}}\begin{bmatrix} 1 \\ -1 \end{bmatrix} = \begin{bmatrix} \frac{1}{\sqrt{2}} \\ \frac{-1}{\sqrt{2}} \end{bmatrix} \qquad [21]$$

It follows from equation 21 that $\mathbf{u}_1$ and $\mathbf{u}_2$ satisfy the equations $\mathbf{u}_1 \cdot \mathbf{u}_2 = 0$ and $\mathbf{u}_1 \cdot \mathbf{u}_1 = \mathbf{u}_2 \cdot \mathbf{u}_2 = 1$ and therefore form an orthonormal set of vectors in $\mathbf{R}^2$.

Moving on to Step 4, we construct the orthogonal matrix T whose columns are $\mathbf{u}_1$ and $\mathbf{u}_2$. Thus, we have

$$T = \begin{bmatrix} \dfrac{1}{\sqrt{2}} & \dfrac{1}{\sqrt{2}} \\ \dfrac{1}{\sqrt{2}} & \dfrac{-1}{\sqrt{2}} \end{bmatrix} \quad \text{and} \quad T^{-1} = T^t = T$$

A simple check shows that $T^{-1}AT$ is in fact a diagonal matrix. We have

$$T^{-1}AT = \begin{bmatrix} \dfrac{1}{\sqrt{2}} & \dfrac{1}{\sqrt{2}} \\ \dfrac{1}{\sqrt{2}} & \dfrac{-1}{\sqrt{2}} \end{bmatrix} \begin{bmatrix} 1 & -1 \\ -1 & 1 \end{bmatrix} \begin{bmatrix} \dfrac{1}{\sqrt{2}} & \dfrac{1}{\sqrt{2}} \\ \dfrac{1}{\sqrt{2}} & \dfrac{-1}{\sqrt{2}} \end{bmatrix} = \begin{bmatrix} 0 & 0 \\ 0 & 2 \end{bmatrix}$$

Example 6 Let $A = \begin{bmatrix} 1 & -1 & -1 \\ -1 & 1 & -1 \\ -1 & -1 & 1 \end{bmatrix}$. Find an orthogonal matrix T such that $T^{-1}AT$ is a diagonal matrix.

Note that A is symmetric.

Solution The characteristic polynomial of A is given by

$$f(\lambda) = \begin{bmatrix} 1-\lambda & -1 & -1 \\ -1 & 1-\lambda & -1 \\ -1 & -1 & 1-\lambda \end{bmatrix} \qquad [22]$$

Expanding the determinant in equation 22 and simplifying yields

$$f(\lambda) = -\lambda^3 + 3\lambda^2 - 4 = -(\lambda^3 - 3\lambda^2 + 4) \qquad [23]$$

Note: We use the rational root test for the polynomial equation $f(\lambda) = 0$. In this case the possible rational roots are ± 1, ± 2, ± 4.

We obtain the factorization of $\lambda^3 - 3\lambda^2 + 4$ by using a well-known algebraic technique for finding rational solutions of polynomial equations. We have

$$\lambda^3 - 3\lambda^2 + 4 = (\lambda + 1)(\lambda - 2)^2 \qquad [24]$$

Combining equations 23 and 24, we obtain

$$f(\lambda) = -(\lambda + 1)(\lambda - 2)^2 \qquad [25]$$

It follows from equation 25 that $\lambda_1 = -1$ is an eigenvalue of A of multiplicity 1 and $\lambda_2 = 2$ is an eigenvalue of A of multiplicity 2. Solving the system of equations $(A - \lambda_1 I)\mathbf{v} = \mathbf{0}$, we obtain the eigenvector $\mathbf{v}_1 = \begin{bmatrix} 1 \\ 1 \\ 1 \end{bmatrix}$. Solving the system of equations $(A - \lambda_2 I)\mathbf{v} = \mathbf{0}$, we obtain two linearly independent eigenvectors, $\mathbf{v}_2 = \begin{bmatrix} 1 \\ -1 \\ 0 \end{bmatrix}$ and $\mathbf{v}_3 = \begin{bmatrix} 1 \\ 0 \\ -1 \end{bmatrix}$. Normalizing $\mathbf{v}_1 = \begin{bmatrix} 1 \\ 1 \\ 1 \end{bmatrix}$, we obtain the eigenvector $\mathbf{u}_1 = \frac{1}{\sqrt{3}} \begin{bmatrix} 1 \\ 1 \\ 1 \end{bmatrix} = \begin{bmatrix} \frac{1}{\sqrt{3}} \\ \frac{1}{\sqrt{3}} \\ \frac{1}{\sqrt{3}} \end{bmatrix}$. We now apply the Gram-Schmidt process to the set $\{\mathbf{v}_2, \mathbf{v}_3\}$ in order to obtain an orthonormal set $\{\mathbf{u}_2, \mathbf{u}_3\}$. First, we normalize $\mathbf{v}_3 = \begin{bmatrix} 1 \\ 0 \\ -1 \end{bmatrix}$, obtaining the eigenvector $\mathbf{u}_3 = \frac{1}{\sqrt{2}} \begin{bmatrix} 1 \\ 0 \\ -1 \end{bmatrix} = \begin{bmatrix} \frac{1}{\sqrt{2}} \\ 0 \\ \frac{-1}{\sqrt{2}} \end{bmatrix}$. Second, we construct the vector

See Section 5-3

$$\tilde{\mathbf{v}}_2 = \mathbf{v}_2 - (\mathbf{v}_2 \cdot \mathbf{u}_3)\mathbf{u}_3 = \begin{bmatrix} 1 \\ -1 \\ 0 \end{bmatrix} - \frac{1}{\sqrt{2}} \begin{bmatrix} \frac{1}{\sqrt{2}} \\ 0 \\ \frac{-1}{\sqrt{2}} \end{bmatrix} = \begin{bmatrix} \frac{1}{2} \\ -1 \\ \frac{1}{2} \end{bmatrix}$$

Normalizing $\tilde{\mathbf{v}}_2$, we obtain $\mathbf{u}_2 = \begin{bmatrix} \frac{1}{\sqrt{6}} \\ \frac{-2}{\sqrt{6}} \\ \frac{1}{\sqrt{6}} \end{bmatrix}$. A simple check shows that we have

$$\mathbf{u}_1 \cdot \mathbf{u}_1 = \mathbf{u}_2 \cdot \mathbf{u}_2 = \mathbf{u}_3 \cdot \mathbf{u}_3 = 1$$

and

$$\mathbf{u}_1 \cdot \mathbf{u}_2 = \mathbf{u}_1 \cdot \mathbf{u}_3 = \mathbf{u}_2 \cdot \mathbf{u}_3 = 0$$

Thus, the set of eigenvectors $\{\mathbf{u}_1,\mathbf{u}_2,\mathbf{u}_3\}$ is orthonormal.

Using $\mathbf{u}_1$, $\mathbf{u}_2$, and $\mathbf{u}_3$ as columns for the orthogonal matrix T, we obtain

$$T = \begin{bmatrix} \frac{1}{\sqrt{3}} & \frac{1}{\sqrt{2}} & \frac{1}{\sqrt{6}} \\ \frac{1}{\sqrt{3}} & 0 & \frac{-2}{\sqrt{6}} \\ \frac{1}{\sqrt{3}} & \frac{-1}{\sqrt{2}} & \frac{1}{\sqrt{6}} \end{bmatrix}$$

The reader should check that the matrix product $T^{-1}AT$ yields the result

$$T^{-1}AT = \begin{bmatrix} -1 & 0 & 0 \\ 0 & 2 & 0 \\ 0 & 0 & 2 \end{bmatrix}$$

Exercises

1 Determine which of the following matrices are symmetric:

a $\begin{bmatrix} 2 & 4 \\ -4 & 0 \end{bmatrix}$ **b** $\begin{bmatrix} 3 & 1 \\ 1 & 5 \end{bmatrix}$ **c** $\begin{bmatrix} 6 & -1 & 2 \\ -1 & 0 & 3 \\ 2 & 3 & 5 \end{bmatrix}$

d $\begin{bmatrix} 7 & 2 & 1 \\ 2 & 1 & 8 \\ -1 & 8 & 0 \end{bmatrix}$ **e** $\begin{bmatrix} 5 & 0 & 0 \\ 0 & 7 & 0 \\ 0 & 0 & 9 \end{bmatrix}$

Hint: Take the transposes of AA^t and $A + A^t$.

2 Let A be a square matrix. Show that the matrices AA^t and $A + A^t$ are symmetric matrices.

3 A square matrix B is called **skew-symmetric** if it satisfies the condition $B^t = -B$. Show that if B is skew-symmetric then the entries of its main diagonal must all be zeros.

4 Determine which of the following matrices are skew-symmetric:

a $\begin{bmatrix} 0 & 2 \\ -2 & 0 \end{bmatrix}$ **b** $\begin{bmatrix} 0 & 1 \\ -1 & 1 \end{bmatrix}$ **c** $\begin{bmatrix} 0 & 1 & 2 \\ -1 & 0 & 3 \\ 2 & -3 & 0 \end{bmatrix}$

d $\begin{bmatrix} 0 & -5 & 7 \\ 5 & 0 & -8 \\ -7 & 8 & 0 \end{bmatrix}$ **e** $\begin{bmatrix} 0 & 0 & 1 \\ 0 & 0 & 0 \\ -1 & 0 & 1 \end{bmatrix}$

Hint: Take the transpose of $A - A^t$.

Hint: Use problems 2 and 5.

5 Let A be a square matrix. Show that $A - A^t$ is a skew-symmetric matrix.

6 Let A be a square matrix. Show that A can be expressed as a sum $B + C$ where B is a symmetric matrix and C is a skew-symmetric matrix.

7 a Let A be a symmetric 2×2 matrix with real entries. Show (without using Theorem 7-6) that A has only real eigenvalues.

b Give an example of a skew-symmetric 2×2 matrix B with real entries such that B has complex eigenvalues which are not real.

8 Determine which of the following matrices are orthogonal:

a $\begin{bmatrix} 1 & -1 \\ 1 & 1 \end{bmatrix}$ **b** $\begin{bmatrix} \frac{1}{2} & \frac{\sqrt{3}}{2} \\ \frac{-\sqrt{3}}{2} & \frac{1}{2} \end{bmatrix}$ **c** $\begin{bmatrix} \frac{4}{5} & \frac{-3}{5} \\ \frac{3}{5} & \frac{4}{5} \end{bmatrix}$

d $\begin{bmatrix} 1 & 0 & -1 \\ 0 & 1 & 0 \\ 1 & 0 & 1 \end{bmatrix}$ **e** $\begin{bmatrix} \frac{1}{3} & \frac{2}{3} & \frac{2}{3} \\ \frac{-2}{\sqrt{5}} & \frac{1}{\sqrt{5}} & 0 \\ \frac{-2}{3\sqrt{5}} & \frac{-4}{3\sqrt{5}} & \frac{5}{3\sqrt{5}} \end{bmatrix}$

9 Find an inverse for each of the matrices in problem 8.

10 Let A and B be $n \times n$ orthogonal matrices. Show that AB is also orthogonal.

11 Let A be an $n \times n$ matrix. Show that if $A^{-1} = A^t$, then A is orthogonal.

12 Let A be an $n \times n$ matrix. Show that if the row vectors of A form an orthonormal set of vectors in $\mathbf{R}^n$, then A is orthogonal.

13 Can you determine x and y such that the matrix $\begin{bmatrix} 1 & x \\ y & 2 \end{bmatrix}$ will have orthogonal eigenvectors $\mathbf{v}_1$ and $\mathbf{v}_2$ without directly computing $\mathbf{v}_1$ and $\mathbf{v}_2$? Explain your answer.

14 For each of the following matrices A, find an orthogonal matrix T such that $T^{-1}AT$ is a diagonal matrix.

a $\begin{bmatrix} 4 & -2 \\ -2 & 4 \end{bmatrix}$ **b** $\begin{bmatrix} -8 & 6 \\ 6 & 8 \end{bmatrix}$ **c** $\begin{bmatrix} -23 & 36 \\ 36 & -2 \end{bmatrix}$

d $\begin{bmatrix} 0 & 1 & 0 \\ 1 & 0 & 0 \\ 0 & 0 & 0 \end{bmatrix}$ **e** $\begin{bmatrix} 1 & 1 & 0 \\ 1 & 1 & 0 \\ 0 & 0 & 0 \end{bmatrix}$ **f** $\begin{bmatrix} 0 & 1 & 1 \\ 1 & 0 & 1 \\ 1 & 1 & 0 \end{bmatrix}$

g $\begin{bmatrix} 0 & 1 & 0 & 0 \\ 1 & 0 & 0 & 0 \\ 0 & 0 & 0 & 0 \\ 0 & 0 & 0 & 0 \end{bmatrix}$ **h** $\begin{bmatrix} 1 & 1 & 1 & 1 \\ 1 & 1 & 1 & 1 \\ 1 & 1 & 1 & 1 \\ 1 & 1 & 1 & 1 \end{bmatrix}$

15 Show that if D is an $n \times n$ diagonal matrix which satisfies the equation $D^2 = 0$, then $D = 0$.

16 Show that if A is a symmetric matrix (with real entries) such that $A^2 = 0$, then $A = 0$.

Hint: Show that A is similar to a diagonal matrix whose main diagonal entries are 1 or -1.

17 Show that if A is a symmetric matrix which is also orthogonal, then its eigenvalues are 1 or -1.

$|A| = \det A$.

18 Show that if A is an orthogonal matrix, then $|A| = \pm 1$.

19 Show that the matrices $\begin{bmatrix} \cos\varphi & -\sin\varphi \\ \sin\varphi & \cos\varphi \end{bmatrix}$ and $\begin{bmatrix} \cos\varphi & \sin\varphi \\ \sin\varphi & -\cos\varphi \end{bmatrix}$ are orthogonal.

20 Show that if A is an orthogonal 2×2 matrix, then there exists a real number φ such that either $A = \begin{bmatrix} \cos\varphi & -\sin\varphi \\ \sin\varphi & \cos\varphi \end{bmatrix}$ or $A = \begin{bmatrix} \cos\varphi & \sin\varphi \\ \sin\varphi & -\cos\varphi \end{bmatrix}$.

Review of Chapter 7

1 Let A and B be similar matrices and let λ be an eigenvalue of A. Is λ an eigenvalue of B too?

2 What are the eigenvalues of the matrix $A = \begin{bmatrix} 0 & 1 & 2 \\ 0 & 3 & 4 \\ 0 & 0 & 5 \end{bmatrix}$?

3 What are the eigenvalues of the matrix $B = \begin{bmatrix} 3 & 0 & 0 \\ 2 & 4 & 0 \\ 1 & 5 & 6 \end{bmatrix}$?

4 Are the matrices $\begin{bmatrix} 0 & 1 & 2 \\ 0 & 3 & 4 \\ 0 & 0 & 5 \end{bmatrix}$ and $\begin{bmatrix} 3 & 0 & 0 \\ 2 & 4 & 0 \\ 1 & 5 & 6 \end{bmatrix}$ similar?

5 Let $A = \begin{bmatrix} 1 & 3 & 4 \\ 0 & 2 & 6 \\ 0 & 0 & 5 \end{bmatrix}$. What are the eigenvalues of A^4?

6 Let A and B be similar matrices and let $\mathbf{v}$ be an eigenvector of A. Is $\mathbf{v}$ an eigenvector of B too?

7 Let A be a 4×4 matrix whose eigenvalues λ_1, λ_2, λ_3, and λ_4 are distinct. Is A diagonalizable?

8 Let A be a 5×5 matrix. What is the maximum number of different eigenvalues that A can have?

9 Show that if A is a diagonalizable matrix then so is A^2.

10 Let A be an $n \times n$ matrix and let A have n linearly independent eigenvectors. If $A^k = 0$ for some positive integer k, show that $A = 0$.

11 Find all possible values of a and b such that the matrix $\begin{bmatrix} 1 & 0 \\ a & b \end{bmatrix}$ is orthogonal.

12 Find all possible values of a, b, and c such that the matrix $\begin{bmatrix} \frac{1}{\sqrt{2}} & \frac{1}{\sqrt{2}} & 0 \\ \frac{2}{3} & -\frac{2}{3} & \frac{1}{3} \\ a & b & c \end{bmatrix}$ is orthogonal.

13 Let A and B be $n \times n$ symmetric matrices. Is AB symmetric too?

14 How do you find the inverse of an orthogonal matrix?

15 Determine (without computation) whether the matrix $\begin{bmatrix} 1 & 2 & 2 \\ 2 & 1 & -2 \\ 2 & -2 & 1 \end{bmatrix}$ is diagonalizable.

16 Let A and B be $n \times n$ symmetric matrices with the same characteristic polynomial. Show that there exists an orthogonal matrix T such that $TAT^{-1} = B$.

Some Applications

8

Historical Introduction

The study of conic sections dates back to the early Greeks (c. 300 B.C.). Euclid and Apollonius wrote about these plane curves, and it was the latter who introduced the terms parabola, hyperbola, and ellipse. Johannes Kepler (1571–1630) pointed out that the various conic sections can be obtained by varying the inclination of the plane that produces the section out of a cone having two nappes.

In the eighteenth century, a number of mathematicians studied the problem of reducing equations of the conic sections to **standard** or **canonical** form—that is, transforming quadratics into simpler form by choosing the coordinate axes to coincide with the principal axes. Cauchy, Sylvester, and Jacobi all worked on this problem. According to the theory they and others refined, a quadratic equation in two variables, $ax^2 + 2by + cy^2 = d$, has associated with it the determinant $\det A = \begin{vmatrix} a & b \\ c & d \end{vmatrix}$. The **characteristic polynomial** (the term was introduced by Cauchy) of the quadratic form is $f(\lambda) = \det(A - \lambda I)$, and the characteristic roots of $f(\lambda) = 0$ yield the length of the principal axes. It was Lagrange and Laplace, in their work on linear differential equations, who introduced the notion of characteristic polynomials. Lagrange used the characteristic roots to find the solutions of the equations of motion of the six planets then known; and Laplace showed that if the six planets moved in the same direction, the characteristic roots of the equations of motion are real and unequal. In 1826, Cauchy proved the invariance of the characteristic equation of any change of rectangular axes. In 1858, Weierstrass was interested in the problem of small oscillations about a position of equilibrium of a dynamical system. In dealing with this problem, he showed how two quadratic forms are reduced simultaneously to sums of squares. Euler's work on the notion of characteristic polynomials involved the reduction to standard form of quadratic equations in three variables.

Applications of conic sections abound in both pure and applied mathematics. Under ideal conditions the path of a projectile is a parabola, and automobile headlights, some bridges, and certain equipment for radio and television involve the parabola in their design. Hyperbolas are used on the battlefield to locate an enemy gun position, and a ship receiving signals from two unknown stations on shore can locate its position as the intersection of two hyperbolas. The paths of planets and artificial satellites are ellipses.

8-1 An Application to Geometry

Objective
Use matrices in the reduction of quadratics to standard form.

The general equation of a central conic in the plane with center at the origin of a Cartesian coordinate system is given by

$$ax^2 + 2bxy + cy^2 = d \qquad [1]$$

where a, b, c, and d are constants. The expression $Q(x,y) = ax^2 + 2bxy + cy^2$ is called the **general quadratic form** in the two variables x and y.

In drawing the graph of a quadratic equation the task is easier if the coordinate axes and the principal axes coincide. If $b \neq 0$ in equation 1, the principal and coordinate axes do not coincide. However, the cross product term can be eliminated by a simple transformation into another coordinate system in which the principal axes are along the coordinate axes. We shall illustrate how matrices are used to accomplish this task, first considering equation 1 and then extending our discussion to quadratic equations in more than two variables.

If $\mathbf{v} = \begin{bmatrix} x \\ y \end{bmatrix}$, we can express the general quadratic form as

$$Q(\mathbf{v}) = ax^2 + 2bxy + cy^2 \qquad [2]$$

Using matrices, we can see that if

$$A = \begin{bmatrix} a & b \\ b & c \end{bmatrix}$$

To see how equation 3 is derived, the reader should visualize $Q(\mathbf{v})$ in the form $Q(\mathbf{v}) = ax + bxy + bxy + cy^2$, which suggests the product of the matrices shown in equation 3.

then equation 2 can be rewritten in the form

$$Q(\mathbf{v}) = \mathbf{v}^t A \mathbf{v} \qquad [3]$$

The matrix A, which is a symmetric matrix, is called the **matrix of the form.**

Our interest at this point is to effect a change in variables to rid equation 1 of the cross product term.

If we consider $\mathbf{v}$ as a point in a coordinate space, then a change of variables is the effect of a nonsingular linear transformation on the space. Thus, if $\mathbf{v}$ is the point in the xy-coordinate system and $\mathbf{u}$ is the point in the x_1y_1-coordinate system, then for some matrix S

$$\mathbf{v} = S\mathbf{u}$$

denotes the relationship between the two coordinate systems.

Definition 8-1 Two quadratic forms

$$\mathbf{v}^t A \mathbf{v} \qquad \text{and} \qquad \mathbf{u}^t B \mathbf{u}$$

are said to be **equivalent** if there exists a linear transformation of variables

$$\mathbf{v} = S\mathbf{u}$$

where S is a nonsingular matrix that transforms the first form into the second. We find that if $Q(\mathbf{v})$ is a quadratic form in $\mathbf{v}$, then

The reader should justify the various steps involved.

$$Q(\mathbf{v}) = \mathbf{v}^t A\mathbf{v} = (S\mathbf{u})^t A(S\mathbf{u}) = (\mathbf{u}^t S^t)A(S\mathbf{u})$$
$$= \mathbf{u}^t(S^t A S)\mathbf{u} = \mathbf{u}^t B\mathbf{u} = Q'(\mathbf{u})$$

Prove this statement!

Thus, the two quadratic forms are equivalent if and only if there is a nonsingular matrix S such that

$$B = S^t AS$$

Definition 8-2 Two square matrices A and B are said to be **congruent** if there exists a nonsingular matrix S such that

$$B = S^t AS$$

Clearly, if the matrix A is congruent to the symmetric matrix B, then A is also symmetric. (The student should prove this statement.)

Also, we have the following relation:

Definition 8-3 Two quadratic forms

$$\mathbf{v}^t A\mathbf{v} \qquad \text{and} \qquad \mathbf{u}^t B\mathbf{u}$$

are said to be **orthogonally equivalent** if there is an orthogonal transformation of variables

$$\mathbf{v} = P\mathbf{u}$$

where P is an orthogonal matrix that transforms the first quadratic form into the second.

The student can easily show that two quadratic forms $\mathbf{v}^t A\mathbf{v}$ and $\mathbf{u}^t B\mathbf{u}$ are orthogonally equivalent if and only if there is an orthogonal matrix P such that

$$B = P^t AP = P^{-1}AP$$

Note: Since P is orthogonal, $P^t = P^{-1}$

In fact, here B is both congruent and similar to A.

Finally, we state the following theorem:

Theorem 8-1 Every quadratic form $\mathbf{v}^t A\mathbf{v}$ is orthogonally equivalent to a quadratic form

$$\lambda_1 x_1^2 + \lambda_2 y_1^2$$

where λ_1 and λ_2 are the eigenvalues of the matrix A.

The reader is encouraged to prove this theorem.

Example 1 Find an orthogonal matrix P to remove the cross product term in the equation

$$5x^2 + 4xy + 5y^2 = 42 \qquad [4]$$

Write the transformed equation and identify the graph.

Solution Let

$$A = \begin{bmatrix} 5 & 2 \\ 2 & 5 \end{bmatrix} \quad \text{and} \quad \mathbf{v} = \begin{bmatrix} x \\ y \end{bmatrix}$$

Then equation 4 may be written as

$$\mathbf{v}^t A \mathbf{v} = 42$$

The characteristic polynomial of A is

$$\begin{aligned} f(\lambda) &= \det(A - \lambda I) \\ &= \lambda^2 - 10\lambda + 21 \end{aligned}$$

with $\lambda_1 = 3$ and $\lambda_2 = 7$ being the eigenvalues. Solving the equation

$$(A - \lambda I)\mathbf{X} = 0$$

for $\lambda = 3$ and $\lambda = 7$, we obtain the respective solutions

$$\mathbf{X}_1 = \begin{bmatrix} 1 \\ -1 \end{bmatrix} \quad \text{and} \quad \mathbf{X}_2 = \begin{bmatrix} 1 \\ 1 \end{bmatrix}$$

Normalizing these to unit length and using the resulting vectors as columns for the matrix P, we have

Observe that P is an orthogonal matrix.

$$P = \begin{bmatrix} \frac{1}{\sqrt{2}} & \frac{1}{\sqrt{2}} \\ \frac{-1}{\sqrt{2}} & \frac{1}{\sqrt{2}} \end{bmatrix}$$

From this we obtain

$$\begin{aligned} B = P^{-1}AP &= P^tAP \\ &= \begin{bmatrix} \frac{1}{\sqrt{2}} & \frac{-1}{\sqrt{2}} \\ \frac{1}{\sqrt{2}} & \frac{1}{\sqrt{2}} \end{bmatrix} \begin{bmatrix} 5 & 2 \\ 2 & 5 \end{bmatrix} \begin{bmatrix} \frac{1}{\sqrt{2}} & \frac{1}{\sqrt{2}} \\ \frac{-1}{\sqrt{2}} & \frac{1}{\sqrt{2}} \end{bmatrix} \\ &= \begin{bmatrix} 3 & 0 \\ 0 & 7 \end{bmatrix} \end{aligned}$$

Thus, if $\mathbf{u} = \begin{bmatrix} x_1 \\ y_1 \end{bmatrix}$ then equation 4 is transformed into the equation

$$\mathbf{u}^t B \mathbf{u} = 42$$

or

$$3x_1^2 + 7y_1^2 = 42$$

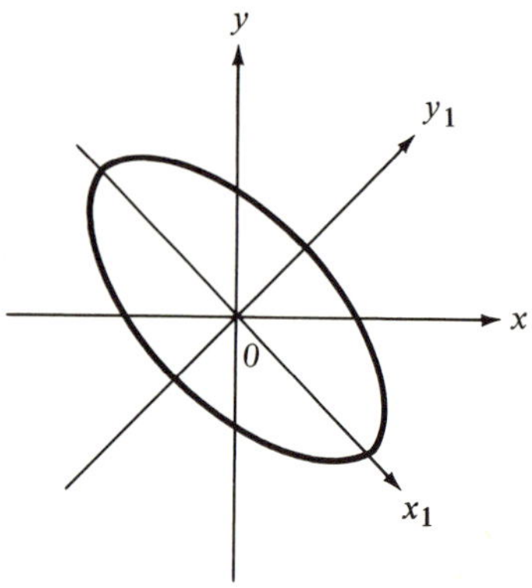

Figure 8-1

or equivalently

$$\frac{x_1^2}{14} + \frac{y_1^2}{6} = 1 \quad [5]$$

The quadratic forms in equations 4 and 5 are orthogonally equivalent, under the coordinate transformations given by

$$\begin{bmatrix} x_1 \\ y_1 \end{bmatrix} = P^{-1}\begin{bmatrix} x \\ y \end{bmatrix}$$

The graph of equation 5 is easily recognizable as an ellipse and is shown in Figure 8-1.

We note that by using the eigenvalues and Theorem 8-1 we can arrive at equation 5 without finding P and B. However, to sketch the graph we need to know the relationship between the old and new coordinate systems.

We recall that a curve defined by the equation

$$\lambda_1 x_1^2 + \lambda_2 y_1^2 = d \quad [6]$$

describes

Observation: If we take $d > 0$ and $B = \begin{bmatrix} \lambda_1 & 0 \\ 0 & \lambda_2 \end{bmatrix}$, then the nature of the curve defined by equation 6 may be obtained from $\det B$. That is, if $\det B > 0$ then the graph of equation 6 is an ellipse; if $\det B < 0$, then the graph of equation 6 is a hyperbola.

a an ellipse if λ_1, λ_2, and D are of the same sign;

b a hyperbola if λ_1 and λ_2 have opposite signs ($d \neq 0$);

c an imaginary ellipse (i.e., the graph is empty) if λ_1 and λ_2 have the same sign but the sign of d is different.

Now, if $\mathbf{v}^t A\mathbf{v}$ and $\mathbf{u}^t B\mathbf{u}$ are orthogonally equivalent where A is the matrix of the form $ax^2 + 2bxy + cy^2$, then for an orthogonal matrix P where $B = P^{-1}AP$ we have

$$\det B = \det(P^{-1}AP) = \det A \qquad \text{(Why?)}$$

Note: In a given equation, $ax^2 + 2bxy + cy^2 = d$, if $d < 0$. We multiply the equation by -1 and then find A and $\det A$.

From this we conclude that the graph of $\mathbf{v}^t A\mathbf{v} = d$, $d > 0$, is

a an ellipse or is empty if $\det A > 0$

b a hyperbola if $\det A < 0$

Example 2 Identify the graph of the equation

$$-3x^2 + 2xy + 5y^2 = 10 \quad [7]$$

Solution Here the matrix of the form is

$$A = \begin{bmatrix} -3 & 1 \\ 1 & 5 \end{bmatrix}$$

Since $d = 10 > 0$ and $\det A = -16 < 0$, the graph of equation 7 is a hyperbola.

We now extend the above procedures to the general quadratic surface in 3-space. We consider the equation of the quadric with center at the origin of a Cartesian coordinate system,

$$a_1x^2 + a_2y^2 + a_3z^2 + 2a_4xy + 2a_5xz + 2a_6yz = b \qquad [8]$$

where $a_i (i = 1, 2, \ldots, 6)$ and b are constants. If $\mathbf{v} = \begin{bmatrix} x \\ y \\ z \end{bmatrix}$ then the general quadratic form

$$Q(\mathbf{v}) = a_1x^2 + a_2y^2 + a_3z^2 + 2a_4xy + 2a_5xz + 2a_6yz$$

can be written as

$$\begin{aligned} Q(\mathbf{v}) = \quad & a_1x^2 + a_4xy + a_5xz \\ & + a_4xy + a_2y^2 + a_6yz \\ & + a_5xz + a_6yz + a_3z^2 \end{aligned}$$

suggesting the product

$$Q(\mathbf{v}) = \mathbf{v}^t A\mathbf{v}$$

where A is the symmetric matrix

$$A = \begin{bmatrix} a_1 & a_4 & a_5 \\ a_4 & a_2 & a_6 \\ a_5 & a_6 & a_3 \end{bmatrix}$$

We note that Theorem 8-1 can be extended to $\mathbf{R}^n$. Thus we find that every quadratic form $\mathbf{v}^t A\mathbf{v}$ is orthogonally equivalent to a quadratic form

$$\lambda_1 y_1^2 + \lambda_2 y_2^2 + \cdots + \lambda_n y_n^2$$

where $\mathbf{v}^t = [x_1, x_2, \ldots, x_n]$ and $\lambda_1, \lambda_2, \ldots, \lambda_n$ are the eigenvalues of the matrix A.

Example 3 Identify the graph of the equation

$$2x^2 + y^2 + 2z^2 + 2xy + 2yz = 12 \qquad [9]$$

Solution The matrix of the form is

$$A = \begin{bmatrix} 2 & 1 & 0 \\ 1 & 1 & 1 \\ 0 & 1 & 2 \end{bmatrix}$$

and its characteristic polynomial is

$$f(\lambda) = \det(A - \lambda I)$$
$$= \det\begin{bmatrix} 2-\lambda & 1 & 0 \\ 1 & 1-\lambda & 1 \\ 0 & 1 & 2-\lambda \end{bmatrix}$$
$$= \lambda(2-\lambda)(\lambda-3)$$

with eigenvalues $\lambda_1 = 0$, $\lambda_2 = 2$, and $\lambda_3 = 3$.

Solving the equation

$$(A - \lambda I)\mathbf{X} = 0$$

for $\lambda = 0$, $\lambda = 2$, and $\lambda = 3$, we obtain the respective solutions $\begin{bmatrix} 1 \\ -2 \\ 1 \end{bmatrix}$, $\begin{bmatrix} 1 \\ 0 \\ -1 \end{bmatrix}$, and $\begin{bmatrix} 1 \\ 1 \\ 1 \end{bmatrix}$. Normalizing these to unit length and using the resulting vectors as columns for the matrix P, we obtain

Observe that P is an orthogonal matrix.

$$P = \begin{bmatrix} \frac{1}{\sqrt{6}} & \frac{1}{\sqrt{2}} & \frac{1}{\sqrt{3}} \\ \frac{-2}{\sqrt{6}} & 0 & \frac{1}{\sqrt{3}} \\ \frac{1}{\sqrt{6}} & \frac{-1}{\sqrt{2}} & \frac{1}{\sqrt{3}} \end{bmatrix}$$

From this we obtain

The reader can easily verify this result.

$$B = P^{-1}AP = P^tAP$$
$$= \begin{bmatrix} 0 & 0 & 0 \\ 0 & 2 & 0 \\ 0 & 0 & 3 \end{bmatrix}$$

Thus if $\mathbf{u} = \begin{bmatrix} x_1 \\ y_1 \\ z_1 \end{bmatrix}$ then equation 9 is transformed into the equation

$$\mathbf{u}^tB\mathbf{u} = 12$$

or

$$2y_1^2 + 3z_1^2 = 12 \qquad \textbf{[10]}$$

or equivalently

$$\frac{y_1^2}{6} + \frac{z_1^2}{4} = 1 \qquad \textbf{[11]}$$

where the linear transformation of variables from the old coordinate system to the new coordinate system is given by

$$\mathbf{u} = P^{-1}\mathbf{v} \qquad [12]$$

The graph of the quadric surface is an elliptic cylinder.

Example 4 Identify the graph of the equation

$$x^2 - 6yz = 1 \qquad [13]$$

Solution The matrix of the form is

$$A = \begin{bmatrix} 1 & 0 & 0 \\ 0 & 0 & -3 \\ 0 & -3 & 0 \end{bmatrix}$$

and its characteristic polynomial is

$$\begin{aligned} f(\lambda) &= \det(A - \lambda I) \\ &= (1 - \lambda)(\lambda^2 - 9) \end{aligned}$$

with eigenvalues $\lambda_1 = 1$, $\lambda_2 = 3$, and $\lambda_3 = -3$. Now, we know that the quadratic form in equation 13 can be diagonalized to

This is an application of the generalization of Theorem 8-1.

$$x_1^2 + 3y_1^2 - 3z_1^2$$

Thus, equation 13 is transformed into the new coordinate system as the quadric

$$x_1^2 + 3y_1^2 - 3z_1^2 = 1 \qquad [14]$$

The graph of the quadric surface is an hyperboloid of one sheet.

Graphs of Quadric Surfaces (Figure 8-2)

Elliptic cylinder
$$\frac{x^2}{a^2} + \frac{y^2}{b^2} = 1$$

Elliptic cone
$$\frac{x^2}{a^2} + \frac{y^2}{b^2} = z^2$$

Elliptic paraboloid
$$\frac{x^2}{a^2} + \frac{y^2}{b^2} = z$$

Ellipsoid
$$\frac{x^2}{a^2} + \frac{y^2}{b^2} + \frac{z^2}{c^2} = 1$$

Hyperboloid of two sheets
$$\frac{y^2}{b^2} - \frac{x^2}{a^2} - \frac{z^2}{c^2} = 1$$

Hyperboloid of one sheet
$$\frac{x^2}{a^2} + \frac{y^2}{b^2} - \frac{z^2}{c^2} = 1$$

Hyperbolic paraboloid
$$\frac{y^2}{b^2} - \frac{x^2}{a^2} = cz$$

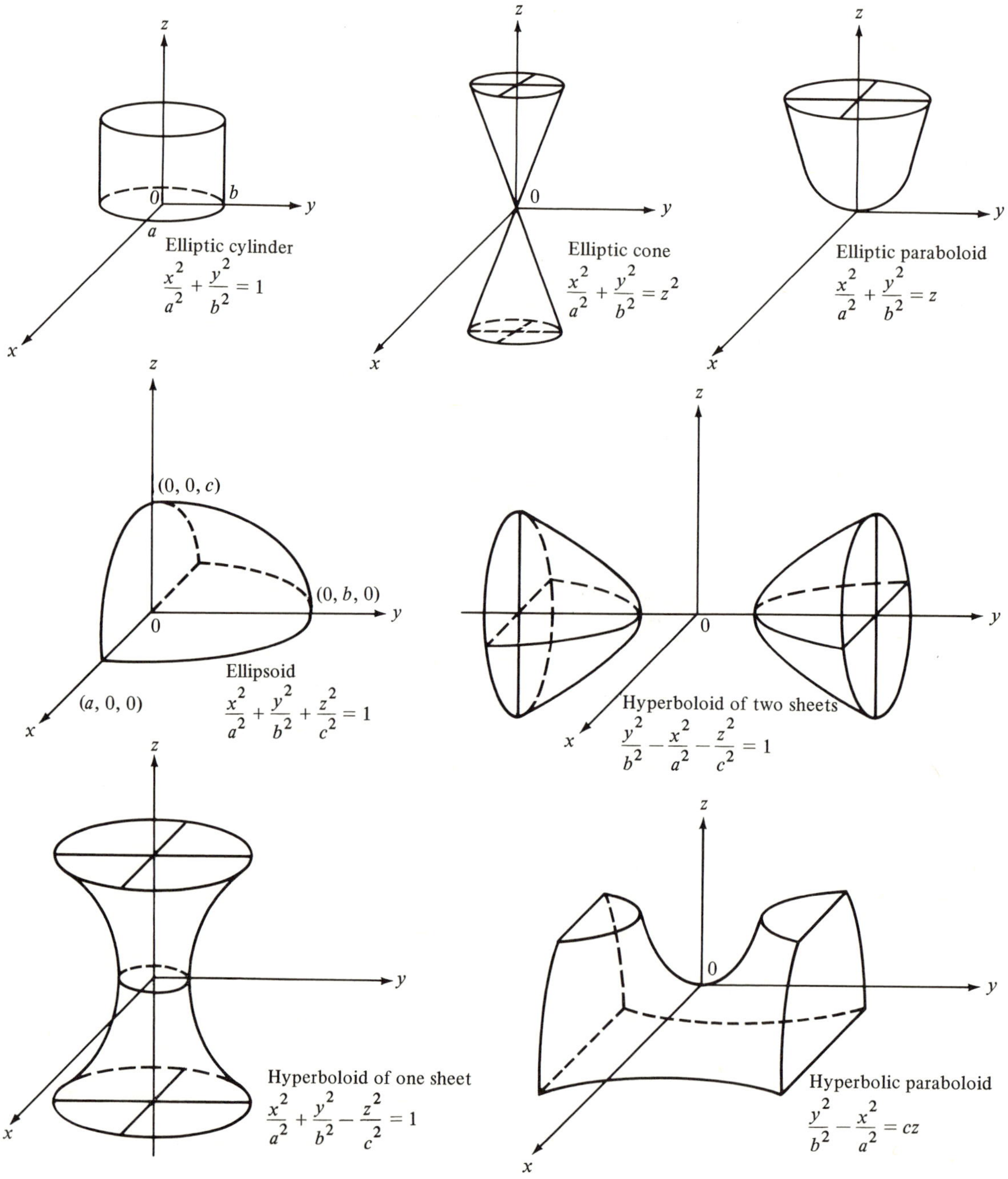

Figure 8-2

Exercises In problems 1 through 6, determine an orthogonal matrix P so as to remove the cross product terms. Using P, find the transformed equation and identify the central conic.

1 $xy = 2$

2 $5x^2 + 4xy + 8y^2 = 9$

3 $2x^2 + 4xy - y^2 = 12$

4 $-3x^2 + 8xy + 3y^2 = 20$

5 $3x^2 + 2xy + 3y^2 = 8$

6 $x^2 + \sqrt{3}xy + 2y^2 = 1$

In problems 7 through 10, using the matrix of the form, identify the graph of the given equation without transforming the coordinates.

7 $x^2 + 5xy + y^2 = 2$

8 $-x^2 + 6xy + 3 = 0$

9 $3x^2 + 2xy + 2y^2 = 4$

10 $4x^2 - 6xy + 3y^2 = -2$

In problems 11 through 15, find an orthogonal matrix P so as to remove the cross product term. Using P, find the transformed equation and identify the quadric surface.

11 $x^2 - 4yz = 16$

12 $3x^2 + 4y^2 + 2\sqrt{3}yz + 6z^2 = 42$

13 $2x^2 + 4y^2 + 3z^2 + 4xz - 4yz = 6$

14 $xy + yz + xz = 1$

15 $2x^2 + 2y^2 + 5z^2 - 4xy - 2yz + 2xz = 20$

16 Show that if one of the eigenvalues of the matrix of the form is zero, then the central conic $ax^2 + 2bxy + cy^2 = d$ $(d > 0)$ is a pair of parallel lines.

17 Show that if $\mathbf{v}^t A\mathbf{v}$ and $\mathbf{v}^t B\mathbf{v}$ are identical and A and B are symmetric, then $A = B$.

18 **a** Consider the central conic $ax^2 + 2bxy + cy^2 = 1$. If r is the largest positive eigenvalue of the matrix of the form, show that the minimum distance from the origin to the curve is $1/\sqrt{r}$.

b Let $\mathbf{v}^t A\mathbf{v} = 1$ be a central quadric in $\mathbf{R}^n$. If r is the largest positive eigenvalue of A, show that the minimum distance from the origin to the quadric surface is $1/\sqrt{r}$.

19 Find the points closest to the origin on the graph of $5x^2 + 4xy + 8y^2 = 1$.

20 Repeat problem 19 for the equation $-3x^2 + 6xy + 2y^2 = 4$.

Historical Note

In the seventeenth century, the need to solve many physical problems led to the study of a class of equations called differential equations. Galileo and Robert Hooke studied the behavior of beams under vertical and horizontal loads. Such problems arose in building some of the great medieval cathedrals. In the eighteenth century, Newton, among others, in his investigation of the shape of the earth and the inverse square law of gravitational attraction, studied the problem of the swinging pendulum, which also involved a differential equation. It was Leibnitz who coined the term differential equation. The famous Bernoulli family used differential equations to formulate and solve problems arising in mechanics.

In the eighteenth century, astronomy was the dominant science, and many problems arising in this field of study led to differential equations. Newton, for instance, used differential equations to study the motion of a planet around the sun (the two-body problem). Navigational problems led to the study of the motion of the moon, and this in turn led to the study of certain differential equations.

Today, differential equations arise in many branches of the physical sciences, engineering, the social sciences, and other disciplines. From the study of motion of celestial objects to the motion of atomic particles, and from the study of population growth to the decomposition of radioactive substances—all involve the study of differential equations. In studying differential equations we rely heavily on topics in linear algebra.

8-2 Applications to Linear Differential Equations

Objective
Use methods of linear algebra to solve linear differential equations of order 1 that arise in the sciences.

Note: Some knowledge of elementary calculus is required in this section.

In this section we shall investigate equations that describe such problems as population growth, radioactive decay, social diffusion, and related phenomena. The Italian mathematician Volterra was a pioneer in using mathematics to study population growth processes, the subject of our first discussion, which illustrates how a problem involving population growth in bacteria is formulated and solved.

VITO VOLTERRA (1860–1940) *was born at Ancona, Italy, and spent most of his childhood in Florence. Volterra's interest in mathematics started at an early age; while still a university student, he wrote his first original paper. In 1883 he was appointed full professor at the University of Pisa.*

Volterra did outstanding work in integral equations, potential theory, and differential equations. His interests were practical as well as theoretical: he designed the mounting of guns on airships. When Mussolini took power in Italy, Volterra refused to sign a required oath of loyalty, and in protest resigned his post at the University of Pisa.

The Problem of Population Growth

Example 1

In an experiment it is found that the population of bacteria grows from 100 units to 400 units in 2 hours. If the rate of change of the population is assumed to be proportional to the number present at any time t, what is the size of the population at the end of the first hour?

We shall first set up a mathematical model describing this situation and then show how the problem is solved.

Let $p(t)$ denote the population of the bacteria at time t. As time changes from t to $t + h$ $(h > 0)$, the population growth is proportional to $p(t)$. That is, for some constant $k > 0$, we have approximately

$$\frac{p(t + h) - p(t)}{h} = kp(t) \qquad [1]$$

If we assume p to be a continuous and differentiable function of t, we obtain

$$p'(t) = kp(t) \qquad [2]$$

where p' is the derivative function describing the rate of change of population.

Equation 2 is called a **differential equation.** In general, an (ordinary) differential equation is an equation involving an independent variable x, a dependent variable y, and certain derivatives of y with respect to x. This may be expressed by

$$f(x,y,y',\ldots,y^{(n)}) = 0$$

Note: A differential equation in which the highest derivative involved is the nth derivative is called an nth-**order** differential equation.

The relationship effectively involves some derivative of y. A **solution** of a differential equation is a function defined on some interval such that on that interval the function and its derivatives satisfy the given equation.

In the example above we find that at $t = 0$ the population is 100 units. We express this as

$$p(0) = 100 \qquad [3]$$

This is called an **initial condition,** and equation 2 together with equation 3 is called an **initial-value problem.** In this case we find that we have another supplementary condition to satisfy: namely, that at $t = 2$ the population is 400 units. Thus, the problem here is to find a function p that satisfies equations 2 and 3 as well as

$$p(2) = 400 \qquad [4]$$

Note: D is a linear operator, and is defined on the space of all differentiable functions.

Let D denote the differentiation operator. Then equation 2 may be written as

$$(D - kI)p = 0 \qquad [5]$$

Recall from the calculus:

$De^{kt} = ke^{kt}$

$Du = 0$ if $u = c$

The product rule:

$D(uv) = uDv + vDu$

Solving equation 2 is equivalent to describing the null space of the operator $L \equiv D - kI$. We find that $Le^{kt} = 0$. Thus, e^{kt} is in the null space of L. We shall now show that if p is in the null space of L, then $p = ce^{kt}$ where c is a constant.

Set $u = e^{-kt}p$. Then

$$\begin{aligned} Du &= e^{-kt}p' - ke^{-kt}p \\ &= e^{-kt}(p' - kp) \\ &= e^{-kt}(D - kI)p \\ &= 0 \end{aligned}$$

Thus, $u = c$ (a constant) and so

$$c = e^{-kt}p$$

or

$$p = ce^{kt} \qquad [6]$$

The function defined by equation 6 is called the **general solution** of equation 2. That is, for every real number c, $p = ce^{kt}$ is a solution of equation 2. Also, every solution of equation 2 is of the form $p = ce^{kt}$ for some constant c.

Now if there is a function p that satisfies equation 2 and the supplementary conditions in equations 3 and 4, it must be of the form given in equation 6. Thus, we need to choose c and k to satisfy equations 3 and 4. From equation 3 we find

$$p(0) = 100 = ce^0$$

or

$$c = 100$$

and from equation 4 we have

$$p(2) = 400 = 100\, e^{2k}$$

or

$$k = \ln 2.$$

Note:

$4 = e^{2k} \Longrightarrow \ln 4 = 2k \Longrightarrow$
$k = \frac{1}{2}\ln 4 = \ln 2$

Therefore, the solution to the problem described by equations 2, 3, and 4 is given by

$$p(t) = 100\, e^{(\ln 2)t} \qquad [7]$$

From equation 7 we find the size of the population at the end of the first hour:

$$p(1) = 100\, e^{\ln 2} = 200.$$

The Problem of Radioactive Decay

WILLARD F. LIBBY (1908-) *is presently a professor at the University of California. He developed the method of carbon dating used to estimate the age of ancient objects. Libby was awarded the 1960 Nobel Prize in chemistry for his method.*

It has been determined that carbon 14 is a radioactive substance that decays continuously. This phenomenon was used by Professor Libby to estimate the age of ancient objects. All living organisms contain carbon 14 and carbon 12. As an organism dies the carbon 14 decays continuously and is no longer replaced, while carbon 12 remains unchanged. Comparing the amounts of carbon 14 and carbon 12 left in an organism makes it possible to estimate the time at which the organism lived.

Radioactive substances decrease in quantity at a rate proportional to the amount of substance present. If $A(t)$ is the amount of substance present at time t, then

$$\frac{dA}{dt} = -kA \tag{8}$$

where $k > 0$ is a constant of proportionality and $\frac{dA}{dt}$ is the rate of change of A with respect to t.

We shall now illustrate how the age of an object is estimated using this formula.

Example 2 Some ancient artifacts were discovered in the Upper Nile region containing organisms with 40 percent of the original of carbon 14 left in them. If the half-life of carbon 14 is taken to be 5,500 years, estimate the age of the objects.

Note: The **half-life** of carbon 14 is the time taken for half the amount of the original substance to decay.

Solution Let $A(t)$ be the amount of carbon 14 at time t satisfying differential equation 8. Let A_0 be the amount of carbon 14 originally present in the organism. That is, at $t = 0$,

$$A(0) = A_0 \tag{9}$$

We also have that at $t = 5{,}500$ years the amount of carbon 14 present is $A_0/2$. That is,

$$A(5{,}500) = A_0/2 \tag{10}$$

Thus the problem is to find $A(t)$ that satisfies equation 8 and the supplementary conditions 9 and 10. Rewriting equation 8 in the form

$$(D + kI)A = 0$$

the student can readily obtain the general solution

$$A(t) = Ce^{-kt} \tag{11}$$

where C is an arbitrary constant. From equation 9 we find that

$$A(0) = A_0 = Ce^0$$

or

$$C = A_0$$

Thus equation 11 becomes

$$A(t) = A_0 e^{-kt} \qquad [12]$$

From equation 10 we obtain

$$A(5{,}500) = A_0/2 = A_0 e^{-k(5{,}500)}$$

or

$$\tfrac{1}{2} = e^{-5{,}500k}$$

or

$$k = \frac{\ln 2}{5{,}500}$$

Therefore, the solution of equation 8 satisfying the supplementary conditions 9 and 10 is given by

$$A(t) = A_0 e^{\frac{-\ln 2}{5{,}500} t} \qquad [13]$$

Since 40 percent of the original carbon 14 is present in the organism, we estimate the age by solving for t in equation 13:

$$\frac{40}{100} A_0 = A_0 e^{\frac{-\ln 2}{5{,}500} t}$$

or

The student is encouraged to include the steps involved to solve for t.

$$t = -5{,}500 \left(\frac{\ln 4 - \ln 10}{\ln 2} \right)$$

$$\approx 7{,}300$$

That is, the objects found are about 7,300 years old.

A Problem in Social Diffusion

Let us consider the modeling of a problem of how some information diffuses through the inhabitants of a country. We illustrate such a problem in Example 3.

Example 3

Remarks:
1 Note that on the first day of the commercials 10 percent of the population was reached. On the second day, however, only

Consider a new car company which is trying to market its latest version of an "electrocar." Suppose as a result of the company's advertising efforts on television a fixed percentage (say 10 percent) of the population is reached per day. If the information spreads from the TV commercials only and we assume that the rate of spread of

some of the 10 percent were reached for the first time.

2 We shall assume that the size of the population receiving the information is increasing continuously.

information is proportional to the number of people who have not yet heard of the "electrocar," find the number of people who would have heard of the car at time t.

Solution

Let $p(t)$ denote the number of people who have heard of the car at time t. Let N represent the number of people in this country. Then $N - p(t)$ denotes the number of people who have not yet heard of the "electrocar." Since we have assumed that the rate of spread of the information $\left(\text{i.e., } \frac{dp}{dt}\right)$ is proportional to the number of people who have not yet heard about the car, we have

$$\frac{dp}{dt} = k(N - p) \qquad [14]$$

where k is a constant. We can easily assume that

$$p(0) = 0 \qquad [15]$$

That is, just before the commercials were aired not one member of the public knew of the "electrocar." Thus we have an initial-value problem given by equations 14 and 15.

It is easy to see that differential equation 14 does not fit the form given in equation 5. We shall use a simple technique to solve equation 14. Set

$$x(t) = N - p(t) \qquad [16]$$

Then

$$\frac{dx}{dt} = \frac{-dp}{dt}$$

and from equation 14 we obtain

$$\frac{dx}{dt} = -kx \qquad [17]$$

At $t = 0$, we find

$$x(0) = N - p(0) = N \qquad [18]$$

Thus the initial-value problem posed in equations 14 and 15 has been transformed into the initial-value problem in equations 17 and 18. The student can readily show that the solution is

$$x(t) = Ne^{-kt} \qquad [19]$$

Substituting equation 16 into equation 19, we get

$$p(t) = N(1 - e^{-kt}) \qquad [20]$$

The student can show that p defined by equation 20 is in fact a solution of the initial-value problem in equations 14 and 15. In this model we are also given that 10 percent of the population is reached per day. Thus at the end of the first day we have

$$p(1) = \frac{10}{100} N = 0.1N \qquad \textbf{[21]}$$

Using this condition, we evaluate k in equation 20:

$$p(1) = 0.1N = N(1 - e^{-k})$$

or

$$-k = \ln 0.9$$

Therefore, the size of the population that has been reached at time t is given by

$$p(t) = N[1 - e^{(\ln 0.9)t}] \qquad \textbf{[22]}$$

Exercises

1 Find the general solution of each of the following differential equations:

a $y' = y$ b $y' + 2y = 0$

c $y' + \sqrt{3}y = 0$ d $2y' + 3y = 0$

2 Solve each of the following initial-value problems:

a $y' = 2y,\quad y(0) = 3$

b $2y' + y = 0,\quad y(0) = 2$

c $y' + 3y = 0,\quad y(1) = -4$

d $3y' - 4y = 0,\quad y(2) = 5$

3 Use the technique of Example 3 to solve each of the following differential equations:

a $y' + 2y = 3$

b $y' - y = 10$

c $2y' + 3y = 5$

In problems 4 through 6, assume that the rate of change in population p at any time t is proportional to p.

4 In a culture the population of bacteria grows from 10,000 to 40,000 units in 10 hours. Find the size of the population at the end of the first 5 hours.

5 The population of the city of Greenville is 100,000. Ten years ago its population was 60,000. What will the population of Greenville be in 10 years from now?

6 The population of a country increases at 5 percent per year. If the present population is 200 million, what will it be in 10 years from now? How long will it be before the population has doubled in size?

7 A radioactive substance decays continuously at a rate proportional to the amount present. If 40 lbs. are stored now, 24 lbs. will be left at the end of 100 years. Find the amount present at time t.

8 A radioactive substance decays continuously and has a half-life of 1,400 years. Estimate the amount of substance left in 1,000 years.

9 If 30 percent of the original carbon 14 is still present in an ancient object, find the approximate age of the object. (Assume that carbon 14 decays continuously and that its half-life is 5,500 years.)

10 The governor of a state with a population A goes on television and announces his resignation. Suppose N people hear the news on television and they transmit this information by word of mouth. Suppose that the rate of spread of this information (per hour) is proportional to the number of people who have heard the news. How many people will have heard the news at time t?

11 It is found experimentally that if an object with temperature T is immersed in a medium with constant temperature C°, the rate of change of the temperature of the object is proportional to the difference between the temperatures of the medium and the object; that is,

$$\frac{dT}{dt} = k(C - T)$$

If the object is at a temperature of 180° and is left exposed to the air at a constant temperature of 60°, it is found that the temperature of the object will decrease to 120° in 1 minute. How long will it take before the object is at a temperature of 90°?

12 A thermometer is placed outdoors where the temperature is fixed at 20°. Initially the thermometer read 75° and in 4 minutes it read 30°. Find the thermometer reading 6 minutes after it was placed outdoors. How long will it take for the reading to drop to 22°?

8-3 On Systems of Differential Equations

Objectives

1. Show how systems of differential equations arise in the sciences.
2. Show how to solve a linear first-order system of differential equations.

Many problems arising in the sciences lead to a system of differential equations. We shall now illustrate how such systems arise and show how linear algebra is used to solve them.

Example 1 In order for a drug to pass from the blood to a cell in the human body it undergoes diffusion twice. First the drug passes from the blood into the membrane that surrounds the cell; then it passes from the membrane into the cell.

Note: Inside each fluid (the blood, fluids in the membrane, and fluids in the cell) the substance spreads out very quickly. Thus the concentration of the substance throughout each of the fluids is uniformly distributed; differences exist only at the boundaries between the blood and the membrane and between the membrane and the cell.

Let c denote the constant concentration of a drug in the blood and let $x(t)$ and $y(t)$ denote the concentration of the drug in the membrane and the cell, respectively, at time t. By the law of diffusion, the amount of drug passing from one fluid to another is a constant multiplied by the difference in concentrations. Thus, the amounts of drug passing from the blood into the membrane and from the membrane into the cell are, respectively,

$$k_1(c - x) \qquad \text{and} \qquad k_2(x - y)$$

where k_1 and k_2 are constants.

Now, let A_1 and A_2 be the amounts of drug in the membrane and the cell, respectively. Then, the rate of change $\left(\frac{dA_1}{dt}\right)$ in the amount of drug A_1 in the membrane is given by

$$\frac{dA_1}{dt} = k_1(c - x) - k_2(x - y) \tag{1}$$

Note: In equation 1 we have the rate of change in the amount of drug A_1 in the membrane as the amount of drug entering the membrane minus the amount leaving it.

and the rate of change $\frac{dA_2}{dt}$ in the amount of drug in the cell is given by

$$\frac{dA_2}{dt} = k_2(x - y) \tag{2}$$

Now if v_1 is the volume of fluid in the membrane, then $A_1 = v_1x$; and similarly, $A_2 = v_2y$ where v_2 is the volume of fluid in the cell.

Assuming v_1 and v_2 to be constant, we find

$$\frac{dA_1}{dt} = v_1\frac{dx}{dt} \qquad \text{and} \qquad \frac{dA_2}{dt} = v_2\frac{dy}{dt}$$

Hence, equations 1 and 2 become

$$v_1\frac{dx}{dt} = k_1(c - x) - k_2(x - y) \tag{3}$$

$$v_2\frac{dy}{dt} = k_1(x - y) \tag{4}$$

Equations 3 and 4 represent a set of two differential equations in the two unknown functions $x(t)$ and $y(t)$. The remaining terms are constants. Such a set of equations is called a **system of differential equations.** A **solution** of such a system consists of an ordered pair

A linear first-order system in the n unknowns, $x_1, x_2, \ldots, x_n$ is of the form

$$\frac{dx_1}{dt} = a_{11}x_1 + a_{12}x_2 + \cdots + a_{1n}x_n$$
$$\frac{dx_2}{dt} = a_{21}x_1 + a_{22}x_2 + \cdots + a_{2n}x_n$$
$$\vdots$$
$$\frac{dx_n}{dt} = a_{n1}x_1 + a_{n2}x_2 + \cdots + a_{nn}x_n$$

of functions (x,y) defined on a common interval that satisfy both the equations. A system of differential equations for unknown functions $x_1, x_2, \ldots, x_n$ in which the highest derivative involved is the first derivative is called a **first-order system.**

A differential equation of order greater than 1 may be reduced to a system of first-order equations. For example, consider the second-order equation

$$y'' = ty' - 4y + 3 \qquad [5]$$

An equivalent system can be defined where

$x_1 = y$ and $x_2 = y'$

This gives the system of equations

$$x_1' = x_2$$
$$x_2' = tx_2 - 4x_1 + 3$$

Using matrices, we write

$$\mathbf{x}' = A\mathbf{x} + \mathbf{b} \qquad [6]$$

Note: $\mathbf{x} = \begin{bmatrix} x_1 \\ x_2 \end{bmatrix}$, $A = \begin{bmatrix} 0 & 1 \\ -4 & t \end{bmatrix}$, and $\mathbf{b} = \begin{bmatrix} 0 \\ 3 \end{bmatrix}$.

Thus, systems of differential equations are also useful in solving differential equations of order greater than 1.

In general a linear first-order system of differential equations involving n unknown functions $x_1, x_2, \ldots, x_n$ may be written as

$$\mathbf{x}' = A\mathbf{x} + \mathbf{b} \qquad [7]$$

where A is an $n \times n$ matrix function and

$$\mathbf{x} = \begin{bmatrix} x_1 \\ x_2 \\ \vdots \\ x_n \end{bmatrix} \quad \text{and} \quad \mathbf{b}(t) = \begin{bmatrix} b_1(t) \\ b_2(t) \\ \vdots \\ b_n(t) \end{bmatrix}$$

If $\mathbf{b}(t) \equiv \mathbf{0}$ then system 7 is called a **homogeneous** system; otherwise the system is called **nonhomogeneous.**

In this section we shall consider the homogeneous system

$$\mathbf{x}' = A\mathbf{x} \qquad [8]$$

where A is an $n \times n$ constant matrix.

A solution of this system is an n-dimensional vector function. The student can easily verify that the set of all real solutions forms a real vector space. Clearly, the zero vector function $\mathbf{0}$ is a solution of equation 8; it is called the **trivial solution.** The student can also prove that the dimension of the solution space of equation 8 is n where n is the number of equations in the system.

To solve equation 8 we need to find a basis for the space of solutions. A basis for the solution space of equation 8 is called a **fundamental set** of solutions. The general solution of equation 8 is of the form

$$c_1\mathbf{v}_1 + c_2\mathbf{v}_2 + \cdots + c_n\mathbf{v}_n$$

where $\mathbf{v}_1, \mathbf{v}_2, \ldots, \mathbf{v}_n$ are linearly independent solutions and $c_1, c_2, \ldots, c_n$ are real constants.

We shall now illustrate how equation 8 is solved. We *assume* a solution of the form

$$\mathbf{x} = e^{\lambda t}\mathbf{a} \qquad [9]$$

where λ and $\mathbf{a}$ (a constant vector) are to be determined. Then the vector function in equation 9 is a solution of equation 8 if and only if

$$\lambda e^{\lambda t}\mathbf{a} = e^{\lambda t}A\mathbf{a} \qquad [10]$$

or

Note: I is the identity matrix.

$$(A - \lambda I)\mathbf{a} = \mathbf{0} \qquad [11]$$

Thus $\mathbf{x}$ is a nontrivial solution of equation 8 if and only if λ is an eigenvalue of the matrix A and $\mathbf{a}$ is a corresponding eigenvector.

Question: Can we find a fundamental set of solutions of the form given by equation 9? The answer is yes, and we shall illustrate the procedure involved in the following example.

Example 2 Solve the system

$$\mathbf{x}' = A\mathbf{x} \qquad [12]$$

where $\mathbf{x} = \begin{bmatrix} x_1 \\ x_2 \end{bmatrix}$ and $A = \begin{bmatrix} 1 & 4 \\ 1 & 1 \end{bmatrix}$.

Solution Assume a solution of the form

$$\mathbf{x} = \mathbf{a}e^{\lambda t}$$

and substitute that solution into equation 12 to obtain

$$(A - \lambda I)\mathbf{a} = 0 \qquad [13]$$

Now, nontrivial solutions of algebraic system 13 exist if and only if λ is chosen so that the determinant of coefficients is zero—that is,

$$\det(A - \lambda I) = 0 \qquad [14]$$

or

$$\begin{vmatrix} 1-\lambda & 4 \\ 1 & 1-\lambda \end{vmatrix} = \lambda^2 - 2\lambda - 3 = 0 \qquad [15]$$

Equation 15 is called the **characteristic equation** for system 12, and it is the condition which determines the values of λ. The roots of equation 15 are $\lambda_1 = -1$ and $\lambda_2 = 3$.

Note: λ_1 and λ_2 are the eigenvalues of the matrix A in equation 14; $\mathbf{a}_1$ and $\mathbf{a}_2$ are the corresponding eigenvectors.

Setting $\lambda_1 = -1$ in equation 13, we obtain $\mathbf{a}_1 = c_1\begin{bmatrix} 2 \\ -1 \end{bmatrix}$ where c_1 is an arbitrary constant. A corresponding solution of equation 12 is

$$\mathbf{v}_1 = \begin{bmatrix} 2 \\ -1 \end{bmatrix} e^{-t} \qquad [16]$$

Setting $\lambda_2 = 3$ in equation 13, we obtain $\mathbf{a}_2 = c_2\begin{bmatrix} 2 \\ 1 \end{bmatrix}$ where c_2 is an arbitrary constant; and a corresponding solution of equation 12 is given by

$$\mathbf{v}_2 = \begin{bmatrix} 2 \\ 1 \end{bmatrix} e^{3t} \qquad [17]$$

The student can readily show that $\mathbf{v}_1$ and $\mathbf{v}_2$ are indeed solutions of equation 12 and that $\mathbf{v}_1$ and $\mathbf{v}_2$ are linearly independent. Under these conditions the general solution of equation 12 is given by

$$\mathbf{x} = c_1\begin{bmatrix} 2 \\ -1 \end{bmatrix} e^{-t} + c_2\begin{bmatrix} 2 \\ 1 \end{bmatrix} e^{3t} \qquad [18]$$

where c_1 and c_2 are constants.

In general, if A in equation 8 has n real and unequal eigenvalues, then the corresponding eigenvectors are linearly independent and yield the general solution of the system. It may be that A has some complex eigenvalues. In such a case it is still possible to obtain a fundamental set of real solutions using the method described in Example 2. We illustrate this in the following example.

Note: It is possible that some of the eigenvalues of A are equal. It is still possible to find n linearly independent solutions. Here, however, we shall not discuss this case.

Example 3 Find the general solution of the equation

$$\mathbf{x}' = \begin{bmatrix} 6 & 8 \\ -1 & 2 \end{bmatrix} \mathbf{x} \qquad [19]$$

Solution Assume a solution of the form $\mathbf{x} = \mathbf{a}e^{\lambda t}$ and substitute that solution into equation 19 to obtain

$$(A - \lambda I)\mathbf{a} = 0 \qquad [20]$$

where

$$A = \begin{bmatrix} 6 & 8 \\ -1 & 2 \end{bmatrix}$$

Note: Nontrivial solutions of algebraic system 20 exist if and only if λ is chosen so that equation 21 is satisfied.

To obtain nontrivial solutions of equation 20, set

$$\det(A - \lambda I) = 0 \qquad \textbf{[21]}$$

or

$$\begin{vmatrix} 6-\lambda & 8 \\ -1 & 2-\lambda \end{vmatrix} = \lambda^2 - 8\lambda + 20 = 0 \qquad \textbf{[22]}$$

Therefore, the eigenvalues of A are

$\lambda_1 = 4 + 2i$ and $\lambda_2 = 4 - 2i$

Setting $\lambda = \lambda_1$ in equation 20 yields an eigenvector

$$\mathbf{a}_1 = c_1 \begin{bmatrix} -2 - 2i \\ 1 \end{bmatrix}$$

where c_1 is an arbitrary constant. A corresponding solution of equation 19 is given by

Recall:
a $e^{a+b} = e^a e^b$
b Euler's identity:
$e^{xi} = \cos x + i \sin x$

$$\mathbf{v}_1 = \begin{bmatrix} -2 - 2i \\ 1 \end{bmatrix} e^{(4+2i)t}$$

$$= e^{4t}\left\{\begin{bmatrix} -2 \\ 1 \end{bmatrix} + i\begin{bmatrix} -2 \\ 0 \end{bmatrix}\right\} e^{2it}$$

$$= e^{4t}\left\{\begin{bmatrix} -2 \\ 1 \end{bmatrix} + i\begin{bmatrix} -2 \\ 0 \end{bmatrix}\right\} (\cos 2t + i \sin 2t)$$

We find that $\mathbf{v}_1$ is a complex solution of equation 19. Also, the real and imaginary components of $\mathbf{v}_1$ are

$$\mathbf{u} = e^{4t}\left\{\begin{bmatrix} -2 \\ 1 \end{bmatrix} \cos 2t - \begin{bmatrix} -2 \\ 0 \end{bmatrix} \sin 2t\right\}$$

and

$$\mathbf{w} = e^{4t}\left\{\begin{bmatrix} -2 \\ 1 \end{bmatrix} \sin 2t + \begin{bmatrix} -2 \\ 0 \end{bmatrix} \cos 2t\right\}$$

See problem 11.

respectively. The student can readily verify that the functions $\mathbf{u}$ and $\mathbf{w}$ are linearly independent real solutions of equation 19. Therefore the general solution of equation 19 is given by

$$\mathbf{x} = c_1\mathbf{u} + c_2\mathbf{w}$$

where c_1 and c_2 are arbitrary constants. Note that the general solution could have been obtained by using λ_2.

Exercises In problems 1 through 8, find the general solution of the given system of equations.

1 $\mathbf{x}' = \begin{bmatrix} 1 & -3 \\ -2 & 2 \end{bmatrix} \mathbf{x}$

2 $\mathbf{x}' = \begin{bmatrix} 3 & 4 \\ 2 & 1 \end{bmatrix} \mathbf{x}$

3 $\mathbf{x}' = \begin{bmatrix} 5 & 3 \\ 4 & 1 \end{bmatrix} \mathbf{x}$

4 $\mathbf{x}' = \begin{bmatrix} 6 & -3 \\ 2 & 1 \end{bmatrix} \mathbf{x}$

5 $\mathbf{x}' = \begin{bmatrix} -3 & -1 \\ 2 & -1 \end{bmatrix} \mathbf{x}$

6 $\mathbf{x}' = \begin{bmatrix} 2 & 4 \\ -2 & -2 \end{bmatrix} \mathbf{x}$

7 $\mathbf{x}' = \begin{bmatrix} 1 & -1 & 0 \\ 1 & 2 & 1 \\ -2 & 1 & -1 \end{bmatrix} \mathbf{x}$

8 $\mathbf{x}' = \begin{bmatrix} 1 & 1 & -1 \\ 2 & 3 & -4 \\ 4 & 1 & -4 \end{bmatrix} \mathbf{x}$

9 Solve problem 1 with the following initial condition:

$$\mathbf{x}(0) = \begin{bmatrix} 0 \\ 5 \end{bmatrix}.$$

10 Solve problem 5 with the following initial condition:

$$\mathbf{x}(0) = \begin{bmatrix} -1 \\ 1 \end{bmatrix}.$$

11 Let $\mathbf{v} = \mathbf{u} + i\mathbf{w}$ be a complex vector function which is a solution of equation 8. Show that the real vector functions $\mathbf{u}$ and $\mathbf{w}$ are also solutions of that equation.

In problems 12 through 16, write the given equation as a system of first-order differential equations and solve.

12 $y'' - 3y' + 2y = 0$

13 $y'' - 5y' + 6y = 0$

14 $y'' + y = 0$

15 $y'' - 6y' + 25y = 0$

16 $y''' - 4y'' + y' + 6y = 0$

17 Solve problem 15 with initial conditions $y(0) = -3$ and $y'(0) = -1$.

18 As shown in Figure 8-3, an object of mass m is attached to a spring on a smooth surface and is set in motion by pulling the object a distance y and released. Applying Newton's second law of motion and Hooke's law (that the tension in the spring is proportional to the distance of the object from the point of equilibrium), we find that the equation of motion of the object is

$$y'' = \frac{-k}{m} y$$

where k (a positive constant) is called the **spring constant.** Write this equation as a system of first-order equations and find a solution that satisfies the initial conditions $y(0) = c_0$ and $y'(0) = 0$.

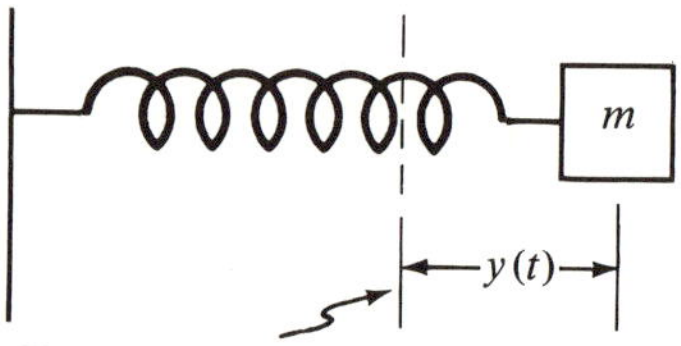

Figure 8-3

ROBERT HOOKE (1635-1703), *an English physicist, studied at Oxford and was an assistant to Robert Boyle. Hooke invented many useful instruments, including the wheel barometer, but he bitterly resented those who took his ideas and expressed them in mathematical models.*

19 Consider the second-order system of differential equations

$$\mathbf{x}'' = A\mathbf{x}$$

where A is an $n \times n$ matrix of constants. Assume that A has n distinct eigenvalues and find the solution of the equation.

Review of Chapter 8

1 What is the general quadratic form in the two variables x and y?

2 What is the matrix of the form for $Q(x, y) = ax^2 + 2bxy + cy^2$?

3 Complete the statement: Two quadratic forms $\mathbf{v}^t A\mathbf{v}$ and $\mathbf{u}^t B\mathbf{u}$ are equivalent if and only if ______________.

4 Define congruent matrices.

5 Complete the statement: Two quadratic forms $\mathbf{v}^t A\mathbf{v}$ and $\mathbf{u}^t A\mathbf{u}$ are said to be orthogonally equivalent if ______________.

6 Identify the graph of the equation $-2x^2 + 4xy + 3y^2 = 8$.

7 Complete the statement: Every quadratic form $ax^2 + 2bxy + cy^2$ is orthogonally equivalent to a quadratic form $\lambda_1 x_1^2 + \lambda_2 y_1^2$ where λ_1 and λ_2 are ______________.

8 The curve defined by the equation $\lambda_1 x^2 + \lambda_2 y^2 = d$ describes

a an ellipse if ________________.

b a hyperbola if ________________.

9 Define ordinary differential equation.

10 Consider the third-order differential equation $y''' - ty'' + 3t^2y' + y = 4 - t$ and write an equivalent first-order system. Is the system homogeneous or nonhomogeneous?

Answers to Problems

Section 0-1

1 b, d, e, f

2 $A = \{2,4,6,8\}$; $B = \{\text{f,l,o,w}\}$;
$C = \{$Mercury, Venus, Earth, Mars, Jupiter, Saturn, Uranus, Neptune, Pluto$\}$;
$D = \{1,2,3, \ldots, 11,12\}$; $E = \{1,2,5,10,25,50\}$; $F = \{2,3,5,7,11,13\}$

3 $\{2,4,6,8\}$ **4** $\{5,6,7,8,9\}$ **5** $\{1,3,5,7,9\}$ **6** $\{1,2,3,4,5,7,9\}$ **7** $\{1,2,3,4,6,8\}$
8 $\{1,2,3,4,5,6,7,8,9\}$ **9** $\varnothing$ **10** $\{2,4\}$ **11** $\varnothing$ **12** $\{2,4,6,8\}$ **13** $\{6,8\}$
14 $\{2,4,5,6,7,8,9\}$ **15** $\{2,4\}$ **16** $\varnothing$ **17** $\{1,3,5,6,7,8,9\}$ **18** $\{1,2,3,4,5,6,7,8,9\}$
19 $\{1,3,5,7,9\}$ **20** $\varnothing$ **21** $\{i\}, \{o\}, \{u\}, \{i,o\}, \{i,u\}, \{o,u\}, \{i,o,u\}$ **22** Yes.
23 **a** $x = 2$ and $y = 9$ or $x = 3$ and $y = 6$ **b** $x = 0$ **c** $x = \frac{10}{4}$ **d** $x = 9$ and $y = \sqrt{3}$; $x = 9$ and $y = -\sqrt{3}$; $x = 3$ and $y = 3$; $x = 3$ and $y = -3$ **e** $x = 3$
f no solution

24 $A = \{2,4\}$, $B = \{1,2,3,4\}$, $C = \{1,3,5,7\}$

25 **a** $A \subset B$ **b** $A = B = \varnothing$ **c** $A = U$ **d** $B = \varnothing$ **e** $B \subset A$ **f** $A = B$

26 **a** 10 **b** 17 **c** 12 **d** 12

Section 0-2

1 $\{x : x \in R\}$ **2** $\{x : x \in R, x \neq 1\}$ **3** $\{t : t \in R\}$ **4** $\{u : u \in R\}$
5 $\{t : t \in R, t \neq 0, -1\}$ **6** $\{r : r \in R, r \neq \pm 1\}$ **7** $\{x : x \geq -5\}$ **8** $\{x : x \in R\}$
9 $\{x : -3 \leq x \leq 3\}$ **10** $\{u : u \leq 0\} \cup \{u : u \geq 2\}$ **11** $\{t : t \in \mathbf{R}, t \neq -\frac{1}{3}\}$
12 $\{s : s \geq 4\} \cup \{s : s \leq -1\}$ **13** $\{x : x \in \mathbf{R}, x < 2\}$ **14** $\{x : -2 < x < 2\}$
15 $\{x : x > 2\} \cup \{x : x < -1\}$

16 **a** -2 **b** -3 **c** 1 **d** $-\frac{11}{4}$ **e** $r^2 + 2r - 2$ **f** $2ah + h^2$

17 **a** 1 **b** $\sqrt{7}$ **c** 0 **d** $\sqrt{6x + 3}$ **18** **a** $\pm\sqrt{\frac{5}{2}}$ **b** ± 2 **c** $\pm\sqrt{5}$

19 **a** $x^2 + x - 3$; $\{x : x \in \mathbf{R}\}$ **b** $x - x^2 - 5$; $\{x : x \in \mathbf{R}\}$
c $x^3 - 4x^2 + x - 4$; $\{x : x \in \mathbf{R}\}$ **d** $\dfrac{x - 4}{x^2 + 1}$; $\{x : x \in \mathbf{R}\}$
e $x^2 - 3$; $\{x : x \in \mathbf{R}\}$ **f** $x^2 - 8x + 17$; $\{x : x \in \mathbf{R}\}$

20 **a** $x^2 + 2x + 1$; $\{x : x \in \mathbf{R}\}$ **b** $2x - x^2 + 1$; $\{x : x \in \mathbf{R}\}$
c $2x^3 + x^2$; $\{x : x \in \mathbf{R}\}$ **d** $\dfrac{2x + 1}{x^2}$; $\{x : x \in \mathbf{R}, x \neq 0\}$
e $2x^2 + 1$, $\{x : x \in \mathbf{R}\}$ **f** $4x^2 + 4x + 1$; $\{x : x \in \mathbf{R}\}$

21 **a** $x^2 + \sqrt{x} + 4$; $\{x : x \geq 0\}$ **b** $x^2 - \sqrt{x} + 4$; $\{x : x \geq 0\}$
c $x^{5/2} + 4\sqrt{x}$; $\{x : x \geq 0\}$ **d** $\dfrac{x^2 + 4}{\sqrt{x}}$; $\{x : x > 0\}$
e $x + 4$; $\{x : x \geq 0\}$ **f** $\sqrt{x^2 + 4}$; $\{x : x \in \mathbf{R}\}$

22 **a** $\dfrac{x^2 + x + 2}{x(x + 1)}$; $\{x : x \in \mathbf{R}, x \neq 0, -1\}$ **b** $\dfrac{x^2 - 3x - 2}{x(x + 1)}$; $\{x : x \in \mathbf{R}, x \neq 0, -1\}$
c $\dfrac{2(x - 1)}{x(x + 1)}$; $\{x : x \in \mathbf{R}, x \neq 0, -1\}$ **d** $\dfrac{x^2 - x}{2(x + 1)}$; $\{x : x \in \mathbf{R}, x \neq 0, -1\}$
e $\dfrac{2 - x}{2 + x}$; $\{x : x \in \mathbf{R}, x \neq 0, -2\}$ **f** $\dfrac{2(x + 1)}{x - 1}$; $\{x : x \in \mathbf{R}, x \neq \pm 1\}$

23 **a** $\dfrac{2}{x} + \sqrt{x - 3}$; $\{x : x \geq 3\}$ **b** $\dfrac{2}{x} - \sqrt{x - 3}$; $\{x : x \geq 3\}$

c $\dfrac{2\sqrt{x-3}}{x}$; $\{x : x \geq 3\}$ **d** $\dfrac{2}{x\sqrt{x-3}}$; $\{x : x > 3\}$

e $\dfrac{2}{\sqrt{x-3}}$; $\{x : x > 3\}$ **f** $\sqrt{\dfrac{2}{x} - 3}$; $\{x : 0 < x \leq \frac{2}{3}\}$

24 **a** $\sqrt{x^2-4} + \sqrt{x-2}$; $\{x : x \geq 2\}$ **b** $\sqrt{x^2-4} - \sqrt{x-2}$; $\{x : x \geq 2\}$

c $\sqrt{(x^2-4)(x-2)}$; $\{x : x \geq 2\}$ **d** $\sqrt{x+2}$; $\{x : x > 2\}$

e $\sqrt{x-6}$; $\{x : x \geq 6\}$ **f** $[\sqrt{x^2-4} - 2]^{1/2}$; $\{x : x \geq 2\sqrt{2}$ or $x \leq -2\sqrt{2}\}$

25 even **26** odd **27** even **28** neither **29** even **30** neither

Section 1-1

1 $\begin{bmatrix} -1 & 3 \\ 6 & 1 \end{bmatrix}$ **2** $\begin{bmatrix} 1 & 5 & 1 \\ 1 & 8 & 5 \end{bmatrix}$ **3** $\begin{bmatrix} 7 & 7 \\ 7 & 7 \\ 7 & 7 \end{bmatrix}$ **4** $\begin{bmatrix} 12 & 3 & 6 \\ 2 & 3 & 2 \\ 4 & -1 & 7 \end{bmatrix}$ **5** $\begin{bmatrix} 4 & 7 \\ -4 & 0 \end{bmatrix}$

6 $A = \begin{bmatrix} 3 & -2 \\ 2 & 1 \end{bmatrix}$ **7** $A = \begin{bmatrix} 5 & -4 & -1 \\ 1 & 1 & -4 \end{bmatrix}$

8 $a_{11} = 2,\ a_{12} = 3,\ a_{21} = 3,\ a_{22} = 4,\ a_{31} = 4,\ a_{32} = 5$

9 $b_{11} = 0,\ b_{12} = -1,\ b_{13} = -2,\ b_{21} = 1,\ b_{22} = 0,\ b_{23} = -1,\ b_{31} = 2,\ b_{32} = 1,\ b_{33} = 0$

10 $c_{11} = 5,\ c_{12} = 8,\ c_{21} = 7,\ c_{22} = 10$

11 $d_{11} = \delta_{11} = 1,\ d_{12} = \delta_{12} = 0,\ d_{13} = \delta_{13} = 0,\ d_{14} = \delta_{14} = 0$
$d_{21} = \delta_{21} = 0,\ d_{22} = \delta_{22} = 1,\ d_{23} = \delta_{23} = 0,\ d_{24} = \delta_{24} = 0$
$d_{31} = \delta_{31} = 0,\ d_{32} = \delta_{32} = 0,\ d_{33} = \delta_{33} = 1,\ d_{34} = \delta_{34} = 0$
$d_{41} = \delta_{41} = 0,\ d_{42} = \delta_{42} = 0,\ d_{43} = \delta_{43} = 0,\ d_{44} = \delta_{44} = 1$

12 $A = \begin{bmatrix} 2 & 4 \\ 4 & 8 \end{bmatrix}$, $B = \begin{bmatrix} 2 & 5 \\ 5 & 8 \end{bmatrix}$, $A + B = \begin{bmatrix} 4 & 9 \\ 9 & 16 \end{bmatrix}$, $A - B = \begin{bmatrix} 0 & -1 \\ -1 & 0 \end{bmatrix}$

13 $\alpha(A + B) = \begin{bmatrix} -6 & 2 \\ 4 & -18 \end{bmatrix}$, $\alpha A + \alpha B = \begin{bmatrix} -6 & 2 \\ 4 & -18 \end{bmatrix}$

14 $(\alpha + \beta)A = \begin{bmatrix} 7 & -3 \\ 8 & 1 \end{bmatrix}$, $\alpha A + \beta A = \begin{bmatrix} 7 & -3 \\ 8 & 1 \end{bmatrix}$, $\alpha(\beta A) = \begin{bmatrix} -42 & 18 \\ -48 & -6 \end{bmatrix}$,

$(\alpha\beta)A = \begin{bmatrix} -42 & 18 \\ -48 & -6 \end{bmatrix}$

16 $A = \begin{bmatrix} -1 & -4 & -7 \\ 1 & -2 & -5 \\ 3 & 0 & -3 \end{bmatrix}$

Section 1-2

1 **a** $[7]$ **b** $[-1]$ **c** $\begin{bmatrix} -7 & 1 \\ 3 & 6 \end{bmatrix}$ **d** $\begin{bmatrix} 5 & 0 \\ -13 & -6 \end{bmatrix}$ **e** $\begin{bmatrix} 1 & -2 \\ -1 & -7 \end{bmatrix}$

f $\begin{bmatrix} -7 & 4 \\ 13 & 7 \end{bmatrix}$ **g** $\begin{bmatrix} -4 & 3 & 2 \\ 1 & 0 & 1 \\ -5 & 6 & 0 \end{bmatrix}$ **h** $\begin{bmatrix} 1 & 4 & 7 \\ 2 & 5 & 8 \\ 3 & 6 & 9 \end{bmatrix}$

2 $AB = \begin{bmatrix} -5 & -3 \\ -3 & -10 \end{bmatrix}$, $BA = \begin{bmatrix} -10 & -3 & -2 \\ -3 & -1 & 0 \\ -2 & 0 & -4 \end{bmatrix}$

3 AB, BA, AD, DA, BC, CD, and DC are undefined.

$AC = \begin{bmatrix} -3 & 1 \\ 12 & -7 \end{bmatrix}$, $CA = \begin{bmatrix} -6 & 7 & -2 \\ -1 & 2 & 0 \\ -5 & -5 & -6 \end{bmatrix}$, $CB = \begin{bmatrix} 5 & -2 & -2 & -1 \\ 1 & 0 & -1 & 0 \\ 2 & -6 & 7 & -3 \end{bmatrix}$

4 $(AB)C = \begin{bmatrix} -2 & 0 & 13 & 0 \\ -4 & 0 & 26 & 0 \end{bmatrix}$, $A(BC) = \begin{bmatrix} -2 & 0 & 13 & 0 \\ -4 & 0 & 26 & 0 \end{bmatrix}$

5 $(\alpha A)B = \begin{bmatrix} -50 & 40 \\ -10 & 0 \\ -40 & 80 \end{bmatrix}$, $A(\alpha B) = \begin{bmatrix} -50 & 40 \\ -10 & 0 \\ -40 & 80 \end{bmatrix}$, $\alpha(AB) = \begin{bmatrix} -50 & 40 \\ -10 & 0 \\ -40 & 80 \end{bmatrix}$

7 $(A + B)C = \begin{bmatrix} 4 & -2 \\ 4 & -2 \end{bmatrix}$, $AC + BC = \begin{bmatrix} 4 & -2 \\ 4 & -2 \end{bmatrix}$, $C(A + B) = \begin{bmatrix} 4 & 0 & -4 \\ -4 & 0 & 4 \\ 2 & 0 & -2 \end{bmatrix}$,

$CA + CB = \begin{bmatrix} 4 & 0 & -4 \\ -4 & 0 & 4 \\ 2 & 0 & -2 \end{bmatrix}$

9 Let $A = \begin{bmatrix} 1 & 3 \\ 2 & 4 \end{bmatrix}$ and $B = \begin{bmatrix} 1 & 3 \\ 2 & 1 \end{bmatrix}$. Then $AB = \begin{bmatrix} 7 & 6 \\ 10 & 10 \end{bmatrix}$ and $BA = \begin{bmatrix} 7 & 15 \\ 4 & 10 \end{bmatrix}$.

Thus $AB \neq BA$.

10 **a** $(A + B)(A + B) = A^2 + AB + BA + B^2$. Thus, $(A + B)(A + B) = A^2 + 2AB + B^2$ if and only if $AB = BA$.

b $(A + B)(A - B) = A^2 - AB + BA - B^2$. Thus, $(A + B)(A - B) = A^2 - B^2$ if and only if $AB = BA$.

11 Let $A = \begin{bmatrix} 0 & 1 \\ 0 & 0 \end{bmatrix}$ and $B = \begin{bmatrix} 1 & 0 \\ 0 & 0 \end{bmatrix}$. We have $A \neq B$, $A \neq 0$, $B \neq 0$, and

$$AB = \begin{bmatrix} 0 & 1 \\ 0 & 0 \end{bmatrix}\begin{bmatrix} 1 & 0 \\ 0 & 0 \end{bmatrix} = \begin{bmatrix} 0 & 0 \\ 0 & 0 \end{bmatrix} = 0.$$

12 Let $A = \begin{bmatrix} 0 & 1 \\ 0 & 0 \end{bmatrix}$. Then $A^2 = \begin{bmatrix} 0 & 1 \\ 0 & 0 \end{bmatrix}\begin{bmatrix} 0 & 1 \\ 0 & 0 \end{bmatrix} = \begin{bmatrix} 0 & 0 \\ 0 & 0 \end{bmatrix} = 0.$

13 Let $B = \begin{bmatrix} 0 & 1 & 1 \\ 0 & 0 & 1 \\ 0 & 0 & 0 \end{bmatrix}$. Then $B^2 = \begin{bmatrix} 0 & 0 & 1 \\ 0 & 0 & 0 \\ 0 & 0 & 0 \end{bmatrix} \neq 0$ and $B^3 = \begin{bmatrix} 0 & 0 & 0 \\ 0 & 0 & 0 \\ 0 & 0 & 0 \end{bmatrix} = 0.$

14 Let $C = \begin{bmatrix} 0 & 1 & 1 & 1 \\ 0 & 0 & 1 & 1 \\ 0 & 0 & 0 & 1 \\ 0 & 0 & 0 & 0 \end{bmatrix}$. Then $C^2 = \begin{bmatrix} 0 & 0 & 1 & 2 \\ 0 & 0 & 0 & 1 \\ 0 & 0 & 0 & 0 \\ 0 & 0 & 0 & 0 \end{bmatrix} \neq 0$, $C^3 = \begin{bmatrix} 0 & 0 & 0 & 1 \\ 0 & 0 & 0 & 0 \\ 0 & 0 & 0 & 0 \\ 0 & 0 & 0 & 0 \end{bmatrix} \neq 0$, and

$$C^4 = \begin{bmatrix} 0 & 0 & 0 & 0 \\ 0 & 0 & 0 & 0 \\ 0 & 0 & 0 & 0 \\ 0 & 0 & 0 & 0 \end{bmatrix} = 0.$$

16 Let $A = \begin{bmatrix} 2 & 3 \\ 4 & 5 \end{bmatrix}$, $B = \begin{bmatrix} 2 & 2 \\ 4 & 5 \end{bmatrix}$, and $C = \begin{bmatrix} 1 & 2 \\ 0 & 0 \end{bmatrix}$. Then $A \neq B$ and $AC = \begin{bmatrix} 2 & 4 \\ 4 & 8 \end{bmatrix} = BC$.

Section 1-3

1 **a** $E = \begin{bmatrix} 0 & 0 & 1 \\ 0 & 1 & 0 \\ 1 & 0 & 0 \end{bmatrix}$ **b** $G = \begin{bmatrix} 1 & 0 & 0 \\ 0 & 1 & 0 \\ 0 & 0 & -3 \end{bmatrix}$ **c** $H = \begin{bmatrix} 1 & 0 & 0 \\ 4 & 1 & 0 \\ 0 & 0 & 1 \end{bmatrix}$

2 **a, c, d, g, h, j,** and **l** are elementary matrices.

3 $D = \begin{bmatrix} 0 & 0 & 1 & 0 \\ 0 & 1 & 0 & 0 \\ 1 & 0 & 0 & 0 \\ 0 & 0 & 0 & 1 \end{bmatrix}$, $E = \begin{bmatrix} 1 & 0 & 0 & 0 \\ 0 & 1 & 0 & 0 \\ 0 & 0 & \frac{1}{5} & 0 \\ 0 & 0 & 0 & 1 \end{bmatrix}$, $F = \begin{bmatrix} 1 & 0 & 0 & 0 \\ 0 & 1 & 0 & 0 \\ 0 & -3 & 1 & 0 \\ 0 & 0 & 0 & 1 \end{bmatrix}$

4 **a** $E_1 = \begin{bmatrix} 1 & 0 & 0 \\ 0 & 1 & 0 \\ 0 & 0 & -1 \end{bmatrix}$ **b** $E_2 = \begin{bmatrix} 1 & 0 & 0 \\ 0 & 1 & 0 \\ 0 & 0 & -1 \end{bmatrix}$ **c** $E_3 = \begin{bmatrix} 1 & 0 & 0 \\ 0 & 0 & 1 \\ 0 & 1 & 0 \end{bmatrix}$

d $E_4 = \begin{bmatrix} 1 & 0 & 0 \\ 0 & 0 & 1 \\ 0 & 1 & 0 \end{bmatrix}$ **e** $E_5 = \begin{bmatrix} 1 & 0 & 0 \\ 1 & 1 & 0 \\ 0 & 0 & 1 \end{bmatrix}$ **f** $E_6 = \begin{bmatrix} 1 & 0 & 0 \\ -1 & 1 & 0 \\ 0 & 0 & 1 \end{bmatrix}$

g $E_7 = \begin{bmatrix} 1 & 0 & 0 \\ 0 & 0 & 1 \\ 0 & 1 & 0 \end{bmatrix}$, $E_8 = \begin{bmatrix} 1 & 0 & 0 \\ 0 & 1 & 0 \\ 0 & 0 & -1 \end{bmatrix}$ **h** $E_9 = \begin{bmatrix} 1 & 0 & 0 \\ 1 & 1 & 0 \\ 0 & 0 & 1 \end{bmatrix}$,

$E_{10} = \begin{bmatrix} 1 & 0 & 0 \\ 0 & 1 & 0 \\ 0 & 0 & -1 \end{bmatrix}$

5 **a** $AI_4 = A$ and $I_3A = A$ **b** $BI_n = B$ and $I_mB = B$ **6** $B = 5I_n + A$

7 $E^7 = \begin{bmatrix} 0 & 0 & 1 \\ 0 & 1 & 0 \\ 1 & 0 & 0 \end{bmatrix} = E$, $G^7 = \begin{bmatrix} 1 & 0 & 0 \\ 0 & (2^7) & 0 \\ 0 & 0 & 1 \end{bmatrix}$, $H^7 = \begin{bmatrix} 1 & 0 & 0 \\ 35 & 1 & 0 \\ 0 & 0 & 1 \end{bmatrix}$

Section 1-4

1 **a** $A = \begin{bmatrix} 2 & -4 & -1 \\ 1 & 5 & 0 \\ 0 & 2 & 3 \end{bmatrix}$, $\begin{bmatrix} 2 & -4 & -1 \\ 1 & 5 & 0 \\ 0 & 2 & 3 \end{bmatrix} \begin{bmatrix} x \\ y \\ z \end{bmatrix} = \begin{bmatrix} 1 \\ 6 \\ 7 \end{bmatrix}$

b $A = \begin{bmatrix} 1 & -1 \\ 2 & 3 \\ 1 & 2 \end{bmatrix}$, $\begin{bmatrix} 1 & -1 \\ 2 & 3 \\ 1 & 2 \end{bmatrix} \begin{bmatrix} x \\ y \end{bmatrix} = \begin{bmatrix} 2 \\ 7 \\ 4 \end{bmatrix}$

c $A = \begin{bmatrix} 5 & -2 & -4 & -1 \\ -1 & 1 & -2 & 0 \\ 2 & -7 & 1 & 6 \\ 0 & 5 & -2 & -1 \end{bmatrix}$, $\begin{bmatrix} 5 & -2 & -4 & -1 \\ -1 & 1 & -2 & 0 \\ 2 & -7 & 1 & 6 \\ 0 & 5 & -2 & -1 \end{bmatrix} \begin{bmatrix} x_1 \\ x_2 \\ x_3 \\ x_4 \end{bmatrix} = \begin{bmatrix} 0 \\ 0 \\ 0 \\ 0 \end{bmatrix}$

d $A = \begin{bmatrix} 2 & 3 & -4 & -5 & 7 \\ -1 & -5 & 6 & -1 & -1 \end{bmatrix}, \begin{bmatrix} 2 & 3 & -4 & -5 & 7 \\ -1 & -5 & 6 & -1 & -1 \end{bmatrix} \begin{bmatrix} x_1 \\ x_2 \\ x_3 \\ x_4 \\ x_5 \end{bmatrix} = \begin{bmatrix} 12 \\ 7 \end{bmatrix}$

2 A^{-1} does not exist. **3** B^{-1} does not exist. **4** C^{-1} does not exist.
5 D^{-1} does not exist. **6** E^{-1} does not exist. **7** A^{-1} does not exist.

Section 1-5

1 **a** $\begin{bmatrix} 1 & 2 \\ 2 & 1 \end{bmatrix}^{-1} = \begin{bmatrix} -\frac{1}{3} & \frac{2}{3} \\ \frac{2}{3} & -\frac{1}{3} \end{bmatrix}$ **b** The matrix $\begin{bmatrix} 2 & 6 \\ 3 & 9 \end{bmatrix}$ is singular.

c The matrix $\begin{bmatrix} 1 & 2 & 7 \\ -1 & 3 & 3 \\ 0 & 1 & 2 \end{bmatrix}$ is singular. **d** $\begin{bmatrix} 1 & -3 & 4 \\ 2 & -5 & 8 \\ 1 & 1 & 5 \end{bmatrix}^{-1} = \begin{bmatrix} -33 & 19 & -4 \\ -2 & 1 & 0 \\ 7 & -4 & 1 \end{bmatrix}$

2 $A^{-1} = \begin{bmatrix} \frac{3}{7} & \frac{1}{7} \\ -\frac{1}{7} & \frac{2}{7} \end{bmatrix}$. One possibility for expressing A^{-1} as a product of elementary matrices is given by: $A^{-1} = \begin{bmatrix} 1 & -3 \\ 0 & 1 \end{bmatrix} \begin{bmatrix} 1 & 0 \\ 0 & -\frac{1}{7} \end{bmatrix} \begin{bmatrix} 1 & 0 \\ -2 & 1 \end{bmatrix} \begin{bmatrix} 0 & 1 \\ 1 & 0 \end{bmatrix}$; similarly, we may express A by: $A = \begin{bmatrix} 0 & 1 \\ 1 & 0 \end{bmatrix} \begin{bmatrix} 1 & 0 \\ 2 & 1 \end{bmatrix} \begin{bmatrix} 1 & 0 \\ 0 & -7 \end{bmatrix} \begin{bmatrix} 1 & 3 \\ 0 & 1 \end{bmatrix}$

3 $B^{-1} = \begin{bmatrix} 11 & -6 & 2 \\ -2 & 1 & 0 \\ 1 & -\frac{2}{3} & \frac{1}{3} \end{bmatrix}$. We may write B^{-1} and B as:

$$B^{-1} = \begin{bmatrix} 1 & 0 & 0 \\ 0 & 1 & 0 \\ 0 & 0 & \frac{1}{3} \end{bmatrix} \begin{bmatrix} 1 & 0 & 2 \\ 0 & 1 & 0 \\ 0 & 0 & 1 \end{bmatrix} \begin{bmatrix} 1 & 0 & 0 \\ 0 & 1 & 0 \\ 0 & -2 & 1 \end{bmatrix} \begin{bmatrix} 1 & -2 & 0 \\ 0 & 1 & 0 \\ 0 & 0 & 1 \end{bmatrix} \begin{bmatrix} 1 & 0 & 0 \\ 0 & 1 & 0 \\ -1 & 0 & 1 \end{bmatrix} \begin{bmatrix} 1 & 0 & 0 \\ -2 & 1 & 0 \\ 0 & 0 & 1 \end{bmatrix}$$

$$B = \begin{bmatrix} 1 & 0 & 0 \\ 2 & 1 & 0 \\ 0 & 0 & 1 \end{bmatrix} \begin{bmatrix} 1 & 0 & 0 \\ 0 & 1 & 0 \\ 1 & 0 & 1 \end{bmatrix} \begin{bmatrix} 1 & 2 & 0 \\ 0 & 1 & 0 \\ 0 & 0 & 1 \end{bmatrix} \begin{bmatrix} 1 & 0 & 0 \\ 0 & 1 & 0 \\ 0 & 2 & 1 \end{bmatrix} \begin{bmatrix} 1 & 0 & -2 \\ 0 & 1 & 0 \\ 0 & 0 & 1 \end{bmatrix} \begin{bmatrix} 1 & 0 & 0 \\ 0 & 1 & 0 \\ 0 & 0 & 3 \end{bmatrix}$$

4 $p = -3$ leads to a singular matrix A.
5 **a** $x_1 = \frac{17}{14}, x_2 = \frac{1}{7}$ **b** $x_1 = \frac{9}{14}, x_2 = \frac{3}{7}$ **c** $x_1 = 0, x_2 = 0$
d $x_1 = -2, x_2 = 1$ **e** $x_1 = \frac{1}{5}, x_2 = -\frac{1}{2}$ **f** $x_1 = 0, x_2 = 0$
6 **a** $x_1 = \frac{1}{3}, x_2 = -3, x_3 = \frac{1}{3}$ **b** $x_1 = 0, x_2 = 1, x_3 = 2$
c $x_1 = 0, x_2 = 0, x_3 = 0$
7 **a** $x_1 = -\frac{3}{2}, x_2 = -\frac{1}{2}, x_3 = \frac{1}{3}, x_4 = -\frac{1}{4}$ **b** $x_1 = 2, x_2 = -1, x_3 = -2, x_4 = 3$
8 $a = \frac{1}{5}, b = \frac{1}{5}$ **9** $a = \frac{1}{3}, b = \frac{1}{3}, c = \frac{2}{3}$

11 $A^{-1} = \begin{bmatrix} 1 & -1 & 0 & 0 \\ 0 & 1 & -1 & 0 \\ 0 & 0 & 1 & -1 \\ 0 & 0 & 0 & 1 \end{bmatrix}, \quad (A^2)^{-1} = (A^{-1})^2 = \begin{bmatrix} 1 & -2 & 1 & 0 \\ 0 & 1 & -2 & 1 \\ 0 & 0 & 1 & -2 \\ 0 & 0 & 0 & 1 \end{bmatrix}$

Section 1-6

1 **a, c, e, g, h, j,** and **k** are matrices in row-reduced echelon form.

2 One such example would be $A = \begin{bmatrix} 0 & 0 & 0 & 0 \\ 1 & 2 & 3 & 4 \\ 0 & 1 & 2 & 3 \end{bmatrix}$.

3 **a** $x_1 = 3 + x_3 - 2x_4$ $\quad x_3$ and x_4 are arbitrary real numbers.
$x_2 = 4 + 2x_3 - x_4$

b $x_1 = 1 + x_2 + 2x_4$ $\quad x_2$ and x_4 are arbitrary real numbers.
$x_3 = 1 - 3x_4$
$x_5 = 1$

c No solution.

4 **a** $x_1 = -8 + 3x_3$ $\quad x_3$ is an arbitrary real number.
$x_2 = 17 - 5x_3$

b $x_1 = 0$ $\quad x_4$ is an arbitrary real number.
$x_2 = \frac{17}{14} + \frac{1}{7}x_4$
$x_3 = -\frac{13}{7} + \frac{10}{7}x_4$

c $x_1 = \frac{1}{2}x_4$ $\quad x_4$ is an arbitrary real number.
$x_2 = -x_3 + \frac{1}{2}x_4$

5 **b** and **d**

6 **a** System has a solution for any real number k. **b** For all $k \neq -1$ and $k \neq 4$ the system is inconsistent and thus has no solution.

9 If the system $\begin{matrix} ax + by = 0 \\ cx + dy = 0 \end{matrix}$ has a unique solution, it must be $x = 0, y = 0$, and therefore L_1 and L_2 are two different lines intersecting at the origin. If the system has more than one solution, then the lines L_1 and L_2 have at least two points in common and thus they coincide.

10 If the system
$$\begin{aligned} ax + by + cz &= 0 \\ dx + ey + fz &= 0 \\ gx + hy + kz &= 0 \end{aligned}$$
has a unique solution, then the planes p_1, p_2, and p_3 intersect at the origin. If the system has more than one solution, then the planes p_1, p_2, and p_3 intersect in a line through the origin. (If the system has solutions representing three noncollinear points, then p_1, p_2, and p_3 are three coinciding planes).

Section 2-1

3 S_2 is a vector space under the operations of addition and scalar multiplication as defined.

4 S_2 (under the operations as defined) is not a vector space. The commutative and associative laws are not satisfied.

5 P (under the operations defined) is not a vector space because it fails to satisfy the closure properties and does not contain a zero vector.

6 P_2 (under the operations defined) is a vector space.

Section 2-2

1 The set of all pairs of real numbers of the form $\begin{bmatrix} x \\ 0 \end{bmatrix}$ with the operations

$\begin{bmatrix} x_1 \\ 0 \end{bmatrix} + \begin{bmatrix} x_2 \\ 0 \end{bmatrix} = \begin{bmatrix} x_1 + x_2 \\ 0 \end{bmatrix}$ and $\alpha \begin{bmatrix} x \\ 0 \end{bmatrix} = \begin{bmatrix} \alpha x \\ 0 \end{bmatrix}$ is a vector space.

2 The set of all pairs of real numbers $\begin{bmatrix} x \\ y \end{bmatrix}$ with the operations $\begin{bmatrix} x_1 \\ y_1 \end{bmatrix} + \begin{bmatrix} x_2 \\ y_2 \end{bmatrix} = \begin{bmatrix} x_1 + x_2 \\ y_1 + y_2 \end{bmatrix}$ and $\alpha \begin{bmatrix} x \\ y \end{bmatrix} = \begin{bmatrix} \alpha x \\ y \end{bmatrix}$ is not a vector space. It fails to satisfy the axiom $(\alpha + \beta)\mathbf{v} = \alpha\mathbf{v} + \beta\mathbf{v}$ for a vector space.

3 The set of all pairs of real numbers of the form $\begin{bmatrix} x \\ 2x \end{bmatrix}$ with the standard operations on $\mathbf{R}^2$ is a vector space.

4 The set of all pairs of real numbers $\begin{bmatrix} x \\ y \end{bmatrix}$, where $x < 0$, with the standard operations on $\mathbf{R}^2$ is not a vector space. The set does not have a zero vector (and no element has an additive inverse). The set is not closed under scalar multiplication, since $(-1)\begin{bmatrix} x \\ y \end{bmatrix} = \begin{bmatrix} -x \\ -y \end{bmatrix}$ and $-x > 0$ if $x < 0$.

5 The set of all triples of real numbers of the form $\begin{bmatrix} x \\ x \\ x \end{bmatrix}$ with the standard operations on $\mathbf{R}^3$ is a vector space.

6 The set of all triples of real numbers $\begin{bmatrix} x \\ y \\ z \end{bmatrix}$ with the operations $\begin{bmatrix} x_1 \\ y_1 \\ z_1 \end{bmatrix} + \begin{bmatrix} x_2 \\ y_2 \\ z_2 \end{bmatrix} = \begin{bmatrix} x_1 + x_2 \\ y_1 + y_2 \\ z_1 + z_2 \end{bmatrix}$ and $\alpha \begin{bmatrix} x \\ y \\ z \end{bmatrix} = \begin{bmatrix} x \\ \alpha y \\ z \end{bmatrix}$ is not a vector space. It fails to satisfy the axiom $(\alpha + \beta)\mathbf{v} = \alpha\mathbf{v} + \beta\mathbf{v}$ for a vector space.

7 The set of all triples of real numbers of the form $\begin{bmatrix} x \\ y \\ 0 \end{bmatrix}$ with the standard operations on $\mathbf{R}^3$ is a vector space.

8 The set of all real-valued functions defined on the real line whose graph contains the origin, with the operations as defined in Example 4, is a vector space.

9 The set of all real-valued functions defined on the real line whose graph contains the point $(1,1)$ in the x-y plane, with operations as defined in Example 4, is not a vector space. It fails to satisfy the closure properties and does not have a zero element.

10 The set of all pairs of real numbers $\begin{bmatrix} x \\ y \end{bmatrix}$ with the operations $\begin{bmatrix} x_1 \\ y_1 \end{bmatrix} + \begin{bmatrix} x_2 \\ y_2 \end{bmatrix} = \begin{bmatrix} (x_1^5 + x_2^5)^{\frac{1}{5}} \\ (y_1^5 + y_2^5)^{\frac{1}{5}} \end{bmatrix}$ and $\alpha \begin{bmatrix} x \\ y \end{bmatrix} = \begin{bmatrix} \alpha^{\frac{1}{5}} x \\ \alpha^{\frac{1}{5}} y \end{bmatrix}$ is a vector space.

11 The set of all pairs of real numbers $\begin{bmatrix} x \\ y \end{bmatrix}$ with the operations $\begin{bmatrix} x_1 \\ y_1 \end{bmatrix} + \begin{bmatrix} x_2 \\ y_2 \end{bmatrix} = \begin{bmatrix} (x_1^2 + x_2^2)^{\frac{1}{2}} \\ (y_1^2 + y_2^2)^{\frac{1}{2}} \end{bmatrix}$ and $\alpha \begin{bmatrix} x \\ y \end{bmatrix} = \begin{bmatrix} |\alpha|^{\frac{1}{2}} x \\ |\alpha|^{\frac{1}{2}} y \end{bmatrix}$ is not a vector space. The set does not

contain a zero element. For all $\begin{bmatrix} x \\ y \end{bmatrix} \neq \begin{bmatrix} 0 \\ 0 \end{bmatrix}$ there do not exist additive inverses. The axiom $(\alpha + \beta)\mathbf{v} = \alpha\mathbf{v} + \beta\mathbf{v}$ is not satisfied here.

12 The set of all matrices of the form $\begin{bmatrix} 0 & a & b \\ -a & 0 & c \\ -b & -c & 0 \end{bmatrix}$ with the usual addition and scalar multiplication defined on matrices is a vector space.

Section 2-3

1 **a, b,** and **d** are subspaces of $\mathbf{R}^2$. **c** is not a subspace of $\mathbf{R}^2$ because the set of all vectors $\begin{bmatrix} x \\ y \end{bmatrix}$, where $x + y = 1$, is not closed under addition and scalar multiplication defined on $\mathbf{R}^2$.

2 **a** and **b** are subspaces of $\mathbf{R}^3$. **c** and **d** are *not* subspaces of $\mathbf{R}^3$ because closure axioms are not satisfied.

3 **a** and **b** are subspaces of P_4 while **c** and **d** are not.

4 **b** and **c** are subspaces of M_2 while **a** and **d** are not.

5 **a, c,** and **d** are subspaces of the function space F while **b** is not.

6 **c** Let V be the vector space $\mathbf{R}^2$. If S is the subspace of $\mathbf{R}^2$ consisting of all pairs of the form $\begin{bmatrix} x \\ 0 \end{bmatrix}$ and T is the subspace of $\mathbf{R}^2$ consisting of pairs of the form $\begin{bmatrix} 0 \\ y \end{bmatrix}$ then $S \cup T$ contains all elements of the form $\begin{bmatrix} x \\ 0 \end{bmatrix}$ or $\begin{bmatrix} 0 \\ y \end{bmatrix}$. It follows that $S \cup T$ is *not* closed under addition defined on $\mathbf{R}^2$ and thus $S \cup T$ fails to be a subspace of $\mathbf{R}^2$. $\left(\begin{bmatrix} 1 \\ 0 \end{bmatrix} \in S, \begin{bmatrix} 0 \\ 1 \end{bmatrix} \in T \text{ but } \begin{bmatrix} 1 \\ 0 \end{bmatrix} + \begin{bmatrix} 0 \\ 1 \end{bmatrix} = \begin{bmatrix} 1 \\ 1 \end{bmatrix} \notin S \cup T.\right)$

Section 3-1

1 **a** Let $\alpha_1 \begin{bmatrix} 2 \\ 3 \end{bmatrix} + \alpha_2 \begin{bmatrix} 1 \\ 4 \end{bmatrix} + \alpha_3 \begin{bmatrix} -1 \\ 2 \end{bmatrix} = \begin{bmatrix} 0 \\ 0 \end{bmatrix}$. One possible solution is $\alpha_1 = 6$, $\alpha_2 = -7$, $\alpha_3 = 5$. Thus, the set $\left\{\begin{bmatrix} 2 \\ 3 \end{bmatrix}, \begin{bmatrix} 1 \\ 4 \end{bmatrix}, \begin{bmatrix} -1 \\ 2 \end{bmatrix}\right\}$ is linearly dependent.

b Let $\beta_1 \begin{bmatrix} 5 \\ 6 \end{bmatrix} + \beta_2 \begin{bmatrix} 10 \\ 12 \end{bmatrix} = \begin{bmatrix} 0 \\ 0 \end{bmatrix}$. One possible solution is $\beta_1 = -2$, $\beta_2 = 1$. Thus, the set $\left\{\begin{bmatrix} 5 \\ 6 \end{bmatrix}, \begin{bmatrix} 10 \\ 12 \end{bmatrix}\right\}$ is linearly dependent.

2 **a** Reader must show that the vector equation $\alpha_1 \begin{bmatrix} 1 \\ -1 \end{bmatrix} + \alpha_2 \begin{bmatrix} 3 \\ 4 \end{bmatrix} = \begin{bmatrix} 0 \\ 0 \end{bmatrix}$ has the unique solution $\alpha_1 = 0$, $\alpha_2 = 0$. **b** Reader must show that the vector equation $\beta_1 \begin{bmatrix} 3 \\ -1 \end{bmatrix} + \beta_2 \begin{bmatrix} 1 \\ 3 \end{bmatrix} = \begin{bmatrix} 0 \\ 0 \end{bmatrix}$ has the unique solution $\beta_1 = 0$, $\beta_2 = 0$.

3 **a, b,** and **d** are linearly dependent. **c** is linearly independent.
4 Reader should use the identity $\cos 2x = \cos^2 x - \sin^2 x$.
6 **b** and **c** are linearly dependent. **a** is linearly independent.

Section 3-2

1 **b** $\begin{bmatrix} 5 \\ -2 \\ 5 \end{bmatrix} = 3\mathbf{v}_1 + 1\mathbf{v}_2$ **d** $\begin{bmatrix} 0 \\ 0 \\ 0 \end{bmatrix} = 0\mathbf{v}_1 + 0\mathbf{v}_2$

The vectors in **a, c,** and **e** are not linear combinations of $\mathbf{v}_1$ and $\mathbf{v}_2$.

2 **a** $\begin{bmatrix} 1 \\ 0 \\ 0 \end{bmatrix} = (-2)\mathbf{v}_1 + 2\mathbf{v}_2 + 1\mathbf{v}_3$ **b** $\begin{bmatrix} 0 \\ 1 \\ 0 \end{bmatrix} = 1\mathbf{v}_1 + (-1)\mathbf{v}_2 + 0\mathbf{v}_3$

c $\begin{bmatrix} 0 \\ 0 \\ 1 \end{bmatrix} = 2\mathbf{v}_1 + (-1)\mathbf{v}_2 + (-1)\mathbf{v}_3$ **d** $\begin{bmatrix} 0 \\ 0 \\ 0 \end{bmatrix} = 0\mathbf{v}_1 + 0\mathbf{v}_2 + 0\mathbf{v}_3$

e $\begin{bmatrix} 1 \\ 2 \\ 3 \end{bmatrix} = 6\mathbf{v}_1 + (-3)\mathbf{v}_2 + (-2)\mathbf{v}_3$

4 **a** $-x^2 - 2x - 1 = \frac{2}{5}p_1 + (-2)p_2 + \frac{1}{5}p_3$
b $3x^2 + 2x - 4 = (-1)p_1 + 9p_2 + (-4)p_3$
c $5x^2 - x + 3 = 2p_1 + 1p_2 + 0p_3$
d $2x^2 + 3x = (-\frac{4}{5})p_1 + 5p_2 + (-\frac{7}{5})p_3$

7 The polynomial $p = x^2 + 3x + 1$ is *not* a linear combination of $p_1 = 2x^2 - x + 1$ and $p_2 = x^2 + x + 1$; thus, $p \notin S$ but $p \in P_2$.

8 **b** $A_5 = \frac{5}{2}A_1 + (-3)A_2 + \frac{3}{2}A_3 + (-\frac{7}{2})A_4$.

10 **b** Add to the set $S = \left\{\begin{bmatrix} 1 \\ 2 \end{bmatrix}, \begin{bmatrix} 3 \\ 2 \end{bmatrix}\right\}$ in **a** any other vector in $\mathbf{R}^2$, say the vector $\begin{bmatrix} 1 \\ 1 \end{bmatrix}$.

c Add to the set S any three vectors in $\mathbf{R}^2$, say, $\begin{bmatrix} 1 \\ 1 \end{bmatrix}, \begin{bmatrix} 0 \\ 1 \end{bmatrix}$ and $\begin{bmatrix} 1 \\ 0 \end{bmatrix}$.

12 **a** The set $S = \left\{\begin{bmatrix} 1 \\ 0 \end{bmatrix}, \begin{bmatrix} 0 \\ 1 \end{bmatrix}\right\}$ spans $\mathbf{R}^2$. The deletion of any vector from S leaves a set which does not span $\mathbf{R}^2$. **b** The set $S = \{1,x,x^2\}$ spans P_2. If we delete 1, x, or x^2 from S we get a set which does not span P_2.

Section 3-3

1 **a, c,** and **d** are bases for $\mathbf{R}^2$.

2 **a** $\left\{\begin{bmatrix} 1 \\ 2 \end{bmatrix}, \begin{bmatrix} 1 \\ 3 \end{bmatrix}, \begin{bmatrix} 1 \\ 4 \end{bmatrix}\right\}$ is linearly dependent.

b $\left\{\begin{bmatrix} 2 \\ 3 \end{bmatrix}, \begin{bmatrix} 4 \\ 6 \end{bmatrix}, \begin{bmatrix} 6 \\ 9 \end{bmatrix}\right\}$ is linearly dependent and does not span $\mathbf{R}^2$.

c $\left\{\begin{bmatrix}1\\0\end{bmatrix}\right\}$ does not span $\mathbf{R}^2$.

d $\left\{\begin{bmatrix}1\\0\end{bmatrix}, \begin{bmatrix}0\\0\end{bmatrix}\right\}$ is linearly dependent and does not span $\mathbf{R}^2$.

3 **a** and **c** are bases for $\mathbf{R}^3$.

4 **a** $\left\{\begin{bmatrix}1\\2\\3\end{bmatrix}\right\}$ does not span $\mathbf{R}^3$. **b** $\left\{\begin{bmatrix}1\\2\\3\end{bmatrix}, \begin{bmatrix}1\\1\\1\end{bmatrix}\right\}$ does not span $\mathbf{R}^3$.

c $\left\{\begin{bmatrix}1\\2\\3\end{bmatrix}, \begin{bmatrix}1\\1\\1\end{bmatrix}, \begin{bmatrix}0\\0\\0\end{bmatrix}\right\}$ does not span $\mathbf{R}^3$ and is linearly dependent.

d $\left\{\begin{bmatrix}1\\2\\3\end{bmatrix}, \begin{bmatrix}1\\1\\1\end{bmatrix}, \begin{bmatrix}1\\4\\1\end{bmatrix}, \begin{bmatrix}-1\\-1\\0\end{bmatrix}\right\}$ is linearly dependent.

5 **b** and **c** are bases for P_2.

6 **a** and **b** do not span P_2. **c** and **d** are linearly dependent sets.

9 Any bases for $M_{2\times 3}$ must contain exactly 6 vectors.

10 **a** Let M_{ij} be the 5×5 matrix such that its (i,j) entry is 1 while all other entries are zero. The set of these 25 matrices form a basis for M_5. **b** A subset S of M_5 which contains 26 elements must be linearly dependent.

Section 3-4

3 $\dim S = 2$. **4** $\dim T = 1$. Notice that $\mathbf{u}_2 = (-2)\mathbf{u}_1$. **5** $\dim S = k$.

6 Let $\bar{S} = \left\{\begin{bmatrix}1\\2\end{bmatrix}, \begin{bmatrix}2\\3\end{bmatrix}\right\}$. Then $\bar{S}$ is a basis for $\mathbf{R}^2$.

7 Let $\bar{S} = \left\{\begin{bmatrix}1\\-1\\2\end{bmatrix}, \begin{bmatrix}1\\-2\\6\end{bmatrix}, \begin{bmatrix}0\\1\\2\end{bmatrix}\right\}$. Then $\bar{S}$ is a basis for $\mathbf{R}^3$.

8 Let $\bar{S} = \{2x^2 - x + 1, x^2 + 2x, -3x^2 + 1\}$. Then $\bar{S}$ is a basis for P_2.

10 $\dim M_2 = 4$. **12** Let $S = \left\{\begin{bmatrix}-2\\3\end{bmatrix}, \begin{bmatrix}1\\0\end{bmatrix}\right\}$. Then S is a basis for $\mathbf{R}^2$.

13 Let $S = \left\{\begin{bmatrix}1\\1\\1\end{bmatrix}, \begin{bmatrix}1\\2\\1\end{bmatrix}, \begin{bmatrix}1\\0\\0\end{bmatrix}\right\}$. Then S is a basis for $\mathbf{R}^3$.

14 Let $S = \{2x^2 + 3x - 4, 1, x\}$. Then S is a basis for P_2.

15 Let $S = \left\{\begin{bmatrix}1 & 3\\2 & 4\end{bmatrix}, \begin{bmatrix}1 & 0\\0 & 0\end{bmatrix}, \begin{bmatrix}0 & 1\\0 & 0\end{bmatrix}, \begin{bmatrix}0 & 0\\1 & 0\end{bmatrix}\right\}$. Then S is a basis for M_2.

16 Let $S = \{2x^3 + x^2 - 3x, 2x^2 - 4x, 1, x\}$. Then S is a basis for P_3.

17 Let M_{ij} be a 3×3 matrix such that its (i,j) entry is 1 while all the remaining entries are zero. The set of these 9 matrices form a basis for M_3. We have $\dim M_3 = 9$.

18 **a** Let $S = \left\{\begin{bmatrix}a\\b\end{bmatrix} : a = b\right\}$. A basis for the subspace S is given by $\left\{\begin{bmatrix}1\\1\end{bmatrix}\right\}$.
We have $\dim S = 1$.

b Let $S = \left\{\begin{bmatrix} a \\ b \end{bmatrix} : a + b = 0\right\}$. A basis for S is given by $\left\{\begin{bmatrix} 1 \\ -1 \end{bmatrix}\right\}$. We have $\dim S = 1$.

c Let $S = \left\{\begin{bmatrix} a \\ b \end{bmatrix} : a = 5b\right\}$. A basis for S is given by $\left\{\begin{bmatrix} 5 \\ 1 \end{bmatrix}\right\}$. We have $\dim S = 1$.

19 **a** A basis for the subspace is $\left\{\begin{bmatrix} 1 \\ 1 \\ 1 \end{bmatrix}\right\}$ and the dimension is 1.

b A basis is given by $\left\{\begin{bmatrix} 1 \\ 0 \\ -1 \end{bmatrix}, \begin{bmatrix} 0 \\ 1 \\ 0 \end{bmatrix}\right\}$ and the dimension is 2.

c A basis is given by $\left\{\begin{bmatrix} 1 \\ 0 \\ 1 \end{bmatrix}, \begin{bmatrix} 0 \\ 1 \\ 1 \end{bmatrix}\right\}$ and the dimension is 2.

d A basis is given by $\left\{\begin{bmatrix} 1 \\ 0 \\ 0 \end{bmatrix}, \begin{bmatrix} 0 \\ 0 \\ 1 \end{bmatrix}\right\}$. The dimension is 2.

e A basis is given by $\left\{\begin{bmatrix} 15 \\ 5 \\ 1 \end{bmatrix}\right\}$. The dimension is 1.

20 **a** A basis is given by $\{x^2, x\}$. The dimension is 2.
b A basis is given by $\{x^2 + 1, x - 1\}$. The dimension is 2.
c A basis is given by $\{x^2 + x - 1\}$. The dimension is 1.
d A basis is given by $\{3x^2 + 1, 3x + 1\}$. The dimension is 2.
e A basis is given by $\{3x^2 + 6x + 2\}$. The dimension is 1.

21 $\dim S = 2$. A basis for S is given by $\left\{\begin{bmatrix} 1 & 1 \\ 1 & 1 \end{bmatrix}, \begin{bmatrix} 3 & 0 \\ 2 & 1 \end{bmatrix}\right\}$.

23 **a** A basis is given by $\left\{\begin{bmatrix} 1 & 0 \\ 0 & 0 \end{bmatrix}, \begin{bmatrix} 0 & 1 \\ 1 & 0 \end{bmatrix}, \begin{bmatrix} 0 & 0 \\ 0 & 1 \end{bmatrix}\right\}$. The dimension is 3.

b A basis is given by $\left\{\begin{bmatrix} 0 & 1 \\ -1 & 0 \end{bmatrix}\right\}$. The dimension is 1.

24 **a** A basis is given by

$$\left\{\begin{bmatrix} 1 & 0 & 0 \\ 0 & 0 & 0 \\ 0 & 0 & 0 \end{bmatrix}, \begin{bmatrix} 0 & 0 & 0 \\ 0 & 1 & 0 \\ 0 & 0 & 0 \end{bmatrix}, \begin{bmatrix} 0 & 0 & 0 \\ 0 & 0 & 0 \\ 0 & 0 & 1 \end{bmatrix}, \begin{bmatrix} 0 & 1 & 0 \\ -1 & 0 & 0 \\ 0 & 0 & 0 \end{bmatrix}, \begin{bmatrix} 0 & 0 & 1 \\ 0 & 0 & 0 \\ 1 & 0 & 0 \end{bmatrix}, \begin{bmatrix} 0 & 0 & 0 \\ 0 & 0 & 1 \\ 0 & 1 & 0 \end{bmatrix}\right\}$$

The dimension is 6.

b A basis is given by $\left\{\begin{bmatrix} 0 & 1 & 0 \\ -1 & 0 & 0 \\ 0 & 0 & 0 \end{bmatrix}, \begin{bmatrix} 0 & 0 & 1 \\ 0 & 0 & 0 \\ -1 & 0 & 0 \end{bmatrix}, \begin{bmatrix} 0 & 0 & 0 \\ 0 & 0 & 1 \\ 0 & -1 & 0 \end{bmatrix}\right\}$. The dimension is 3.

Section 3-5

1 **a** One possible basis for the row space is given by $\mathbf{u}_1 = [1,0]$ and $\mathbf{u}_2 = [0,1]$. The dimension of the row space is 2.

b One possible basis for the row space is given by $\mathbf{u}_1 = [1,-3,-4]$ and $\mathbf{u}_2 = [0,7,8]$. The dimension of the row space is 2.

c One possible basis for the row space is given by $\mathbf{u}_1 = [1,0,0]$, $\mathbf{u}_2 = [0,1,-2]$, and $\mathbf{u}_3 = [0,0,1]$. The dimension of the row space is 2.

d One possible basis for the row space is given by $\mathbf{u}_1 = [1,2,-3,2,3]$ and $\mathbf{u}_2 = [0,5,-9,4,10]$. The dimension of the row space is 2.

2 For the matrix in **1a** we have

$[3,4] = 3[1,0] + 4[0,1] = 3\mathbf{u}_1 + 4\mathbf{u}_2$
$[1,-1] = 1[1,0] + (-1)[0,1] = 1\mathbf{u}_1 + (-1)\mathbf{u}_2$
$[5,0] = 5[1,0] + 0[0,1] = 5\mathbf{u}_1 + 0\mathbf{u}_2$

For the matrix in **1b** we have

$[2,1,0] = 2[1,-3,-4] + 1[0,7,8] = 2\mathbf{u}_1 + 1\mathbf{u}_2$
$[-1,3,4] = (-1)[1,-3,-4] + 0[0,7,8] = (-1)\mathbf{u}_1 + 0\mathbf{u}_2$
$[2,8,8] = 2[1,-3,-4] + 2[0,7,8] = 2\mathbf{u}_1 + 2\mathbf{u}_2$

For the matrix in **1c** we have

$[5,-1,2] = 5[1,0,0] + (-1)[0,1,0] + 2[0,0,1] = 5\mathbf{u}_1 + (-1)\mathbf{u}_2 + 2\mathbf{u}_3$
$[4,4,-8] = 4[1,0,0] + 4[0,1,0] + (-8)[0,0,1] = 4\mathbf{u}_1 + 4\mathbf{u}_2 + (-1)\mathbf{u}_3$
$[3,1,-2] = 3[1,0,0] + 1[0,1,0] + (-2)[0,0,1] = 3\mathbf{u}_1 + 1\mathbf{u}_2 + (-2)\mathbf{u}_3$
$[2,-2,4] = 2[1,0,0] + (-2)[0,1,0] + 4[0,0,1] = 2\mathbf{u}_1 + (-2)\mathbf{u}_2 + 4\mathbf{u}_3$

For the matrix in **1d** we have

$[2,-1,3,0,-4] = 2[1,2,-3,2,3] + (-1)[0,5,-9,4,10] = 2\mathbf{u}_1 + (-1)\mathbf{u}_2$
$[7,-1,6,2,-9] = 7[1,2,-3,2,3] + (-3)[0,5,-9,4,10] = 7\mathbf{u}_1 + (-3)\mathbf{u}_2$
$[5,5,-6,6,5] = 5[1,2,-3,2,3] + (-1)[0,5,-9,4,10] = 5\mathbf{u}_1 + (-1)\mathbf{u}_2$
$[3,1,0,2,-1] = 3[1,2,-3,2,3] + (-1)[0,5,-9,4,10] = 3\mathbf{u}_1 + (-1)\mathbf{u}_2$

3 **a** $\begin{bmatrix} 2 & 3 \\ 1 & 5 \end{bmatrix}^t = \begin{bmatrix} 2 & 1 \\ 3 & 5 \end{bmatrix}$ **b** $\begin{bmatrix} 1 \\ 2 \\ 3 \end{bmatrix}^t = [1 \quad 2 \quad 3]$ **c** $\begin{bmatrix} 1 & 2 & 3 & 4 \\ 5 & 6 & 7 & 8 \end{bmatrix}^t = \begin{bmatrix} 1 & 5 \\ 2 & 6 \\ 3 & 7 \\ 4 & 8 \end{bmatrix}$

d $\begin{bmatrix} 0 & 1 & 2 & 3 \\ -1 & 0 & 4 & 5 \\ -2 & -4 & 0 & 6 \\ -3 & -5 & -6 & 0 \end{bmatrix}^t = \begin{bmatrix} 0 & -1 & -2 & -3 \\ 1 & 0 & -4 & -5 \\ 2 & 4 & 0 & -6 \\ 3 & 5 & 6 & 0 \end{bmatrix}$ **e** $\begin{bmatrix} 1 & 5 & 6 & 7 \\ 5 & 2 & 8 & 9 \\ 6 & 8 & 3 & 0 \\ 7 & 9 & 0 & 4 \end{bmatrix}^t = \begin{bmatrix} 1 & 5 & 6 & 7 \\ 5 & 2 & 8 & 9 \\ 6 & 8 & 3 & 0 \\ 7 & 9 & 0 & 4 \end{bmatrix}$

4 **a** One possible basis for the column space is given by $\mathbf{u}_1 = \begin{bmatrix} 1 \\ 0 \end{bmatrix}$ and $\mathbf{u}_2 = \begin{bmatrix} 0 \\ 1 \end{bmatrix}$.
The dimension of the column space is 2.

b One possible basis for the column space is given by $\mathbf{u}_1 = \begin{bmatrix} 1 \\ 0 \\ 0 \end{bmatrix}$, $\mathbf{u}_2 = \begin{bmatrix} 0 \\ 1 \\ 0 \end{bmatrix}$,

and $\mathbf{u}_3 = \begin{bmatrix} 0 \\ 0 \\ 1 \end{bmatrix}$. The dimension of the column space is 3.

c One possible basis for the column space is given by $\mathbf{u}_1 = \begin{bmatrix} 1 \\ -2 \\ 3 \end{bmatrix}$. The dimension of the column space is 1.

d One possible basis for the column space is given by $\mathbf{u}_1 = \begin{bmatrix} 1 \\ 2 \\ 3 \\ 4 \end{bmatrix}$ and $\mathbf{u}_2 = \begin{bmatrix} 0 \\ 10 \\ 11 \\ 11 \end{bmatrix}$.

The dimension of the column space is 2.

5 For the matrix in **4a** we have

$$\begin{bmatrix} 1 \\ 4 \end{bmatrix} = 1\begin{bmatrix} 1 \\ 0 \end{bmatrix} + 4\begin{bmatrix} 0 \\ 1 \end{bmatrix} = 1\mathbf{u}_1 + 4\mathbf{u}_2$$

$$\begin{bmatrix} 2 \\ 5 \end{bmatrix} = 2\begin{bmatrix} 1 \\ 0 \end{bmatrix} + 5\begin{bmatrix} 0 \\ 1 \end{bmatrix} = 2\mathbf{u}_1 + 5\mathbf{u}_2$$

$$\begin{bmatrix} 3 \\ 6 \end{bmatrix} = 3\begin{bmatrix} 1 \\ 0 \end{bmatrix} + 6\begin{bmatrix} 0 \\ 1 \end{bmatrix} = 3\mathbf{u}_1 + 6\mathbf{u}_2$$

For the matrix in **4b** we have

$$\begin{bmatrix} 3 \\ -1 \\ -3 \end{bmatrix} = 3\begin{bmatrix} 1 \\ 0 \\ 0 \end{bmatrix} + (-1)\begin{bmatrix} 0 \\ 1 \\ 0 \end{bmatrix} + (-3)\begin{bmatrix} 0 \\ 0 \\ 1 \end{bmatrix} = 3\mathbf{u}_1 + (-1)\mathbf{u}_2 + (-3)\mathbf{u}_3$$

$$\begin{bmatrix} 1 \\ 2 \\ 4 \end{bmatrix} = 1\begin{bmatrix} 1 \\ 0 \\ 0 \end{bmatrix} + 2\begin{bmatrix} 0 \\ 1 \\ 0 \end{bmatrix} + 4\begin{bmatrix} 0 \\ 0 \\ 1 \end{bmatrix} = 1\mathbf{u}_1 + 2\mathbf{u}_2 + 4\mathbf{u}_3$$

$$\begin{bmatrix} 2 \\ 0 \\ 1 \end{bmatrix} = 2\begin{bmatrix} 1 \\ 0 \\ 0 \end{bmatrix} + 0\begin{bmatrix} 0 \\ 1 \\ 0 \end{bmatrix} + 1\begin{bmatrix} 0 \\ 0 \\ 1 \end{bmatrix} = 2\mathbf{u}_1 + 0\mathbf{u}_2 + 1\mathbf{u}_3$$

For the matrix in **4c** we have

$$\begin{bmatrix} 2 \\ -4 \\ 6 \end{bmatrix} = 2\begin{bmatrix} 1 \\ -2 \\ 3 \end{bmatrix} = 2\mathbf{u}_1, \quad \begin{bmatrix} -3 \\ 6 \\ -9 \end{bmatrix} = (-3)\begin{bmatrix} 1 \\ -2 \\ 3 \end{bmatrix} = (-3)\mathbf{u}_1, \quad \begin{bmatrix} 1 \\ -2 \\ 3 \end{bmatrix} = 1\mathbf{u}_1,$$

$$\begin{bmatrix} -1 \\ 2 \\ -3 \end{bmatrix} = (-1)\mathbf{u}_1$$

For the matrix in **4d** we have

$$\begin{bmatrix} 1 \\ 2 \\ 3 \\ 4 \end{bmatrix} = 1\mathbf{u}_1 + 0\mathbf{u}_2, \quad \begin{bmatrix} -2 \\ 6 \\ 5 \\ 3 \end{bmatrix} = (-2)\begin{bmatrix} 1 \\ 2 \\ 3 \\ 4 \end{bmatrix} + 1\begin{bmatrix} 0 \\ 10 \\ 11 \\ 11 \end{bmatrix} = (-2)\mathbf{u}_1 + 1\mathbf{u}_2$$

$$\begin{bmatrix} -3 \\ 4 \\ 2 \\ -1 \end{bmatrix} = (-3)\begin{bmatrix} 1 \\ 2 \\ 3 \\ 4 \end{bmatrix} + 1\begin{bmatrix} 0 \\ 10 \\ 11 \\ 11 \end{bmatrix} = (-3)\mathbf{u}_1 + 1\mathbf{u}_2$$

$$\begin{bmatrix} 5 \\ 0 \\ 4 \\ 9 \end{bmatrix} = 5\begin{bmatrix} 1 \\ 2 \\ 3 \\ 4 \end{bmatrix} + (-1)\begin{bmatrix} 0 \\ 10 \\ 11 \\ 11 \end{bmatrix} = 5\mathbf{u}_1 + (-1)\mathbf{u}_2,$$

$$\begin{bmatrix} 2 \\ -6 \\ -5 \\ -3 \end{bmatrix} = 2\begin{bmatrix} 1 \\ 2 \\ 3 \\ 4 \end{bmatrix} + (-1)\begin{bmatrix} 0 \\ 10 \\ 11 \\ 11 \end{bmatrix} = 2\mathbf{u}_1 + (-1)\mathbf{u}_2$$

6 **a** $\begin{bmatrix} 0 & 0 & 0 \\ 0 & 0 & 0 \end{bmatrix}$ **b** $\begin{bmatrix} 1 & 0 \\ 0 & 0 \\ 0 & 0 \end{bmatrix}$ **c** $\begin{bmatrix} 1 & 0 & 0 & 0 \\ 0 & 1 & 0 & 0 \\ 0 & 0 & 0 & 0 \end{bmatrix}$

d $\begin{bmatrix} 1 & 0 & 0 & 0 \\ 0 & 1 & 0 & 0 \\ 0 & 0 & 1 & 0 \\ 0 & 0 & 0 & 0 \end{bmatrix}$ **e** $\begin{bmatrix} 1 & 0 & 0 & 0 & 0 \\ 0 & 1 & 0 & 0 & 0 \\ 0 & 0 & 1 & 0 & 0 \\ 0 & 0 & 0 & 1 & 0 \\ 0 & 0 & 0 & 0 & 1 \end{bmatrix} = I_5$

7 **a** $\begin{bmatrix} 3 & 2 \\ -6 & -4 \end{bmatrix}$ has rank 1. (Matrix is singular.)

b $\begin{bmatrix} 1 & 0 \\ 0 & 1 \\ 3 & 5 \end{bmatrix}$ has rank 2.

c $\begin{bmatrix} 0 & 2 & 3 \\ 3 & 1 & 0 \\ -2 & 0 & 1 \end{bmatrix}$ has rank 2. (Matrix is singular.)

d $\begin{bmatrix} 0 & 1 & 1 & 1 \\ -1 & 0 & 1 & 1 \\ -1 & -1 & 0 & 1 \\ -1 & -1 & -1 & 0 \end{bmatrix}$ has rank 4. (Matrix is nonsingular.)

8 The rank of all elementary matrices of order 3×3 is 3. Every elementary matrix is nonsingular. **9** $r(A) = r(A^t)$.

Section 3-6

1 **a** One possible basis for the null space of the matrix $[2 \quad 1 \quad -3]$ is given by $\begin{bmatrix} -\frac{1}{2} \\ 1 \\ 0 \end{bmatrix}$ and $\begin{bmatrix} \frac{3}{2} \\ 0 \\ 1 \end{bmatrix}$. The dimension of the null space is 2.

b One possible basis for the null space of the matrix $\begin{bmatrix} 3 & 0 & 1 \\ 1 & 2 & 0 \end{bmatrix}$ is given by $\begin{bmatrix} -\frac{1}{3} \\ \frac{1}{6} \\ 1 \end{bmatrix}$. The dimension of the null space is 1.

c One possible basis for the null space of the matrix $\begin{bmatrix} 2 & 4 \\ 1 & 2 \\ -3 & -6 \end{bmatrix}$ is given by $\begin{bmatrix} -2 \\ 1 \end{bmatrix}$. The dimension of the null space is 1.

d One possible basis for the null space of the matrix $\begin{bmatrix} 1 & 2 & 3 & 4 & 5 \\ 1 & 1 & 1 & 1 & 1 \\ 2 & -1 & 2 & -1 & 2 \end{bmatrix}$

is given by $\begin{bmatrix} 1 \\ -1 \\ -1 \\ 1 \\ 0 \end{bmatrix}$ and $\begin{bmatrix} 1 \\ 0 \\ -2 \\ 0 \\ 1 \end{bmatrix}$. The dimension of the null space is 2.

2 a $\begin{bmatrix} x_1 \\ x_2 \\ x_3 \end{bmatrix} = \begin{bmatrix} \frac{13}{7} \\ -\frac{3}{7} \\ 0 \end{bmatrix} + x_3 \begin{bmatrix} -\frac{1}{7} \\ -\frac{10}{7} \\ 1 \end{bmatrix}$ (x_3 is arbitrary)

b $\begin{bmatrix} x_1 \\ x_2 \\ x_3 \\ x_4 \end{bmatrix} = \begin{bmatrix} \frac{22}{7} \\ \frac{10}{7} \\ 0 \\ 0 \end{bmatrix} + x_3 \begin{bmatrix} \frac{11}{7} \\ \frac{5}{7} \\ 1 \\ 0 \end{bmatrix} + x_4 \begin{bmatrix} -\frac{1}{7} \\ \frac{4}{7} \\ 0 \\ 1 \end{bmatrix}$ (x_3 and x_4 are arbitrary)

3 a The rank of the matrix of coefficients is 1. The rank of the augmented matrix is 2. System has no solution.

b The rank of the matrix of coefficients is 2. The rank of the augmented matrix is 2. System has a unique solution.

c The rank of the matrix of coefficients is 2. The rank of the augmented matrix is 3. System has no solution.

d The rank of the matrix of coefficients is 2. The rank of the augmented matrix is 2.

System has solutions given by $\begin{bmatrix} x_1 \\ x_2 \\ x_3 \end{bmatrix} = \begin{bmatrix} 2 \\ 4 \\ 0 \end{bmatrix} + x_3 \begin{bmatrix} 1 \\ -2 \\ 1 \end{bmatrix}$ (x_3 is arbitrary).

e The rank of the matrix of coefficients is 3. The rank of the augmented matrix is 3.

System has a unique solution given by $\begin{bmatrix} x_1 \\ x_2 \\ x_3 \end{bmatrix} = \begin{bmatrix} 1 \\ -1 \\ 2 \end{bmatrix}$.

Section 4-1

1 a $M_{12} = \begin{vmatrix} 4 & 0 & -2 \\ 0 & 1 & 2 \\ -1 & 3 & 0 \end{vmatrix}, \quad A_{12} = (-1)^{1+2} M_{12}$

b $M_{23} = \begin{vmatrix} 1 & 0 & 3 \\ 0 & -3 & 2 \\ -1 & -2 & 0 \end{vmatrix}, \quad A_{23} = (-1)^{2+3} M_{23}$

c $M_{34} = \begin{vmatrix} 1 & 0 & 2 \\ 4 & 1 & 0 \\ -1 & -2 & 3 \end{vmatrix}, \quad A_{34} = (-1)^{3+4} M_{34}$

d $M_{42} = \begin{vmatrix} 1 & 2 & 3 \\ 4 & 0 & -2 \\ 0 & 1 & 2 \end{vmatrix}, \quad A_{42} = (-1)^{4+2} M_{42}$

2 **a** -2 **b** 30 **c** 0 **d** 0 **e** $3-2x$
3 **a** $x=3$ **b** $x=\pm\sqrt{2}$ **c** $x=0, x=1, x=-1$
4 **a** 16 **b** 68 **c** 1
5 **a** $x=1, x=-\frac{14}{13}$ **b** $x=1, x=2, x=-3$ **c** $x=3, x=2, x=-2$
6 **a** -6 **b** -66 **c** 4
7 **a** $\det A = a_{11}A_{11} + a_{12}A_{12} + a_{13}A_{13} + a_{14}A_{14} + a_{15}A_{15}$
b $\det A = a_{13}A_{13} + a_{23}A_{23} + a_{33}A_{33} + a_{43}A_{43} + a_{53}A_{53}$

Section 4-2

1 **a** 24 **b** 0 **c** 0 **d** 0 **e** 4 **f** 0 **g** 0
2 **a** -3 **b** 1 **c** 1 **d** $-1{,}024$ **e** 180 **f** 64
3 **a** singular **b** nonsingular **c** nonsingular **d** singular
4 **a** $x=2$ and $x=-1$ **b** $x=2$ and $x=3$
5 **a** $\det(A^{-1})=\frac{1}{9}$ **b** $\det(A^t)=9$ **c** $\det((A^t)^{-1})=\frac{1}{9}$
8 Let $A=\begin{bmatrix}1 & 2\\ 3 & 4\end{bmatrix}$ and $B=\begin{bmatrix}1 & 0\\ 0 & 0\end{bmatrix}$. Then $A+B=\begin{bmatrix}2 & 2\\ 3 & 4\end{bmatrix}$. We have $\det(A+B)=2$, $\det A=-2$, and $\det B=0$. Thus, $\det(A+B)\neq \det A+\det B$.
10 **a** -6 **b** 24 **c** $n!(-1)^{\frac{n(n-1)}{2}}$

Section 4-3

1
$$a_{11}A_{21}+a_{12}A_{22}+a_{13}A_{23} = 1\cdot(-1)^{2+1}\begin{vmatrix}4 & 7\\ 6 & 9\end{vmatrix} + 4\cdot(-1)^{2+2}\begin{vmatrix}1 & 7\\ 3 & 9\end{vmatrix} + 7\cdot(-1)^{2+3}\begin{vmatrix}1 & 4\\ 3 & 6\end{vmatrix}$$
$$= (-1)(-6) + 4\cdot(-12) + 7\cdot(-1)(-6) = 0$$

2
$$a_{11}A_{13}+a_{21}A_{23}+a_{31}A_{33} = 5\cdot(-1)^{1+3}\begin{vmatrix}4 & 2\\ 3 & 7\end{vmatrix} + 4\cdot(-1)^{2+3}\begin{vmatrix}5 & 6\\ 3 & 7\end{vmatrix} + 3\cdot(-1)^{3+3}\begin{vmatrix}5 & 6\\ 4 & 2\end{vmatrix}$$
$$= 5\cdot 22 + 4\cdot(-1)\cdot 17 + 3\cdot(-14) = 0$$

3 **a** $\operatorname{adj} A = \begin{bmatrix}2 & -1\\ -4 & 3\end{bmatrix}$ **b** $A(\operatorname{adj} A) = \begin{bmatrix}3 & 1\\ 4 & 2\end{bmatrix}\begin{bmatrix}2 & -1\\ -4 & 3\end{bmatrix} = \begin{bmatrix}2 & 0\\ 0 & 2\end{bmatrix}$ **c** $\det A = 2$

4 **a** $\operatorname{adj} A = \begin{bmatrix}13 & -4 & 1\\ 4 & 8 & -2\\ 1 & 2 & 7\end{bmatrix}$

b $A(\operatorname{adj} A) = \begin{bmatrix}2 & 1 & 0\\ -1 & 3 & 1\\ 0 & -1 & 4\end{bmatrix}\begin{bmatrix}13 & -4 & 1\\ 4 & 8 & -2\\ 1 & 2 & 7\end{bmatrix} = \begin{bmatrix}30 & 0 & 0\\ 0 & 30 & 0\\ 0 & 0 & 30\end{bmatrix}$ **c** $\det A = 30$

5 **a** $\begin{bmatrix}5 & 2\\ 6 & 3\end{bmatrix}^{-1} = \frac{1}{3}\begin{bmatrix}3 & -2\\ -6 & 5\end{bmatrix} = \begin{bmatrix}1 & -\frac{2}{3}\\ -2 & \frac{5}{3}\end{bmatrix}$ **b** $\begin{bmatrix}5 & 1 & 2\\ -1 & 3 & 0\\ -2 & 0 & 4\end{bmatrix}^{-1} = \frac{1}{76}\begin{bmatrix}12 & -4 & -6\\ 4 & 24 & -2\\ 6 & -2 & 16\end{bmatrix}$

c $\begin{bmatrix}1 & 1 & 1 & 1\\ 1 & 2 & 3 & 4\\ 1 & 3 & 6 & 10\\ 1 & 4 & 10 & 20\end{bmatrix}^{-1} = \begin{bmatrix}4 & -6 & 4 & -1\\ -6 & 14 & -11 & 3\\ 4 & -11 & 10 & -3\\ -1 & 3 & -3 & 1\end{bmatrix}$

7 **a** $x_1=8, x_2=2$ **b** $x_1=3, x_2=1, x_3=2$
c $x_1=1, x_2=-1, x_3=2, x_4=-2$

8 **a** If rank of A is n then rank of $\operatorname{adj} A$ is also n.
b If rank of A is smaller than n then the rank of $\operatorname{adj} A$ is also smaller than n.

12 Let x and y be the number of male and female applicants interviewed, respectively. Then we have

$$15x + 10y = 240$$
$$10x + 20y = 360$$

We get $x = 6$ and $y = 15$.

Section 5-1

1 **a** $\sqrt{13}$ **b** $\sqrt{26}$ **c** $\sqrt{63}\ (=3\sqrt{7})$ **d** $\sqrt{55}$

2 **a** $\sqrt{29}$ **b** $\sqrt{14}$ **c** $\sqrt{54}\ (=3\sqrt{6})$ **d** $\sqrt{13}$

3 **a** -7 **b** 13 **c** -9 **d** 22

4 **a** $-\dfrac{7}{5\sqrt{2}}$ **b** $\dfrac{1}{3}\sqrt{\dfrac{13}{3}}$ **c** $-\dfrac{3}{2\sqrt{5}}$ **d** $\dfrac{22}{3\sqrt{70}}$ **6** $0, \sqrt{\dfrac{6}{41}}, \sqrt{\dfrac{35}{41}}$

8 For $\cos\theta = 1$ we have $\theta = 0$ and for $\cos\theta = -1$ we have $\theta = -\pi(\pi$ radians $= 180°)$.

Section 5-2

1 **a** $k = \frac{10}{3}$ **b** $k = -1, k = -2$ **c** $k = 0, k = 2, k = -3$

2 **a** $(2,-3,1)$ is a basis for the subspace. **b** The dimension of the subspace is 1.
c $\dfrac{1}{\sqrt{14}}(2,-3,1)$ and $\dfrac{1}{\sqrt{14}}(-2,3,-1)$

3 **a** and **d** are orthogonal sets.

4 No, because any linearly independent set in $\mathbf{R}^4$ contains *at most* 4 vectors and any orthonormal set is linearly independent.

5 **a** and **c** are orthonormal. **6** **a** $\frac{18}{5}$ **b** $\frac{2}{3}$ **c** $\frac{8}{3}$

7 **a** $(-\frac{8}{25}, \frac{6}{25})$ **b** $(\frac{14}{9}, -\frac{14}{9}, \frac{7}{9})$ **c** $(\frac{8}{9}, 0, \frac{16}{9}, -\frac{16}{9})$

8 **b** $(3,2) = \dfrac{7}{\sqrt{5}}\left(\dfrac{1}{\sqrt{5}}, \dfrac{2}{\sqrt{5}}\right) + \dfrac{4}{\sqrt{5}}\left(\dfrac{2}{\sqrt{5}}, -\dfrac{1}{\sqrt{5}}\right)$

9 **b** $(1,2,3) = 0\left(\dfrac{1}{\sqrt{3}}, \dfrac{1}{\sqrt{3}}, -\dfrac{1}{\sqrt{3}}\right) + \dfrac{4}{\sqrt{2}}\left(\dfrac{1}{\sqrt{2}}, 0, \dfrac{1}{\sqrt{2}}\right) + \dfrac{6}{\sqrt{6}}\left(-\dfrac{1}{\sqrt{6}}, \dfrac{2}{\sqrt{6}}, \dfrac{1}{\sqrt{6}}\right)$

Section 5-3

1 **b** $\mathbf{v}_1 = \left(\dfrac{3}{\sqrt{10}}, \dfrac{1}{\sqrt{10}}\right), \quad \mathbf{v}_2 = \left(\dfrac{1}{\sqrt{10}}, \dfrac{-3}{\sqrt{10}}\right)$

2 $\mathbf{v}_1 = \left(\dfrac{1}{3}, \dfrac{2}{3}, \dfrac{2}{3}\right), \quad \mathbf{v}_2 = \left(\dfrac{10}{\sqrt{153}}, \dfrac{-7}{\sqrt{153}}, \dfrac{2}{\sqrt{153}}\right)$

3 **b** $\mathbf{v}_1 = \left(\dfrac{1}{\sqrt{3}}, \dfrac{1}{\sqrt{3}}, \dfrac{1}{\sqrt{3}}\right), \quad \mathbf{v}_2 = \left(\dfrac{-1}{\sqrt{6}}, \dfrac{-1}{\sqrt{6}}, \dfrac{2}{\sqrt{6}}\right), \quad \mathbf{v}_3 = \left(\dfrac{-1}{\sqrt{2}}, \dfrac{1}{\sqrt{2}}, 0\right)$

4 $\mathbf{v}_1 = \left(\frac{1}{\sqrt{2}}, 0, \frac{1}{\sqrt{2}}, 0\right), \quad \mathbf{v}_2 = \left(\frac{1}{\sqrt{3}}, 0, \frac{-1}{\sqrt{3}}, \frac{1}{\sqrt{3}}\right), \quad \mathbf{v}_3 = \left(\frac{-1}{\sqrt{6}}, 0, \frac{1}{\sqrt{6}}, \frac{2}{\sqrt{6}}\right)$

5 b $\mathbf{v}_1 = \left(\frac{1}{2}, \frac{1}{2}, \frac{1}{2}, \frac{1}{2}\right), \quad \mathbf{v}_2 = \left(\frac{-3}{\sqrt{20}}, \frac{-1}{\sqrt{20}}, \frac{1}{\sqrt{20}}, \frac{3}{\sqrt{20}}\right),$

$\mathbf{v}_3 = \left(\frac{1}{2}, -\frac{1}{2}, -\frac{1}{2}, \frac{1}{2}\right), \quad \mathbf{v}_4 = \left(\frac{-1}{\sqrt{20}}, \frac{3}{\sqrt{20}}, \frac{-3}{\sqrt{20}}, \frac{1}{\sqrt{20}}\right)$

Section 6-1

4 From $T(x,y) = (1,2)$ it follows that $T(1,1) = (1,2)$, $T(2,2) = (1,2)$ and also $T(3,3) = (1,2)$. But $T[(1,1) + (2,2)] = T(3,3) = (1,2)$ and $T(1,1) + T(2,2) = (1,2) + (1,2) = (2,4)$. Thus, T is not linear.

5 The transformation $L: \mathbf{R}^2 \longrightarrow \mathbf{R}^2$ such that L maps each point in the plane into its reflection about the y-axis is given by $L(x,y) = (-x,y)$.

8 **a, c,** and **e** are linear while **b, d,** and **f** are not linear.

9 **a, b,** and **d** are linear while **c** is not linear.

Section 6-2

1 a $\begin{bmatrix} 1 & 1 \\ 1 & -1 \end{bmatrix}$ **b** $\begin{bmatrix} 0 & 1 \\ 1 & 0 \end{bmatrix}$ **c** $\begin{bmatrix} 1 & 0 \\ 0 & 1 \end{bmatrix}$ **d** $\begin{bmatrix} 0 & 0 \\ 0 & 0 \end{bmatrix}$

2 a $\begin{bmatrix} \frac{\sqrt{3}}{2} & -\frac{1}{2} \\ \frac{1}{2} & \frac{\sqrt{3}}{2} \end{bmatrix}$ **b** $\begin{bmatrix} \frac{1}{\sqrt{2}} & -\frac{1}{\sqrt{2}} \\ \frac{1}{\sqrt{2}} & \frac{1}{\sqrt{2}} \end{bmatrix}$ **c** $\begin{bmatrix} \frac{1}{2} & -\frac{\sqrt{3}}{2} \\ \frac{\sqrt{3}}{2} & \frac{1}{2} \end{bmatrix}$

d $\begin{bmatrix} 0 & -1 \\ 1 & 0 \end{bmatrix}$ **e** $\begin{bmatrix} -\frac{1}{2} & -\frac{\sqrt{3}}{2} \\ \frac{\sqrt{3}}{2} & -\frac{1}{2} \end{bmatrix}$ **f** $\begin{bmatrix} -1 & 0 \\ 0 & -1 \end{bmatrix}$

3 a $\begin{bmatrix} 1 & 1 & 1 \\ 1 & -2 & -1 \\ 3 & 1 & -2 \\ 2 & -3 & 4 \end{bmatrix}$ **b** $\begin{bmatrix} 1 & 2 & 0 & 0 \\ 0 & 0 & 2 & -1 \end{bmatrix}$ **c** $[2 \quad -1 \quad 3]$

4 $\begin{bmatrix} \frac{9}{25} & \frac{12}{25} \\ \frac{12}{25} & \frac{16}{25} \end{bmatrix}$ **5** $P\begin{bmatrix} x \\ y \end{bmatrix} = \begin{bmatrix} (u_1)^2 & u_1u_2 \\ u_1u_2 & (u_2)^2 \end{bmatrix}\begin{bmatrix} x \\ y \end{bmatrix}$ and $\begin{vmatrix} (u_1)^2 & u_1u_2 \\ u_1u_2 & (u_2)^2 \end{vmatrix} = 0$

6 $\begin{bmatrix} \gamma & 0 & 0 & 0 \\ 0 & \gamma & 0 & 0 \\ 0 & 0 & \gamma & 0 \\ 0 & 0 & 0 & \gamma \end{bmatrix}$ **8** $T\begin{bmatrix} x \\ y \end{bmatrix} = [1 \quad -1]\begin{bmatrix} x \\ y \end{bmatrix}$ **9** $T\begin{bmatrix} x \\ y \end{bmatrix} = \begin{bmatrix} 2 & 1 \\ -1 & 3 \end{bmatrix}\begin{bmatrix} x \\ y \end{bmatrix}$

Section 6-3

4 $T\begin{bmatrix} x \\ y \end{bmatrix} = \begin{bmatrix} 0 & -1 \\ -1 & 0 \end{bmatrix}\begin{bmatrix} x \\ y \end{bmatrix}$ 5 $T\begin{bmatrix} x \\ y \end{bmatrix} = \begin{bmatrix} 2(u_1)^2 - 1 & 2u_1u_2 \\ 2u_1u_2 & 2(u_2)^2 - 1 \end{bmatrix}\begin{bmatrix} x \\ y \end{bmatrix}$

6 $\mathbf{u} = \begin{bmatrix} \frac{1}{\sqrt{5}} \\ \frac{2}{\sqrt{5}} \end{bmatrix}$ or $\mathbf{u} = \begin{bmatrix} -\frac{1}{\sqrt{5}} \\ -\frac{2}{\sqrt{5}} \end{bmatrix}$

7 $(ST)\begin{bmatrix} x \\ y \end{bmatrix} = \begin{bmatrix} 0 \\ 2x + 5y \end{bmatrix}$, $(TS)\begin{bmatrix} x \\ y \end{bmatrix} = \begin{bmatrix} -x - 2y \\ 3x + 6y \end{bmatrix}$, $T\begin{bmatrix} x \\ y \end{bmatrix} = \begin{bmatrix} 2 & -1 \\ 0 & 3 \end{bmatrix}\begin{bmatrix} x \\ y \end{bmatrix}$,

$S\begin{bmatrix} x \\ y \end{bmatrix} = \begin{bmatrix} 0 & 0 \\ 1 & 2 \end{bmatrix}\begin{bmatrix} x \\ y \end{bmatrix}$, $(ST)\begin{bmatrix} x \\ y \end{bmatrix} = \begin{bmatrix} 0 & 0 \\ 2 & 5 \end{bmatrix}\begin{bmatrix} x \\ y \end{bmatrix}$, $(TS)\begin{bmatrix} x \\ y \end{bmatrix} = \begin{bmatrix} -1 & -2 \\ 3 & 6 \end{bmatrix}\begin{bmatrix} x \\ y \end{bmatrix}$

Section 6-4

1 **a** The null space of T consists of all vectors $\begin{bmatrix} x \\ y \end{bmatrix}$ where $x = y$. A basis for the null space of T is given by the set $\left\{\begin{bmatrix} 1 \\ 1 \end{bmatrix}\right\}$.

b The range of T consists of all vectors of the form $\alpha\begin{bmatrix} 1 \\ 1 \end{bmatrix}$ where α is an arbitrary scalar (real number). Thus, a basis for the range of T is given by the set $\left\{\begin{bmatrix} 1 \\ 1 \end{bmatrix}\right\}$.

c The nullity of T is equal to 1. The rank of T is equal to 1.

2 **a** The null space of T consists of the zero vector $\begin{bmatrix} 0 \\ 0 \\ 0 \end{bmatrix}$ of $\mathbf{R}^3$.

b The range of T consists of all linear combinations of the vectors $\begin{bmatrix} 2 \\ 3 \\ 0 \end{bmatrix}$, $\begin{bmatrix} 4 \\ 1 \\ 1 \end{bmatrix}$, and $\begin{bmatrix} 6 \\ 2 \\ -1 \end{bmatrix}$. Since these vectors are linearly independent, they form a basis for $\mathbf{R}^3$. Thus, the range of T is $\mathbf{R}^3$.

c The nullity of T is equal to 0. The rank of T is equal to 3.

3 **a** The null space of T consists of all vectors $\begin{bmatrix} x \\ y \\ z \\ w \end{bmatrix}$ such that $x = -w$ and $y = -z$.

It follows that a basis for the null space of T is given by the set $\left\{\begin{bmatrix} 0 \\ -1 \\ 1 \\ 0 \end{bmatrix}, \begin{bmatrix} -1 \\ 0 \\ 0 \\ 1 \end{bmatrix}\right\}$.

b The range of T consists of all linear combinations of the vectors $\begin{bmatrix} 1 \\ 1 \end{bmatrix}$ and $\begin{bmatrix} 1 \\ -1 \end{bmatrix}$.

c The nullity of T is equal to 2. The rank of T is equal to 2.

4 b The null space of T consists of the zero polynomial.

c The range of T consists of all polynomials of the form $ax^2 + bx$ (constant term is zero). It follows that a basis for the range of T is given by the set $\{x^2,x\}$.

d The nullity of T is equal to 0. The rank of T is equal to 2.

6 $T\begin{bmatrix} x \\ y \end{bmatrix} = 2\begin{bmatrix} x \\ y \end{bmatrix}$ **7** $T\begin{bmatrix} x \\ y \\ z \end{bmatrix} = \begin{bmatrix} \frac{37}{44} & \frac{3}{44} & -\frac{4}{11} \\ \frac{37}{44} & \frac{3}{44} & -\frac{4}{11} \end{bmatrix}\begin{bmatrix} x \\ y \\ z \end{bmatrix}$

9 b The null space of D consists of all constant polynomials $p(x) = a_0$. A basis for this null space is given by the set $\{1\}$.

c The range of D consists of all polynomials having the form $a_1x + a_0$. Thus, the range of D coincides with the vector space P_1. A basis for P_1 is given by the set $\{x,1\}$.

d The nullity of D is equal to 1. The rank of D is equal to 2.

10 b The null space of D^2 consists of all the polynomials of the form $a_1x + a_0$. Thus, the null space of D^2 is the vector space P_1.

c The range of D^2 is the same as its null space.

d The nullity of D^2 is equal to 2. The rank of D^2 is equal to 2.

Section 6-5

1 $[\mathbf{p}]_B = \begin{bmatrix} 3 \\ -3 \\ 1 \end{bmatrix}$, $[\mathbf{p}]_{B'} = \begin{bmatrix} 3 \\ 2 \\ 1 \end{bmatrix}$ **2** $A = \begin{bmatrix} 1 & 0 & 0 \\ 1 & \frac{1}{2} & \frac{1}{2} \\ 1 & \frac{1}{2} & -\frac{1}{2} \end{bmatrix}$, $A' = \begin{bmatrix} 1 & 0 & 0 \\ -2 & 1 & 1 \\ 0 & 1 & -1 \end{bmatrix}$

4 $[\mathbf{v}]_B = \begin{bmatrix} -3 \\ 1 \end{bmatrix}$, $[\mathbf{v}]_{B'} = \begin{bmatrix} \frac{4}{3} \\ -\frac{5}{3} \end{bmatrix}$ **5 a** $A = \begin{bmatrix} -\frac{1}{3} & \frac{1}{3} \\ \frac{2}{3} & \frac{1}{3} \end{bmatrix}$ **b** $A' = \begin{bmatrix} -1 & 1 \\ 2 & 1 \end{bmatrix}$

7 $[\mathbf{p}]_B = \begin{bmatrix} \frac{7}{2} \\ \frac{3}{2} \end{bmatrix}$, $[\mathbf{p}]_{B'} = \begin{bmatrix} \frac{7}{3} \\ \frac{1}{3} \end{bmatrix}$ **8 a** $A = \begin{bmatrix} \frac{2}{3} & 0 \\ -\frac{1}{3} & 1 \end{bmatrix}$ **b** $A' = \begin{bmatrix} \frac{3}{2} & 0 \\ \frac{1}{2} & 1 \end{bmatrix}$

10 $[\mathbf{v}]_B = \begin{bmatrix} -1 \\ -1 \\ 3 \end{bmatrix}$, $[\mathbf{v}]_{B'} = \begin{bmatrix} 0 \\ 2 \\ 1 \end{bmatrix}$ **11 a** $A = \begin{bmatrix} -1 & -2 & -1 \\ 0 & 1 & 1 \\ 1 & 1 & 1 \end{bmatrix}$ **b** $A' = \begin{bmatrix} 0 & -1 & 1 \\ -1 & 0 & -1 \\ 1 & 1 & 1 \end{bmatrix}$

Section 6-6

1 $\begin{bmatrix} 3 & 0 \\ -1 & 1 \\ 1 & 2 \end{bmatrix}$ **2** $[\mathbf{v}]_B = \begin{bmatrix} 5 \\ -2 \end{bmatrix}$, $[T(\mathbf{v})]_{B'} = \begin{bmatrix} 15 \\ -7 \\ 1 \end{bmatrix}$ **3** $\begin{bmatrix} 0 & 0 \\ -1 & 1 \\ 1 & 1 \end{bmatrix}$

4 $[\mathbf{p}]_B = \begin{bmatrix} 2 \\ 0 \end{bmatrix}$, $[T(\mathbf{p})]_{B'} = \begin{bmatrix} 0 \\ -2 \\ 2 \end{bmatrix}$ **5** $\begin{bmatrix} 3 & 1 \\ -2 & 1 \end{bmatrix}$

6 $[\mathbf{v}]_B = \begin{bmatrix} 3 \\ -2 \end{bmatrix}$, $[T(\mathbf{v})]_B = \begin{bmatrix} 7 \\ -8 \end{bmatrix}$ **7** $\begin{bmatrix} -1 & 0 & 0 \\ 0 & -1 & 0 \\ 2 & 2 & 2 \end{bmatrix}$

8 $[\mathbf{v}]_B = \begin{bmatrix} 3 \\ -2 \\ 4 \end{bmatrix}, \qquad [T(\mathbf{v})]_B = \begin{bmatrix} -3 \\ 2 \\ 10 \end{bmatrix}$

11 a $\begin{bmatrix} 0 & 1 & 0 & 0 \\ 0 & 0 & 2 & 0 \\ 0 & 0 & 0 & 3 \\ 0 & 0 & 0 & 0 \end{bmatrix}$ **b** $\begin{bmatrix} -\frac{1}{2} & \frac{1}{2} & -1 & -1 \\ -\frac{1}{2} & \frac{1}{2} & 1 & 1 \\ 0 & 0 & -\frac{3}{2} & \frac{3}{2} \\ 0 & 0 & -\frac{3}{2} & \frac{3}{2} \end{bmatrix}$

Section 6-7

1 $A' = \begin{bmatrix} 4 & 1 \\ -7 & -1 \end{bmatrix}, \quad P = \begin{bmatrix} 1 & 2 \\ -1 & -3 \end{bmatrix}, \quad P^{-1} = \begin{bmatrix} 3 & 2 \\ -1 & -1 \end{bmatrix},$

$A = P^{-1}A'P = \begin{bmatrix} -3 & -7 \\ 3 & 6 \end{bmatrix}$

2 $A' = \begin{bmatrix} 5 & 1 \\ -5 & 1 \end{bmatrix}, \quad A = P^{-1}A'P = \begin{bmatrix} 0 & -5 \\ 2 & 6 \end{bmatrix}$

3 $A' = \begin{bmatrix} -2 & -1 & -4 \\ 0 & -1 & 1 \\ 1 & 2 & 2 \end{bmatrix}, \quad P = \begin{bmatrix} -1 & -2 & -1 \\ 1 & 1 & 0 \\ 1 & 1 & 1 \end{bmatrix}, \quad P^{-1} = \begin{bmatrix} 1 & 1 & 1 \\ -1 & 0 & -1 \\ 0 & -1 & 1 \end{bmatrix},$

$A = P^{-1}A'P = \begin{bmatrix} 0 & 1 & 0 \\ 0 & -1 & 1 \\ 3 & 2 & 0 \end{bmatrix}$

4 $A' = \begin{bmatrix} -1 & -5 & -3 \\ 1 & 3 & 2 \\ -1 & 0 & 1 \end{bmatrix}, \quad A = P^{-1}A'P = \begin{bmatrix} -1 & 0 & 1 \\ 5 & 3 & 0 \\ -2 & 0 & 1 \end{bmatrix}$

10 $X = \begin{bmatrix} 3 & 0 \\ 0 & 2 \end{bmatrix}, \quad X = \begin{bmatrix} -3 & 0 \\ 0 & 2 \end{bmatrix}, \quad X = \begin{bmatrix} 3 & 0 \\ 0 & -2 \end{bmatrix},$ and $X = \begin{bmatrix} -3 & 0 \\ 0 & -2 \end{bmatrix}$

11 $X = \begin{bmatrix} -1 & 0 \\ 0 & 2 \end{bmatrix}$

12 Any of the following 8 matrices is a solution of the matrix equation $X^4 = A$:

$\begin{bmatrix} 2 & 0 & 0 \\ 0 & 1 & 0 \\ 0 & 0 & 3 \end{bmatrix}, \begin{bmatrix} -2 & 0 & 0 \\ 0 & 1 & 0 \\ 0 & 0 & 3 \end{bmatrix}, \begin{bmatrix} 2 & 0 & 0 \\ 0 & -1 & 0 \\ 0 & 0 & 3 \end{bmatrix}, \begin{bmatrix} 2 & 0 & 0 \\ 0 & 1 & 0 \\ 0 & 0 & -3 \end{bmatrix}$

$\begin{bmatrix} -2 & 0 & 0 \\ 0 & -1 & 0 \\ 0 & 0 & 3 \end{bmatrix}, \begin{bmatrix} -2 & 0 & 0 \\ 0 & 1 & 0 \\ 0 & 0 & -3 \end{bmatrix}, \begin{bmatrix} 2 & 0 & 0 \\ 0 & -1 & 0 \\ 0 & 0 & -3 \end{bmatrix}, \begin{bmatrix} -2 & 0 & 0 \\ 0 & -1 & 0 \\ 0 & 0 & -3 \end{bmatrix}$

13 b The 4 solutions are $\begin{bmatrix} 1 & 2 \\ -1 & 4 \end{bmatrix}, \begin{bmatrix} -7 & 10 \\ -5 & 8 \end{bmatrix}, \begin{bmatrix} 7 & -10 \\ 5 & -8 \end{bmatrix},$ and $\begin{bmatrix} -1 & -2 \\ 1 & -4 \end{bmatrix}.$

14 b The 8 solutions are:

$$\tfrac{1}{3}\begin{bmatrix} 9 & 0 & -3 \\ 4 & 5 & -2 \\ -2 & 2 & 4 \end{bmatrix}, \tfrac{1}{3}\begin{bmatrix} 17 & -8 & -7 \\ 20 & -11 & -10 \\ 6 & -6 & 0 \end{bmatrix}, \tfrac{1}{3}\begin{bmatrix} 7 & 2 & -5 \\ 4 & 5 & -2 \\ -6 & 6 & 0 \end{bmatrix}, \tfrac{1}{3}\begin{bmatrix} -15 & 6 & 9 \\ -20 & 11 & 10 \\ -2 & 2 & 4 \end{bmatrix}$$

$$\tfrac{1}{3}\begin{bmatrix} 15 & -6 & -9 \\ 20 & -11 & -10 \\ 2 & -2 & -4 \end{bmatrix}, \tfrac{1}{3}\begin{bmatrix} -7 & -2 & 5 \\ -4 & -5 & 2 \\ 6 & -6 & 0 \end{bmatrix}, \tfrac{1}{3}\begin{bmatrix} -17 & 8 & 7 \\ -20 & 11 & 10 \\ -6 & 6 & 0 \end{bmatrix}, \tfrac{1}{3}\begin{bmatrix} -9 & 0 & 3 \\ -4 & -5 & 2 \\ 2 & -2 & -4 \end{bmatrix}$$

15 Let $A = \begin{bmatrix} 1 & 0 \\ 0 & 1 \end{bmatrix}$ and $B = \begin{bmatrix} 1 & 1 \\ 0 & 1 \end{bmatrix}$. We have $\det A = \det B = 1$, but A is not similar to B since an identity matrix is similar only to itself.

Section 7-1

1 The eigenvalue associated with $\mathbf{v} = \begin{bmatrix} -2 \\ 1 \end{bmatrix}$ is $\lambda = 3$.

2 The eigenvalue associated with the vector $\begin{bmatrix} 1 \\ 1 \end{bmatrix}$ is $\lambda = 3$; that associated with the vector $\begin{bmatrix} 3 \\ 2 \end{bmatrix}$ is $\lambda = 4$.

3 **a** $(2 - \lambda)(3 - \lambda) = 0$, $\lambda = 2$, and $\lambda = 3$. $\begin{bmatrix} 1 \\ 0 \end{bmatrix}$ is an eigenvector associated with $\lambda = 2$. $\begin{bmatrix} 1 \\ 1 \end{bmatrix}$ is an eigenvector associated with $\lambda = 3$.

b $(4 - \lambda)^2 = 0$, $\lambda = 4$. $\begin{bmatrix} 1 \\ 0 \end{bmatrix}$ and $\begin{bmatrix} 0 \\ 1 \end{bmatrix}$ are linearly independent eigenvectors associated with $\lambda = 4$.

c $\lambda^2 = 0$, $\lambda = 0$. $\begin{bmatrix} 1 \\ 0 \end{bmatrix}$ and $\begin{bmatrix} 0 \\ 1 \end{bmatrix}$ are linearly independent eigenvectors associated with $\lambda = 0$.

d $(\lambda - 5)(\lambda + 3) = 0$, $\lambda = 5$, and $\lambda = -3$. $\begin{bmatrix} 1 \\ 1 \end{bmatrix}$ is an eigenvector associated with $\lambda = 5$. $\begin{bmatrix} 1 \\ -3 \end{bmatrix}$ is an eigenvector associated with $\lambda = -3$.

e $(\lambda - 4)(\lambda + 1) = 0$, $\lambda = 4$, and $\lambda = -1$. $\begin{bmatrix} 2 \\ 3 \end{bmatrix}$ is an eigenvector associated with $\lambda = 4$. $\begin{bmatrix} 1 \\ -1 \end{bmatrix}$ is an eigenvector associated with $\lambda = -1$.

4 $\lambda^2 - 4\lambda + 13 = 0$

5 **a** $(2 - \lambda)^2(1 - \lambda) = 0$, $\lambda = 2$, and $\lambda = 1$. $\begin{bmatrix} 1 \\ 0 \\ 2 \end{bmatrix}$ is an eigenvector associated with $\lambda = 2$. $\begin{bmatrix} 0 \\ 0 \\ 1 \end{bmatrix}$ is an eigenvector associated with $\lambda = 1$.

b $(\lambda-1)(\lambda-2)(\lambda-5)=0$, $\lambda=1$, $\lambda=2$, and $\lambda=5$. $\begin{bmatrix}2\\1\\4\end{bmatrix}$ is an eigenvector associated with $\lambda=1$. $\begin{bmatrix}1\\1\\2\end{bmatrix}$ is an eigenvector associated with $\lambda=2$. $\begin{bmatrix}0\\1\\1\end{bmatrix}$ is an eigenvector associated with $\lambda=5$.

c $(\lambda-1)(\lambda-4)(\lambda-6)=0$, $\lambda=1$, $\lambda=4$, and $\lambda=6$. $\begin{bmatrix}1\\0\\0\end{bmatrix}$ is an eigenvector associated with $\lambda=1$. $\begin{bmatrix}2\\3\\0\end{bmatrix}$ is an eigenvector associated with $\lambda=4$. $\begin{bmatrix}16\\25\\10\end{bmatrix}$ is an eigenvector associated with $\lambda=6$.

d $(\lambda-2)(\lambda-3)(\lambda+2)=0$, $\lambda=2$, $\lambda=3$, and $\lambda=-2$. $\begin{bmatrix}1\\0\\-1\end{bmatrix}$ is an eigenvector associated with $\lambda=2$. $\begin{bmatrix}1\\-1\\-1\end{bmatrix}$ is an eigenvector associated with $\lambda=3$. $\begin{bmatrix}1\\-1\\4\end{bmatrix}$ is an eigenvector associated with $\lambda=-2$.

6 **a** $\begin{bmatrix}1&-1&-1\\1&3&1\\-3&1&-1\end{bmatrix}$ **b** $\lambda=2$, $\lambda=3$, and $\lambda=-2$

c x^2-1 is an eigenvector of T associated with $\lambda=2$. x^2-x-1 is an eigenvector of T associated with $\lambda=3$. x^2-x+4 is an eigenvector of T associated with $\lambda=-2$.

9 $\lambda=1$, $\lambda=-1$, $\lambda=32$, $\lambda=-32$.

11 The vector $\mathbf{v}=\begin{bmatrix}-2\\1\end{bmatrix}$ is an eigenvector for the matrix $A=\begin{bmatrix}5&4\\-1&1\end{bmatrix}$. We have $A^t\mathbf{v}=\begin{bmatrix}5&-1\\4&1\end{bmatrix}\begin{bmatrix}-2\\1\end{bmatrix}=\begin{bmatrix}-11\\-7\end{bmatrix}$. Since $\begin{bmatrix}-11\\-7\end{bmatrix}$ is not a scalar multiple of the vector $\begin{bmatrix}-2\\1\end{bmatrix}$, it follows that $\begin{bmatrix}-2\\1\end{bmatrix}$ is not an eigenvector of A^t.

Section 7-2

1 $P=\begin{bmatrix}1&-1\\1&1\end{bmatrix}$ **2** $P=\begin{bmatrix}1&1\\-3&1\end{bmatrix}$

3 The matrix $A=\begin{bmatrix}3&-1\\4&-1\end{bmatrix}$ has an eigenvalue $\lambda=1$ of multiplicity 2. This leads to an eigenvector $\mathbf{v}=\begin{bmatrix}1\\2\end{bmatrix}$. Since A does *not* have two linearly independent eigenvectors, it is not diagonalizable.

4 The matrix $A=\begin{bmatrix}5&4\\-1&1\end{bmatrix}$ has an eigenvalue $\lambda=3$ of multiplicity 2. This leads to an eigenvector $\mathbf{v}=\begin{bmatrix}2\\-1\end{bmatrix}$. Since A does *not* have two linearly independent eigenvectors, it is not diagonalizable.

5 The characteristic equation of A is given by $(\lambda - 2)^2(\lambda + 1) = 0$. The eigenvalue $\lambda = 2$ (with multiplicity 2) leads to two linearly independent eigenvectors $\begin{bmatrix} 1 \\ 0 \\ -1 \end{bmatrix}$ and $\begin{bmatrix} 0 \\ 1 \\ -1 \end{bmatrix}$. The eigenvalue $\lambda = -1$ leads to an eivenvector $\begin{bmatrix} 1 \\ 1 \\ 1 \end{bmatrix}$. The matrix A is diagonalizable by the matrix P given by $P = \begin{bmatrix} 1 & 0 & 1 \\ 0 & 1 & 1 \\ -1 & -1 & 1 \end{bmatrix}$.

$P^{-1}AP = \begin{bmatrix} 2 & 0 & 0 \\ 0 & 2 & 0 \\ 0 & 0 & -1 \end{bmatrix}$.

6 The eigenvalue of $A = \begin{bmatrix} 0 & 1 & 1 \\ 0 & 0 & 1 \\ 0 & 0 & 0 \end{bmatrix}$ is given by $\lambda = 0$. This eigenvalue leads to an eigenvector $\begin{bmatrix} 1 \\ 0 \\ 0 \end{bmatrix}$. Since A does *not* have three linearly independent eigenvectors it is not diagonalizable.

7 The characteristic equation of A is given by $(\lambda - 2)(\lambda - 3)(\lambda + 2) = 0$. The eigenvalue $\lambda = 2$ leads to an eigenvector $\begin{bmatrix} 1 \\ 0 \\ -1 \end{bmatrix}$. The eigenvalue $\lambda = 3$ leads to an eigenvector $\begin{bmatrix} 1 \\ -1 \\ -1 \end{bmatrix}$. The eigenvalue $\lambda = -2$ leads to an eigenvector $\begin{bmatrix} 1 \\ -1 \\ 4 \end{bmatrix}$. The matrix A is diagonalizable by the matrix P given by

$P = \begin{bmatrix} 1 & 1 & 1 \\ 0 & -1 & -1 \\ -1 & -1 & 4 \end{bmatrix}$. $P^{-1}AP = \begin{bmatrix} 2 & 0 & 0 \\ 0 & 3 & 0 \\ 0 & 0 & -2 \end{bmatrix}$.

10 **a** $Q = \begin{bmatrix} 1 & 3 \\ -1 & -4 \end{bmatrix}$ **b** $A^8 = \begin{bmatrix} 256 & 0 \\ 0 & 256 \end{bmatrix}$

14 **a** The matrix $\begin{bmatrix} 0 & 1 \\ 0 & 0 \end{bmatrix}$ is nilpotent because $\begin{bmatrix} 0 & 1 \\ 0 & 0 \end{bmatrix}^2 = \begin{bmatrix} 0 & 0 \\ 0 & 0 \end{bmatrix}$, and thus it is not diagonalizable.

b The matrix $\begin{bmatrix} 0 & 1 & 1 \\ 0 & 0 & 1 \\ 0 & 0 & 0 \end{bmatrix}$ is nilpotent because $\begin{bmatrix} 0 & 1 & 1 \\ 0 & 0 & 1 \\ 0 & 0 & 0 \end{bmatrix}^3 = \begin{bmatrix} 0 & 0 & 0 \\ 0 & 0 & 0 \\ 0 & 0 & 0 \end{bmatrix}$, and thus it is not diagonalizable.

c The matrix $\begin{bmatrix} 0 & 1 & 1 & 1 \\ 0 & 0 & 1 & 1 \\ 0 & 0 & 0 & 1 \\ 0 & 0 & 0 & 0 \end{bmatrix}$ is nilpotent because $\begin{bmatrix} 0 & 1 & 1 & 1 \\ 0 & 0 & 1 & 1 \\ 0 & 0 & 0 & 1 \\ 0 & 0 & 0 & 0 \end{bmatrix}^4 = \begin{bmatrix} 0 & 0 & 0 & 0 \\ 0 & 0 & 0 & 0 \\ 0 & 0 & 0 & 0 \\ 0 & 0 & 0 & 0 \end{bmatrix}$, and thus it is not diagonalizable.

19 We have $T(4+x) = 4+x$ and $T(-3x) = (-3)(-3x)$. Let $B = \{4+x, -3x\}$. Then B is an ordered basis for P_1. The matrix of T with respect to B is a diagonal matrix A given by $A = \begin{bmatrix} 1 & 0 \\ 0 & -3 \end{bmatrix}$.

Section 7-3

1 **b, c,** and **e** are symmetric matrices. **a** and **d** are *not* symmetric matrices.

4 **a** and **d** are skew-symmetric matrices. **b, c,** and **e** are *not* skew-symmetric matrices.

7 **b** Let $B = \begin{bmatrix} 0 & 1 \\ -1 & 0 \end{bmatrix}$. The characteristic equation for B is $\lambda^2 + 1 = 0$. The solutions are $\lambda = \pm i$.

8 **b, c,** and **e** are orthogonal matrices. **a** and **d** are *not* orthogonal matrices.

9 **a** $\begin{bmatrix} 1 & -1 \\ 1 & 1 \end{bmatrix}^{-1} = \frac{1}{2}\begin{bmatrix} 1 & 1 \\ -1 & 1 \end{bmatrix}$ **b** $\begin{bmatrix} \frac{1}{2} & \frac{\sqrt{3}}{2} \\ \frac{-\sqrt{3}}{2} & \frac{1}{2} \end{bmatrix}^{-1} = \begin{bmatrix} \frac{1}{2} & \frac{-\sqrt{3}}{2} \\ \frac{\sqrt{3}}{2} & \frac{1}{2} \end{bmatrix}$

c $\begin{bmatrix} \frac{4}{5} & -\frac{3}{5} \\ \frac{3}{5} & \frac{4}{5} \end{bmatrix}^{-1} = \begin{bmatrix} \frac{4}{5} & \frac{3}{5} \\ -\frac{3}{5} & \frac{4}{5} \end{bmatrix}$ **d** $\begin{bmatrix} 1 & 0 & -1 \\ 0 & 1 & 0 \\ 1 & 0 & 1 \end{bmatrix}^{-1} = \frac{1}{2}\begin{bmatrix} 1 & 0 & 1 \\ 0 & 2 & 0 \\ -1 & 0 & 1 \end{bmatrix}$

e $\begin{bmatrix} \frac{1}{3} & \frac{2}{3} & \frac{2}{3} \\ \frac{-2}{\sqrt{5}} & \frac{1}{\sqrt{5}} & 0 \\ \frac{-2}{3\sqrt{5}} & \frac{-4}{3\sqrt{5}} & \frac{5}{3\sqrt{5}} \end{bmatrix}^{-1} = \begin{bmatrix} \frac{1}{3} & \frac{-2}{\sqrt{5}} & \frac{-2}{3\sqrt{5}} \\ \frac{2}{3} & \frac{1}{\sqrt{5}} & \frac{-4}{3\sqrt{5}} \\ \frac{2}{3} & 0 & \frac{5}{3\sqrt{5}} \end{bmatrix}$

13 Take $x = y$; then $\begin{bmatrix} 1 & x \\ y & 2 \end{bmatrix}$ is a symmetric matrix whose eigenvalues are given by $\lambda_1 = \dfrac{3 + \sqrt{1+4x^2}}{2}$ and $\lambda_2 = \dfrac{3 - \sqrt{1+4x^2}}{2}$. Since $\lambda_1 \neq \lambda_2$ it follows by Theorem 7-11 that the corresponding eigenvectors $\mathbf{v}_1$ and $\mathbf{v}_2$ are perpendicular.

14 **a** $T = \begin{bmatrix} \frac{1}{\sqrt{2}} & \frac{1}{\sqrt{2}} \\ \frac{1}{\sqrt{2}} & \frac{-1}{\sqrt{2}} \end{bmatrix}$, $\quad T^{-1}AT = \begin{bmatrix} 2 & 0 \\ 0 & 6 \end{bmatrix}$

b $T = \begin{bmatrix} \frac{1}{\sqrt{10}} & \frac{3}{\sqrt{10}} \\ \frac{3}{\sqrt{10}} & \frac{-1}{\sqrt{10}} \end{bmatrix}$, $\quad T^{-1}AT = \begin{bmatrix} 10 & 0 \\ 0 & -10 \end{bmatrix}$

c $T = \begin{bmatrix} \frac{3}{5} & \frac{4}{5} \\ \frac{4}{5} & \frac{-3}{5} \end{bmatrix}$, $\quad T^{-1}AT = \begin{bmatrix} 25 & 0 \\ 0 & -50 \end{bmatrix}$

d $T = \begin{bmatrix} 0 & \frac{1}{\sqrt{2}} & \frac{1}{\sqrt{2}} \\ 0 & \frac{1}{\sqrt{2}} & \frac{-1}{\sqrt{2}} \\ 1 & 0 & 0 \end{bmatrix}$, $T^{-1}AT = \begin{bmatrix} 0 & 0 & 0 \\ 0 & 1 & 0 \\ 0 & 0 & -1 \end{bmatrix}$

e $T = \begin{bmatrix} 0 & \frac{1}{\sqrt{2}} & \frac{1}{\sqrt{2}} \\ 0 & \frac{-1}{\sqrt{2}} & \frac{1}{\sqrt{2}} \\ 1 & 0 & 0 \end{bmatrix}$, $T^{-1}AT = \begin{bmatrix} 0 & 0 & 0 \\ 0 & 0 & 0 \\ 0 & 0 & 2 \end{bmatrix}$

f $T = \begin{bmatrix} \frac{-1}{\sqrt{2}} & \frac{-1}{\sqrt{6}} & \frac{1}{\sqrt{3}} \\ \frac{1}{\sqrt{2}} & \frac{-1}{\sqrt{6}} & \frac{1}{\sqrt{3}} \\ 0 & \frac{2}{\sqrt{6}} & \frac{1}{\sqrt{3}} \end{bmatrix}$, $T^{-1}AT = \begin{bmatrix} -1 & 0 & 0 \\ 0 & -1 & 0 \\ 0 & 0 & 2 \end{bmatrix}$

g $T = \begin{bmatrix} 0 & 0 & \frac{1}{\sqrt{2}} & \frac{1}{\sqrt{2}} \\ 0 & 0 & \frac{1}{\sqrt{2}} & \frac{-1}{\sqrt{2}} \\ 1 & 0 & 0 & 0 \\ 0 & 1 & 0 & 0 \end{bmatrix}$, $T^{-1}AT = \begin{bmatrix} 0 & 0 & 0 & 0 \\ 0 & 0 & 0 & 0 \\ 0 & 0 & 1 & 0 \\ 0 & 0 & 0 & -1 \end{bmatrix}$

h $T = \begin{bmatrix} \frac{-1}{\sqrt{2}} & \frac{-1}{\sqrt{6}} & \frac{-1}{\sqrt{12}} & \frac{1}{2} \\ \frac{1}{\sqrt{2}} & \frac{-1}{\sqrt{6}} & \frac{-1}{\sqrt{12}} & \frac{1}{2} \\ 0 & \frac{2}{\sqrt{6}} & \frac{-1}{\sqrt{12}} & \frac{1}{2} \\ 0 & 0 & \frac{3}{\sqrt{12}} & \frac{1}{2} \end{bmatrix}$, $T^{-1}AT = \begin{bmatrix} 0 & 0 & 0 & 0 \\ 0 & 0 & 0 & 0 \\ 0 & 0 & 0 & 0 \\ 0 & 0 & 0 & 4 \end{bmatrix}$

Section 8-1

1 $x^2 - y^2 = 4$ (hyperbola); **2** $9x^2 + 4y^2 = 9$ (ellipse)
3 $3x^2 - 2y^2 = 12$ (hyperbola); **4** $x^2 - y^2 = 4$ (hyperbola)
5 $x^2 + 2y^2 = 4$ (ellipse); **6** $5x^2 + y^2 = 2$ (ellipse)
7 hyperbola **8** hyperbola **9** ellipse
10 ellipse **11** $x^2 + 2y^2 - 2z^2 = 16$; hyperboloid of one sheet
12 $3x^2 + 3y^2 + 7z^2 = 42$; ellipsoid **13** $2x^2 + y^2 = 1$; elliptic cylinder
14 $x^2 + y^2 - 2z^2 = 4$; hyperboloid of one sheet **15** $5x^2 + 4y^2 = 20$; elliptic cylinder

Section 8-2

1 **a** $y = ce^{t}$ **b** $y = ce^{-2t}$ **c** $y = ce^{-\sqrt{3}t}$ **d** $y = ce^{-3/2t}$

2 **a** $y = 3e^{2t}$ **b** $y = 2e^{-1/2t}$ **c** $y = -4e^{3-3t}$ **d** $y = 5e^{4/3(t-2)}$

3 **a** $y = ce^{-2t} + \frac{3}{2}$ **b** $y = ce^{t} - 10$ **c** $y = ce^{-3/2t} + \frac{5}{3}$

4 19,937 **5** 168,000 **6** 326.5; 14.2 **7** $A = 40e^{-0.0051t}$

8 0.6 of original amount **9** 9,700 **10** $N = N_0e^{kt}$ **11** 2 min.

12 24.25; 7.8 min.

Section 8-3

1 $\mathbf{x} = c_1\begin{bmatrix}1\\-1\end{bmatrix}e^{4t} + c_2\begin{bmatrix}3\\2\end{bmatrix}e^{-t}$ **2** $\mathbf{x} = c_1\begin{bmatrix}1\\-1\end{bmatrix}e^{-t} + c_2\begin{bmatrix}2\\1\end{bmatrix}e^{5t}$

3 $\mathbf{x} = c_1\begin{bmatrix}3\\2\end{bmatrix}e^{7t} + c_2\begin{bmatrix}1\\-2\end{bmatrix}e^{-t}$ **4** $\mathbf{x} = c_1\begin{bmatrix}1\\1\end{bmatrix}e^{3t} + c_2\begin{bmatrix}3\\2\end{bmatrix}e^{4t}$

5 $\mathbf{x} = c_1e^{-2t}\left\{\begin{bmatrix}-1\\1\end{bmatrix}\cos t - \begin{bmatrix}0\\1\end{bmatrix}\sin t\right\} + c_2e^{-2t}\left\{\begin{bmatrix}-1\\1\end{bmatrix}\sin t + \begin{bmatrix}0\\1\end{bmatrix}\cos t\right\}$

6 $\mathbf{x} = c_1\left\{\begin{bmatrix}1\\-1\end{bmatrix}\cos 2t - \begin{bmatrix}1\\0\end{bmatrix}\sin 2t\right\} + c_2\left\{\begin{bmatrix}1\\0\end{bmatrix}\cos 2t + \begin{bmatrix}1\\-1\end{bmatrix}\sin 2t\right\}$

7 $\mathbf{x} = c_1\begin{bmatrix}1\\2\\-7\end{bmatrix}e^{-t} + c_2\begin{bmatrix}1\\0\\-1\end{bmatrix}e^{t} + c_3\begin{bmatrix}1\\-1\\-1\end{bmatrix}e^{2t}$

8 $\mathbf{x} = c_1\begin{bmatrix}1\\1\\1\end{bmatrix}e^{t} + c_2\begin{bmatrix}1\\2\\1\end{bmatrix}e^{2t} + c_3\begin{bmatrix}1\\7\\11\end{bmatrix}e^{-3t}$

9 $\mathbf{x} = \begin{bmatrix}-3\\3\end{bmatrix}e^{4t} + \begin{bmatrix}3\\2\end{bmatrix}e^{-t}$ **10** $\mathbf{x} = e^{-2t}\left\{\begin{bmatrix}-1\\1\end{bmatrix}\cos t - \begin{bmatrix}0\\1\end{bmatrix}\sin t\right\}$

12 $\mathbf{x}' = \begin{bmatrix}0 & 1\\-2 & 3\end{bmatrix}\mathbf{x};\quad \mathbf{x} = c_1\begin{bmatrix}1\\1\end{bmatrix}e^{t} + c_2\begin{bmatrix}1\\2\end{bmatrix}e^{2t}$

13 $\mathbf{x}' = \begin{bmatrix}0 & 1\\-6 & 5\end{bmatrix}\mathbf{x};\quad \mathbf{x} = c_1\begin{bmatrix}1\\2\end{bmatrix}e^{2t} + c_2\begin{bmatrix}1\\3\end{bmatrix}e^{3t}$

14 $\mathbf{x}' = \begin{bmatrix}0 & 1\\-1 & 0\end{bmatrix}\mathbf{x};\quad \mathbf{x} = c_1\left\{\begin{bmatrix}0\\-1\end{bmatrix}\cos t - \begin{bmatrix}1\\0\end{bmatrix}\sin t\right\} + c_2\left\{\begin{bmatrix}0\\-1\end{bmatrix}\sin t + \begin{bmatrix}1\\0\end{bmatrix}\cos t\right\}$

15 $\mathbf{x}' = \begin{bmatrix}0 & 1\\-25 & 6\end{bmatrix}\mathbf{x};$

$\mathbf{x} = c_1e^{3t}\left\{\begin{bmatrix}1\\3\end{bmatrix}\cos 4t - \begin{bmatrix}0\\4\end{bmatrix}\sin 4t\right\} + c_2e^{3t}\left\{\begin{bmatrix}0\\4\end{bmatrix}\cos 4t + \begin{bmatrix}1\\3\end{bmatrix}\sin 4t\right\}$

16 $\mathbf{x}' = \begin{bmatrix} 0 & 1 & 0 \\ 0 & 0 & 1 \\ -6 & -1 & 4 \end{bmatrix}\mathbf{x}; \qquad \mathbf{x} = c_1 \begin{bmatrix} 1 \\ -1 \\ 1 \end{bmatrix} e^{-t} + c_2 \begin{bmatrix} 1 \\ 2 \\ 4 \end{bmatrix} e^{2t} + c_3 \begin{bmatrix} 1 \\ 3 \\ 9 \end{bmatrix} e^{3t}$

17 $y = e^{3t}\,(2 \sin 4t - 3 \cos 4t)$

18 $\mathbf{x}' = \begin{bmatrix} 0 & 1 \\ -k/m & 0 \end{bmatrix}\mathbf{x}; \qquad \mathbf{x} = c_0 \begin{bmatrix} \cos \sqrt{k/m} \\ -\sqrt{k/m}\, \sin(\sqrt{k/m}\; t) \end{bmatrix}$

Index